Current Progress in Plant Science

Current Progress in Plant Science

Edited by **Clive Koelling**

Callisto Reference
New York

Published by Callisto Reference,
106 Park Avenue, Suite 200,
New York, NY 10016, USA
www.callistoreference.com

Current Progress in Plant Science
Edited by Clive Koelling

International Standard Book Number: 978-1-63239-693-8 (Hardback)

Printed in China.

Contents

Preface

In my initial years as a student, I used to run to the library at every possible instance to grab a book and learn something new. Books were my primary source of knowledge and I would not have come such a long way without all that I learnt from them. Thus, when I was approached to edit this book; I became understandably nostalgic. It was an absolute honor to be considered worthy of guiding the current generation as well as those to come. I put all my knowledge and hard work into making this book most beneficial for its readers.

Plant Science, also called botany or plant biology, refers to the scientific study of plants. This field includes the study of the smallest of plants like algae, fungi to the largest in this category like trees. The main aim of botany is to improve the supply of various plant products like medicines, foods, fibres, building materials, etc. and to manage wilderness. There are many sub-fields of plant biology like anatomy which refers to study of cells and tissues of plants, ecology which is the study of plant and their relation with environment, paleobotany which is the study of evolution of fossil plants, biophysics which refers to the study of physical aspects of plant life processes, etc. This book unravels the recent studies in the field of plant science by presenting the researches conducted by international experts. It will also provide interesting topics for research that readers can take up. This book will provide in-depth knowledge of this topic and its sub-fields. It will elucidate the new technologies and concepts related to this area. It will serve as a reference guide to broad spectrum of readers including botanists, environmentalists, ecologists, researchers and students associated with this field.

I wish to thank my publisher for supporting me at every step. I would also like to thank all the authors who have contributed their researches in this book. I hope this book will be a valuable contribution to the progress of the field.

Editor

1

Study on the Pollen Thermo-Sensitive and Stigma Exsertion Male Sterile Line Tomato (*Lycopersicon esculentum* Mill.) cv. Da107

Qinzhi Yu[1], Congyao Liang[2], Xianyu Wang[3*], Yongcheng Du[4], Masaharu Masuda[5], Kenji Murakami[5]

[1]Guilin Economic Crops Technology Extending Station, Guilin, China
[2]Nanzhang County Agricultural Technology Promotion Center, Xiangyang, China
[3]Agricultural College, Guangxi University, Nanning, China
[4]Institute of Vegetables and Flowers, Chinese Academy of Agricultural Sciences, Beijing, China
[5]Okayama University, Okayama, Japan
Email: fengdi053@163.com, liangcongyao@163.com, [*]wang12261962@163.com

Abstract

To accelerate the breeding process, male sterile line is used to leave out the troublesome procedures of the artificial emasculation in tomato breeding. However, the fertility of the pollen thermo-sensitivity male sterile line (PTMSL) and the stigma exsertion male sterile line (SEMSL) are affected easily by the environments when used alone. The trial materials were Da107 and the control was First. This study was conducted to create a new male sterile line of tomato characterized by pollen thermo-sensitivity and long style by genic recombination through the hybridizing of the PTMSL and SEMSL. Research on the statistics of the pollen germination rate, the contamination rate in F_1 and the flower organics indicated that Da107 was an ideal TS and SE male sterile line with the sterility of 95%, as well as it also could be used as fertile line at low temperature. Meanwhile, the results showed that hybrid-seed contamination risk with selfed seeds from residual fertility in Da107 was low.

Keywords

Lycopersicon esculentum **Mill., Pollen-Thermo-Sensitivity Male Sterility, Stigma-Exsertion Male Sterility, Pollen Germination, Seed Contamination Rate**

[*]Corresponding author.

1. Introduction

Tomato (*Lycopersicon esculentum* Mill.) belongs to the *Solanaceae* and *Lycopersicon* family, a self-pollinated plant with obvious heterosis. And heterosis breeding is the general method of obtaining high production which depends on the elaboration of effective methods of hybrid seed production [1]. So it is critical to make wide use of male sterility for hybrid seed production [2] and identify genetically stable lines that can be used in hybrid seed-breeding programs [3]. Plant breeders paid great attentions to the male sterility of tomato since the first one was discovered by Crane in 1915, and work had been done one after another [4]. The different temperature grads and time grads, could affect the elongation of exserted-stigma [5], the functional style male sterility line of stigma exsertion tomato could reach 95% of the infertility rate in the field of more than 20˚C temperature [6]. And many researches related to select breeding, cytology, physiology and biochemistry have been reported [7] [8]. Although more than 40 ms genes are known in tomato, most of which are not applicable for a practical hybrid seed production, because of the difficulty in maintaining their pure line by seeds [9]. In this study, we took the high pure hybrid-seed rate for the objection, and tried to take full advantage of male sterile materials.

2. Materials and Methods

2.1. Materials

This study was carried out under natural conditions at College of Agriculture Experiment Base of Guangxi University in late June. The temperature was between 10˚C and 30˚C. Experimental plants Da107 used as the test group were derived from successive self-pollination for six generations after the hybridization of T-4 (a genic male sterile line from Okayama University) and stigma exsertion mutant type (WANYU), and the fertile line First was used as the control.

The experimental materials were treated with 3 steps as follow: 1) The seeds were disinfected in water solution with Trisodium phosphate dodecahydrate's concentration of 10% for 20 minutes; flushed with water; and then soaked in normal temperature water for 12 hours; finally, accelerated germination with gauze in normal temperature, and each day during germination with water washing 1 - 2 times. 2) Culture of seedling was in seedling matrix with seedling-raising disk. 3) Transplanted when the seedlings were 25 days old in a glass greenhouse; and other management was according to the normal level. The experimental dates of the tomato trial were tabulated.

2.2. Method

2.2.1. The Statistics of Pollen Germination Rate (PGR)

The pollens were collected from each flower blooming that day of Da107 and First at anthesis, and marked with Da107 and First (adding different temperature treatments and the date of sampling) respectively, then cultivated with the pollen medium (0.005 g boric acid, o. o 1 g calcium chloride, 10 g glucose and xml distilled water per 100 ml solution) at 25˚C for 2 h. Finally, the pollen germination rate of 5 eye shots in the microscope was calculated.

2.2.2. The Study of Floral Organs

The flowers were gathered from Da107 and First when bloomed completely that day. A ruler was used to measure the length of the styluses and the stamens, as well as the size of the styluses' diameter. And the photos of the typical flowers were taken.

2.2.3. Investigation of F_1 Seed Contamination Rate

F_1 seeds whose female parent was Da107 were sowed in pots. The contamination rate was observed, as the seedlings grew up with three main leaves, characterized by leaf shapes compared with the female parent's.

3. Results and Discussion

3.1. The Pollen Germination Rate and Temperature

There were no significant difference in the pollen germination rate between Da107 and First grew at night temperature(NT) between 14˚C - 18˚C (**Figure 1**). On the contrary, the results showed significant difference between those grew at 20˚C - 28˚C (NT) (**Figure 2**). The two pictures (as below) indicated that temperature was a

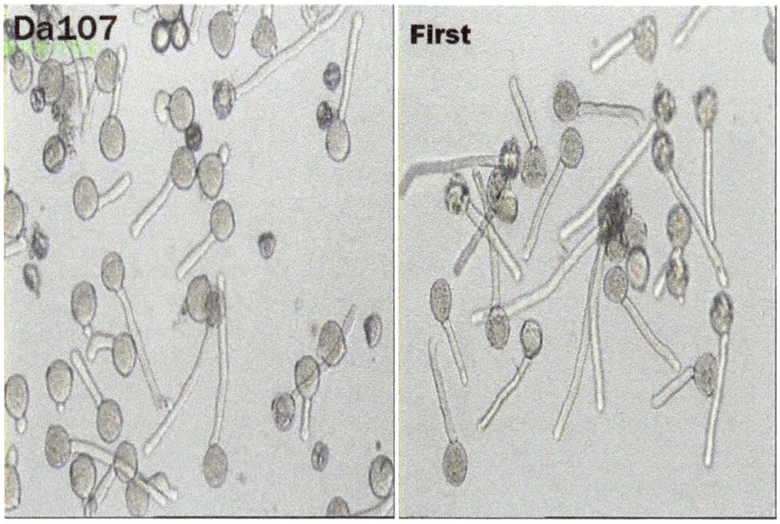

Figure 1. The comparison chart of PGR [grew at 14˚C - 18˚C (NT)].

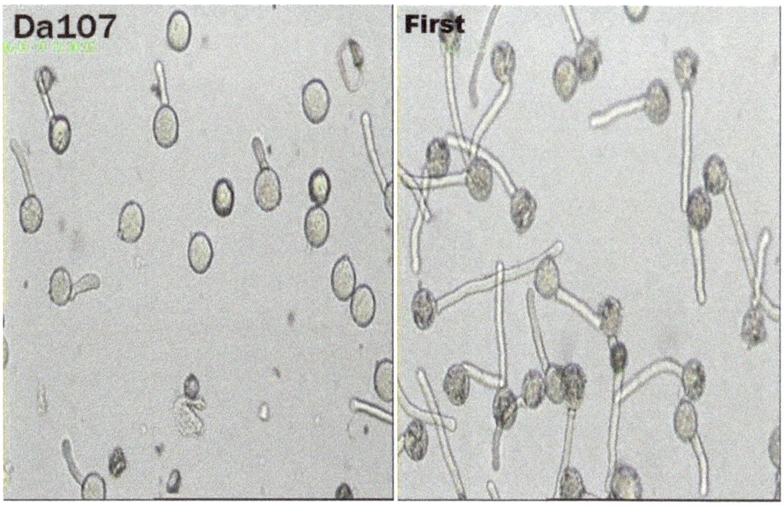

Figure 2. The comparison chart of PGR [planted at 20˚C - 28˚C (NT)].

major factor affecting the PGR of Da107.

The PGR of Da107 was low to 4.34% at 26.43˚C night temperature, but up to 56.21% at 16.42˚C. It can be seen that correlation coefficient between PGR and temperature was −0.97 (**Table 1**). It meant that RGR of Da107 was significantly negatively related to temperature. With the increase of temperature, the PGR of Da107 was decreased. The PGR of 56.21% of Da107 had no significant difference with First when the temperature was 16.42˚C. In other words, Da107 restored fertility under the conditions of low temperature. The T-4 mutant plant's PGR was four times higher in autumn (low temperature) than that in spring (high temperature) and it was restored when exposed to low temperatures [10]. Masaharu Masuda reported that the fertility of T-4 mutant type was affected by night temperature and the male-sterile line which restored fertility could be propagated under low night temperature conditions (above 12˚C and below 18˚C) [11]. SPSS analysis test showed that the analysis of variance at 1% level of significance indicated that the differences of the PGR between Da107 and the control first were statistically significant. So the results manifested that with the environment conditions changed, Da107 was a perfect MSL whose sterility was up to 95% and maintenance line also.

3.2. The Floral Organ

Normal fertile line First (**Figure 3.3**) showed big corolla, full-fledged filament, full stylus and yellow plump anther, and tetradynamous stamen was as high as or slightly higher than the stylus. Compared with the First, small coralla, browning or whiting thin less anther, and tetradynamous stamen much higher than the stylus was characterized as Da107 (**Figure 3.1** vs **Figure 3.2**).

In natural conditions, that the length of Da107's stylus was about 75% longer than the stamen (**Table 2**) posed an obstacle to self-pollination, and the trait stood for a functional male sterile phenotype (FMSP) [12]. So Da107 was a double male sterility line. In other words, Da107 was not only the thermo-sensitivity male-sterile mentioned above, but also FMSP. Studies on the genetic variability of exserted stigma-sterility showed that this male-sterile was not to be evaluated based only on the length of the style that protruded from the anther cone, but also depended on the length of both style and anther, and all these were affected mainly by the temperature [13] [14]). The length of the stylus and stamen increased with the temperature rising. That was to say, the sterility of Da107 increased with temperature rising. It could be concluded that Da107 was a perfect male sterile line under the condition of higher temperature. Furthermore, Da107's stylus had the diameter almost one bigger than First's (**Table 2**), which was in favor of the hand pollination immensely. In any case, Da107 was used as MSL at high temperature or maintenance line at low temperature

3.3. F_1 Seeds Contamination

Table 3 showed that the F_1 hybrid-seed contamination rate with the selfed-seed was inapparent, so, Da107 was a valuable sterile material for breeding when taken as female parent.

4. Conclusion

The new type MSL Da107 characterized with pollen temperature-sensitivity and stigma exsertion can be used as

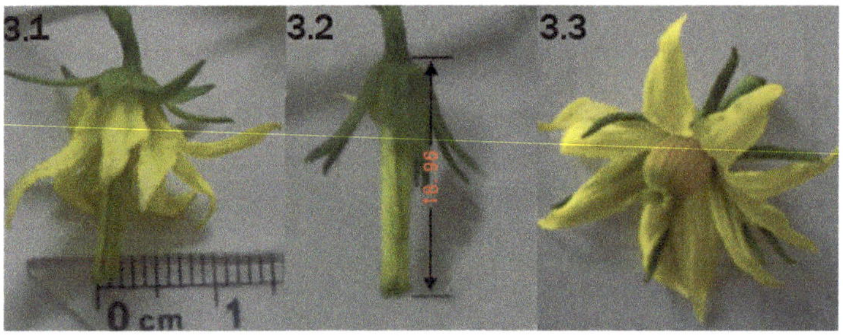

Figure 3. The comparison charts of about floral organs of da107 (3.1 vs 3.2) and first (3.3).

Table 1. Correlation analysis of the pollen germination rate at different temperatures.

Period (month/day)	Average night temperature (°C)	PGR of Da107 (%)	error	PGR of first (%)	Error
8/7 - 8/9	26.43	4.34	0.95	67.45	1.65
10/5 - 10/7	25.91	11.06	0.9	79.12	1.79
2/11 - 2/13	23.05	33.61	1.65	68.70	1.07
12/9 - 12/11	16.42	56.21	2.41	65.60	1.68
Correlation coefficient with temperature		−0.97**		0.56	

**Significant at 0.01 probability level.

Table 2. Correlation analysis of floral organ at different temperature.

Period (month/day)	Night average temperature (°C)	Length of stylus (mm)	Length of stamen (mm)	Difference of stylus and stamen (mm)	Stylus diameter (Da107) (mm)	Stylus diameter (first) (mm)
9/24 - 9/28	27.9 ± 0.3	20.33	12.37	7.96	2.31	1.37
10/15 - 10/19	24.9 ± 0.3	18.85	11.86	6.99	2.27	1.11
12/10 - 12/14	17.9 ± 0.4	17.17	11.13	6.04	2.13	1.01
1/6 - 1/9	9.38 ± 0.12	15.85	9.86	5.99	1.95	0.97
Correlation coefficient with temperature		0.98**	0.99**	0.88**	1.00**	0.84**

Table 3. Rate of F_1 seeds contamination in different male parents cross combinations, whose female parent was Da107.

Cross combination	No. of trial seedlings (a)	No. of selfed-seedlings (b)	Contamination rate (b/a × 100)
Da107 × Dongsheng	430	3	0.6977
Da107 × 12	562	7	1.6279
Da107 × Niufanqie	624	16	3.7209
Da107 × A3	437	12	2.7907

fertility line in autumn (low temperature), and the hybrid seeds are produced by the hybridization with maintenance line in the way of artificial pollination in spring. What's more, the length of Da107's stylus was much longer than that of another, which ensured the purity of the hybrid seeds. Meanwhile, Da107, leaving a perfect male-sterile trait, overcame the problems associated with the labor-intensive process for emasculation during hybrid-seed production.

Funding

The national ministry of science and technology and international cooperation funding project (2006DFA33380); funding project of Guilin comprehensive experimental station, national commodity vegetable industry technology system (CARS-25-X-XX).

References

[1] Kilchevsky, A. and Dobrodkin, M. (2000) Use of Functional Male Sterility in Seed Production of Tomato Hybrids. *Acta Physiologiae Plantarum*, **22**, 232-234. http://dx.doi.org/10.1007/s11738-000-0018-1

[2] Masuda, M., Uchida, K., Kato, K. and Agong, S.G. (1999) Characterization and Genetic Analysis of Male Sterile Mutant Induced in Tomato cv. First, Having Mature Pollen Stainable with Acetocarmine. *Journal of the Japanese Society for Horticultural Science*, **68**, 566-568. http://dx.doi.org/10.2503/jjshs.68.566

[3] Gorman, S.W. and McCormick, S. (1997) Male Sterility in Tomato. *Critical Reviews in Plant Sciences*, **161**, 31-53. http://dx.doi.org/10.1080/713608142

[4] Zhu, G.-M., Li, J.F. and Li, P. (1982) A Preliminary Investigation on Morphological and Anatomical Characters of Sterile Flowers in Tomato (*Lycopersicon esculentum* Mill.). *Transactions on Journal of Northeast Agricultural College*, **2**, 46-52.

[5] Chen, L., Tan, L.L. and Wang, Y. (2009) Preliminary Study on Physiological Mechanism of Temperature Sensitive Male Sterile Tomato. *Transactions on Journal of Northeast Agricultural College*, **1**, 31-35.

[6] Yan Z.-M. (2010) Analysis of ESTs Related to Temperature Sensitivity of Longstigma Mutant Tomato T431 and the Preliminary Analysis of LeMBF1 in the Flower Bud Differentiation. Thesis, Northest Agricultural University, Harbin.

[7] Wang, X.-Y., Liang, C.-Y. and Deng, L.-F. (2011) Studies on Physiological and Biochemical Characters in Thermo-Sensitive Male Sterile Line Tomato *Solanum lycopersicum* L. T-4. *Transactions on Northern Horticulture*, **15**, 29-31.

[8] Guo, Y.-H., Yu F.-D. and Gan, G.-Y. (2013) Studies on Protein Content, POD and EST Isozyme of Thermo-Sensitive Genic Male-Sterile Tomato T-4. *Transactions on China Vegetable*, **6**, 58-62.

[9] Georgiev, H. (1991) Heterosis in Tomato Breeding. In: Kalloo, G., Ed., *Genetic Improvement of Tomato*, *Monographs on Theoretical and Applied Genetics*, Springer-Verlag, Berlin, 83-98.

[10] Masuda, M., Uchida, K., Kato, K. and Agong, S.G. (2000) Restoration of Male Fertility in Seasonally Dependent Male Sterile Mutant Tomato, *Lycopersicon esculentum* cv. First. *Journal of the Japanese Society for Horticultural Science*, **69**, 557-562. http://dx.doi.org/10.2503/jjshs.69.557

[11] Masuda, M., Kato, K., Murakami, K., Nakamura, H., Ojiewo, C.O. and Masinde, P.W. (2007) Partial Fertility Restoration as Affected by Night Temperature in a Male-Sterile T-4 Mutant Tomato, *Lycopersicon esculentum* L. *Journal of the Japanese Society for Horticultural Science*, **76**, 41-46. http://dx.doi.org/10.2503/jjshs.76.41

[12] Rick, C.M. and Robinson, J. (1951) Inherited Defects of Floral Structure Affecting Fruitfulness in *Lycopersicon esculenturn*. *American Journal of Botany*, **38**, 639-652. http://dx.doi.org/10.2307/2437776

[13] Takhur, M.R, (1970) Environmental and Genetic Control of Heterostyly in Tomato (*Lycopersicon esculentum* Mill.). Dissertation, The Ohio State University, Columbus.

[14] Bistra, A. (2000) Functional Male Sterility in Tomato (*Lycopersicon esculentum* Mill.) and Its Application in Hybrid Seed Production. *Acta Physiologiae Plantarum*, **22**, 221-225. http://dx.doi.org/10.1007/s11738-000-0015-4

2

Studies on Vegetative and Reproductive Ecology of *Paris polyphylla* Smith: A Vulnerable Medicinal Plant

Chitta Ranjan Deb*, Sakutemsu L. Jamir, Nangshimeren Sakutemsu Jamir

Department of Botany, Nagaland University, Lumami, India
Email: *debchitta@rediffmail.com, debchitta@gmail.com

Abstract

Paris polyphylla **Smith is an important medicinal plant that is considered vulnerable in many parts of the world. The species exhibits poor adaptability to changed environment and susceptible to climatic conditions for reproductive vigor. During the present study the species was studied in four different natural niches as well as synthetic habitats to check the adaptability to artificial conditions, its morphological and physiological characters and its interaction in its community so that conservation strategies could be worked out. The present study indicates that *P. polyphylla* exhibits healthy growth as well as reproduction in undisturbed area with canopy covers over 80%. The plant grows in well-drained rich humus soil and grows well with certain key associated species such as *Quercus* species, *Taxus baccata*, *Aconitum* species, *Eupatorium adenophorum*, *Smilax* species etc. Many plants were non-flowering and more than 60% were found non-flowering in its habitat during its flowering season in any particular year. There are clear morphological changes as the flowers stages from pre-fertilization to post-fertilization. Arrangement of anthers in two whorls shows a distinct and unique floral display. Light is an important factor during seed setting and less than 50% shade lowers the seed productivity very significantly. Predators such as slugs and snails are found on the plant. Urgent conservation and new innovative ways are necessary to propagate this economically important plant to ensure its continuity.**

Keywords

Associated Species, Floral Biology, Niche Characterization, Primary Habitat, Reproductive Biology

1. Introduction

Plant reproductive biology and phenological studies are important for understanding and knowing the plant in its

*Corresponding author.

various stages of its life cycle from the time of its sprouting till its senescence along with the various barriers associated the plant so that this knowledge helps in knowing how plants interact within its community and what actions can be done for the conservation of the plants [1]-[2]. The word "*Paris*" is derived from Latin word "*pars*" meaning equal often referring to the symmetry of the plant for its equal number of floral parts and leaves and the work "*Polyphylla*" refers to the many (poly) and leaves (phylla). *Paris* L. (Melanthiaceae) genus consists of about 24 species which is distributed from Europe to Asia [3]. *Paris polyhylla* is categorized as sparse [4] in India but it is listed under vulnerable category in Nepal [5]. *Paris polyphylla* is an important medicinal plant containing saponin steroids polyphyllin D, dioscin and balanitin 7 [6]. Rhizome of the plant is used as antihelmintic, antispasmodic, digestive stomachic, expectorant and vermafuge [5], scabies, rashes, or itching problems [7], to treat liver cancer [6]. The roots are also fed to cattle with diarrhea and dysentery [5].

In Nagaland, forests are lost or fragmented due to intensive "Slash and Burn" cultivation or "Jhum" cultivation, logging and developmental activities [8], incidences of forest fires along with large scale landslides effects and threaten the biodiversity of a forest. Plants which are at constant threat need immediate conservation. *Paris polyphylla* is under constant threat because of rampant harvesting for local medicine as well as illegal trade to the neighboring states like Manipur and countries like Myanmar [9]. The plant which was quite abundant in the last decade has decreased drastically and therefore conservation of this plant is highly important in the wild as well as educating the locals on the importance of this plant and its conservation in the state.

Plants with its certain morphological limitation factors like habitat specific, slow growth and time taken to become reproductively mature along with reproductive limitations such as non-reproductive plants and long term seed dormancy adds to the rapid reduction in its species. Thus studies on floral phenology, reproductive biology are very important for determining conservation strategies of such economically important threatened plant. Present study focuses on different aspects of habitat ecology, adaptability to artificial habitat, floral biology, reproductive biology and seed biology of *Paris polyphylla*.

2. Materials and Methods

Field collection and visits were conducted in selected four parts of Nagaland (India) where *Paris polyphylla* are found. The selected four areas were based on the abundance of the plant species through primary data from the locals, high gap ranges of elevation differences between the selected areas so as to check any differences on the plant species, vegetation type and landscape features and human role played on the selected forest area. The four areas are forest area of Pangsha village (Tuensang district) at an elevation of 2278 m (ASL), 26°14'27.2"N and 95°07'05.7"E, Chida and its surrounding area (Phek district) at an elevation of 1874 m ASL, 25°30'11"N and 94°13'37"E, Aradura Hill (Kohima district) at an elevation of 1396 m (ASL), 25°38'49.5" N and 94°06'25.8"E and Longkum village forest area (Mokokchung district) at an elevation of 1404.51 m (ASL), 29°15'98.8"N and 94°24'03.5"E.

Field visits conducted for this perennial herb for four years to monitor the plant formation, sprouting of shoots till its senescence. Morphological aspects of the plants, their phenology and reproductive biology were studied. Along with this, characteristics ecology of the plant, its associated species, canopy cover, level of anthropogenic disturbances were studied.

Adaptability of the plant species outside its natural habitat was studied through establishment in artificial habitat at Penli ward (Mokokchung) at 1259 m (ASl) where the plants collected from the four studied areas were kept in different potting mixture at different shaded conditions:

1) Manure soil by mixing decayed wood, sand and top black soil at a ratio of 1:1:3.
2) Mixture of cow dung, sand and top black soil at a ratio of 1:1:3.
3) Only to black soil.

Plants were monitored constantly for its growth and development. Changes in the soil were done every year to ensure nutrient availability to the plants.

The experimental beds were provided with different shaded conditions as below:

a) Zero shade.
b) Plants kept in association with domestic garden plants and climbers which provides partial shade.
c) Plants kept in polyhouse ca. 75% shade.

Plants grown in different soil mixture, their adaptation outside the natural habitats, effect of temperature and shaded conditions on the plant growth and development were studied.

3. Results

3.1. Habitat

Paris polyphylla is an important medicinal perennial plant. The present investigation study was conducted in 4 natural habitats with a total of 8 populations for 4 years. Plants usually grow inside deep forest where human interference is minimal and under shade (**Table 1**, **Figure 1(a)**). But with increase in disturbance level the population was found to become thinner. *Paris* is a shade loving plants and grows under the canopy closure of more than 80% at an altitude of 1300 - 2500 m ASL in Sub-tropical broad leaved hill forest to moist temperate broad leaf forest. It grows in moist and well-drained soil and usually on slope areas with north facing with soils that are usually covered with dry and decay organic matters. The plants are erect herbs growing up to a meter tall (**Figure 1(b)**). Plants that lack inflorescence are usually shorter in height. The plants grow well in well-drained humus soil and water logging is lethal to this group of plants.

Of the four sites studied during the present investigation, high level of disturbances was found in the unprotected forest area of Pangsha area while village protected forest areas like in Longkhum forest area and Chida area shows limited disturbances (**Table 1**). These disturbances are not due to deforestation or any other anthropogenic activities but due to over-exploitation of *P. polyphylla* rhizome for local medicinal use as well as illegal exports outside the state. *Paris polyphylla* found in and around the area of Chida (Khezakenoma) area were found growing the best in slope areas with well-drainage and the plants were found growing for even more than 1 m in this area (**Figure 1(c)**).

Besides the natural habitats, plants were also rehabilitated in artificial habitats with similar altitudes and temperatures. It was found that in these artificial habitats though plants grew normally in the first year, failed to flower. In some cases wherever they flowered, failed to form fruit and seed setting and degenerated subsequently.

During the present study it was observed that in all the study areas, there were some common associated species (**Table 2**). Healthy growth of *P. polyphylla* was observed wherever these associated species were present

Table 1. Niche characterization of *Paris polyphylla* in Nagaland.

Site name	Locality	Elevation	Canopy Closure*	Disturbances
Tuensang	Pangsha area	2278 m ASL (26˚14'27.2"N and 95˚07'05.7"E)	86%	High human interference
Phek	Chida and its surrounding area	1874 m ASL (25˚30'11"N and 94˚13'37"E)	88%	Limited human interference
Kohima	Aradura hill	1490 m ASL, (26˚14'27.2"N and 95˚07'05.7"E)	84%	Protected area
Mokokchung	Longkum village	1404.51 m ASL (29˚15'988"N and 094˚24'035"E)	87%	Limited human interference

Table 2. Associated herbaceous plants found near *Paris polyphylla*.

Sl. No.	Plant species	Family
1.	*Ageratum conyzoides*	Asteraceae
2.	*Bidens pilosa*	Asteraceae
3.	*Breynia* sp.	Euphorbiaceae
4.	*Curculigo capitulata*	Hypoxidaceae
5.	*Dioscorea* sp.	Dioscoreaceae
6.	*Drymeria cordata*	Schrophulariaceae
7.	*Eupatorium adenophorum*	Asteraceae
8.	*Fagopyrum esculentum*	Polygonaceae
9.	*Hottuynia cordata*	Saurauiaceae
10.	*Lycopodium* sp.	Lycopodiaceae
11.	*Smilax* sp.	Smilaceae
12.	*Spilanthes acmella*	Asteraceae

Note: The above table gives generalized common plants which are found within 5 m perimeter where *Paris polyphylla* is found in its natural habitat.

Figure 1. (a) *Paris polyphylla* growing in the natural habitat; (b) *P. polyphylla* plants with rhizome; (c) over 1 m tall plants growing in Chida village (at higher altitude); (d) common associated species growing with *P. polyphylla* in natural habitat; (e) a *P. polyphylla* rhizome; (f) sprouting of new shoots from rhizome in the mid season resulted in non-flowering plants.

within 5 m perimeter compared to the areas where these species were absent. Some of the common associated species identified in all the niches are: **Trees**—*Schima wallichi*, *Alnus nepalensis*, *Bauhinia Purpurea*, *Albizia Procera*, *Pinus Roxburghii*, *Rhodedendrons arboreus*, *Quercus sps. Taxus Baccata*, *Bamboos*; **Herbs**—*Berberis sps*, *Aconitium sps* , *Impatiens sps*, *Ageratum conyzoides*, *Bidens pilosa*, *Curculigo capitulata*, *Dioscorea species*, *Drymeria cordata*, *Eupatorium adenophorum*, *Fagopyrum esculentum*, *Hottuynia cordata*, *Spilanthes acmella*, *Lycopodium* sps. Ferns, Bryophytes; **Climbers**—*Smilax species* (**Figure 1(d)**).

3.2. Rhizome Characteristics

Paris polyphylla rhizomes are horizontal and creeping with transverse rings/bud scale scars on its rhizome (**Figure 1(e)**). One ring or bud scale scar are added usually every year from the senescence of the above ground part during late autumn and therefore one can determine the age of the plant by simply counting the rings but sometimes a plant can give rise to compound bud, therefore counting the age of the plant by the rings/bud scale scars is inapplicable and not accurate. Further it was found that in some cases a rhizome which already has a matured flowering plant gives rise to new buds in the middle of the year usually during the month of April and May (**Figure 1(f)**).

3.3. Phenology and Reproductive Biology of *P. polyphylla*

The degree of plant survival is related to its reproductive output and success. Therefore, considering this, *P. polyphylla* plants were carefully studied and tagged to see their inflorescence development and non-development in the studied plant population. The method applied was simple and traditional. Plants in a given population during the March and April were counted and tagged as reproductive (flowering) and non-reproductive (non-flowering) (**Figure 2(a)**). The bud in *Paris polyphylla* remains dormant for 4 - 5 months during the winter season until the next sprouting season. *Paris polyphylla* starts sprouting from February onwards till the month of March gives rise to new above ground stems, leaves and inflorescence. Majority of the plants found all the natural habitats and also grown under controlled conditions were found to be non-reproductive or in other words absence of inflorescence. Over 60% of the studied plants were found to be non-reproductive (**Table 3**). During the sprouting of the above ground plant body, the leaves and the inflorescence usually develops synchronously

Table 3. Percentage of non-reproductive *Paris polyphylla* plants.

Location	Total number of plants observed	Non-reproductive plants	% non-reproductive plants
Tuensang	732	453	61.88
Phek	602	318	52.82
Kohima	167	113	67.66
Mokokchung	244	152	62.29
Total	**1745**	**1036**	**61.16**

Note: Plants were studied in different locations within the whole district; non-flowered plants were very high in *Paris polyphylla* plants and it accounted for more than 60% in total plant populations.

Figure 2. (a) Development of flowering and non-flowering plants in the same cluster; (b) solitary flower of *P. polyphylla*; (c) a flower showing roundish ovary with stigmatic lobes; (d) anthers of *P. polyphylla* arranged in 2 whorls; (e) a common pollinator engaged in pollination; (f) dispersal of pollens on the other parts of flowers; (g) post fertilization of modification of ovary (ovary turned dull purple); (h) mature fruit showing red seeds; (i) fruit with fewer seeds from plants growing under higher light intensity; (j) predator on the plant causing death to enlarging ovary.

and therefore non-reproductive plants are easily identified when the inflorescence lacks in *P. polyphylla*. The non-reproductive plants are independent in relation to the age of its rhizome in all matured plants. Plants which are non-reproductive can remain the same in the next season or becomes reproductive which is vice-versa for reproductive plants.

In *P. polyphylla* which bears inflorescence, flowering begins in March-April where the flowers bloom at the terminal and they are solitary (**Figure 2(b)**). At the initial growth, the inflorescence forms a closed whorl covered by the sepals enclosing the tepals, the anthers and the stigma. The ovary is roundish in shape situated in the middle with usually 5 stigmatic lobes which is wet type in nature (**Figure 2(c)**). Usually the stigma appears deep purple in color during the pollination period. In *Paris polyphylla*, the male function first than the female, therefore the flowers are protrandous. Flowers are perfect flowers as they contain both the male and female gametes within the same flower. Outer tepals are leafy and green, while the inner tepals are yellowish green and filiform. (Usually the numbers of outer and inner tepals are same). The number of stamens is usually more than double that of outer tepals. The anther displays a unique character. In *P. polyphylla*, the anther occurs in two whorls (**Figure 2(d)**). Closing and opening of the anthers take place. Usually the outer whorl has two additional anther filaments compare to inner whorl. The inner whorl of the anther filament lies in between the two outer anther filaments. However, in the flower there will be two outer anther filaments on adjoining side that has one inner anther whorl filament within them. This anther display is same in many flowers of *Paris polyphylla.* Stigma is wet and the stigma lobes are usually 4 or 5 and re-curved at the tips. Plants generally remain at bloom for a period of 2 - 3 weeks. The timing of anthesis is between 6:30 - 7:30 A.M. and the anther dehisces by 7:30 - 8:30 A.

M. The timing of anther dehiscence of *P. poyphylla* varies even within the same individual species population in a particular area. Variation in difference of timing can be even up to 2 - 3 weeks. Usually the flowers are self pollinated. However chances of pollination can occur by small insects. The plants are visited by different insects such as different types of flies and bees especially during the middle of daytime from 11:00 A.M. till 3:00 P.M. (**Figure 2(e)**, **Table 4**). These agents might help pollination indirectly. The buzzing of their wings of the insect flies especially bees might facilitate the pollen dispersal to the stigma within the flower itself or even to the surrounding flowers as the pollens are very light. However loss of large amount of pollens occurs within the plant. Since the pollen grains are very light, wind also acts as a pollinating agent. The plants are self compatible and they produce large amount of pollens. Huge amount of pollens are wasted which is indicated by the falling of pollens all over the outer tepals (**Figure 2(f)**). Stigmatic lobes are usually bright purple but turns dull purple once the pollination takes place. After pollination, the anthers shriveled up and the ovary color also changes from bright purple to dull purple which finally fades away. After fertilization, the ovary enlarges forming a globular capsule (**Figure 2(g)**). The anther, the filaments and the tepals slowly disintegrates off from the end towards the inflorescence. It was observed that a significant *per cent* of *P. polyphylla* though successful in yielding inflorescence and flowers (considered as fertile plants) failed to produce seeds.

3.4. Fruiting and Seed Formation

In *P. polyphylla* viable seed formation is very low and climatic factors play an important role. The fruit of *Paris polyphylla* which is a globular capsule bears the seeds inside them. The seeds are reddish orange in color and

Table 4. Reproductive phenology of *Paris polyphylla*.

Parameters	*Observations*
Height of the plant	Up to 1 m
Rhizome type	Horizontal and creeping
Sprouting of plant from rhizome	February-March
Dormancy of rhizome	4 - 5 Months
Leaves	Lanceolate or sometimes oblong
Outer tepals	Leafy green
Inner tepals	Yellowish green and filiform
Flowering	March-April
Flowers	Solitary and terminal
Time of anthesis	6:30 - 7:30 AM
Time of anther dehiscence	7:30 - 8:30 AM
Stigma	Wet type
Stigma lobes	(4) or 5
Pollination	Self-pollination
Capsule before and during pollination	Bright and deep purple
Capsule after pollination	Dull purple
Fruiting	May
Temperature for seed production	18˚C - 23˚C
seed	Reddish orange color
Seed maturation	September-October
Dispersal of seeds	Gravity
Senescence	October-November

when the fruit matures (**Figure 2(h)**). It was observed that the seed production is affected by the sunlight and heat. During the development phase, if the plants are exposed to direct sunlight, the seed production is very low. If the plants are maintained under <80% shade compare to normal day light seed production is affected, but if again the shade cover is less than 50%, than either the fruits are seedless or will have very few seeds (**Figure 2(i)**). It can be at times as few as 1 - 2 seeded fruits. Generally seeds are produce in large quantity in well-suited plant habitats. The outer covering of the capsule slits open exposing the seeds. The occurrence of a slit on the side of the globular capsule fruit indicates the enlargement of fruits and its maturity. The seeds are dispersed by gravity and dispersal is localized in and around the mother plant. Seeds do not germinate during the first year after its dispersal but stays dormant.

3.5. *Paris polyphylla* Susceptibility to Climatic Conditions and Predators

Plants are highly competitive with other herbaceous plants but at the same time they were found to be susceptible to the attack by various predators. In most of the habitats it was observed that snails and slugs were the predators of the fruits and the presence of them causes the death of the enlarging ovary (**Figure 2(j)**). Besides the biological agents, the plants are highly sensitive to the heat and sun light. Prolonged exposure to light and heat causes the plant to droop down and later dies off. Moisture content in the soil was also found to be very crucial for the growth of the plant and water retention in the soil causes the rotting of the stem base of the plant as well as rotting of rhizomes.

4. Discussion

Paris polyphylla is an important perennial medicinal plant of North East India in general and Nagaland in particular. The plant species is widely distributed in the high altitude areas across the state but its population is decreasing at an alarming rate. At present the populations are scattered in some small patches. The main dispersal agent is gravity and the young plantlets are found in and around the mother plants. This factor also contributes to loss of the plant species. This is because the local medicine men, illegal harvesters collect the rhizomes from the natural populations as there is no commercial cultivation of the species. Random collection of plants without taking into consideration the flowering and non-flowering plants and uprooting the rhizomes without taking care of the younger plants which sprouted from either from seeds or fragmented rhizomes is one of the key factors for downsizing the population of *P. polyphylla*. But on the positive note, the loss of huge amount of pollens from plants can be dispersed easily among the plants within this population groups. Thus conservation of the species is very important and the first step is identifying and understanding the habitat of the plant. In order to locate the species whose population is very thin, knowledge of the associated species is very important to locate the species of interest whose population otherwise is very difficult to locate. There are very reports available on this approach of population inventory of RET species. Madhu *et al.* [5] used *Artisaema* species for determining the presence or absence of *P. polyphylla* in Nepal while Langhu and Deb [10] successfully used the associated species for locating *Aconitum nagarum* in Nagaland from various habitats. In the present study, this approach was found to very helpful for locating the *Paris polyphylla* population. Besides this, the information on associated species could be important tools for characterization of the niches for the species, rehabilitation of the threatened species from conservation view point. There might be a possibility that wherever the associated species are growing, the key species is absent. This could be due to sweeping species from that area due to various anthropogenic activities, but could be a place where the key species may be introduced for propagation and or conservation.

During the present study it was found that there was great amount of variations/polymorphism in phenology of the plants. From literatures available, it reflects that the polymorphism in the genus *Paris* is universal. According to Zhang *et al.* [3] *P. polyphylla* var. *yunnanensis* also exhibits phenotypic diversity which correlates with the morphological diversity of *P. polyphylla* found in Nagaland as in the present study.

It was found that anthers are arranged in two whorls and large numbers of pollens are wasted. Presence of large number of anthers and wastage of large number of pollens indicates that the species is not efficient in energy economy. In *Paris polyphylla* var. *yunnanensis* also reported the anthers are arranged in two whorls where they also studied the closing and opening of anthers [11]. According to them conservation of pollens in the flowers and enhanced the male fitness by its closing during night and during rains which was not observed in the present study. However, the occurrence of large amount of pollen waste during pollination might give heavy loss and expenditure of the plant energy, which shows plants poor effective mechanism in pollen dispersal to the stigma of the flower.

In *Paris polyphylla* var. *yunnanensis* it was found that fly was the main pollinating agent [11] which is in agreement with the present study. In addition, in the present study bees were also acting as pollinators. Seed production is very important for sexually reproduced plants for new generations. In *P. polyphylla* seeds are produced during warm weather usually mid-summer but if the plants are exposed to direct sunlight, aborts the seed or seed production is affected significantly. As the plant are shade loving, drooping of the plants usually occurs if they are exposed to direct sunlight even for 1 - 2 hours indicates that for good reproductive output the commercial growers should ensure proper care especially during flowering and seed setting.

5. Conclusion

Paris polyphylla is an economically important medicinal plant and the species is under severe threat. The findings of the present study provide us a clear picture about the natural habitat and their primary requirements for survival, flowering, fruiting, seed setting. Further the findings pinpoint why the species cannot be cultivated easily away from the natural habitat. The findings from the seed biology study will help the commercial growers for seed propagation of the seed. The local authority, government agency should join hands to develop *Paris* park/garden in the natural habitats to propagate and conserve this economically important species.

Acknowledgements

This work is financially supported by the Department of Biotechnology, Ministry of Science and Technology, Govt. of India, New Delhi, India through a research grant No. BT/ENV/BC/01/2010 to Prof. Chitta Ranjan Deb.

References

[1] Fenner, M. (1998) The Phenology of Growth and Reproduction in Plants. *Perspectives in Plant Ecology, Evolution and Systematics*, **1**, 78-91. http://dx.doi.org/10.1078/1433-8319-00053

[2] Tandon, R., Shivana, K.R. and Mohan Ram, H.Y. (2003) Reproductive Biology of *Butea monosperma* (Fabaceae). *Annals of Botany*, **92**, 715-723. http://dx.doi.org/10.1093/aob/mcg193

[3] Zhang, J., Wang, Y., Zhang, J., Ding, Y., Yu, H. and Jin, H. (2011) Evaluation of Mineral Element Contents in *Paris polyphylla* var. *yunnanensis* from Southwest China. *African Journal of Pharmacy and Pharmacology*, **5**, 1792-1796. http://dx.doi.org/10.5897/AJPP11.433

[4] Yonzone, R., Bhujel, R.B. and Rai, S. (2012) Genetic Resources, Current Ecological Status and Altitude Wise Distribution of Medicinal Plants Diversity of Darjeeling Himalaya of West Bengal, India. *Asian Pacific Journal of Tropical Biomedicine*, **12**, S439-S445. http://dx.doi.org/10.1016/s2221-1691(12)60203-2

[5] Madhu, K.C., Phoboo, S. and Jha, P.K. (2010) Ecological Study of *Paris polyphylla* Sm. *Ecoprint*, **17**, 87-93.

[6] Li, F., Jiao, P., Yao, S., Sang, H., Qin, S., Zhang, W., Zhang, Y. and Gao, L. (2012) *Paris polyphylla* Smith Extract Induces Apoptosis and Activates Cancer Suppressor Gene Connexin26 Expression. *Asian Pacific Journal of Cancer Prevention*, **13**, 205-209. http://dx.doi.org/10.7314/APJCP.2012.13.1.205

[7] Jamir, N.S., Lanusunep and Pongener, N. (2012) Medico-Herbal Medicine Practiced by the Naga Tribes in the State of Nagaland (India). *Indian Journal of Fundamental and Applied Life Sciences*, **2**, 328-333.

[8] Reddy, C.S., Dutta, K. and Jha, C.S. (2013) Analysing the Gross and Net Deforestation Rates in India. *Current Science*, **105**, 1492-1500.

[9] Mao, A.A., Hynniewta, T.M. and Sanjappa, M. (2009) Plant Wealth of Northeast India with Reference to Ethnobotany. *Indian Journal of Traditional Knowledge*, **8**, 96-103.

[10] Langhu, T. and Deb, C.R. (2014) Studies on the Reproductive Biology and Seed Biology of *Aconitum nagarum* Stapf: A Threatened Medicinal Plant of North East India. *Journal of Research in Biology*, **4**, 1465-1474.

[11] Dingkang, W., Guifang, S., Lifang, W., Shuhua, Z. and Xiaojiang, C. (2009) A Novel Mechanism Controls Anther Opening and Closing in *Paris polyphylla* var. *yunnanensis*. *Chinese Science Bulletin*, **54**, 244-248. http://dx.doi.org/10.1007/s11434-008-0507-y

3

Chemistry and Biological Activity of Condamineeae Tribe: A Chemotaxonomic Contribution of Rubiaceae Family

Vinicius F. Moreira*, **Ivo J. C. Vieira**, **Raimundo Braz-Filho**

Sector of Natural Products Chemistry, Universidade Estadual do Norte Fluminense Darcy Ribeiro, Campos dos Goytacazes, Brazil
Email: *vfmoreira@gmail.com

Abstract

This study is a review of the Condamineeae tribe, through the acquisition of data from phytochemical studies and evaluation of genera activities that constitute the tribe, in order to contribute to chemotaxonomic classification of this tribe in the Rubiaceae family. This review also states the scarcity of phytochemical investigations of several genera and consequently a lot of species of family.

Keywords

Biological Activity, Condamineeae, Chemotaxomy, Phytochemistry, Rubiaceae

1. Introduction

The Rubiaceae family has approximately 660 genus and around 11,150 species [1]. Based on molecular phylogenetic studies [2] [3] this family is partitioned in three subfamilies: Cinchonoideae, Ixoroideae and Rubioideae.

It is widely distributed, mainly on tropical and subtropical regions but also on cold and temperate regions in Europe and north of Canada [4]. In America this family is represented by approximately 229 genus and 5200 species [5]. Nowadays it has around 118 genus and 1347 species in Brazil, corresponding to one of the main families of Brazilian flora [6].

This vast family is composed by classes of secondary metabolites with highly significant pharmacological potential. Into these classes the iridoids, anthraquinones, triterpenes and indole alkaloids, which are considered chemotaxonomic markers of the family [7], are highlighted. Many of chemical constituents as flavonoids, other

*Corresponding author.

phenolic derivates and terpenoids (diterpens) are also found on this family.

In Ixoroideae subfamily the iridoids are presented as chemotaxonomic markers, while in Cinchonoideae there is a predominance of indolic alkaloids. In Rubioideae, the antraquinoms are the main class of secondary metabolites found [8]. Although the chemotaxonomic markers of Ixoroidae, it is also observed the occurrence of iridoids in the remaining subfamilies.

2. Methods

The data survey was done through academic researches and database sites and portals: www.ibict.br, www.scielo.org, www.sciencedirect.com, https://scifinder.cas.org, and www.periodicos.capes.gov.br. The terms used on the search strategy of scientific publications were restricted to species, genera, subfamilies, families and tribes. Aiming to provide greater scope to the bibliographic survey there were also selected some specific terms related to phytochemical and biological activities.

The academic investigations and scientific publications obtained for each genus are classified and presented in this study into four items: phytochemical studies, biological activities, ethno pharmacological activities and other studies, according to the available information.

3. The Genera of Condamineeae Tribe—Ixoroideae Subfamily

The Ixoroideae subfamily is represented by trees or scrubs and is characterized by levogyrous or valvular aestivation in most tribes. It presents pantropical e pan subtropical distribution and is composed by 15 tribes, including the Condamineeae, tribe that *Simira genus* belongs [9].

Phylogenetic molecular investigations revealed several relations on Rubiaceae systematic [10]. In a study conducted by [2] an initially divergent clade of the Ixoroideae subfamily was composed by genus previously classified into several distinct tribes, including *Calycophylleae*, *Cinchoneae*, *Condamineeae*, *Hippotideae*, and *Rondeletieae*. This clade was considered asan unique tribe and referred as Condamineeae, composed by the genus *Alseis*, *Calycophyllum*, *Capirona*, *Chimarrhis*, *Condaminea*, *Emmenopterys*, *Pentagonia*, *Pinckneya*, *Pogonopus*, *Rustia*, *Simira* and *Warszewiczia* [11].

Phylogenetic molecular studies associates several other genus to Condamineeae, such as *Bathysa*, *Bothriospora*, *Dioicodendron*, *Dolichodelphys*, *Ferdinandusa*, *Hippotis*, *Macbrideina*, *Macrocnemum*, *Mastixiodendron*, *Parachimarrhis*, *Picardaea*, *Sommera* and *Wittmackanthus* [3], and *Mussaendopsis* [12]. The *Dialypetalanthus* genus seemed to belong to Ixoroideae subfamily [13] and later classified as part of Condamineeae tribe [9].

The Condamineeae tribe is constituted by species very different morphologically wherein rare morphological characteristics of Rubiaceae are found into the tribe. The genus that compose the tribe are presented on **Table 1** [11].

The result of bibliographic survey showed that only genres *Alseis*, *Bathysa*, *Calycophyllum*, *Chimarrhis*, *Condaminea*, *Elaeagia*, *Emmenopterys*, *Pentagonia*, *Pogonopus*, *Rustia*, *Simira*, *Sommera* and *Warszewiczia* had some type of phytochemical study. Presented studies evaluating biological activity, genres *Alseis*, *Bathysa*, *Calycophyllum*, *Capirona*, *Chimarrhis*, *Elaeagia*, *Emmenopterys*, *Macrocnemum*, *Pentagonia*, *Pogonopus*, *Simira*, *Sommera* and *Warszewiczia*. E ethnopharmacological activity reports to the generos A *lseis*, *Bathysa*, *Bothriospora*, *Calycophyllum*, *Capirona*, *Condaminea*, *Hippotis*, *Macrocnemum*, *Pentagonia*, *Pogonopus*, *Rustia*, *Simira*, *Sommera* and *Warszewiczia*.

The results of the literature survey data related to the genres of Condamineeae tribe is presented below, with the exception of *Simira* genre, which has already been presented in our previous research [14].

3.1. Alseis

3.1.1. Phytochemical Studies

It was found only one screening with 299 samples representing 133 species from 48 genera of plants of Panama Rubiaceae family. Samples of each extract were diluted in ammonia and chloroform for acid-base extraction to produce the alkaloids extracts. The extractions were applied on filter paper to comparison effects with emetine, with Dragendorff reagents. The samples were also chromatographed in silica gel plates for TLC with standards and pulverized with Dragendorff and $FeCl_3/HClO_4$ reagents to detect the presence of alkaloids. The *A. blackiana* species presented positive results for alkaloids on these chromatographic tests [15].

Table 1. Genera that constitutes the Condamineeae tribe.

Condamineeae		
Alseis	*Bathysa*	*Bothriospora*
Calycophyllum	*Capirona*	*Chimarrhis*
Condaminea	*Dialypetalanthus*	*Dioicodendron*
Dolichodelphys	*Dicholobium*	*Elaeagia*
Emmenopterys	*Ferdinandusa*	*Hippotis*
Holtonia	*Macbrideina*	*Macrocnemum*
Mastixiodendron	*Mussaendopsis*	*Parachimarrhis*
Pentagonia	*Picardaea*	*Pinckneya*
Pogonopus	*Rustia*	*Schizocalyx*
Semaphyllanthe	*Simira*	*Sommera*
Tammsia	*Warszewiczia*	*Wittmackanthus.*

3.1.2. Biological Activity

A. yucatanensis extracts were tested to it effect upon the induced contraction by norepinephrine (NE) 0.3 μM or 55 ml KCl in cropped aorta of free endothelial rings. The extract of *A. yucatanensis* shell presented vasoactive activity (relaxed vase) being capable of relax induced contractions by NE and KCL with effective dose for 50% (ED50) of 0.12 and 1.73 mg/mL; respectively [16] [17].

3.1.3. Ethnopharmacological Activity

It was found only one ethnopharmacological investigation citing *A. floribunda* in a survey of plants named "bitter plants" used as substitutes of *Chinchona* spp. on Brazilian traditional medicine [18].

3.2. Bathysa

3.2.1. Phytochemical Study

It is noteworthy the phytochemical study that isolated the paeonol phenolic compound **(1)** of *B. meridionalis* roots (**Figure 1**) [19].

Positive results are found for indolic alkaloids and flavonoids on chromatographic tests with *B. australis* extracts analyzed by HPLC [20]. For *B. cuspidata* the presence of alkaloids was evaluated by thin layer chromatography [21] and phytochemical analyzes were done through chromatographic reagent tests to the presence of flavonoids, phenolic compounds, alkaloids, tannins, cumarins and triterpenes [22].

Studies with stem extracts of *B. veraguensis* species, as well as to *B. panamensis* [15], also presented positive results on TLC chromatographic tests (with Dragendorff reagents) to alkaloids [23].

3.2.2. Biological Activity

Special emphasis is given to studies evaluating the antifungal activity of secretory proteins of *B. nicholsonii* against *Fusarium oxysporum*, *Colletotrichum musae* e *Colletotrichum lindemuthianum*. The inhibition of fungal growth was tested with conidia (20,000 cells/mL in 1mL of saline solution) incubated under 28°C in microplates of 200 μL, followed by addition of the exudates (secretory proteins) to a final concentration of 100 μg/mL, revealing an inhibitory effect on the growth around 15%, 20% and 64%, respectively, after 48 h [24].

Inhibitory activity evaluation of *B. australis* against *Mycobacterium fortuitum* and *M. malmoense* using the resazurin as a viable cell indicator, the crops were placed on microplates in contact with the extracts to be evaluated, in raw and partition forms. It was incubated under 37°C and after 48 hours, for *M. malmoense*, and four days, for *M. fortuitum*, it was added the resazurin. Firstly it was done the susceptibility screening to determine the activity of extracts and partitions of vegetable extracts in fixed concentrations (100 μg/mL). The *B. australis* extracts did not inhibited the growth of the microorganisms involved [25].

The ethanolic extract of *B. australis* was evaluated on the activity of the protein Pdr5p ATPase of plasmatic

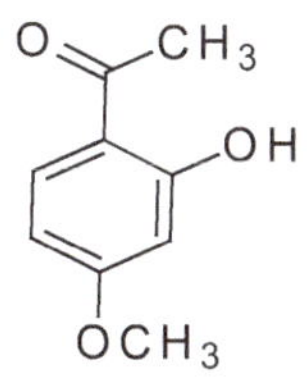

(1)

Figure 1. Chemical Structure of paeonol.

membranes of yeasts, on concentration ranges of 0 to 400 μg/mLwith IC_{50} of 35.3 μg/mL. Based on this result the ethanolic extract was fractionated by liquid-liquid partition resulting in hexane, dichloromethane, ethyl acetate and butanol fractions. After the solubilization in 20% of dimethylsulfoxid, the fractions were tested in Pdr5p ATPase on fixed concentration of 200 μg/mL and it was observed that the ethyl acetate, dichloromethane and hexane fractions presented and inhibition on the Pdr5p ATPase activity very similar (around 80%) compared with the ethanolic raw extract, while the butanol fraction produced a very small inhibition (around 20%) [20].

The antifungal evaluation of *B. meridionalis*, against *Alternaria antexnata*, that causes diseases on tangerines. The methanolic extracts of leaves, bark, flowers and stems were dissolved in 500 μL of 1% (v/v) of Tween 80 and mixed with 100 μL of suspended conidae of *A. antexnata* in 2.6 - 3.0 × 10^5 conidae/ml and the conidae suspension resultant was poured in pits (Pfizer, 0.55 mg/ml de PDA). After three days at 25°C and a 12 hours photoperiod, the extracts that hindered the fungal growth were considered active. The *B. meridionalis* extracts did not presented any significant result [26].

On a study of the effect of the bark extract of *B. cuspidata* in pulmonary injuries induced by paraquat (PQ) herbicide in rats, briefly, PQ was dissolved in 0.5 mL of NaCl saline solution at 0.9% (10 mg/mL) and injected intraperitonially in a unique dose of 30 mg/kg of corporal weight. The ethanolic extract of *B. cuspidata* bark of the stem was resuspended in 700 μL of DMSO vehicle and administrated for animals through wash. In the investigation the intoxicated rats' treatments, with 200 and 400 mg/kg of ethanolic extract, reduced the animal mortality to 30% and 20% respectively [27].

The hepatoprotective effect of the *B. cuspidata* bark extract on hepatic injury is induced by carbon tetrachloride. The hepatic injury was induced by intraperitoneal injection with administration of CCl_4 (60% v/v, 1ml/kg) at each 48h during 12 days, the administration of the extract in 200 mg/kg and 400 mg/kg was started six days before the first application of CCl_4 and progressed to be administrated to animals during the 12 days. A significant reduction on serum aspartate transaminase, alanine and gamma-glutamyl transferase and a significant increase on the activity of the antioxidant enzyme superoxide dismutase, low proportions of cellular necrosis and lipid droplets were observed, results that confirm the hepatoprotective activity pronounced of the bark extract of *B. cuspidata* and suggests this effect can be associated with the inhibition of oxidative damages [28].

Antioxidant effect of *B. cuspidata* extract on hepatic damage and glucose kinetic on rats' blood exposed to paraquat. On rats treated with *B. cuspidata* (400 mg/kg), the bark extract was capable of inhibit large glucose variations on blood and reduce the hepatic damage [22].

Yet for *B. cuspidata* studies are found evaluating the mutagenic potential of leaves and stems extracts tested by Ames test on TA98 and TA100 lines of *Salmonella typhimurium* on absence and presence of metabolic activation, revealing that the leaves ethanolic extract did not exert effect on plasmid DNA while the stem extract revealed mutagenic activity of metabolic activation on *S. typhimurium*. The extract effects directly on DNA were assessed by plasmodial cleavage, indicating the absence the genotoxic activity for both stem and leave extracts [21].

And finally, a study investigating the applicability potential of ethanolic extracts ointment from *B. cuspidata's* peel, for wounds treatment in Rats. The circular wounds of 12 mm diameter were done by surgical incision on the back of animals, which were divided in four groups, according to the received application. One group received a 0.9% saline solution, the second group received lanolin, and a third group received 2.5% of the lanolin emulsified extract. After 21 treatment days, the groups with the extract had given better results according to the skin regeneration and wound healing. This study suggests that the healing effects observed in ointment with the *B. cuspidata*'s extract are related to polyphenols compounds, including flavonoids, found in these extracts and which had their presence identified by phytochemical tests [29].

To the *B. veraguensis* stem extracts were prepared and tested for inhibitory activity of acetylcholinesterase. A concentration of 100 μg was applied on plates of TLC and eluted with a mix of $CHCl_3$-MeOH-H_2O (65:35:5) to polar extracts and AcOEt-n-hexane (1:1) to non-polar, respectively. The extracts revealed moderate inhibitory activity to acetylcholinesterase when compared to the HBr galatamine inhibitor tested at 0.01 μg on TLC. On this same study the extracts of *B. veraguensis* were tested for inhibition activity to DPPH free moiety. The extracts revealed moderate activity of inhibiting the DPPH free moiety compared to quercetine, tested as reference compound (0.1 μg) [23].

3.2.3. Ethnopharmacological Activity

For *B. australis* and *B. cuspidata* it was found an ethnophamacological study with a data survey of plants named "bitter plants" used as substitutes of *Chinchona* spp. on Brazilian traditional medicine [18].

3.3. Bothriospora

Ethnopharmacological Activity

It is noteworthy the ethnopharmacological study in biodynamic plants employed by the indians of amazonic region with medicines, poisons and narcotics. To *B. corymbosa* it is stated the use of fruits against intestine parasites [30].

3.4. Calycophyllum

3.4.1. Phytochemical Study

In a plant screening in Costa Rica the presence of alkaloids is found on leaves and stems of *C. candidissimum* through the chromatographic tests [31].

The phytochemical investigation of *C. spruceanum* presented the isolation of seco-iridoids 7-methoxydiderroside **(2)**, 6'-O-acetyldiderroside **(3)** and 8-O-tigloyldiderroside **(4)**, kingside **(5)**, diderroside **(6)** and loganetin **(7)** (**Figure 2**), isolated from the wood bark along with the iridoids loganin **(8)** and secoxyloganin **(9)** (**Figure 2**) [32].

3.4.2. Biological Activity

To the species *C. multiflorum*, it was found an investigation of the antifungal activity evaluation of dichloromethane extracts, methanol and aqueous of the bark against *Cryptococcus neoformans*, *Microsporum gypseum* and *Trichophyton mentagrophytes*. The extracts were evaluated through the disc method—agar diffusion. The dichloromethane extract presented a pronounced activity, with 5 mg per disc, diameters to the inhibition zone with

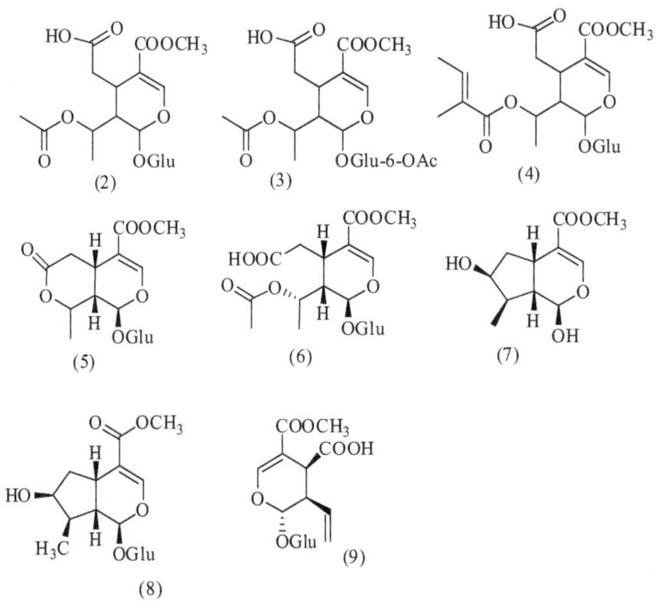

Figure 2. Seco-iridoids and iridoids of *C. spruceanum*.

the values: 21 mm, 18 mm and 18 mm, and to the dichloromethane extract with 10 mg per disc, diameters to the inhibition zone with the values: 24 mm, 20 mm and 20 mm for *C. neoformans*, *M. gypseum* and *T. mentagrophytes* respectively. Besides, the aqueous extracts containing 5 mg and 10 mg per disc present inhibition diameters with the values 23 and 25 respectively to *T. mentagrophytes* [33].

Special emphasis is given to studies evaluating the activities of *C. spruceanum* against leishmania with MTT (3-(4,5-dimethiliazol-2ul)-2,5-difenil bromide of tetrazoline) presenting absence of activity [34].

The antimalarial evaluation of the bark of *C. spruceanum in vivo* determined by the suppression test of four days, and *in vitro*, the infected erythrocytes with malaria were sown on well of microtiter plates, exposed to different concentrations in both tests [35].

Yet, to *C. spruceanum*, the antibacterial evaluation against *Mycobacterium tuberculosis* the radiometric *in vitro* trials using semiautomatic detector for growth monitoring was applied. The raw extracts of dichloromethane of vegetal material were dissolved to the concentration of 1 mg/ml of 100% dimethilsulfoxid (DMSO) and injected on bottles to obtain a final concentration of 50 μg of plant extract per ml of broth culture. The extracts presented less than 50% of inhibition, results considered non-significant [36].

The evaluation of antifungal activity against *Candida albicans* ATCC 90028, *Sporothrix schenckii* IHEM 15,503 and *Trichophyton mantagrophytes* IHEM 0584, *in vitro*, compared with amphotericin B. did not presented significant results [37].

The evaluation of the compounds 7-metoxididerrosideo, 6'-O-acethyldiderrosideo, secoxyloganin **(9)** and diderrode **(6)** exhibited poor activity *in vitro* against *Trypanosoma cruzi* with values of IC_{50} equal to 59, 0, 90, 2, 74, 2 and 84.9 mg/ml, respectively [32].

3.4.3. Ethnopharmacological Activity

The species *C. acreanumi*, *C. obovatum* and *C. spruceanum* are cited on an ethnobotanical investigation as employed by indians of the Amazon region. The shells of *C. acreanum* are scraped and placed in warm water about fungal infections. These shells are employed in infusions to combat intestine parasites. And for *C. spruceanum* a shell decoction is used as black scab (caused by arachnids) and the dry shell can be pulverized and applied on fungal infections [30].

To *C. spruceanum* species, it was found an ethnobotanical and ethnopharmacological investigation of plants used by indians of Peruvian Amazonia against skin infections on shell, fruits and leaves decoction forms [38] [39].

3.5. Capirona

3.5.1 Biological Activity

It is noteworthy an investigation of the antileishmania activity to *C. decorticans*. The leishmanicide activity *in vitro* of ethanolic extracts was evaluated for amastigote and promatigotes cultures by the MTT method, obtaining $IC_{50} > 100$ μg/mL for both forms, no significant results [40].

3.5.2. Ethnopharmacological Activity

The *C. decorticans* species is also used on formulation of medicines against leishmania by the traditional medicine of ethnic groups on Peruvian Amazonia [34]. It is also used for warts, infected wounds, fungal infections and scabies treatments. The shell is scraped and applied as a cataplasm on the affected area, and the broth of the shell is also used on wounds [40].

3.6. *Chimarrhis*

3.6.1. Phytochemical Study

To *C. turbunata* were stated phytochemical investigations where were isolated and indentified the indolic alkaloid monoterpenic glycosylated strictosidine **(10)**, the derived corinanteans 5-α-carboxystrictosidine **(11)**, isovallesiachotamine **(12)**, vallesiachotamine **(13)**, turbinatine **(14)**, 3,4-dehydro-strictosidine **(15)**, 3,4-dehydrostrictosidine acid **(16)**, strictosidine acid **(17)**, and the alkaloids *β*-carbolinic: cordifoline **(18)**, desoxycordifoline **(19)**, and harman-3-carboxylic acid **(20)** [7] [41] [42] (**Figure 3**).

Yet for C. turbinata in phytochemical investigations found the flavonoids chimarrhoside **(21)** quercetin-3-*O*-rutinoside **(22)**, kaempferol-3-*O*-rutinoside **(23)**, kaempferol-3-*O*-*α*-L-rhamnopyranosyl-(1→6)-*β*-D-ga-

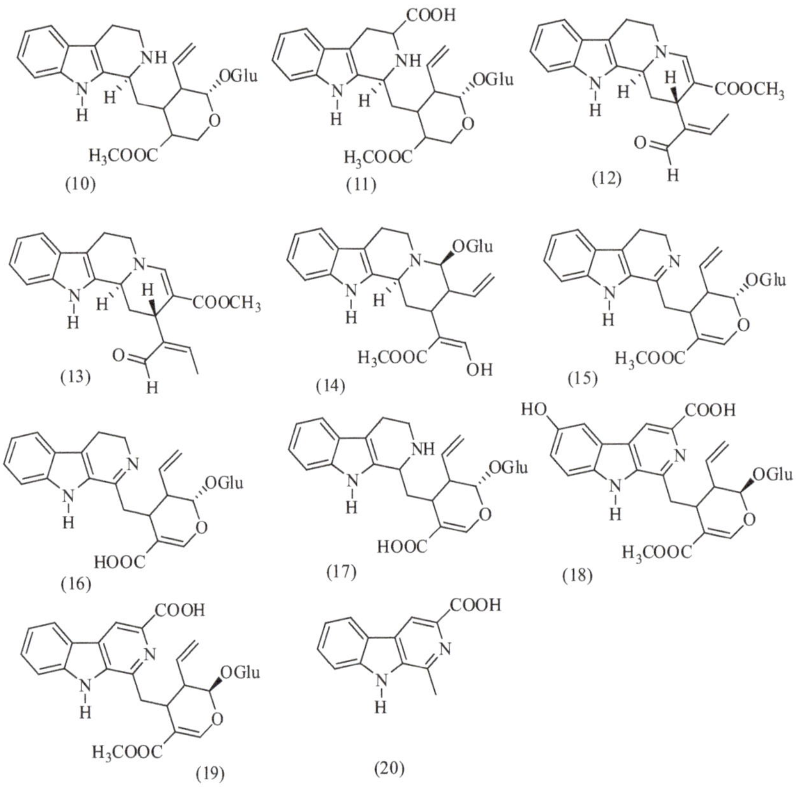

Figure 3. Alkaloids isolated of *C. turbinata.*

lactopyranoside **(24)**, quercetin-3-*O*-*α*-L-rhamnopyranosyl-(1→6)-*β*-D-galactopyranoside **(25)**, 6-hydro-xy-rutin **(26)**, kaempferol-3-*O*-D-galactopyranoside **(27)**, kaempferol-3-*O*-D-glucopyranoside **(28)**, kaempferol-3-*O*-*α*-L-rhamnopyranosyl-(1→6)-*α*-L-rhamnopyranosyl-(1→4)-*β*-D-glucopyranoside **(29)**, catechin **(30)** and catechin-(4*α*→8)-catechin-procyanidin B-3 **(31)** [43] (**Figure 4**).

Also the *C. turbinata* leaves were stated on phytochemical investigations that isolated and identified the derived chlorogenic acid 1-(3',4'-dihydroxycinnamoyl) cyclopentane-2,3-diol **(32)**, *trans*-chlorogenic acid methyl ester **(33)** and *cis*-chlorogenic acid methyl ester **(34)** and the chimarrhinin **(35)** [44] [45] (**Figure 5**).

The *C. parviflora* species also presented positive results for alkaloids on TLC tests with chromatographic reagents [15].

3.6.2. Biological Activity

Special emphasis is given to an investigation of *in vitro* evaluation of medicines based on *C. turbinata* by traditional medicine of French Guiana against *Plasmodium falciparum*, but did not presented significant results, with IC_{50} upper than 11 µg/mL being considered inactive, and for the test *in vivo* (suppressive test in four days) the medicines based on *C. turbinata* did not presented any activity [46].

The isolated compounds **(10)** to **(14)** exhibited poor, but selective, activity with values around $IC_{12} > 250$ and 100 µg/mL or more, on mutant yeast strains RS 188N (RAD+) and RS 322YK (rad 52Y), respectively [41] [47].

The compounds **(10)**, **(11)** and from **(14)** to **(20)** were tested for the moiety 1,1-difenil-2-picrilhidrazil (DPPH). The alkaloid **(18)** presented high scavenging activity to the DPPH radical (IC_{50} 18.3 µg/mL), using the antioxidant rutine pattern (IC_{50} 12.3 µg/mL) as reference, while all other alkaloids isolated revealed poor activity (IC > 40 µM). Besides, these alkaloids were submitted to TLC screening for acehylcolinesterase inhibitors. The compounds **(14)** and **(19)** revealed moderated inhibition to a concentration of 0.1 and 1.0 µM, respectively. A trial *in vitro* with rats' brains **(14)** presented moderated activity (IC_{50} 1.86 µM) compared to the standard compound, galantamine (IC_{50} 0.92 µM) [42].

The isolated *C. turbinata* flavonoids were submitted to an antioxidant activity evaluation on DPPH test, using rutine as standard (IC_{50} 21.34 µg/mL) and BHT (IC_{50} 62.50 µg/mL). The compounds **(22)** with IC_{50} 21.3 µg/mL,

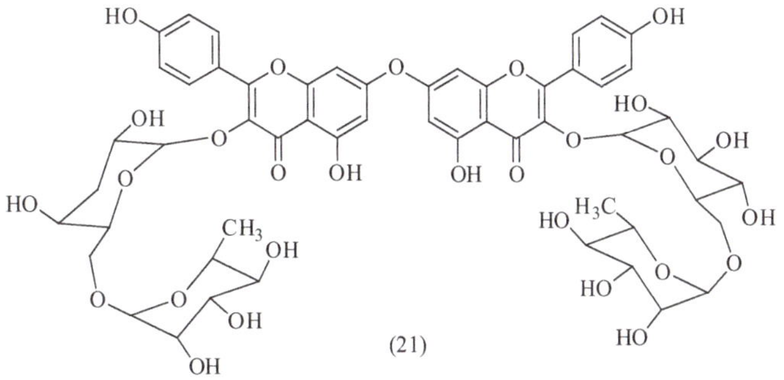

(21)

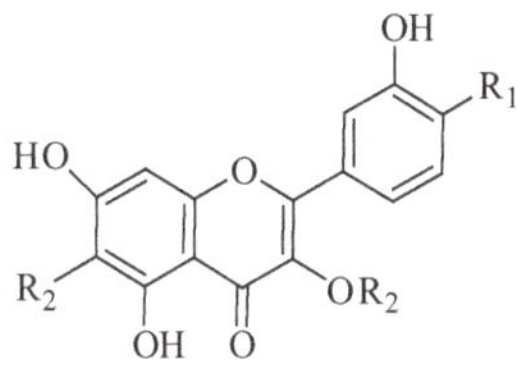

	R_1	R_2	R_3
(22)	OH	Glu-Rha	H
(23)	H	Gal-Rha	H
(24)	H	Glu-Rha	H
(25)	OH	Gal-Rha	H
(26)	H	Glu-Rha	OH
(27)	H	Gal	H
(28)	H	Glu	H
(29)	H	Glu-Rha-Rhu	H

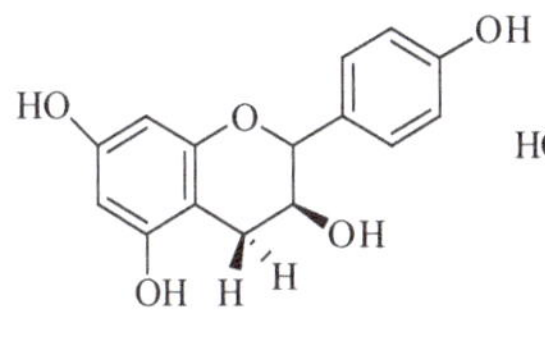

(30)

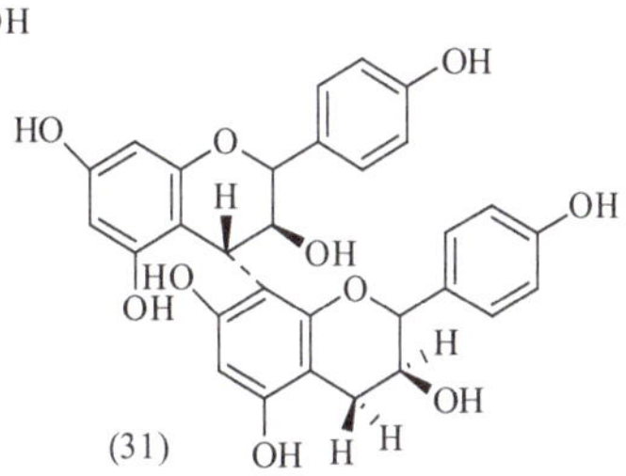

(31)

Figure 4. Flavonoid of *C. turbinata.*

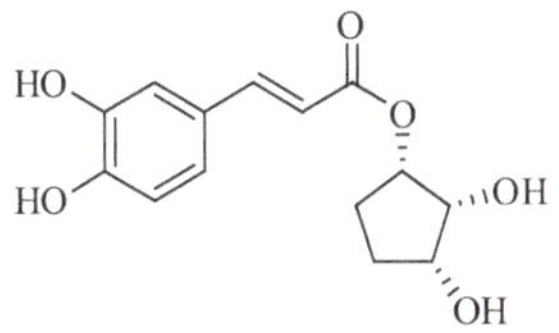

(32)

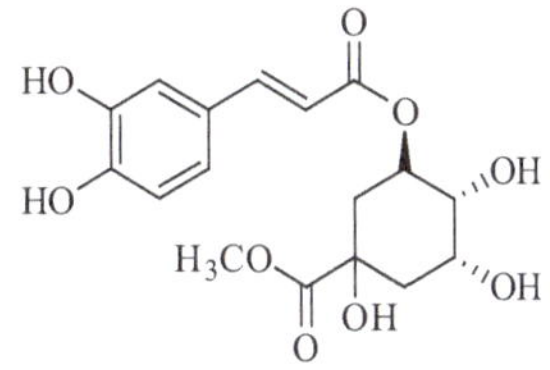

(33)

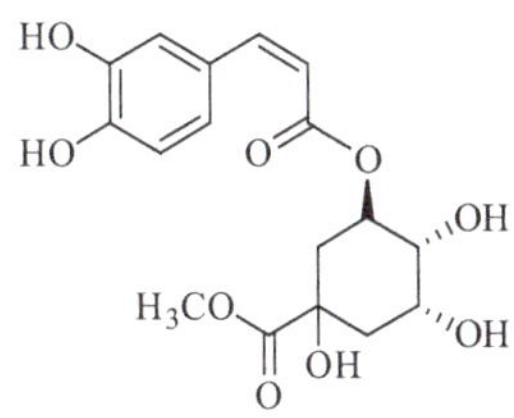

(34)

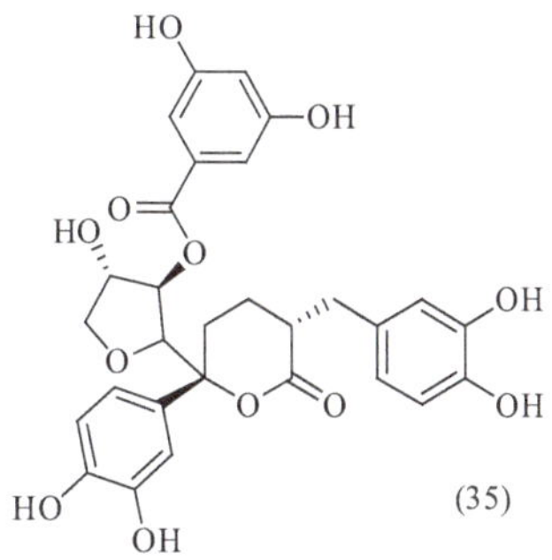

(35)

Figure 5. Other compounds of *C. turbinata.*

(25) with IC_{50} 30.1 µg/mL, **(26)** with IC_{50} 25.4 µg/mL, **(30)** with IC_{50} 33.4 µg/mL, and **(31)** with IC_{50} 11.5 µg/mL presented higher activities [43].

The oxidant activity with DPPH of the compounds **(33)** to **(35)** were evaluated using as standard the antioxidant reference BHT (IC_{50} 62.50 (0.6 μmol·L^{-1}) and chlorogenic acid (IC_{50} 20.0 (0.2 μmol·L^{-1}). The compound **(34)** revealed the best results with IC_{50} value (7.50 ± 0.5 μmol·L^{-1}) below the BHT standard. The compounds **(33)** and **(34)** revealed IC_{50} 18.60 ± 0.4 μmol·L^{-1} and IC_{50} 18.50 ± 0.6 μmol·L^{-1}, respectively. The results indicate the free moiety scavenging activity of these molecules [45].

3.7. Condaminea

3.7.1. Phytochemical Study

Positive result to *Condaminea corymbosa* alkaloids on TLC tests with chromatographic reagents with Rubiaceae species at Panama [15].

3.7.2. Ethnopharmacological Activity

The ethnopharmacological investigation where is cited the use of *C. corymbosa* stem as tonic on Peruvian Amazonia traditional medicine is highlighted [48].

3.8. *Dialypetalanthus*

A taxonomical study of *D. fuscescens* species analyzed the composition of polysaccharides on the cell walls of *D. fuscescens* leaves and compared with *Bathysa meridionalis*. The analyses demonstrated both species have very similar chemical composition of the cell wall corroborating as additional evidence of their taxonomic proximity as is actually accepted [49]. In addition to studies that classify the genus on Ixoroideae subfamily and Condamineeae tribe through molecular phylogenetical investigations [3] [5] [11].

3.9. Elaeagia

3.9.1. Phytochemical Study

The *E. nitidiflora* species presented positive results for alkaloids on tests with chromatographic reagents and on this same investigation the *E. auriculata* species presented negative results [15].

The study in subfraction of ethanol extract obtained from the leaves of *E. utilis*, analyzed by CG/MS. Terpenes, sesquiterpenlactones and simple phenolic compounds were identified in it for comparison with mass and retention time values from literature (**Table 2**) [50].

3.9.2. Biological Activity

The *E. auriculata* species activity was also evaluated for against *Aphis gossypii* and *A. craccivora* (aphids). The trial was realized on microplates covered by parafilm containing leaves extract (60 μg/ml) and proteins of leaves extract (33.3 μg/mL). The aphids fed up of these extracts on holes in the parafilm. The mortality rate was verified at each 24, 72 and 120 hours and the results were expressed in percentage. The leaves extract of *E. auriculata* presented 29.5% of mortality rate while the proteins presented 3.7%, no significant results [51].

E. utilis species' antimicrobial activity was evaluated against the bacteria that cause dental caries: *Streptococcus mutans*, *S. sobrinus* e *Lactobacillus acidophilus*. Two extracts, one from light petroleum and an ethanolic one, were obtained from the leaves of *E. utilis*. Both were fractionated and sub-fractionated, according to the presented activities in the evaluation, done by diffusion method in a concentration of 10 mg/well. In each well, 50 μl of extract were placed, fraction or sub-fraction, dissolved in DMSO. As a positive control, 50 μl of Vancomycin were used with 150 μg/ml concentration and 50 μl of DMSO as negative control. After 24 h of incubation, the inhibition zones were measured and, the minimum inhibitory concentration, calculated (MIC; at least 6

Table 2. Compounds identified by GC-MS in *E. utilis*.

Compounds in *E. utilis*	
Benzyl alcohol	3-methoxy-4-hydroxybenzaldehyde
Benzoic acid	4-vinylphenol
Benzoic acid, 2-hydroxy-, phenyl methyl ester	2,5-dimethyl-(3-methoxymethyl)-p-benzoquinone
2,3-dihydrobenzofuran	1,Z-5,E-7-dodecatrien
4-((1e)-3-hydroxy-1-propenyl)-2-methoxyphenol2	2,6-dimethoxy-4-(2-propenyl) phenol
-methoxy-4-vinylphenol	Loliolide

mm). Between the ethanolic extract fractions, the fraction of dichloromethane presented the greatest level of activity, being sub-fractionated in three fractions, of which the MeOH: H_2O fraction presented MIC = 0.1 µg/well with three tested bacteria. In this fraction, the presence of terpenes, sesquiterpenlactones and phenolic compounds, was confirmed through phytochemical tests, suggesting the relation between these compounds and the presented activity [50].

3.10. Emmenopterys

3.10.1. Phytochemical Study

Special emphasis is given to the phytochemical investigations for *E. henryi* reporting the isolation and investigation of cumarins: scopoletin **(36)**, umbelliferone **(37)** and umbelliferone-7-*β*-D-glucoside **(38)**; the steroids: *β*-sitosterol **(39)** and 3-O-*β*-D-glucopyranosylsitosterol **(40)**; the ursane triterpenes: taraxerol **(41)**, taraxerone **(42)**, ursolic acid acetate **(43)**, 3*β*, 19*α*, 23-trihydroxyurs-12-en-24-al-28-oic acid **(44)**, 3*β*, 19*α*, 24-trihydroxy-23-norurs-12-en-28-oic acid **(45)**, 12*β*-hydroxy-5*α*-pregnane-14.16-dien-3.20-dione **(46)**, pomolic acid **(47)**, 3*β*, 6*β*, 19*α*, 23-tetrahydroxyurs-12-en-28-oic acid **(48)**, 3*β*, 6*β*, 23-trihydroxyolean-12-en-28-oic acid **(49)**, 3*β*, 6*β*, 19*α*, 23-tetrahydroxyolean-12-en-28-oic acid **(50)** and 3*β*, 23.24-trihydroxyolean-12-en-28-oic acid **(51)**; and the pregnanes derivatives: 3*β*, 12*β*-dihydroxy-5*α*-pregnane-14.16-dien-20-one **(52)**, 12*β*-hydroxy-5*α*-pregnane-14, 16-dien-3.20-dione **(53)**, 3*β*, 12*β*-dihydroxy-5*α*-pregnane-16-en-20-one **(54)** and 12*β*-dihydroxy-5*α*-pregnane-16-en-3.20-dione **(55)** [52] [53] (**Figure 6**).

3.10.2 Biological Activity

The petroleum ether extract of the *E. henryi* stem presented antifeedant activity against *Plutella xylostella* on larvae trial treatment. The extract presented 80.09% of mortality rate against *Plutella xylostella* and 62.80% against *Spodoptera litura*, indicating high activity compared to industrially marketed standards on China and suggests further investigations for the use of this species as an insecticide [54].

3.11. Hippotis

In addition to studies which classify the genus on Ixoroideae subfamily and Condamineeae tribe through molecular phylogenetic investigations [3] [11], it was found only one ethnopharmacological study suggesting the *H. tubiflora* stem is used as tonic on Peruvian Amazonia traditional medicine [48].

3.12. Macrocnemum

3.12.1. Biological Activity

Evaluation of *M. roseum* leaves and bracts ethanolic extracts against leishmania using MTT method to determine the extract activities. For bioassays the extracts were dissolved on DMSO (final concentration 1%) at 10 mg/mL concentration, presenting moderate activity with IC_{50} 37 g/mL to bracts extract and IC_{50} 22 g/mL to leaves extract [34].

3.12.2. Ethnopharmacological Activity

An investigation of *M. roseum* states the use of leaves to treat wounds on traditional medicine of Peru. The leaves are boiled on water and the wounds washed the same water, but cold [34] [55].

3.13. Pentagonia

3.13.1. Phytochemical Study

It is noteworthy a phytochemical investigation with *P. gigantifolia* root ethanolic extracts where were isolated the compounds: 6-octadecynoic acid **(56)** and 6-nonadecynoic acid **(57)** [56] (**Figure 7**).

3.13.2. Biological Activity

The investigation with *P. gigantifolia* root ethanolic extract presented antifungal activity against *Candida albicans*, with IC_{50} < 20 µg/mL. Susceptibility tests to antifungal agents were realized using a modified version of NCCLS methods. After being chromatographed one of the fractions was tested and presented the highest activity with IC_{50} < 2 µg/mL [56].

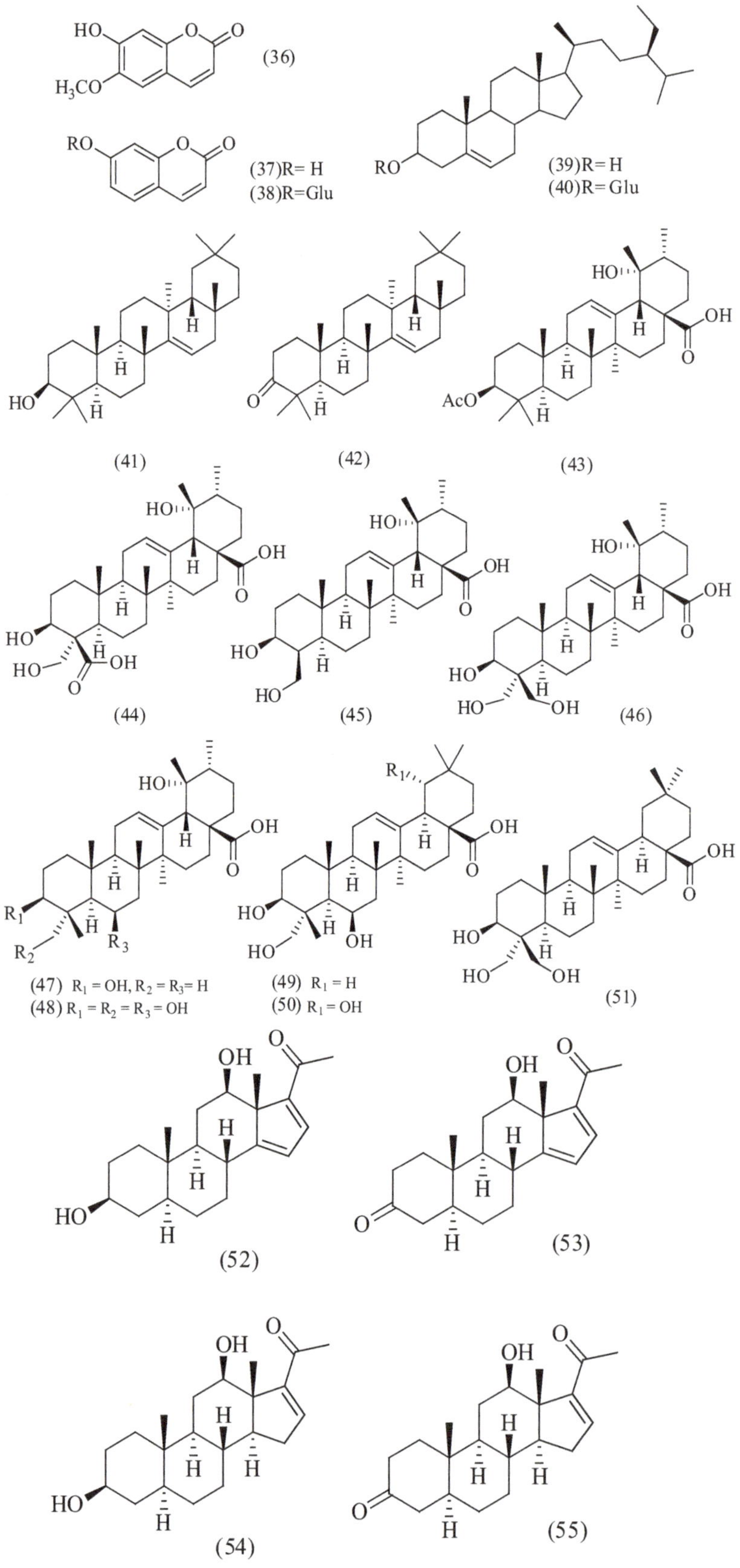

Figure 6. Compounds isolated from *E. henryi*.

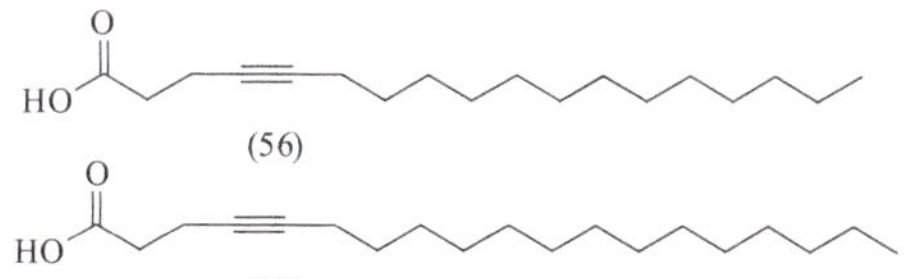
(56)
(57)

Figure 7. Fatty acids of *Pentagonia gigantifolia.*

The evaluation of compounds **(56)** with IC_{50} range of 0.45 to 0.65 μg/mL and **(57)** with IC_{50} range of 0.25 to 0.45 μg/mL presented activity with many *C. albicans* strains [56].

3.13.3. Ethnopharmacological Activity

It was found a study that states the use of *P. pubens* shell boiled on water to combat malaria and also it use as ornamental plant [57].

3.14. *Pogonopus*

3.14.1. Phytochemical Study

Phytochemical study of *P. specious* branches extracts led to the isolation of the alkaloids 1', 2', 3', 4'-tetradehydrotubulosine **(58)**, tubulosine **(59)** and psychotrine **(60)** [58]. The phytochemical investigation of *P. tubulosus* branches extracts, besides the compounds **(59)** and **(60)**, the alkaloid cephaeline **(61)** was also isolated [59] (**Figure 8**).

3.14.2. Biological Activity

On cytotoxicity trial the *P. tubulosus* extract obtained by acid-base extraction revealed IC_{50} = 0.015 μg/mL against epidermoid carcinoma KB cells in culture. The **(58)** and **(59)** alkaloids were evaluated against several human tumor cell lines presenting poor cytotoxicity with ED_{50} = 2.2 to 9.0 μg/mL for **(58)** and ED_{50} = 2.3 to 13.3 μg/mL for **(60)**, activity considered poor compared to the most potent activity observed for tubulosine (59), with ED_{50} < 0.001 to 0.22 μg/mL [58].

The raw ethanolic extract and a *P. tubulosus* extract obtained by acid-base extraction were also evaluated for antiproliferative activity against several lines of human neoplastic cells: ATCC-CCL-23, laryngeal carcinoma, ATCC-HTB-22, breast carcinoma, ATCC-CRL-1932, liver carcinoma, ATCC-HTB-38, colon carcinoma, and murine cancer: ATCC-CRL-6322, murine melanoma. The raw extract presented IC_{100} range of 2.2 to 6.9 μg/mL and the extract of acid-base extraction with IC_{100} range of 0.24 to 2.2 μg/mL presented high antiproliferative activity against all neoplastic cells lines [60].

The alkaloids **(59)**, **(60)** and **(61)** were also tested against sensible strains and resistant to *Plasmodium falsiparumin vitro* and against *P. berghei* and *P. vinckei petteri* (*in vivo*). Tubulosine **(59)** presented *in vitro* IC_{50} = 0.006 μg/mL against sensible *P. falciparum* and IC_{50} = 0.011 μg/mL against resistant *P. falciparum*. And *in vivo* presented antimalarial activity with ED_{50} = 0.05 mg/Kg/day for *P. vinckei petteri* and ED_{50} = 0.45 mg/Kg/day for *P. berghei* against *Plasmodium falciparum in vitro* tests. The *in vivo* tests were done through the classic suppression method during four days [59] [61].

The psychotrine alkaloid **(60)** presented *in vitro* IC_{50} = 0.14 μg/mL against sensible *P. falciparum* and IC_{50} = 0.39 μg/mL against resistant *P. falciparum*. And *in vivo* presented antimalarial activity with ED_{50} > 2 mg/Kg/day for *P. berghei*. The cefaeline alkaloid **(61)** presented *in vitro* IC_{50} = 0.027 μg/mL against sensible *P. falciparum* and IC_{50} = 0.011 μg/mL against resistant *P. falciparum*. And *in vivo* presented antimalarical activity with ED_{50} = 6 mg/Kg/day for *P. berghei* [59].

3.14.3. Ethnopharmacological Activity

Studies of the use of *P. tubulosus* on traditional medicine of Bolivia against malaria in the stem bark decoction form [62].

3.15. *Rustia*

3.15.1. Phytochhemical Study

The study of essential oil extracted from *Rustia formosa* secretory cavities, analyzed by CG/MS, presented a

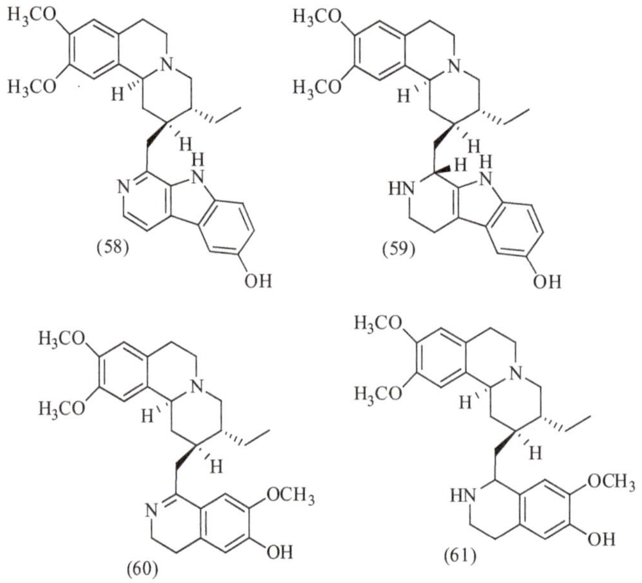

Figure 8. Alkaloids Isolated of *Pogonopus* genus.

complex mix of 75 terpenic components, formed mainly by sesquiterpens, substances solely identified for comparison with mass and retention time values from literature (**Table 3**) [63].

3.15.2. Ethnopharmacological Activity

An investigation states the use of *R. formosa* species on Brazilian traditional medicine as substitute of *Chinchona* spp., based on pharmacological studies [18].

3.16. *Sommera*

3.16.1. Phytochemical Study

The phytochemical investigation of *S. sabiceoides* roots ethanolic extracts presented five fatty acids: 6-hexadecynoic acid **(62)**, 6-heptadecynoic acid **(63)** and 6-icosynoic acid **(64)** (**Figure 9**) and 6-octadecynoic acid **(56)** and 6-nonadecynoic acid **(57)** [63] (**Figure 7**).

3.16.2 Biological Activity

The *S. sabiceoides* dichloromethane extract presented 96% of inhibition against *Mycobacterium tuberculosis*, in 50 μg/mL and presented MIC < 6.25 μg/mL, considered excellent results. The *in vitro* experiments (trials) were done using a radiometric detector semiautomized to monitor the *Mycobacterium tuberculosis* growth [36].

The methanolic extract tested against *Candida albicans* presented 50% inhibition with IC_{50} < 20 μg/mL. The *in vitro* cytotoxocity was determined by the neutral red method. On the *in vivo* cytotoxicity experiment the compounds were dissolved on peanut oil until the desired concentration. Six rat groups received 100 μL intraperitoneal injections, providing compound doses of 3, 4, 17 or 34 μmol/kg of body weight.

The *S. sabiceoides* compounds **(56)** and **(57)** presented positive results against *Trichophyton mentagrophytes* ATCC MYA-4439, *T. mentagrophytes* ATCC 9533, *T. rubrum* ATCC 10218, and *T. rubrum* ATCC MYA-4438 presenting MFC range from 0.7 to 11.1 μM. The *S. sabiceoides* compounds **(62)** and **(64)** along with **(56)** and **(57)** were also active against *Candida krusei* ATCC 6258 with MIC range from 4.4 to 10.6 Mm [64].

The compound **(57)** also presented activity against *C. albicans* (MIC range from 1.3 to 8.0 μM) against *Aspergillus fumigatus* (MIC range from 2.7 to 6.6 μM), against *T. mentagrophytes* e *T. rubrum* strains (MIC range from 0.7 to 3.3 μM) and against fluconazol-resistant *C. albicans* (MIC range from 1.3 to 8.0 μM).

The compounds **(62)** and **(64)** did not presented *in vitro* toxicity to concentrations until 32 μM on cytotoxicity test with SK-MEL cells (melanoma), KB (epidermic carcinoma, oral), BT-549 (ductal carcinoma, breast), SK-OV-3 (ovary carcinoma), and LLCPK-1 (pork kidney epithelial). And on cytotoxicity test **(56)** and **(57)** did not presented toxic effects on doses less than 34 μmol/Kg/day [64].

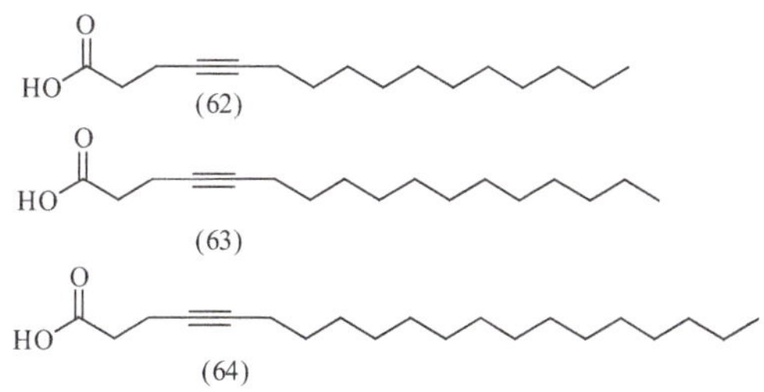

Figure 9. Fatty acids of *Sommera sabiceoides*.

Table 3. Compounds identified in *R. formosa* essential oil.

Compounds in *R. formosa* essential oil					
β-elemene	germacrene D	*α*-humulene	*α*-muurolene	curzerenone	germacrene B
γ-muurolene	*γ*-elemene	*cis*-*β*-guaiene	caryophylleneoxide	*α*-cadinene	*epi*-*α*-muurolol
β-caryophyllene	*β*-selinene	*α*-cadinol	*δ*-cadinene	*epi*-α-cadinol	germacrone

3.17. *Warszewiczia*

3.17.1. Phytochemical Study

The *W. coccinea* shell was tested on a histochemical study for alkaloid presence, presenting negative result [65], however presented positive results on another investigation regarding chromatographic techniques [15].

Still, *W. coccinea* investigations stated the presence of the triterpens 3*β*, 6*β*, 19*α*-trihydroxy-urs-12-en-28-oic acid **(65)** and 3*β*, 6*β*-dihydroxy-olean-12-en-28-oic acid **(66)** [23] and the tiramine methabolites **(67)** and cyanidin 3-glicoside **(68)** [66] (**Figure 10**).

3.17.2. Biological Activities

Evaluation of *W. cordata* antileishmania activity. The experiments were conducted on promastigotes and axenic amastigotes of *Leishmania amazonensis* and the extract activity was determined through MTT method. The *W. cordata* branches ethanolic extract presented IC_{50} = 60.6 μg/ml for the amastigote form, moderated activity, presented no result for promastigote form [67].

An investigation of the *W. coccinea* ethanolic extract activity against leishmania (*Leishmania amazonensis*) revealed poor activity determined through MTT method, presenting IC_{50} > 100 μg/mL to shell extract and IC_{50} = 54.3 μg/mL to branches extract. To the antimalaric (*Plasmodium falciparum* FCR3 *in vitro*) there was no activity [68].

The *W. coccinea* stem extract strongly inhibited acetylcholinesterase in 100 μg on TLC, and inhibited poorly free DPPH moieties on antioxidant test. On the same investigation the compounds **(65)** and **(66)** were active on the inhibition acetylcholinesterase test with 1 μg at each TLC plate [23].

3.17.3. Ethnopharmacological Activity

An ethnobotanic investigation stated the use of *W. coccinea* grated shell in cold water intake for convulsion and epilepsy treatment on Peruvian traditional medicine [68].

There were also found ethnobotanic investigations wherein the *W. coccinea* roots are dried and the powder is applied with oil to fungal skin disorders or the root is boiled and applied as decoction to pain relief [30].

W. coccinea shells are also used as haemostatic by Colombian indians, to nosebleed treatment at Peru and fever at Porto Rico [23].

The *Dioicodendron*, *Dolichodelphys*, *Dolicholobium*, *Ferdinandusa*, *Holtonia*, *Macbrideina*, *Mastixiodendron*, *Mussaendopsis*, *Parachimarrhis*, *Picardaea*, *Pinckneya*, *Schizocalyx*, *Semaphyllanthe*, *Tammsia* e *Wittmackanthus* genus presented absence or irrelevant investigations to this review.

4. Results and Discussion

In relation to the phytochemical studies, from the 33 genus in the Condamineeae Tribe, approximately 13 presented at least one studied species, containing isolated or identified secondary metabolites. In a total of 110

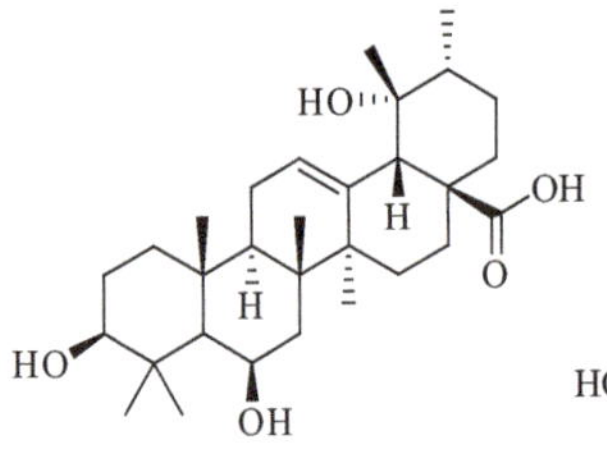

(65)

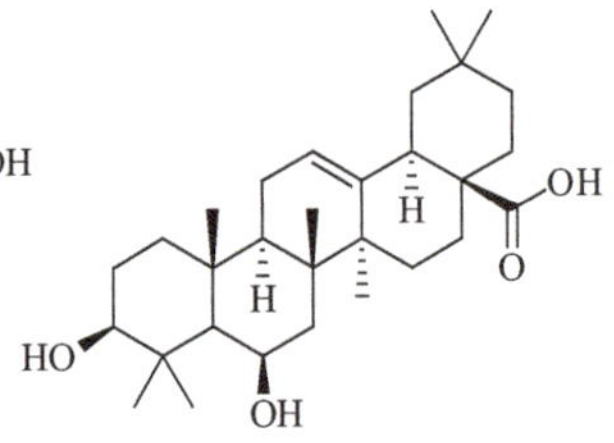

(66)

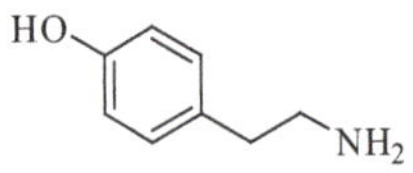

(67)

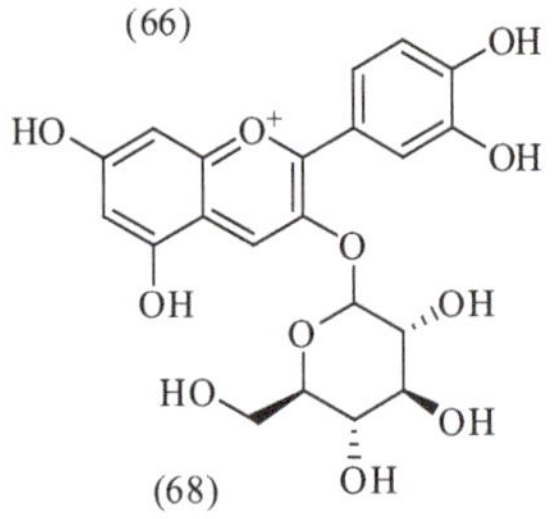

(68)

Figure 10. Compounds isolated of *W. coccinea.*

isolated and identified substances in these species, including the results of the review for the genus *Simira* [14], approximately 24.54% were from the alkaloid's class, and from these substances, approximately 81.18% are indole alkaloids, which are considered as chemotaxonomic markers from family [7]. The class of terpenoids metabolites, including diterpens, triterpens, and steroids, presents a bigger percentage, approximately 27.27% of the substances found in this tribe until this moment (**Graphic 1** and **Graphic 2**)

According to the biological activity, from the 33 genus in the tribe, 13 presented some evaluation of biological activity, and 14 presented reports of ethnopharmacological application, but not necessarily showed biological activity.

From the 13 genus which presented biological activity, 9 genus, including the genus *Simira*, presented studies reporting the presence of alkaloid or a positive result when tested for alkaloids. And from these 9 genus, 7 also presented ethnopharmacological reports (**Graphic 3** and **Graphic 4**).

5. Conclusions

It was observed the species that belongs to the genera of Condamineeae tribe presented a considered number of metabolites in similar biosynthetic origin and a lot of them presenting some kind of biological activity.

The alkaloids found for *Simira* genus are indolic [14] as the alkaloids presented for *Chimarrhis* and *Pogonopus* genus, considered as taxonomic markers of the family, belongs to the most investigated secondary metabolic groups aiming to prove the phyologenetic correlation between Ruibiaceae chemistry and taxonomy. Evidences based on indolic alkaloids skeletons, used for chemical and morphological evolutionary comparisons between Rubiaceae, Apocynaceae and Loganiaceae [8].

The presence of iridoids and it derivatives secoirioids of *Calycophyllum* genus also demonstrates the connection between the genus presenting monoterpenes indolic alkaloids, as *Chimarrhis* and *Pogonopus*, since the secoiridoids are involved on the construction of the basic skeleton of indolic alkaloids [69]. This proximity relationship can extend to the genus that presented positive results for alkaloids on chromatographic tests as *Alseis*, *Bathysa*, *Condaminea*, *Elaeagia*, *Macrocnemum*, *Pentagonia*, *Rustia*, *Sommera*, *Warszewiczia* e também o gênero *Simira*. The *Calycophyllum* genus presented positive results for alkaloids.

For *Chimarrhis* genus it was found the harman-3-carboxilic acid **(20)** (**Figure 3**), structure that could derive from harmana alkaloid, considered as a taxonomic tracer in *Simira* genus [14] corroborating the proximity between the *Simira* and *Chimarrhis* genus.

This review also states the scarcity of phytochemical investigations of several genera and consequently a lot of species of Rubiaceae family. It could contribute more significantly to this family taxonomy.

Acknowledgements

The authors are thankful to CNPq and FAPERJ for financial support.

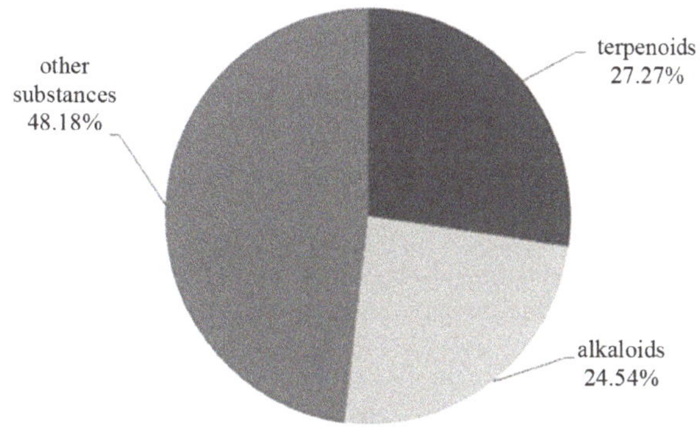

Graphic 1. Percentage of the compounds isolated in Condamineeae tribe.

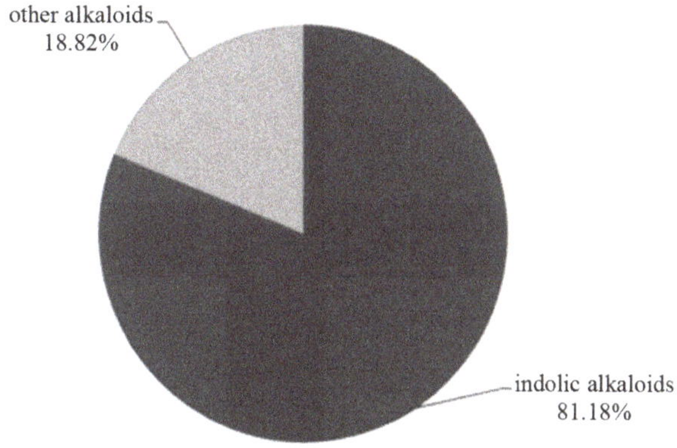

Graphic 2. Percentage of alkaloids isolated in Condamineeae tribe.

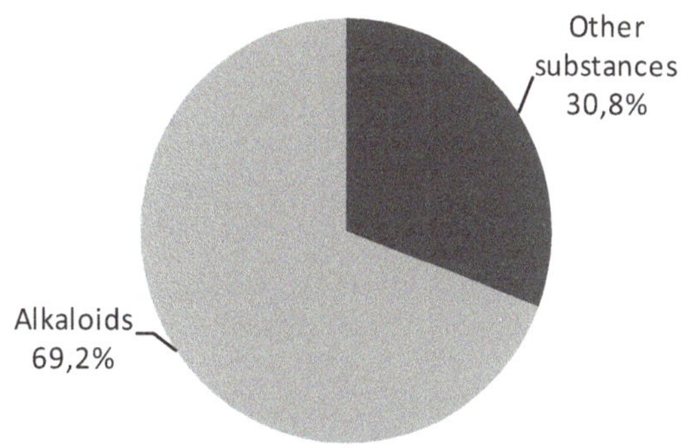

Graphic 3. Alkaloids percentage in genus with biological activity.

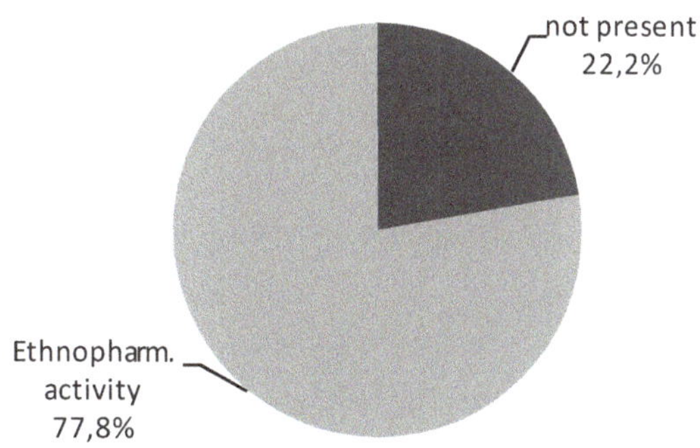

Graphic 4. Ethnopharmacological activity percentage in genus with alkaloids.

Conflict of Interest

The authors confirm that this article content has no conflict of interest.

References

[1] Robbrecht, E. and Manen, J. (2006) The Major Evolutionary Lineages of the Coffee Family (Rubiaceae, Angiosperms). Combined Analysis (nDNA and cpDNA) to Infer the Position of *Coptosapelta* and *Luculia*, and Supertree Construction Based on *rbcL*, *rps*16, *trnL-trnF* and *atpB-rbcL* Data. A New Classification in Two Subfamilies, Cinchonoideae and Rubioideae. *Systematics and Geography of Plants*, **76**, 85-146.

[2] Bremer, B., Jansen, R.K., Oxelman, B., Backlund, M., Lantz, H. and Kim, K.J. (1999) More Characters and More Taxa for a Robust Phylogeny—Case Study from the Coffee Family (Rubiaceae). *Systematic Biology*, **48**, 413-435. http://dx.doi.org/10.1080/106351599260085

[3] Rova, J.H.E., Delprete, P.G., Andersson, L. and Albert, V.A. (2002) A *trnL-F* cpDNA Sequence Study of the Condamineeae-Rondeletieae-Sipaneeae Complex with Implications on the Phylogeny of the Rubiaceae. *American Journal of Botany*, **89**, 145-159. http://dx.doi.org/10.3732/ajb.89.1.145

[4] Judd, W.S., Campbell, C.S., Kellogg, E.A. and Stevens, P.F. (2007) Plant Systematics: A Phylogenetic Approach. 3rd Edition, Sinauer Associates, Inc., Sunderland.

[5] Delprete, P.G. (1999) Rondeletieae (Rubiaceae)—Part I (Rustia, Tresanthera, Condaminea, Picardaea, Pogonopus, Chimarrhis, Dioicodendron, Molopanthera, Dolichodelphys, and Parachimarrhis). Flora Neotropica Monograph, Vol. 77, 1-226.

[6] Barbosa, M.R. *Simira* in List of Species of Flora of Brazil. Botanical Garden of Rio de Janeiro. Accessed 30 March 2014. http://floradobrasil.jbrj.gov.br/jabot/floradobrasil/FB14287

[7] Cardoso, C.L., Silva, D.H.S., Young, M.C.M., Castro-Gamboa, I. and da Silva Bolzani, V. (2008) Indole Monoterpene Alkaloids from *Chimarrhis turbinata* DC Prodr.: A Contribution to the Chemotaxonomic Studies of the Rubiaceae Family. *Brazilian Journal of Pharmacognosy*, **18**, 26-29. http://dx.doi.org/10.1590/s0102-695x2008000100007

[8] Bolzani, V.S., Yong, M.C.M., Furlan, M., Cavalheiro, A.J., Araujo, A.R., Silva, D.H.S. and Lopes, M.N. (2001) Secondary Metabolites from Brazilian Rubiaceae Plant Species: Chemotaxonomical and Biological Significance. *Recent Research Developments in Phytochemistry*, **5**, 19-31.

[9] Bremer, B. and Eriksson, O. (2009) Time Tree of Rubiaceae: Phylogeny and Dating the Family, Subfamily, and Tribes. *International Journal of Plant Science*, **170**, 766-793. http://dx.doi.org/10.1086/599077

[10] Bremer, B. (2009) A Review of Molecular Phylogenetic Studies of Rubiaceae. *Annals of the Missouri Botanical Garden*, **96**, 4-26. http://dx.doi.org/10.3417/2006197

[11] Kainulainen, K., Persson, C., Eriksson, T. and Bremer, B. (2010) Molecular Systematics and Morphological Character Evolution of the Condamineeae (Rubiaceae). *American Journal of Botany*, **97**, 1961-1981. http://dx.doi.org/10.3732/ajb.1000090

[12] Razafimandimbison, S.G. and Bremer, B. (2001) Tribal Delimitation of Nuclease (Cinchonoideae, Rubiaceae): Inference from Molecular and Morphological Data. *Systematics and Geography of Plants*, **71**, 515-538. http://dx.doi.org/10.2307/3668697

[13] Fay, M.F., Bremer, B., Prance, G.T., van der Bank, M., Bridson, D. and Chase, M.W. (2000) Plastid *rbcL* Sequence Data Show *Dialypetalanthus* to Be a Member of Rubiaceae. *Kew Bulletin*, **55**, 853-864. http://dx.doi.org/10.2307/4113630

[14] Moreira, V.F., Vieira, I.J.C. and Braz-Filho, R. (2014) Chemical Constituents and Biological Activities of *Simira* Genus: A Contribution to the Chemotaxonomic of Rubiaceae Family. *The Natural Products Journal*, **4**, 290-298. http://dx.doi.org/10.2174/2210315505666141211225600

[15] Soto-Sobenis, A., Castillo, B., Delgado, A., González, A. and Montenegro R. (2001) Alkaloid Screening of Herbarium Samples of Rubiaceae from Panama. *Pharmaceutical Biology*, **39**, 161-169. http://dx.doi.org/10.1076/phbi.39.3.161.5925

[16] Slish, D.F., Arvigo, R. and Balick, M.J. (2004) *Alseis yucatanensis*: A Natural Product from Belize That Exhibits Multiple Mechanisms of Vasorelaxation. *Journal of Ethnopharmacology*, **92**, 297-302. http://dx.doi.org/10.1016/j.jep.2004.03.003

[17] Slish, D.F., Ueda, H., Arvigo, R. and Balick, M.J. (1999) Ethnobotany in the Search for Vasoactive Herbal Medicines. *Journal of Ethnopharmacology*, **66**, 159-165. http://dx.doi.org/10.1016/S0378-8741(98)00225-6

[18] Cosenza, G.P., Somavilla, N.S., Fagg, C.W. and Brandão, M.G.L. (2013) Bitter Plants Used as Substitute of *Cinchona* spp. (Quina) in Brazilian Traditional Medicine. *Journal of Ethnopharmacology*, **149**, 790-796. http://dx.doi.org/10.1016/j.jep.2013.08.004

[19] Weeks, R.A., Dobberstein, R.H. and Farnsworth, N.R. (1977) Isolation of Paeonol from *Bathysa meridionalis*. *Lloydia*, **40**, 515-516.

[20] Rangel, L.P., de Abreu, L.F., de Andrade, A.R., Leitão, S.G., Leitão, G.G. and Ferreira-Pereira, A. (2008) Effect of Different Extracts from the Brazilian Atlantic Forest on the Pdr5p ATPase Activity. *Brazilian Journal of Pharmacognosy*, **18**, 30-36. http://dx.doi.org/10.1590/s0102-695x2008000100008

[21] Gontijo, D.C., Nunes, L.G., Souza, C.J.A., Fietto, L.G. and Leite, J.P.V. (2012) Evaluation of the Genotoxic Potential of Ethanolic Extracts of Stem Bark and Leaves of *Bathysa cuspidata* (A.St.-Hil.) Hook. *Revista de Ciências Farmacêuticas Básica e Aplicada*, **33**, 355-359.

[22] Novaes, R.D., Gonçalves, R.V., Marques, D.C.S., Cupertino, M.C., Peluzio, M.C.G., Neves, C.A., Leite, J.P.V. and Maldonado, I.R.S.C. (2012) Effect of Bark Extract of *Bathysa cuspidata* on Hepatic Oxidative Damage and Blood Glucose Kinetics in Rats Exposed to Paraquat. *Toxicologic Pathology*, **40**, 62-70. http://dx.doi.org/10.1177/0192623311425059

[23] Calderón, A.I., Simithy, J., Quaggio, G., Espinosa, A., López-Pérez, J.L. and Gupta, M.P. (2009) Triterpenes from *Warszewiczia coccinea* (Rubiaceae) as Inhibitors of Acetylcholinesterase. *Natural Product Communications*, **4**, 1323-1326.

[24] de Castro Miguel, E., Gomes, V.M., de Oliveira, M.A. and da Cunha, M. (2006) Colleters in *Bathysa nicholsonii* K. Schum. (Rubiaceae): Ultrastructure, Secretion Protein Composition, and Antifungal Activity. *Plant Biology*, **8**, 715-722. http://dx.doi.org/10.1055/s-2006-924174

[25] Fonseca, T.L., Von Groll, A., Leitão, G.G., Scaini, C.J., Ramos, D.F. and Silva, P.E.A. (2008) Antimycobacterial Vegetal Extracts against *Mycobacterium fortuitum* and *Mycobacterium malmoense*. *Vittalle*, **20**, 65-71.

[26] Carvalho, D.D.C., Alves, E., Camargos, R.B., Oliveira, D.F., Scolforo, J.R.S., de Carvalho, D.A. and Batista, T.R.S. (2011) Plant Extracts to Control *Alternaria alternata* in Murcott Tangor Fruits. *Revista Iberoamericana de Micología*, **28**, 173-178. http://dx.doi.org/10.1016/j.riam.2011.05.001

[27] Novaes, R.D., Gonçalves, R.V., Cupertino, M.C., Marques, D.C.S., Rosa, D.D., Peluzio, M.C.G., Neves, C.A. and Leite, J.P.V. (2012) Bark Extract of *Bathysa cuspidata* Attenuates Extra-Pulmonary Acute Lung Injury Induced by Paraquat and Reduces Mortality in Rats. *International Journal of Experimental Pathology*, **93**, 225-233. http://dx.doi.org/10.1111/j.1365-2613.2012.00808.x

[28] Gonçalves, R.V., Novaes, R.D., Leite, J.P.V., Vilela, E.F., Cupertino, M.C., Nunes, L.G. and Matta, S.L.P. (2012) Hepatoprotective Effect of *Bathysa cuspidata* in a Murine Model of Severe Toxic Liver Injury. *International Journal of Experimental Pathology*, **93**, 370-376. http://dx.doi.org/10.1111/j.1365-2613.2012.00835.x

[29] Gonçalves, R.V., Novaes, R.D., Cupertino, M.C., Araújo, B.M., Vilela, E.F., Machado, A.T., Leite, J.P.V. and Matta, S.L.P. (2014) *Bathysa cuspidata* Extract Modulates the Morphological Reorganization of the Scar Tissue and Accelerates Skin Wound Healing in Rats: A Time-Dependent Study. *Cells Tissues Organs*, **199**, 266-277. http://dx.doi.org/10.1159/000365504

[30] Schultes, R.E. (1985) De Plantis Toxicariis e Mundo Novo Tropicale Commentationes XXXIV: Biodynamic Rubiaceous Plants of the Northwest Amazon. *Journal of Ethnopharmacology*, **14**, 105-124. http://dx.doi.org/10.1016/0378-8741(85)90082-0

[31] Saenz, R., Nassar, J.A. and Maryssia, C. (1965) Phytochemical Screening of Costa Rican Plants: Alkaloid Analysis II. *Revista de Biologia Tropical*, **13**, 207-212.

[32] Zuleta, L.M.C., Cavalheiro, A.J., Silva, D.H.S., Furlan, M., Young, M.C.M., Albuquerque, S., Castro-Gamboa, I. and da Silva Bolzani, V. (2003) *seco*-Iridoids from *Calycophyllum spruceanum* (Rubiaceae). *Phytochemistry*, **64**, 549-553. http://dx.doi.org/10.1016/S0031-9422(03)00153-5

[33] Portillo, A., Vila, R., Freixa, B., Adzet, T. and Cañigueral, S. (2001) Antifungal Activity of Paraguayan Plants Used in Traditional Medicine. *Journal of Ethnopharmacology*, **76**, 93-98. http://dx.doi.org/10.1016/S0378-8741(01)00214-8

[34] Odonne, G., Bourdy, G., Castillo, D., Estevez, Y., Lancha-Tangoa, A., Alban-Castillo, J., Deharo, E., Rojas, R., Stien, D. and Sauvain, M. (2009) *Ta'ta'*, *Huayani*: Perception of Leishmaniasis and Evaluation of Medicinal Plants Used by the Chayahuita in Peru. Part II. *Journal of Ethnopharmacology*, **126**, 149-158. http://dx.doi.org/10.1016/j.jep.2009.07.015

[35] Deharo, E., Bourdy, G., Quenevo, C., Muñoz, V., Ruiz, G. and Sauvain, M. (2001) A Search for Natural Bioactive Compounds in Bolivia through a Multidisciplinary Approach. Part V. Evaluation of the Antimalarial Activity of Plants Used by the Tacana Indians. *Journal of Ethnopharmacology*, **77**, 91-98. http://dx.doi.org/10.1016/S0378-8741(01)00270-7

[36] Graham, J.G., Pendland, S.L., Prause, J.L., Danzinger, L.H., Vigo, J.S., Cabieses, F. and Farnsworth, N.R. (2003) Antimycobacterial Evaluation of Peruvian Plants. *Phytomedicine*, **10**, 528-535. http://dx.doi.org/10.1078/094471103322331502

[37] Wen, L., Haddad, M., Fernandez, I., Espinoza, G., Ruiz, C., Neyra, E., Bustamante, B. and Rojas, R. (2011) Antifungal Activity of Four Plants Used in Peruvian Traditional Medicine: Isolation of the Active Principle of *Psidium acutangulum*. *Revista de la Sociedad Quimica del Peru*, **77**, 199-204.

[38] Jovel, E.M., Cabanillas, J. and Towers, G.H.N. (1996) An Ethnobotanical Study of the Traditional Medicine of the Mestizo People of Suni Mirano, Loreto, Perú. *Journal of Ethnopharmacology*, **53**, 149-156. http://dx.doi.org/10.1016/0378-8741(96)01437-7

[39] Polesna, L., Polesny, Z., Clavo, M.Z., Hansson, A. and Kokoska, L. (2011) Ethnopharmacological Inventory of Plants Used in Coronel Portillo Province of Ucayali Department, Peru. *Pharmaceutical Biology*, **49**, 125-136. http://dx.doi.org/10.3109/13880209.2010.504927

[40] Castillo, D., Arevalo, J., Herrera, F., Ruiz, C., Rojas, R., Rengifo, E., Vaisberg, A., Lockb, O., Lemesre, J.L., Gornitzka, H. and Sauvain, M. (2007) Spirolactone Iridoids Might Be Responsible for the Antileishmanial Activity of a Peruvian Traditional Remedy Made with *Himatanthus sucuuba* (Apocynaceae). *Journal of Ethnopharmacology*, **112**, 410-414. http://dx.doi.org/10.1016/j.jep.2007.03.025

[41] Cardoso, C.L., Silva, D.H.S., Tomazela, D.M., Verli, H., Young, M.C.M., Furlan, M., Eberlin, M.N. and da Silva Bolzani, V. (2003) Turbinatine, a Potential Key Intermediate in the Biosynthesis of Corynanthean-Type Indole Alkaloids. *Journal of Natural Products*, **66**, 1017-1021. http://dx.doi.org/10.1021/np020547m

[42] Cardoso, C.L., Castro-Gamboa, I., Silva, D.H.S., Furlan, M., Epifanio, R.D.A., da Cunha Pinto, A., de Rezende, C.M., Lima, J.A. and da Silva Bolzani, V. (2004) Indole Glucoalkaloids from *Chimarrhis turbinata* and Their Evaluation as Antioxidant Agents and Acetylcholinesterase Inhibitors. *Journal of Natural Products*, **67**, 1882-1885. http://dx.doi.org/10.1021/np049863m

[43] Cardoso, C.L., Silva, D.H.S., Castro-Gamboa, I. and da Silva Bolzani, V. (2005) New Biflavonoid and Other Flavonoids from the Leaves of *Chimarrhis turbinate* and Their Antioxidant Activities. *Journal of the Brazilian Chemical Society*, **16**, 1353-1359. http://dx.doi.org/10.1590/S0103-50532005000800008

[44] Cardoso, C.L., da Silva Bolzani, V., Silva, D.H.S., Ishii, H., Berova, N. and Nakanishi, K. (2006) The Absolute Configuration of 1-(3',4'-Dihydroxycinnamoyl)Cyclopentane-2,3-diol from the Amazonian Tree *Chimarrhis turbinate*. *Journal of Natural Products*, **69**, 1046-1050. http://dx.doi.org/10.1021/np050522y

[45] Cardoso, C.L., Castro-Gamboa, I., Bergamini, G.M., Cavalheiro, A.J., Silva, D.H.S., Lopes, M.N., Araujo, A.R., Furlan, M., Verli, H. and da Silva Bolzani, V. (2011) An Unprecedented Neolignan Skeleton from *Chimarrhis turbinata*. *Journal of Natural Products*, **74**, 487-491. http://dx.doi.org/10.1021/np1007476

[46] Bertani, S., Bourdy, G., Landau, I., Robinson, J.C., Esterre, P. and Deharo, E. (2005) Evaluation of French Guiana Traditional Antimalarial Remedies. *Journal of Ethnopharmacology*, **98**, 45-54. http://dx.doi.org/10.1016/j.jep.2004.12.020

[47] Gunatilaka, A.A.L., Samaranayake, G. and Kingston, D.G.I. (1992) Bioactive Ergost-5-ene-3ß,7a-diol Derivatives from *Pseudobersama mossambicensis*. *Journal Natural Products*, **55**, 1648-1654. http://dx.doi.org/10.1021/np50089a014

[48] Sanz-Biset, J. and Cañigueral, S. (2011) Plant Use in the Medicinal Practices Known as "Strict Diets" in Chazuta Valley (Peruvian Amazon). *Journal of Ethnopharmacology*, **137**, 271-288. http://dx.doi.org/10.1016/j.jep.2011.05.021

[49] Moraes, F., Hahn, M.G. and Braga, M.R. (2001) Comparative Analysis of Leaf Cell-Wall Polysaccharides of *Dialypetalanthus fuscescens* and *Bathysa meridionalis*: Evidence of Biochemical Similarities between Dialypetalanthaceae and Rubiaceae-Cinchonoideae. *Revista Brasileira de Botânica*, **24**, 289-294. http://dx.doi.org/10.1590/S0100-84042001000300007

[50] Aldana, J., Téllez, N. and Gamboa, F. (2013) Antimicrobial Activity of Fractions and Subfractions of *Elaegia utilis* against Microoganisms of Importance in Dental Caries. *Acta Odontológica Latinoamericana*, **26**, 104-111.

[51] Quiros, D.I., Emmen, D.A., Dominguez, E., Heller, M.V., Coley, P.D. and Kursar, T.A. (2006) A Rapid, Efficient Method for the Bioassay of Extracts, Fractions and Compounds for Activity against Tropical Aphids. *International Journal of Pest Management*, **52**, 333-342. http://dx.doi.org/10.1080/09670870600847525

[52] Ma, Z.W. and He, G.F. (1989) Chemical Constituents of *Emmenopterys henryi* Oliv. Native to China. *Acta Botanica Sinica*, **31**, 620-625.

[53] Wu, X.D., He, J., Li, X.Y., Dong, L.B., Gong, X., Gao, X., Song, L.D., Li Y., Peng, L.Y. and Zhao, Q.S. (2013) Triterpenoids and Steroids with Cytotoxic Activity from *Emmenopterys henryi*. *Planta Medica*, **79**, 1356-1361. http://dx.doi.org/10.1055/s-0033-1350645

[54] Liu, T.X., Xu, H.H. and Luo, W.C. (2006) Opportunities and Potentials of Botanical Extracts and Products for Management of Insect Pests in Cruciferous Vegetables. *Naturally Occurring Bioactive Compounds*, **8**, 171-197. http://dx.doi.org/10.1016/S1572-557X(06)03008-X

[55] Odonne, G., Valadeau, C., Alban-Castillo, J., Stien, D., Sauvain, M. and Bourdy, G. (2013) Medical Ethnobotany of

the Chayahuita of the Paranapura Basin (Peruvian Amazon). *Journal of Ethnopharmacology*, **146**, 127-153. http://dx.doi.org/10.1016/j.jep.2012.12.014

[56] Li, X.C., Jacob, M.R., ElSohly, H.N., Nagle, D.G., Smillie, T.J., Walker, L.A. and Clark, A.M. (2003) Acetylenic Acids Inhibiting Azole-Resistant *Candida albicans* from *Pentagonia gigantifolia. Journal of Natural Products*, **66**, 1132-1135. http://dx.doi.org/10.1021/np030196r

[57] Delprete, P.G. (1996) Systematics, Typification, Distribution, and Reproductive Biology of *Pinckneya bracteata* (Rubiaceae). *Plant Systematics and Evolution*, **201**, 243-261. http://dx.doi.org/10.1007/BF00989065

[58] Ito, A., Lee, Y.H., Chai, H.B., Gupta, M.P., Farnsworth, N.R., Cordell, G.A., Pezzuto, J.M. and Kinghorn, A.D. (1999) 1',2',3',4'-Tetradehydrotubulosine, a Cytotoxic Alkaloid from *Pogonopus speciosus. Journal of Natural Products*, **62**, 1346-1348. http://dx.doi.org/10.1021/np990255u

[59] Sauvain, M., Moretti, C., Bravo, J.A., Callapa, J., Muñoz, V., Ruiz, E., Richard, B. and Le Men-Olivier, L. (1996) Antimalarial Activity of Alkaloids from *Pogonopus tubulosus. Phytotherapy Research*, **10**, 198-201. http://dx.doi.org/10.1002/(SICI)1099-1573(199605)10:3<198::AID-PTR798>3.0.CO;2-Z

[60] Perdomo, R.T. (2011) Evaluation of the Antiproliferative Activity and Inhibition of Topoisomerases by Extracts, Pure and Semi-Synthetic Substances Obtained from Natural Products. Doutorate Degree Thesis, Federal University of Mato Grosso do Sul, Campo Grande.

[61] Desjardins, R.E., Canfield, C.J., Haynes, J.D. and Chulay, J.D. (1979) Quantitative Assessment of Antimalarial Activity *in Vitro* by a Semi-Automated Microdilution Technique. *Antimicrobial Agents and Chemotherapy*, **16**, 710-718. http://dx.doi.org/10.1128/AAC.16.6.710

[62] Frederich, M., Tits, M. and Angenot, L. (2008) Potential Antimalarial Activity of Indole Alkaloids. *Royal Society of Tropical Medicine and Hygiene*, **102**, 11-19. http://dx.doi.org/10.1016/j.trstmh.2007.10.002

[63] Vieira, R.C., Delprete, P.G., Leitão, G.G. and Leitão, S.G. (2001) Anatomical and Chemical Analyses of Leaf Secretory Cavities of *Rustia formosa* (Rubiaceae). *American Journal of Botany*, **88**, 2151-2156. http://dx.doi.org/10.2307/3558376

[64] Li, X.C., Jacob, M.R., Khan, S.I., Ashfaq, M.K., Babu, K.S., Agarwal, A.K., ElSohly, H.N., Manly, S.P. and Clark, A.M. (2008) Potent *in Vitro* Antifungal Activities of Naturally Occurring Acetylenic Acids. *Antimicrobial Agents and Chemotherapy*, **52**, 2442-2448. http://dx.doi.org/10.1128/AAC.01297-07

[65] Loustalot, A.J. and Pagan, C. (1949) Local "Fever" Plants Tested for Presence of Alkaloids. *El Crisol*, **3**, 3-5.

[66] Mohammed, M.J. and Seaforth, C.E. (1981) Extractives of *Warszewiczia coccinea* Klotzch. *Revista Latinoamericana de Quimica*, **12**, 72-73.

[67] Estevez, Y., Castillo, D., Pisango, M.T., Arevalo, J., Rojas, R., Alban, J., Deharo, E. and Bourdy, G. (2007) Sauvain, M. Evaluation of the leishmanicidal Activity of Plants Used by Peruvian Chayahuita Ethnic Group. *Journal of Ethnopharmacology*, **114**, 254-259. http://dx.doi.org/10.1016/j.jep.2007.08.007

[68] Valadeau, C., Pabon, A., Deharo, E., Albán-Castillo, J., Estevez, Y., Lores, F.A., Rojas, R., Gamboa, D., Sauvain, M., Castillo, D. and Bourdy, G. (2009) Medicinal Plants from the Yanesha (Peru): Evaluation of the Leishmanicidal and Antimalarial Activity of Selected Extracts. *Journal of Ethnopharmacology*, **123**, 413-422. http://dx.doi.org/10.1016/j.jep.2009.03.041

[69] Cunha, A.P. (2005) Farmacognosia e fitoquímica. Fundação Calouste Gulbenkian, Lisboa, 403-412.

4

Morphological and Biochemical Changes in Ginseng Seedling Roots Affected with Stripe Symptoms

Zamir K. Punja[1*], Mahfuzur Rahman[2]

[1]Department of Biological Sciences, Simon Fraser University, Burnaby, Canada
[2]WVU Extension Service, West Virginia University, Morgantown, USA
Email: *punja@sfu.ca

Abstract

A unique symptom of longitudinal red stripes on the surface of one-year-old ginseng roots was studied to determine the morphological and biochemical changes taking place. Light and scanning electron microscopy, measurements of phenolic and mineral element content, and enzyme activity, were compared between healthy and stripe-affected root tissues. Light microscopy revealed that the root epidermis had ruptured and fissures extended for 3 - 4 cell layers into the cortex. Phenolic compounds accumulated in the epidermal cells which stained with Toluidine blue O. Total phenolic content was higher in tissues from striped roots compared to healthy roots and HPLC profiles showed increases in a number of specific phenolic compounds. Analysis of epidermal tissues by SEM-EDX for mineral element content showed a marked increase in levels of iron, silicon and aluminum and a decline in potassium in striped root tissues. The activity of the enzymes phenylalanine ammonia lyase and peroxidase were also found to be higher in striped root tissues. Striping of ginseng roots is a physiological condition caused by a rupture of the epidermis due to rapid growth of underlying cells, which results in phenolic accumulation and sequestration of several minerals. Further oxidation causes a visible red striping on the root surface.

Keywords

Cell Disruption, Iron Sequestration, *Panax quinquefolius*, Phenolics, Root Epidermis

1. Introduction

Ginseng (Panax species, Family Araliaceae) is a medicinal plant which is cultivated in various parts of North

*Corresponding author.

America, Asia, Australia and other regions of the world [1]. The root (rhizome) has been used in traditional Chinese medicine and is gaining increasing attention for use as a medicinal herb worldwide. In Canada, American ginseng (*Panax quinquefolius* L.) is cultivated in regions of British Columbia and Ontario and the root is harvested, dried, trimmed, and shipped for export to Asia. Ginseng root quality is determined by shape, size and overall appearance. Blemish-free roots are highly valued and any surface discolorations can significantly reduce market value. Ginseng plants are initiated from stratified seed planted in the fall, and a thick layer of straw mulch is placed on the soil surface. After seedlings emerge in the spring (first-year), the plants are grown for an additional 3 - 4 growing seasons to reach a marketable size [1]. During the four-year growth cycle, ginseng plants can be affected by a number of fungal diseases and abiotic stress factors which can affect root quality [2] [3]. A considerable amount of genetic diversity also exists among ginseng plants in cultivated fields that can lead to variation in morphological traits [4]. Previous studies have shown that physiological (abiotic) and biotic factors can cause root discoloration of the ginseng root surface and distortion of root growth. The abiotic factors include anaerobic soil conditions during root growth [5], ethylene production under stress conditions [6], and soil conditions in which iron accumulates to toxic levels in the tissues [5] [7]. In addition, reddish-brown to orange-brown discolored regions near the crown of the root were shown to be due to infection by fungi that included *Fusarium* spp. [8], *Cylindrocarpon destructans* [9], and *Rhexocercosporidium* sp. [10]. These fungi caused root rot as well as extensive distortion and discoloration of the root.

During annual surveys conducted on ginseng farms in British Columbia to assess the occurrence and incidence of seedling and root diseases, a unique discoloration on the root surface that consisted of longitudinal stripes (striped root) near the crown of the root was observed in different years of sampling (2008-2012). The symptoms did not resemble any of the previously described diseases or disorders associated with ginseng roots [2] [3] [6]-[10]. The objective of this research was to determine the morphological and biochemical changes occurring in the affected roots in an effort to establish the causes of the striping symptom.

2. Materials and Methods

2.1. Root Samples

Commercial ginseng fields from which root samples were obtained were located in Kamloops, BC and were sampled during the months of June to September. Roots showing longitudinal stripes (see **Figure 1**) were selected primarily from one-year-old plantings (in the year following seeding the previous fall season) where the symptoms predominated on seedling roots. Most of the roots were collected in the later part of the growing season (August-September) in two consecutive years. Harvested roots (approx. 100 in total) were placed inside plastic bags and transported to the laboratory in a cooler and stored at 4˚C and used for the experiments below. Field sites where striped roots were observed were marked and re-visited the following year to observe progression of symptom development.

2.2. Light Microscopy

Segments of root tissues approx. 2 mm^2 with visible stripe symptoms (**Figure 1**) were cut from affected roots and placed in a fixative for 72 hr (formalin: acetic acid: alcohol, 1:1:1) and then dehydrated in an ethanol series (70% to 100%, 2 hr in each). The tissue pieces were pre-infiltrated in a 100% ethanol: 2-hydroxyethylmethacrylate mixture (1:1, v/v) (Technovit 7100, Marivac Ltd., Halifax, NS). Samples were then infiltrated in Technovit 7100 (100 ml plus 1 g hardner) for 24 hr at 4˚C and then embedded in Technovit. Cross-sections of the tissue were made using an LKB pyramitome (Diversified Equipment Corp., Lorton, VA) to a thickness of 4 - 8 µm. The sections were stained with Toluidine Blue 0 (0.05%) for 30 sec and rinsed in water. The sections were placed on microscope slides, dried at 50˚C for 10 min, and sealed under a coverslip using Permount (Fisher Scientific Canada, Ottawa, ON). The sections were examined at magnifications of 100 - 400 X under a Zeiss light microscope. Control tissues taken from roots without any striping were included for comparison. A minimum of 5 sections of each root sample was examined.

2.3. Scanning Electron Microscopy

Samples of small root segments (1 mm^2) containing striped tissue were prepared for scanning electron microscopy (SEM) and energy-dispersive spectroscopy analysis (EDX) to observe morphological changes on the root

Figure 1. Symptoms of striping on ginseng seedling roots. (a) Healthy root (extreme left) compared to various stages of striping that develop on the top portion of the root. The roots on the extreme right show splitting of the epidermis and intense red colour development; (b) stripe symptoms extending down to almost two-thirds of the root; (c) top view of a seedling root with stripe symptoms. The white structure at the top of the root is the vegetative bud for the following year's growth. Scale bar = 1 cm.

surface and identify metal compounds in epidermal cells, respectively. Healthy root tissues were included as a control. For SEM, the tissues samples were fixed in 0.1 M cacodylate buffer containing 2% glutaraldehyde, 4% formaldehyde and 4% sucrose using standard microwave processing [11]. Samples were post-fixed in 2% (w/v) tannic acid, followed by fixation in 1% osmium tetroxide in the above buffer for 1 h at room temperature, and 2% aqueous uranyl acetate for 1 h at room temperature in the dark. The samples were then dehydrated in a graded ethanol series and critical point-dried using CO_2. Samples were mounted on stubs using colloidal silver, sputter-coated with gold-palladium in a Nanotech SEMPrep II Sputter Coater and examined in a scanning electron microscope (Hitachi S4700). For EDX, the tissue samples were immediately frozen in liquid nitrogen and freeze-dried for 48 hr, after which the samples were adhered to a metallic stub and carbon-coated to make the sample conductive. The samples were analyzed in a Hitachi S-3000 N, EDX QuartzX one (Tokyo, Japan).

2.4. Mineral Element Determination

The roots selected for this assay were first gently washed in deionized water several times and blotted dry. The striped tissues were gently scraped from the root surface using a sharp scalpel and bulked to obtain a composite sample. Healthy root tissues were included as a control. Air-dried (0.2 g) tissue was digested with 40% (v/v) hydrofluoric acid mixed with nitric acid and made up to 100 ml with 1% nitric acid. Aliquots of this solution were mixed with an internal standard and analyzed by Inductively Coupled Plasma Atomic Emission Spectrometry (ICP-AES) as described in [12]. The results were compared to a calibration curve constructed using multi-element standards and expressed in ppm. Two replicate samples were included for analysis and the procedure was repeated once using a different set of roots.

2.5. Phenolic Compounds

Total phenolic compounds present in striped and healthy root tissues was determined using the Folin-Ciocaleau method [13]. Using a sharp scalpel, the striped areas were gently scraped off the surface of multiple roots and bulked to provide 2 g (fresh weight) which was then frozen in liquid nitrogen and ground using a mortar and pestle. The tissues were mixed with 80% methanol containing 8% formic acid, stirred at 37°C in a water bath for 30 min, and centrifuged at 8000 × g for 10 min. One ml of the supernatant was added to 5 ml of diluted (1:10, v/v) F-C reagent and mixed for 30 sec. After 5 min, 4 ml of 7.5% Na_2CO_3 solution was added and the mixture was incubated for 2 hr in the dark. Absorbance was measured at 740 nm in a Thermo Spectronic Helios UV visible spectrophotometer (Cambridge, UK) by using the reagent mix as a reference. The mean absorbance was compared to a calibration graph prepared using a range of concentrations (10, 25, 50 and 100 $\mu g \cdot ml^{-1}$) of gallic acid and the phenolic concentration was expressed as $\mu g \cdot g^{-1}$ gallic acid equivalents (GAE). Two replicate samples were included for analysis and the procedure was repeated once using a different set of roots.

To determine specific phenolic compounds, the above supernatant extract was evaporated to dryness using a rotary evaporator. The residue was dissolved in 5 ml solvent (methanol, formic acid and water) and centrifuged at 12,000 × g for 10 min and filtered through a 0.2 µm hydrophilic Durapore membrane filter (Millipore Corp., Bedford, MA). Fifty microliters of the extract, which constituted the mobile phase for HPLC, was injected through through an Atlantis™ dC18 column in an LC-module 1 Waters HPLC machine (Mildford, MS) at a flow rate 1.2 ml/min^{-1} and the absorbance was recorded at 280 nm over a 24 min retention time.

2.6. Enzyme Assays

Striped tissues were gently scraped from the surface of multiple roots and bulked to provide 2 g (fresh weight), frozen immediately in liquid N_2, and homogenized with 20 ml of extraction buffer on ice. The buffer was prepared with 80 ml Tris-HCl (0.1 M, pH 7.8), containing 15 ml of 10 mM *β*-mercaptoethanol, 5 ml 0.5% ascorbate and three drops of glycerol. The homogenate was centrifuged at 37,000 × g for 10 min at 4°C. The supernatant was kept on ice until the assay for enzymes was done. Three replicate samples were included in each assay.

Phenylalanine ammonia lyase (PAL). The assay mixture contained 900 µl of 10 mM L-phenylalanine in Tris–HCl (0.1 M, pH 8.8), and 100 µl extract. The mixture was placed in a water bath at 37°C for 60 min. Enzyme activity was determined using a spectrophotometer (Thermo Spectronic Helios UV visible) by the increase in absorbance at 290 nm over a 60 min interval. The change in absorbance was found to be proportional to the amount of enzyme, which produces an increase in absorbance at 290 nm of 1.0 per hour with a 1 cm light path. Specific enzyme activity was expressed as units per milligram protein of the enzyme extract. Protein concentration was measured according to Bradford [14] using bovine serum albumin (BSA) as a standard.

Peroxidase (POD). The tissue extracts were diluted 10-fold with the homogenization buffer. Pyrogallol solution (0.05 M) was prepared in 5 mM $Na2(PO_4)_3$ buffer (pH 6.0). One hundred microliters of the diluted extracts was mixed with 800 µl pyrogallol. The reaction was started by adding 100 µl of 0.6 M H_2O_2 to the mixture, and the rate of increase in absorbance at 470 nm was measured at 20 sec intervals for a minute. Specific enzyme activity was expressed as units per milligram protein of the enzyme extract. One unit is defined as the amount that will form 1.0 mg purpurogallin (PPG) from pyrogallol in 20 sec at pH 6.0 at 20°C by increasing the absorbance by 1.0.

3. Results

3.1 Symptoms and Light Microscopic Observations of Striped Roots

Ginseng roots with stripe symptoms were distinctly different from healthy roots that showed no external discoloration (**Figure 1(a)**). Striped roots had varying degrees of reddish-orange longitudinal stripes that varied in length from 2 cm to up to 8 cm and extended down from the upper crown to more than halfway down the length of the root (**Figure 1(b)**). The stripes consisted of ridges of tissues comprised of reddish-brown material (**Figure 1(c)**). Light microscopic examination of the tissues showed that the cell layers stained a bluish-green with Toluidine blue 0 (**Figure 2(a)**), indicating the presence of phenolic compounds. The underlying cortical cells remained unstained. The ridges appeared to have formed as a result of epidermal fissures that extended down into the cortical cells, causing them to push outwards (**Figure 2(b)**). The fissures could be seen forming over the entire root surface, leading to multiple ridges forming on the root (**Figure 3(a)**). At subsequent later stages of development, the epidermal cell layers that were stained with Toluidine blue and several cortical cell layers were sloughed off by the formation of an abscission zone (**Figure 3(b)**). These cell layers were not seen in control roots (**Figure 3(c)**).

3.2. Scanning Electron Microscopy

Images of the surface of striped root tissues examined under a SEM confirmed that the epidermis had ruptured and cracks (fissures) were formed (**Figure 4(a)**), creating ridges of tissue comprised of the epidermis on either side and exposing the underlying cortical cells (**Figure 4(b)**). These ridges of darkly-pigmented tissues could also be seen as longitudinal lines (stripes) that made their way down the root (**Figure 4(c)**). Examination of the epidermal tissues using SEM-EDX revealed that there was a significant increase in the levels of silicon, aluminum and iron as seen in the spectrum analysis when compared with healthy epidermal tissues, in which only potassium was significantly higher (**Figure 5**). When the compositional levels of all elements were expressed as

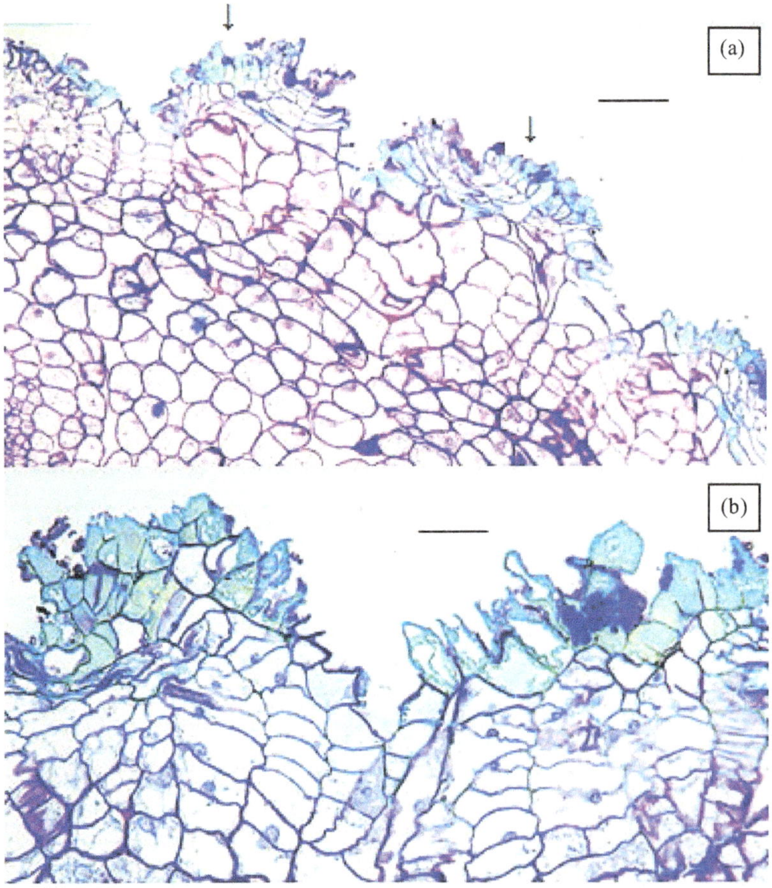

Figure 2. Light microscopic images of thin sections made through roots with stripe symptoms. (a) The epidermal cells which are stained blue with Toluidine blue 0 (arrows) form the striped areas that accumulate phenolic compounds. Note the development of fissures that develop in between the ridges; (b) close-up of blue-stained epidermal cells that form a ridge of cells on either side of a fissure. Scale bar = 25 μm.

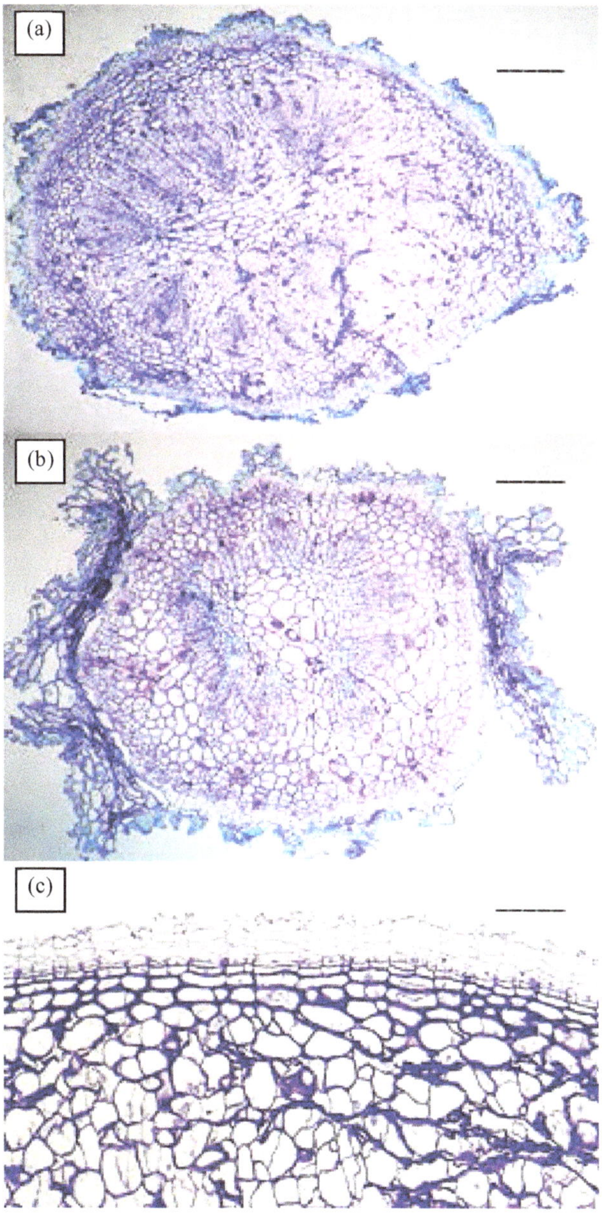

Figure 3. Light microscopic images of thin sections made through whole seedling roots with stripe symptoms. (a) The clusters of epidermal cells that form the ridges are stained blue and can be seen over the entire root surface; (b) a later stage of stripe root in which the epidermal cells are being sloughed off by the formation of an abscission zone; (c) control root showing intact epidermis and absence of fissures. Scale bar = 1 mm.

concentration (wt [%]), the levels of aluminum, calcium, iron and silicon were found to be higher and potassium lower in striped tissues (**Table 1**). As well, magnesium and phosphorus were slightly higher in striped tissues.

3.3. Mineral Element Determination

Analysis of striped root tissues and healthy tissues by Inductively Coupled Plasma Atomic Emission Spectrometry confirmed the results obtained by SEM-EDX. Levels of aluminum and iron were markedly higher in striped tissues, and magnesium and manganese levels were slightly higher (**Table 2**). In healthy tissues, levels of copper and zinc were increased.

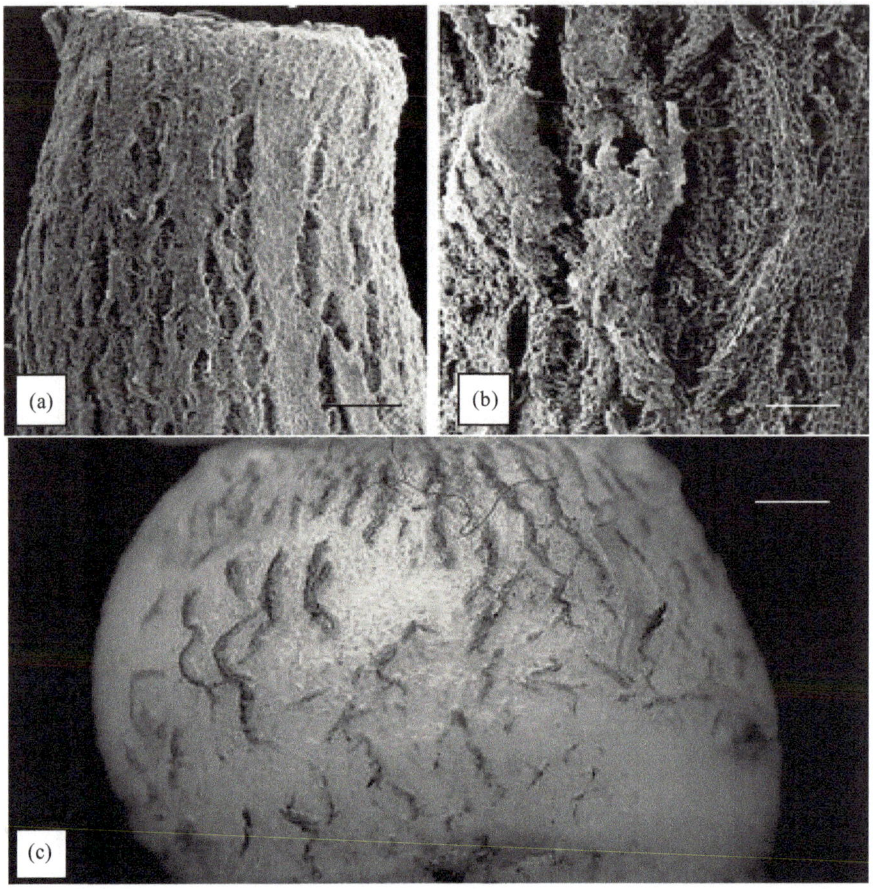

Figure 4. Scanning electron microscopic images of a seedling root with stripe symptoms. (a) Small fissures (cracks) can be seen in the epidermis which expose the underlying cortical cells; (b) close-up of a large fissure showing the formation of ridge tissues on both sides of the crack; (c) seedling root showing ridges of pigmented tissues on the root surface. Scale bar = 0.5 cm.

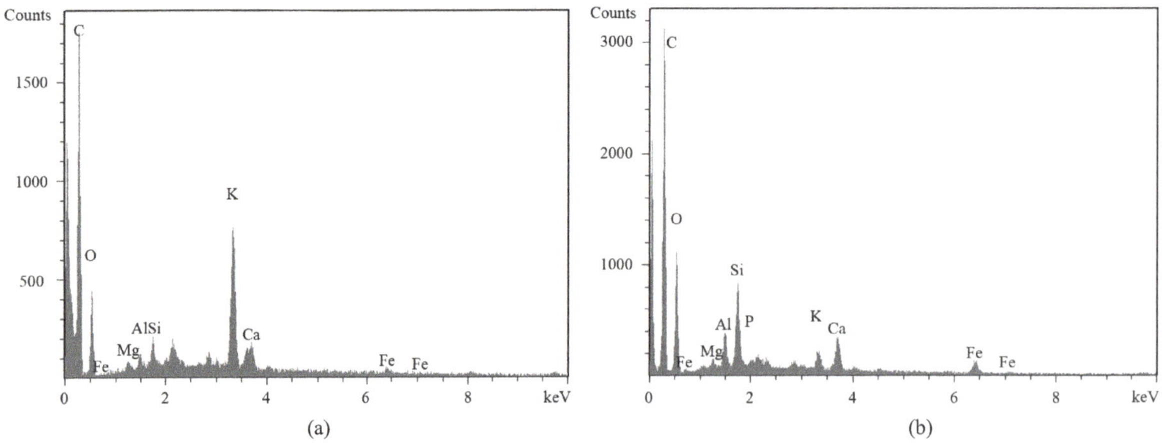

Figure 5. SEM-EDX spectrum analysis of healthy ginseng seedling roots (a) compared to striped roots (b).

3.4. Phenolic Compounds and Enzyme Assays

Total phenolic levels in extracts from striped root tissues were almost 80% higher compared to healthy tissues (69 $\mu g/g^{-1}$ gallic acid equivalents (GAE) compared to 38 $\mu g/g^{-1}$, respectively). When examined by HPLC analysis,

Table 1. Levels of specific mineral elements in healthy ginseng epidermal tissues compared with epidermal tissues showing stripe symptoms as determined by SEM-EDX.

Element	Healthy Root (%)	Stripe Root (%)
Aluminum	0.0	1.13
Calcium	1.24	1.97
Iron	0.71	1.92
Magnesium	0.0	0.28
Manganese	0.0	0.11
Phosphorus	0.0	0.33
Potassium	7.12	0.75
Silicon	0.54	3.13
Sulphur	0.0	0.24

Data are the average from two replicate samples and are expressed as concentration [wt. (%)]. The SEM-EDX profiles are shown in **Figure 5**. Carbon and oxygen were estimated to be 55% and 35%, respectively.

Table 2. Nutrient levels in healthy and stripe-affected ginseng root tissues as determined by ICP-AES.

Element	Healthy Root	Stripe Root
Macronutrients (mg/g)		
Calcium	2.5	2.6
Magnesium	1.8	2.4
Phosphorus	1.4	1.6
Potassium	1.7	1.8
Sodium	0.3	0.3
Sulphur	0.6	0.8
Micronutrients (ug/g)		
Aluminum	115	334
Boron	24.8	27.6
Copper	33.1	15.6
Iron	90.1	232
Manganese	24.7	40.7
Molybdenum	0.8	0.7
Zinc	41.5	32.7

Data are the average of two replicate samples. The analysis was conducted twice using tissues bulked from different roots. Representative data from one analysis are shown.

several phenolic compounds showed higher peaks at specific retention times between 10 and 25 min in extracts from striped root tissues compared to healthy roots (**Figure 6**). By using specific phenolic compounds as standards, some of these higher peaks on the chromatogram were identified as chlorogenic acid, quercetin, protocatechuic acid, O-coumaric acid, catechin and gallic acid (data not shown). The levels of the enzymes phenyalalanine ammonium lyase (PAL) and peroxidase (POD) were also found to be higher in extracts from striped tissues compared to healthy tissues (24.5 and 16.5 units of PAL and 7.2 and 4.6 units of POD, respectively. The data were averaged from 3 replicate samples).

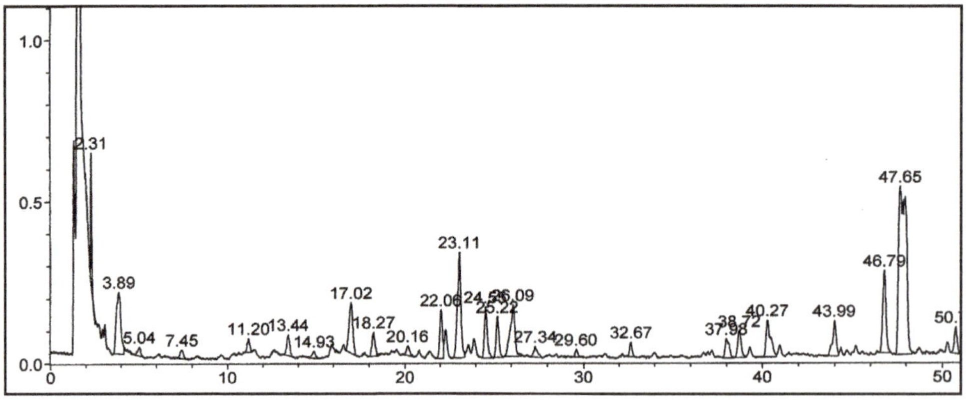

(a)

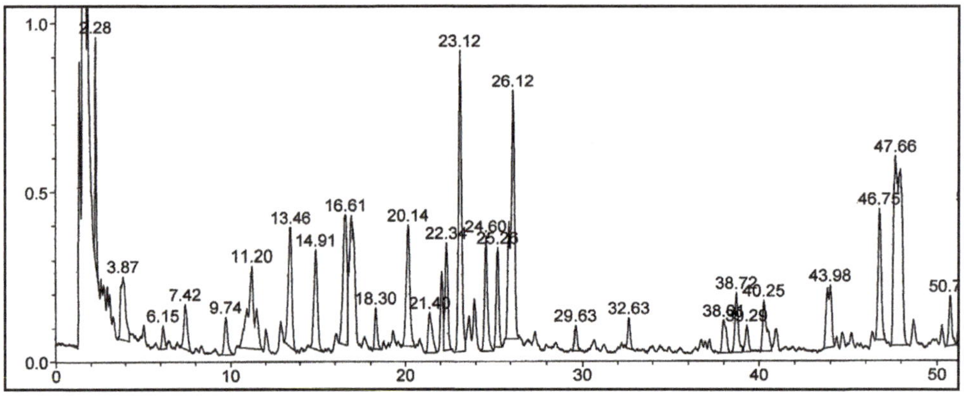

(b)

Figure 6. HPLC chromatograms of phenolic compounds extracted from healthy roots (a) compared to stripe roots (b).

4. Discussion

The results from this study show that the symptom of stripes on ginseng seedling roots is caused by a rupture (fissure) of the epidermal tissues to produce ridges of phenolic-containing cells that accumulate a number of positively-charged elements such as silicon, iron and aluminum. The fissures are likely a result of growth cracks, occurring when roots undergo a rapid increase in cell growth or expansion due to increased nutrient availability or water supply. Most of the symptoms were seen on roots later in the growing season (August-September) when seasonal growth was nearing completion. Warm and dry weather during this period is accompanied by frequent applications of irrigation water. Cracking of roots has been reported to occur in carrots under conditions of excessive nitrogen supply and moisture availability and occurred towards the end of the growing season [15]. A similar phenomenon has been reported to occur in potato tubers, where they are referred to as "growth cracks" [16] [17]. The frequency of growth cracks in potato increases when relatively poor growing conditions, such as prolonged moisture stress or high temperatures, are rapidly followed by relatively good growing conditions, such as excessive irrigation or rainfall, and it is also made worse if nitrogen fertilizer is applied in excess [16] [17]. Cracks in the cuticle of fruits such as cherries and tomatoes were also shown to be due to sudden increases in growth of the fruit or following increased turgor in the underlying pericarp cells due to rapid water uptake that placed a strain on the epidermis [18] [19]. The symptoms on ginseng roots also resembled the large cracks that develop naturally on the bark of certain coniferous tree species such as Douglas fir.

The reddening of the affected tissues was a marked symptom of striping and is a typical response of ginseng roots to mechanical injury or invasion by fungal pathogens [8] [10]. The secretion of phenolic compounds by damaged cells, followed by oxidation and coupled by sequestration of elements such as iron were reported to be involved in "rust spot" symptoms [5]-[7]. The browning reaction seen in wounded or damaged plant tissue, such as in apples and potatoes, is also the result of oxidation of phenolic compounds released from damaged cells by polyphenol oxidase enzymes to form brown-colored products [20]. Similarly, physical abrasion to the surface of peach and nectarine fruits caused brown and/or black spots to appear due to cell disruption in the epidermal and hypodermal cell layers [21]. Phenolic compound secretion by plant cells serves as a defense response to biotic and abiotic stresses [22]. Many phenolic compounds display antimicrobial activity and play an important role in the defense response of plants against diseases [23]. Polyphenols, especially tannins, are also known for their ability to chelate heavy metals, such as Fe, and play a role as potential antioxidants [24].

Phenolic compounds have been shown to be involved with metal chelation both in vitro and in vivo [25] [26]. While studying metallo-phenolic complex formation in peach and nectarine epidermal tissues, Cheng and Crisosto [26] demonstrated that elevated levels of the phenolic compounds caffeic acid, chlorogenic acid, and catechin chelated iron to form an iron-polyphenol complex, which led to tissue discoloration. We observed higher levels of chlorogenic acid, quercetin, catechin, and a number of other compounds, suggesting that these phenolics could be important in sequestering elements such as Fe, Si and Al. The observed increase in elemental concentration in ginseng striped tissues compared to healthy tissues in this study was likely due to metal chelation by these elevated phenolic compounds. The lower levels of potassium observed in striped tissues could be inversely related to the elevated concentrations of di and tribasic cations, which may have replaced K+ ions by cation exchange. Most phenolic compounds have ortho-dihydroxyl groups which are involved in forming metal complexesand also have a low affinity for K+ [27] [28].

One of the key enzymes involved in the metabolic pathway leading towards phenolic synthesis in plants is L-Phenylalanine ammonia-lyase (PAL) [22]. There is a large body of evidence illustrating a strong correlation between increases in PAL gene/protein expression/activity and increased levels of phenolic compounds in response to different stimuli [22]. In addition, peroxidase (POD) plays an important role in wound-healing and repair of damaged plant tissues [23]. The activities of both of these enzymes were higher in ginseng striped tissues compared to healthy tissues, suggesting they were involved in the higher phenolic production and tissue repair following the rupture of the epidermal cells. The sloughing off of the affected epidermal tissues and the formation of an abscission zone involve peroxidases that can catalyze the formation of lignin [29]. In many respects, the biochemical changes observed in the present study following the rupture of the epidermal cells is similar to the generalized defense responses in ginseng roots to pathogen infection [30]. However, root striping appears to be a physiological condition and does not resemble any of the previously described symptoms associated with ginseng pathogens [3] [8] [10].

Acknowledgements

We thank Chai-Na-Ta Ginseng Corp. for providing access to ginseng farms for sample collection, A. Wan for providing technical assistance, and the Natural Sciences and Engineering Research Council of Canada for funding this research through the Discovery Grants Program.

References

[1] Punja, Z.K. (2011) American Ginseng: Research Developments, Opportunities, and Challenges. *Journal of Ginseng Research*, **35**, 368-374. http://dx.doi.org/10.5142/jgr.2011.35.3.368

[2] Proctor, J.T.A. and Bailey, W.G. (1987) Ginseng: Industry, Botany, and Culture. *Horticultural Reviews*, **9**, 188-236. http://dx.doi.org/10.1002/9781118060827.ch6

[3] Punja, Z.K. (1997) Fungal Pathogens of American Ginseng (*Panax quinquefolius* L.) in British Columbia, Canada. *Canadian Journal of Plant Pathology*, **19**, 301-306. http://dx.doi.org/10.1080/07060669709500528

[4] Schluter, C. and Punja, Z.K. (2002) Genetic Diversity among Natural and Cultivated Field Populations and Seed Lots of American Ginseng (*Panax quinquefolius* L.) in Canada. *International Journal of Plant Science*, **163**, 427-439. http://dx.doi.org/10.1086/339512

[5] Lee, T.S., Mok, S.K., Seon, S.K., Yoon, J.H., Baek, N.I. and Choe, J. (2004) Accumulation of Crude Lipids, Phenolic Compounds, and Iron in Rusty Ginseng Root Epidermis. *Journal of Ginseng Research*, **28**, 157-164. http://dx.doi.org/10.5142/JGR.2004.28.3.157

[6] Campeau, C., Proctor, J.T.A., Murr, D.P. and Schooley, J. (2003) Characterization of North American Ginseng Rust-Spots and the Effects of Ethephon. *Journal of Ginseng Research*, **27**, 188-194. http://dx.doi.org/10.5142/JGR.2003.27.4.188

[7] Yingping, W., Zhihong, L., Yanjun, S., Shiwei, G., Shuzhen, T. and Zhorong, L. (1997) Studies on the Genesis of Ginseng Rust Spots. *Korean Journal of Ginseng Science*, **21**, 69-77.

[8] Punja, Z.K., Wan, A., Goswami, R., Verma, N., Rahman, M., Barasubiye, T., Seifert, K. and Lévesque, C.A. (2007) Diversity of *Fusarium* Species Associated With Discolored Ginseng Roots in British Columbia. *Canadian Journal Plant Pathology*, **29**, 340-353. http://dx.doi.org/10.1080/07060660709507480

[9] Reeleder, R.D., Roy, R. and Capell, B.B. (1999) Seed and Root Rots of Ginseng (*Panax quinquefolius*) Caused by *Cylindrocarpon destructans* and *Fusarium* spp. *Journal of Ginseng Research*, **26**, 151-158.

[10] Punja, Z.K., Wan, A., Lieppi, L., Goswami, R.S. and Jayaraj, J. (2013) Growth, Pathogenicity, Infection Behaviour, and Genetic Diversity of *Rhexocercosporidium* Isolates Originating From Ginseng Fields in British Columbia. *Canadian Journal of Plant Pathology*, **35**, 503-513. http://dx.doi.org/10.1080/07060661.2013.843315

[11] Gerrity, R.G. and Forbes, G.W. (2003) Microwave Processing in Diagnostic Electron Microscopy. *Microscopy Today*, **11**, 38-41.

[12] Dahlquist, R.L. and Knoll, J.W. (1978) Inductively Coupled Plasma-Atomic Emission Spectrometry: Analysis of Biological Materials and Soils for Major, Trace, and Ultra-Trace Elements. *Applied Spectroscopy*, **32**, 1-29. http://dx.doi.org/10.1366/000370278774331828

[13] Singleton, V.L., Orthofer, R. and Lamuela-Raventos, R.M. (1999) Analysis of Total Phenols and Other Oxidation Substrates and Antioxidants by Means of Folin-Ciocalteau Reagent. *Methods in Enzymology*, **299**, 152-178. http://dx.doi.org/10.1016/S0076-6879(99)99017-1

[14] Bradford, M.M. (1976) A Rapid and Sensitive Method for the Quantitation of Microgram Quantities of Protein Utilizing the Principle of Protein-Dye Binding. *Analytical Biochemistry*, **72**, 248-254. http://dx.doi.org/10.1016/0003-2697(76)90527-3

[15] Hartz, T.K., Johnstone, P.R. and Nunez, J.J. (2005) Production Environment and Nitrogen Fertility Affect Carrot Cracking. *HortScience*, **40**, 611-615.

[16] Hiller, L.K., Koller, D.C. and Thornton, R.E. (1985) Physiological Disorders of Potatoes. In: Li, P.H., Ed., *Potato Physiology*, Academic Press, New York, 389-455.

[17] Jefferies, R.A. and MacKerron, D.K.L. (1987) Observations on the Incidence of Tuber Growth Cracking in Relation to Weather Patterns. *Potato Research*, **30**, 613-623. http://dx.doi.org/10.1007/BF02367642

[18] Emmons, C.L. and Scott, J.W. (1996) Factors Affecting Cuticle Cracking in Tomato. *Proceedings of the Florida State Horticultural Sciences*, **109**, 178-182.

[19] Peschel, S. and Knoche, M. (2005) Characterization of Microcracks in the Cuticle of Developing Sweet Cherry Fruit. *Journal of the American Society for Horticultural Science*, **130**, 487-495.

[20] Ramirez, E.C., Whitaker, J.R. and Virador, V.M. (2002) Polyphenol Oxidase. In: Whitaker, J.R., Voragen, A.G.J. and Wong, D.W.S., Eds., *Handbook of Food Enzymology*, Marcel Dekker, New York, 509-523. http://dx.doi.org/10.1201/9780203910450.ch39

[21] Crisosto, G.W., Johnson, R.S. and Luza, J. (1993) Incidence of Physical Damage on Peach and Nectarine Skin Discoloration Development: Anatomical Studies. *Journal of the American Society for Horticultural Science*, **118**, 796-800.

[22] Boudet, A.-M. (2007) Evolution and Current Status of Research in Phenolic Compounds. *Phytochemistry*, **68**, 2722-2735. http://dx.doi.org/10.1016/j.phytochem.2007.06.012

[23] Nicholson, R.L. and Hammerschmidt, R. (1998) Phenolic Compounds and Their Role in Disease Resistance. *Annual Review of Phytopathology*, **30**, 369-389. http://dx.doi.org/10.1146/annurev.py.30.090192.002101

[24] Rice-Evans, C.A., Miller, N.J. and Paganga, G. (1996) Structure-Antioxidant Activity Relationship of Flavonoids and Phenolic Acids. *Free Radical Biology in Medicine*, **20**, 933-956. http://dx.doi.org/10.1016/0891-5849(95)02227-9

[25] Brune, M., Rossander, L. and Hallberg, L. (1989) Iron Absorption and Phenolic Compounds: Importance of Different Phenolic Structures. *European Journal of Clinical Nutrition*, **43**, 547-558.

[26] Cheng, G.W. and Crisosto, C.H. (1997) Iron-Polyphenol Complex Formation and Skin Discoloration in Peaches and Nectarines. *Journal of the American Society for Horticultural Science*, **122**, 95-99.

[27] Lavid, N., Schwartz, A., Yarden, O. and Tel-Or, E. (2000) The Involvement of Polyphenols and Peroxidase Activities in Heavy-Metal Accumulation by Epidermal Glands of the Waterlily (Nymphaeceae). *Planta*, **212**, 323-331. http://dx.doi.org/10.1007/s004250000400

[28] Khokhar, S. and Richard, K.A.O. (2003) Iron Binding Characteristics of Phenolic Compounds: Some Tentative Structure-Activity Relations. *Food Chemistry*, **81**, 133-140. http://dx.doi.org/10.1016/S0308-8146(02)00394-1

[29] Poovaiah, B.W. (1973) Peroxidase Activity in the Abscission Zone of Bean Leaves during Abscission. *Plant Physiology*, **52**, 263-267. http://dx.doi.org/10.1104/pp.52.3.263

[30] Goswami, R. and Punja, Z.K. (2008) Molecular and Biochemical Characterization of Defense Responses in Ginseng (*Panax quinquefolius*) Roots Challenged with *Fusarium equiseti*. *Physiological and Molecular Plant Pathology*, **72**, 10-20. http://dx.doi.org/10.1016/j.pmpp.2008.04.006

5

Crude Protein and Proline in Dry Bean Seed Respond to Weeding and Soil Fertility Regimes

Silindile P. Miya, Albert T. Modi*

Crop Science, School of Agricultural, Earth and Environmental Sciences, University of KwaZulu-Natal, Scottsville, USA

Email: *modiat@ukzn.ac.za

Abstract

The objective of this study was to investigate the effect of weeds and fertilizer application on dry bean seed quality. Four dry bean (*Phaseolus vulgaris* L.) cultivars, Caledon (C), Ukulinga (U), Gadra (G) and uMtata (M) were planted for seed production using a field experiment designed as a split, replicated three times. There were three levels of weeding W0 (no weeding), W1 (weeding until 50% flowering) and W2 (weeding all the time until harvest). The weeding treatments were split into no fertilizer application (F0) and optimum fertilizer application (F1) according to soil fertility analysis. At harvest maturity, seeds were compared for quality with respect to size, germination and total protein content. Proline content was determined as a measure of crop response to the weed and fertilizer stresses during crop production. Although seed size was affected by management stress, seed germination was not significantly affected by weeding and fertilizer even when it was explored in terms of seed vigor by determining rate of germination and seedling size. However, weed management and fertilizer application significantly affected proline and total crude protein contents in seeds ($P < 0.05$). The findings of this study show that the biotic stress of weeds and abiotic stress of soil fertility can be used to determine seed physiological quality of dry bean seeds.

Keywords

Abiotic Stress, Biotic Stress, Seed Protein, Proline, Phaseolus Vulgaris

1. Introduction

Dry beans (*Phaseolus vulgaris* L.) are primarily used for worldwide human consumption with an increasing

*Corresponding author.

demand [1]-[3], especially for lower income societies where the crop provides suitable dietary proteins [4]-[6]. The crop is also a rich source of dietary fibre, starch, folate, minerals and vitamins with low amounts of fat and sodium [2]. As is the case for all grain, fodder and vegetable crops, successful dry bean production begins with high quality seed [7]. Seed quality can be influenced by biotic and abiotic environmental conditions during growth and storage before planting. Seed physical and physiological or biochemical qualities are important for seed quality determination. Although seed physiology is an important aspect of seed quality, it is used to a limited extent, mainly in terms of seed viability and germination. Seed viability and germination provide a general indication that the activity of dormancy-breaking enzymes (e.g. alpha-amylase) and phytohormones (e.g. gibberrelic acid) can be confirmed.

Weed management offers more than just eliminating the existing weed population which is the case in weed control. This phenomenon emphasizes weed prevention, reduction of weed emergence after crop planting and minimizing the weed-crop competition [8]. Therefore, weed management is a preventive measure contrary to weed control which is a curative approach. Furthermore, weed management puts great emphasis on the integration of techniques, knowledge and management skills [9]. It has been argued [10] that the phenomenon of weed management depends on varying degrees of crop growth vigor despite the farming practices and environmental diversity where crops are grown. It is crucial to ensure timely weed control before any crop loss, and this should be repeated until late emerging weeds no longer reduce yield [11]. The span time to which crops can tolerate weed interference is very short with variation in pattern of weed growth relative to crops [10]. Preventing yield losses in the field requires the absence of weeds for 5 to 7 weeks after planting [12]. Although there is existing control strategies used to ameliorate weeds problem, there is still a threat in agricultural productivity posed by weeds [13]. Therefore, high level of management is required in dry bean cultivation, more so since the crop is also susceptible to a number of stresses unrelated to weeds [14]. Dry bean yield can be improved several folds with the adoption of appropriate crop management practices [3]. Therefore, implementation of a proper weed management strategy is very important to avoid yield losses [15].

Plant competition and community dynamics are determined and altered by fertilization (nutrient availability) management and timing relative to crop and weed demands upon nutrient supplies [13] [15]. Specific responses of fertilizer application vary depending on the nature of the weed-crop association [16]. However, the interactions between weeds and crops can be influenced by crop fertilization strategy [14]. Therefore, fertility management ensures that crops maximize nutrient uptake and minimize nutrient availability to weeds, thereby reducing weed impact [15]. Moreover, the type of crop and weed species determine the net effect of fertilization on crop yield [17]. Dry matter production of crops increases with an increase in soil fertility unless there is a competition with weeds [18]. Furthermore, the improvement of nutrient uptake by the crop fertilizer application management increases together with its competitiveness for other resources that might otherwise be available for weeds [8] [17].

During initial growth stages, starving weeds may be enhanced by delayed nutrient application, and better matching nutrient supply with crop uptake capacity. However, fertilizer application before or at planting may detrimentally affect the crop through promotion of weed germination whenever weeds absorb nutrients earlier and more rapidly than crops [14]. Under high soil nutrient levels, weeds tend to be more competitive than crops and accumulate more nutrients, resulting in crop loss from weed impact [16] [17]. Evidence of this was shown in a study about the disadvantage of N fertilizer application which led to increased weed density, thus leading to crop yield decreases [19].

Although research has been reported on the effects of weeds and soil fertility on crop performance from the agronomic perspective, there is a need to explain these factors from the seed quality perspective. The aim of this study was to investigate the possible interaction of weed and soil fertility stress on seed quality. The specific objective was on physical seed quality from the point of view of seed size, seed germination and chemical indicators of crop response to weed and soil fertility stress. The results presented here focus on seed size, chemical composition in terms of a stress indicator, proline, and crude protein content in seeds. No data for seed germination are shown since there are no significant effects of treatments on it at any level.

2. Materials and Methods

Four dry bean (*Phaseolus vulgaris* L.) cultivars, Caledon (C), Ukulinga (U), Gadra (G) and uMtata (M) purchased from Pro-seed were planted for seed production using a field experiment designed as a split, replicated three times. U, G and M have a determinate growth habit, whereas, C is indeterminate. There were three levels of weeding W0 (no weeding), W1 (weeding until 50% flowering) and W2 (weeding all the time until harvest). The planting density was 100,000 plants ha^{-1}. Weeding was done culturally using a hand-hoe and hand-pulling. Fertilizer treatments were split into no fertilizer (F0) and optimum fertilizer application (F1) before planting, using 2:3:2 [20] fertilizer according to soil analysis results (80 kg/ha N, 200 kg/ha P and 60 kg/ha K). The crop was grown under irrigation using 580 mm of water over the growing season (13 February-06 June) in 2014 (**Figure 1**).

At harvest maturity [21], seeds were manually harvested to determined seed quality in terms of 1000 seed mass and seed germination [22]. Protein analysis was conducted according to the Bradford method [20], with some amendments. About 2 g DW of seed tissue powder was ground with electric blade grinder was used instead of 0.5 g plant. The sample was extracted in 5 ml 50 mM Tris-HCl buffer (pH 7.4). The Bradford dye (1 ml) was added to test tubes with 20 µl sample extract which was mixed and incubated at room temperature for five minutes in 1 ml plastic cuvettes. The samples were read from the spectrophotometer at 595 nm and the protein concentration was determined through the comparison of results with a standard curve constructed using bovine serum albumin.

Analysis for proline content ($\mu mol \cdot g^{-1}$ DW) was conducted according to [23] with the amendment of using seeds as plant material instead of using roots from field grown dry beans. Plant material was homogenized in 10 ml (3%) aqueous sulfosalicylic acid. Filtrate (2 ml), acid ninhydrin (2 ml) and glacial acetic acid (2ml) were reacted in a test tube for one hour at 100˚C and thereafter cooled in ice bath. Reaction mixture was extracted with toluene which was thoroughly mixed with a test tube stirrer. Absorbance was read at 520 nm using toluene as a blank and proline concentration was determined using a standard curve (and calculated as:

$$\text{Proline} = \frac{\left[\left(\mu\text{g proline/ml} \times \text{ml toluene}\right) \div 115.5\ \mu\text{g}/\mu\text{moles}\right]}{\left[\text{g sample}/5\right]}$$

Data from the study were analyzed by using Genstat® version 16 software. When comparing means, standard error of the difference (SED) at a probability level of 5% was used. Statistical differences were accepted at $P \leq 0.05$.

3. Results and Discussion

There were no significant differences between treatments and cultivars with respect to seed germination. Hence, no seed germination data are shown. However, there were significant ($P < 0.001$) differences between cultivars (**Figure 2**) and weeding regimes (**Figure 3**) with respect to 1000 grain mass. There were no significant interactions with respect to 1000 grain mass. Although the differences between cultivars can be ascribed to genetic

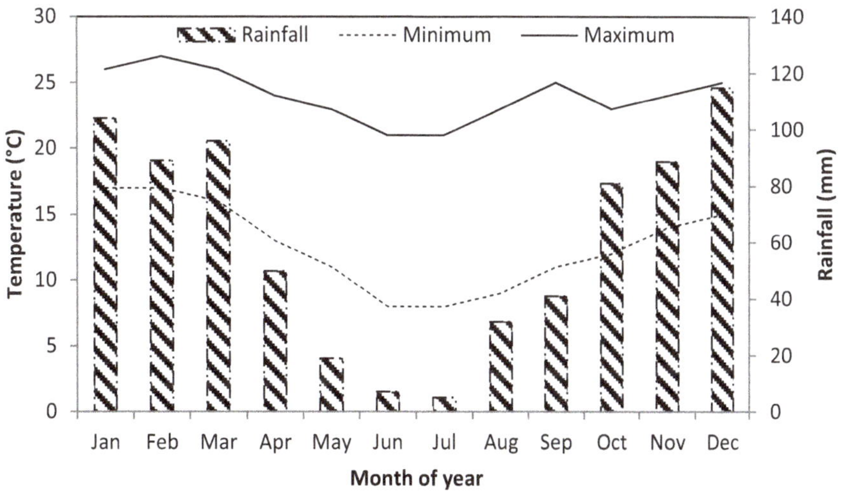

Figure 1. Climatic data for the cropping year (2014), where the crop was grown from February to June.

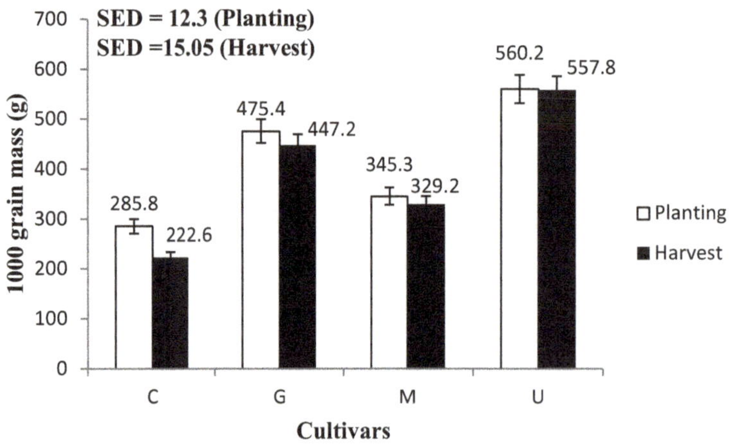

Figure 2. Seed 1000 grain mass at planting and harvest maturity for dry bean cultivars Caledon (C), Ukulinga (U), Gadra (G) and uMtata (M).

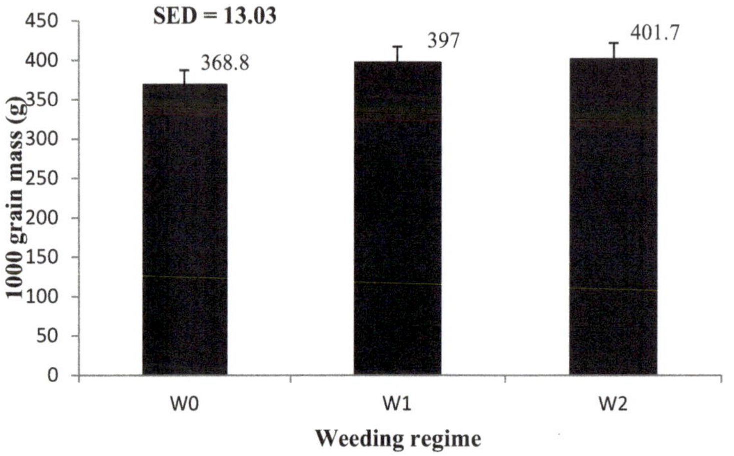

Figure 3. Effect of weeding regime (W0 = no weeding; W1 = Weeding until 50% flowering; W2 = weeding until pod maturation) on dry bean grain mass at harvest maturity.

variation, since there were no significant differences in their harvest maturity seed moisture content of about 15% (data not shown), it is clear from **Figure 1** that cultivar Ukulinga was very resistant to weed stress and Caledon was very sensitive.

There were highly significant ($P < 0.001$) effects of weeding regime on proline seed content (**Figure 4**). Weeding treatment W0 led to about 14% more proline content in seeds compared with W1 and W2 (**Figure 3**). This is an indication that W0 was most stressful than the other two weeding regimes. There was a minor (0.6%), but significant difference between weeding until 50% flowering and keeping the weeds until crop maturity (**Figure 4**) with respect to proline content. Fertiliser application significantly ($P = 0.03$) reduced (4.3%) seed proline content compared to no fertiliser application (**Figure 5**). Fertiliser application also significantly ($P < 0.01$) increased seed crude protein content (**Table 1**). The cultivars differed significantly ($P = 0.02$) with respect to crude protein content (**Table 1**).

Although there were no significant differences among weeding regimes, with respect to total crude protein, it is clear from **Figure 6** that there was a negative relationship between seed protein and proline contents. According to [24], proline is essential for primary metabolism as a proteinogenic amino acid and is exceptionally conformational rigid [25]. This compound is considered to be a compatible solute and a molecular chaperone that protects folded protein structures from denaturation. It is also considered a protein stabilizer for cell membranes by interacting with phospholipids, enhancing different enzymes activities, functioning as a hydroxyl radical

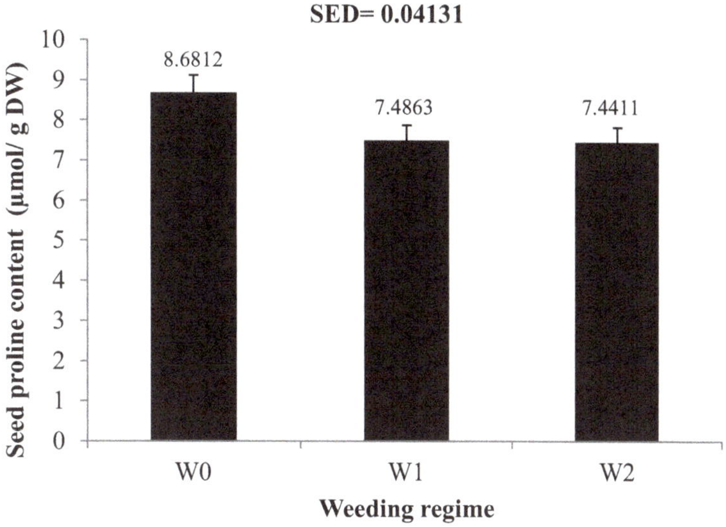

Figure 4. The effect of weeding regime (W0 = no weeding; W1 = Weeding until 50% flowering; W2 = weeding until pod maturation) on seed proline content.

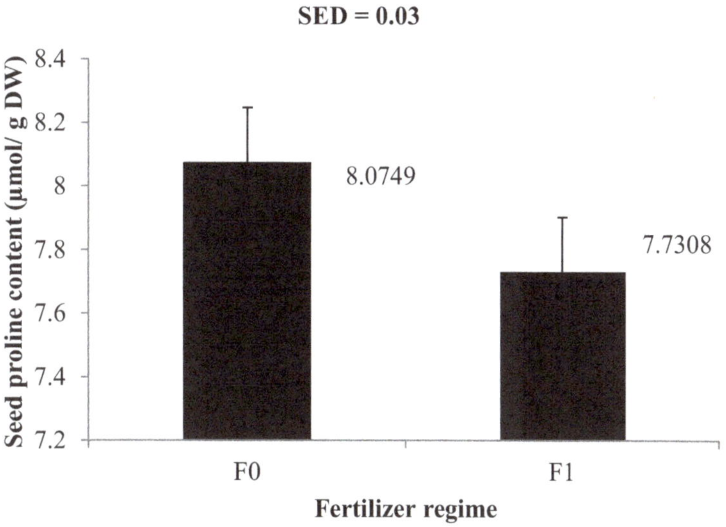

Figure 5. Effect of fertiliser application (F0 = None; F1 optimum) on proline content of mature dry beans.

Table 1. Effect of fertilizer application (F0 = None; F1 optimum) on crude total protein (µg/gDW) of dry bean seeds.

Cultivar	F0	F1
Caledon	0.028a*	0.033b
Ukulinga	0.039c	0.044d
uMtata	0.041c	0.053e
Gadra	0.030a	0.038c
Mean	0.034	0.042

SED (Cultivar × Fertiliser) = 0.002

SED (Fertilizer regime) = 0.003

*Values sharing the same letters are not significantly different at ($P \leq 0.05$).

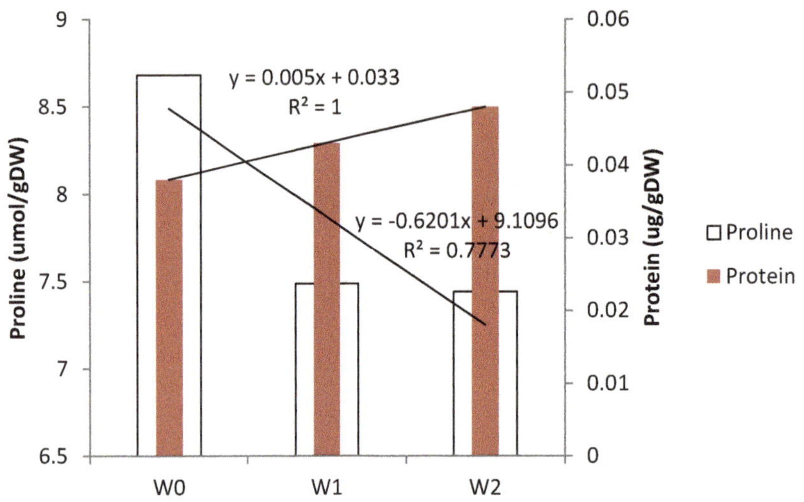

Figure 6. Weeding regime (W0 = no weeding; W1 = Weeding until 50% flowering; W2 = weeding until pod maturation) effects, with respect to proline and crude protein contents in dry been seeds across two fertiliser regimes and four cultivars (see **Table 1**).

scavenger, or serving as an energy and nitrogen source [25] [26]. There have been reports that proline contributes to osmotic adjustment and tolerance of plants exposed to unfavourable environmental conditions [27]. The increase in synthesis of osmoprotectants is a general metabolic adaptation enabling plants to cope with water or osmotic stress, and proline is a well-known osmoprotectant in plants [28] [29]. According to [30] in addition to drought, proline in higher plants tends to increase under different environmental stress conditions such as high salinity, high light and ultraviolet (UV) irradiation, heavy metals, oxidative stress as well as biotic stress [25] [27]. Free proline as an osmoprotectant might eliminate reactive oxygen species (ROS) and promote the antioxidant ability. This compound can also stabilize biological macromolecule structures, decrease cell activity and relieve the toxicity of ammonium (NH_4^+) [31]. Chemical changes in response to biotic and abiotic stresses have been observed in other studies [32], although the plant tissue is different.

4. Conclusion

Crop nutrition and protection are key aspects of agronomy. In cases where grain crops are produced for seed, these factors become even more important, because their implications go beyond the farm product to the market value for future crop establishment. Seed science studies generally focus on seed quality from the basic physical norms, viability, germination and vigour. This study showed that simply relying on the norms for testing seed quality might lead to the conclusion that some important biotic and abiotic stresses had no effect. This study showed that although seed germination was not affected by weed infestation and suboptimal soil fertility, these factors affected an important aspect of seed quality, 1000 grain mass. As expected, 1000 grain mass was negatively affected by weed infestation and this was countered by optimum fertiliser application. The major finding of this study was showing that the biotic and abiotic stresses caused by weeds and poor soil fertility, respectively, could be linked to physiological indicators of seed quality, seed protein and proline. Given that proline has already been shown to be associated with plant protection against water stress, it would be interesting to use the results of this study to develop a model for testing seed quality in the context of agronomic practices and environmental stress.

References

[1] Goodwin, M. (2003) Crop Profile for Dry Beans. http://www.pulsecanada.com

[2] Mkanda, A.V., Minnaar, A. and de Kock, H.L. (2007) Relating Consumer Preferences to Sensory and Physicochemical Properties of Dry Beans (*Phaseolus vulgaris*). *Journal of the Science of Food and Agriculture*, **87**, 2868-2879. http://dx.doi.org/10.1002/jsfa.3046

[3] Singh, S.P., Teran, H., Munoz, C.G., Osorno, J.M., Takegami, J.C. and Thung, M.D. (2003) Low Soil Fertility Toler-

ance in Landraces and Improved Common Bean Genotypes. *Crop Science*, **43**, 110-119. http://dx.doi.org/10.2135/cropsci2003.0110

[4] Escribano, M.R., Santalla, M. and de Ron, A.M. (1997) Genetic Diversity in Pod and Seed Quality Traits of Common Bean Populations form North-Western Spain. *Euphytica*, **93**, 71-81. http://dx.doi.org/10.1023/A:1002908224793

[5] Broughton, W.J., Hernandez, G., Blair, M., Beebe, S., Gepts, P. and Vanderleyden, J. (2003) Beans (*Phaseolus* spp.)—Model Food Legumes. *Plant and Soil*, **252**, 55-128. http://dx.doi.org/10.1023/A:1024146710611

[6] DAFF (Department of Agriculture, Forestry and Fisheries) (2010) Dry Beans. http://www.nda.agric.za/docs/Brochures/brochuresDryBeans.pdf

[7] Black, M. and Halmer, P. (2006) The Encyclopedia of Seeds: Science, Technology and Uses. CABI, Wallingford.

[8] Buhler, D.D. (2002) Challenges and Opportunities for Integrated Weed Management. *Weed Science*, **50**, 273-280. http://dx.doi.org/10.1614/0043-1745(2002)050[0273:AIAAOF]2.0.CO;2

[9] Mashingaidze, A.B. (2004) Improving Weed Management and Crop Productivity in Maize Systems in Zimbabwe. PhD Dissertation, Wageningen University, Wageningen.

[10] McDonald, G.K. and Gill, G.S. (2009) Improving Crop Competitiveness with Weeds: Adaptations and Trade-Offs. Elsevier Inc., Academic Press, San Diego. http://dx.doi.org/10.1016/b978-0-12-374431-9.00018-9

[11] Donald, W.W. (2000) Alternative Ways to Control Weeds between Rows in Weeded Check Plots in Corn (*Zea mays*) and Soybean (*Glycine max*). *Weed Technology*, **14**, 36-44. http://dx.doi.org/10.1614/0890-037X(2000)014[0036:AWTCWB]2.0.CO;2

[12] Wicks, G.A., Johnston, D.N., Nuland, D.S. and Kinbacher, E.J. (1973) Competition between Annual Weeds and Sweet Spanish Onions. *Weed Science*, **21**, 436-439.

[13] Pynenburg, G.M., Sikkema, P.H. and Gillard, C.L. (2011) Agronomic and Economic Assessment of Intensive Pest Management of Dry Bean (*Phaseolus vulgaris*). *Crop Protection*, **30**, 340-348. http://dx.doi.org/10.1016/j.cropro.2010.12.006

[14] Liebman, M. and Davis, A.S. (2000) Integration of Soil, Crop and Weed Management in Low-External-Input Farming Systems. *Weed Research*, **40**, 27- 47. http://dx.doi.org/10.1046/j.1365-3180.2000.00164.x

[15] Barberi, P. (2002) Weed Management in Organic Agriculture: Are We Addressing the Right Issues? *Weed Research*, 42, 177-193. http://dx.doi.org/10.1046/j.1365-3180.2002.00277.x

[16] Buhler, D.D. (2004) Weed Biology, Cropping Systems, and Weed Management. *Journal of Crop Production*, **8**, 245-270. http://dx.doi.org/10.1300/J144v08n01_10

[17] Michigan State University Extension (2015) Integrated Pest Management. http://www.ipm.msu.edu/agriculture/field_crops/identifying_weeds_in_field_crops

[18] Patterson, D.T. (1995) Effects of Environmental Stress on Weed/Crop Interactions. *Weed Science*, **43**, 483-490.

[19] Rahmati, S., Sajedi, N.A. and Gomarian, M. (2013) Effects of Time Cultivation and Weeds Control Methods on Yield and Yield Components of Red Bean (*Phaseolus calcaratus* L.). *International Journal of Agriculture and Crop Science*, **5**, 2795-2803.

[20] Bradford, M.M. (1976) A Rapid and Sensitive Method for the Quantitation of Microgram Quantities of Protein Utilizing the Principle of Protein—Dye Binding. *Analytical Biochemistry*, 72, 248-254. http://dx.doi.org/10.1016/0003-2697(76)90527-3

[21] Sowmya, K.J., Gowda, R., Bhanuprakash, K., Yogeesha, H.S., Puttaraju, T.B. and Channakeshava, B.C. (2013) Enhancement of Seed Quality through Chemopriming in Cucumber (*Cucumis sativus* L.). *Mysore Journal of Agricultural Science*, **47**, 22-30.

[22] ISTA (2015) International Rules for Seed Testing, Vol. 215, Introduction, i-1-6 (10). The International Seed Testing Association, Bassersdorf.

[23] Bates, L.S., Waldren, R.P. and Teare, I.D. (1973) Short Communication: Rapid Determination of Free Proline for Water-Stress Studies. *Plant and Soil*, **39**, 205-207. http://dx.doi.org/10.1007/BF00018060

[24] Wisiol, K. (1979) Clipping of Water-Stressed Blue Grama Affects Proline Accumulation and Productivity. *Journal of Range Management*, **32**, 194-195. http://dx.doi.org/10.2307/3897121

[25] Szabados, L. and Savoure, A. (2009) Proline: A Multifunctional Amino Acid. *Trends in Plant Science*, **15**, 89-97. http://dx.doi.org/10.1016/j.tplants.2009.11.009

[26] Socias, F.X., Pol, A., Aguiw, F., Vadell, J. and Medrano, H. (1997) Effects of Rapidly and Gradually Induced Water Stress on Plant Response in Subterranean Clover Leaves. *Journal of Plant Physiology*, **150**, 212-219. http://dx.doi.org/10.1016/S0176-1617(97)80205-2

[27] Claussen, W. (2005) Proline as a Measure of Stress in Tomato Plants. *Plant Science*, **168**, 241-248.

http://dx.doi.org/10.1016/j.plantsci.2004.07.039

[28] Celik, O. and Unsal, S.G. (2013) Expression Analysis of Proline Metabolism-Related Genes in Salt-Tolerant Soybean Mutant Plants. *Plant Omics Journal*, **6**, 364-370.

[29] Xiao, G.J., Zhang, F.J., Qiu, Z.J., Yao, Y.B., Wang, R.Y. and Huang, J. (2013) Response to Climate Change for Potato Water Use Efficiency in Semi-Arid Areas of China. *Agricultural Water Management*, **127**, 119-123. http://dx.doi.org/10.1016/j.agwat.2013.06.004

[30] Cvikrova, M., Gemperlova, L., Martincova, O. and Vankova, R. (2013) Effect of Drought and Combined Drought and Heat Stress on Polyamine Metabolism in Proline-over-Producing Tobacco Plants. *Plant Physiology and Biochemistry*, **73**, 7-15. http://dx.doi.org/10.1016/j.plaphy.2013.08.005

[31] Xiong, J., Zhang, L., Fu, G.F., Yang, Y.J., Zhu, C. and Tao, L.X. (2012) Drought-Induced Proline Accumulation Is Uninvolved with Increased Nitric Oxide, which Alleviates Drought Stress by Decreasing Transpiration in Rice. *Journal of Plant Research*, **125**, 155-164. http://dx.doi.org/10.1007/s10265-011-0417-y

[32] González-Gallegos, E., Laredo-Alcalá, E., Ascacio-Valdés, J., Jasso de Rodríguez, D. and Hernández-Castillo, F.D. (2015) Changes in the Production of Salicylic and Jasmonic Acid in Potato Plants (*Solanum tuberosum*) as Response to Foliar Application of Biotic and Abiotic Inductors. *American Journal of Plant Sciences*, **6**, 1785-1791. http://dx.doi.org/10.4236/ajps.2015.611179

6

Impacts of Four Invasive Asteraceae on Soil Physico-Chemical Properties and AM Fungi Community

Xinwei Chen[1,2], Yuxiao Liu[3], Hongmei Liu[2], Hui Wang[2], Dianlin Yang[1,2], Chaohe Huangfu[1,2*]

[1]College of Plant Protection, Shenyang Agricultural University, Shenyang, China
[2]Agro-Environmental Protection Institute, Ministry of Agriculture, Tianjin, China
[3]Institute of Information, Tianjin Academy of Agricultural Sciences, Tianjin, China
Email: *huangfu24@163.com

Abstract

Invasive Asteraceae are an important group of plants and might have universal impacts on invaded ecosystems. However, research data on the ecological impacts of specific plants are still lacking. *Chromolaena odorata*, *Ageratina adenophora*, *Flaveria bidentis*, and *Mikania micrantha* are four typical alien Asteraceae in China. The involvement of soil biota, in particular, the community structure of arbuscular mycorrhizal (AM) fungi in their invasion process was tested in present research. It was found that invasion by the four Asteraceae changed, to different extents, the nutrient levels in soils. Invasion by *C. odorata*, *A. adenophora*, and *F. bidentis* followed common patterns: 1) decreasing pH value; 2) increasing the soil AM fungal diversity and species richness. Invasion by all four Asteraceae tested increased nitrogen pools and accelerated nitrogen fluxes with a decrease in the phosphorus pool, especially available phosphorus. It was suggested that mycorrhizal symbiosis might partially promote successful invasion of these Asteraceae by affecting the metabolism of phosphorus in soil. The impacts on soil ecosystems of these Asteraceae tested were also species-specific, and different invasion strategies were exhibited.

Keywords

Invasive Plants, Mycorrhizal Symbiosis, Soil Nutrient Cycle, Enzyme Activity, Plant-Soil Interactions

1. Introduction

Biological invasions have gained increasing attention, and are recognized as a key problem for biodiversity

*Corresponding author.

conservation. As the world's third largest country, China's territories cover several climatic zones from tropical to cold temperate. With its various habitats and environmental conditions, China is especially vulnerable to invasion by alien species [1]. Most of the 270 invasive plants identified in China are annuals; the next largest group is perennial herbs. The invasive species comprise 59 families in China, and the largest is the Asteraceae, which includes approximately 25.3% of all recognized invasive plants [2]. Plants in this family have similar biological traits because of their shared evolutionary history, and these traits benefit their invasive potential [3]. It is worthwhile exploring whether this important group of invasive plants, the Asteraceae, has any common ecological impacts or invasion strategies. *Mikania micrantha*, *Chromolaena odorata*, and *Ageratina adenophora* are typical of the alien Asteraceae that invade southern China. *M. micrantha*, a perennial vine ranked in the top 10 most invasive weeds in the world, has spread widely in China since the 1980s, and it has recently invaded many habitats, including forests, farmlands, wastelands, and orchards in southern China, causing significant damage [4]. *C. odorata*, a perennial forb native to the Americas from the southern United States to northern Argentina, is a major invasive plant of crops, plantations, savannas, and natural forests in many parts of the world, including tropical and subtropical areas such as central and western Africa, tropical America, India, the Philippines, southeast China, South Africa, eastern Indonesia, and Australia. In China, *C. odorata* is spreading rapidly, and can now be found in Yunnan, Guangdong, Guangxi, Hainan, Guizhou, and Taiwan Provinces, and Macao and Hong Kong, thus posing threats to local biodiversity and reducing the productivity of arable lands. *A. adenophora*, which is native to Mexico and Costa Rica, is a worldwide noxious invasive weed. This plant has invaded southern China since the 1940s and has become one of the most economically destructive invasive plants in China.

A new invasive weed, *Flaveria bidentis* (L.) Kuntze, is becoming an increasing problem in China. This species is native to South America, primarily Argentina, and was found for the first time in China in 2001. It outcompetes natural vegetation and forms a dense cover, and is widely distributed in Hebei, Shandong and Henan Provinces and Tianjin. This exotic weed prefers a warm and wet environment, but also has the ability to survive environmental stresses (salinity, barren soil, cold temperature), and evidence suggests that it has increased in abundance and range in some parts of China [5]. Invasion by this weed has brought on the degradation and disappearance of indigenous plant communities, and changed the structure and function of native ecosystems [6], seriously affecting the maintenance and stability of local biodiversity.

Arbuscular mycorrhizal (AM) fungi are ubiquitous root symbionts associated with over 80% of all extant terrestrial plants [7]. Through mutualism, AM fungi enhance nutrient acquisition for the host plant by transferring phosphorus and nitrogen captured from soil in exchange for carbon derived from photosynthate. AM fungi provide a key interface between the host plant and the soil mineral nutrients. AM fungal taxa exhibit host-specific growth responses [8] and further invoke different growth responses in host plants [9]. As host plants can shape distinctive AM communities even when inoculated with the same AM fungus [10], plant invasion seems likely to alter AM fungal communities [11]. The majority of alien invasive plants are mycorrhizal species capable of creating symbioses rapidly in a new habitat containing AM fungi. Researchers have recently proposed scenarios whereby AM fungi play a more significant role in the plant invasion process than have been previously thought [12] [13]. A positive feedback has been reported in mutualism between AM fungi and invasive plants, partly contributing to the successful invasion of alien plants. Until now, only Montesinos-Navarro *et al.* [14] have investigated phylogenetic host specificity in patterns of AM association and found no clear pattern for AM fungi in a semi-arid shrubland in Mexico. However, there are still limited data on the role of AM fungi in the plant invasion process, and much less is known of associations between closely related invasive plants and AM fungi [13].

Denaturing gradient gel electrophoresis (DGGE) has been extensively used to study the diversity and structure of AM fungal communities [15] [16]. The aim of this study was to test how different invasive Asteraceae species might influence the AM fungal status, and whether closely related invasive species hosted more similar AM fungal communities, either in diversity or abundance, than the existing native species. Such research may yield insights into the contribution of AM fungi to plant coexistence, plant diversity, and plant productivity [14] [17] [18]. The aims of the present study are 1) whether selected invasive plants within same genus have similar mycorrhizal symbiosis, and 2) whether their effects on soil ecosystem, especially related to AM fungi-plant interaction, are in common.

2. Materials and Methods

2.1. Study Area

The four Asteraceae species were collected from two wasteland ecosystems in northern and southern China. The study ecosystem invaded by *F. bidentis* was located in Xian County, Hebei Province (38°15'30"N, 115°57'50"E). The climate is a temperate continental monsoon, characterized by an annual average temperature of 12.3°C, annual precipitation of 560 mm, and annual average sunshine of 2800 h. The soil is alluvial, weakly alkaline, relatively compacted, and mildly salinized. Other plants in the ecosystem were herbs such as *Setaria viridis*, *Digitaria sanguinalis*, and *Phragmite saustralis*. Because *C. odorata*, *A. adenophora*, and *M. micrantha* grow in the same ecosystem, only one control soil was selected. Samples were collected from Ruili, Yunnan Province (N24°03'577", E97°85'789"), at an elevation of 794 m. This area has a subtropical monsoon climate, with an annual average temperature of 21°C, annual rainfall of 1394 mm, and annual average sunshine of 2330 h. Soil in this area is red loess and acidic. The native dominant vegetation is composed of *Rabdosia stracheyi* and *Herba taraxaci*, and the three invasive species are highly abundant. A sampling area (500 m^2) was chosen within each ecosystem, to adequately cover the diversity of native plant species and areas completely colonized by the invasive plant species.

2.2. Field Sampling

Field sampling was conducted in the two different ecosystems as described above, and collections were made in August 17 and October 20, 2011 for *F. bidentis* and *C. odorata*, *A. adenophora*, and *M. micrantha*, respectively, corresponding with the period when invasives biomass reaches their peak. To verify the impact pattern found with those invasive plants, sampling was done in following year. Two sampling data was pooled for soil physicochemical parameters analyses and only once PCR-DGGE fingerprinting was presented given the fact that there were no inter-year differences between treatments detected. Two sampling areas were selected in each ecosystem. Sites were characterized as invaded when the invasive Asteraceae were the dominant species (>90% in coverage), while the native sites were sites where only native vegetation was found. A total of 12 sampling plots (1 × 1 m quadrats) were distributed around each study area according to the presence or absence of invasive plants (six quadrats had a history of invasion by targeted plants for at least 5 years or more with a coverage of more than 90%, and six quadrats were dominated by native grasses and devoid of invasive species). Invaded plots were placed inside of invasive Asteraceae patches which covered an area of at least 10 m^2. Paired uninvaded plots were located between 2 and 5 m away so that they were not so close to the invaded plots that they would be invaded, but not so far that there would be obvious differences in light, moisture, or elevation.

The soil (at 0 - 15 cm in depth) in each quadrat was collected and mixed. Soil for microbial and enzyme activity analysis was frozen at −20°C prior to analysis. Soil for measurement of physicochemical properties was kept dry until analysis. The aim of this study was to examine overall differences in AM fungal community composition between sites, so bulk soils were analyzed rather than rhizosphere soil to allow assessment of the potential shift in AM fungal communities caused by plant invasion.

2.3. Analysis of Soil Physicochemical Properties

Soil pH was determined using a 1:5 soil weight/water volume ratio and a pH meter. All soil nutrients were analyzed using conventional methods. Soil organic matter was measured using the $K_2Cr_2O_7$-H_2SO_4 oxidation method. The Kjeldahl method was used for analyzing the total nitrogen (N) content of the soil. Soil mineral N was extracted using 2 mol·L^{-1} KCl, then the concentrations of NO_3^--N in the KCl extracts were determined by hydrazine sulfate colorimetry and the concentrations of NH_4^+-N by indophenol blue colorimetry. Total phosphorus (P) was extracted using the $HClO_4$-H_2SO_4 method, and available P was determined using the sodium bicarbonate method. The urease activity was determined using urea as the substrate, and the soil mixture was incubated at 37°C for 5 h. The residual urea was determined using a colorimetric method. Phosphatase activities were measured using the *p*-nitrophenol method, and potassium permanganate titration was used to determine peroxide dehydrogenase activity.

2.4. Characterization of AM Fungal Community Structure

Genomic DNA was extracted from approximately 0.25 g of soil using the Power Soil DNA isolation kit (MoBio Laboratories, CA, USA). Compositions of AM fungal communities were characterized using polymerase chain reaction (PCR)-DGGE. A nested PCR was used to amplify AM fungal DNA. 2 μ DNA (10 ng) were subjected to an initial PCR as the template using the AM fungal specific primers AML1 and AML2 to amplify an approximately 0.8 kb section of the small subunit rRNA gene. PCR was carried out using 0.2 mMdNTPs, 0.2 μM of each primer, 0.5 U of TaqDNA polymerase (Takara Co. Ltd., China), and the manufacturer-supplied reaction buffer in a total volume of 20 μl. Initial denaturation was at 94˚C for 3 min, followed by 30 cycles at 94˚C for 30 s, 50˚C for 30 s, and 72˚C for 45 s, followed by a final extension period at 72˚C for 7 min. PCR products were analyzed using 1.0% (w/v) agarose gel electrophoresis. The first PCR product was diluted 1/20 with ddH_2O, and further employed as template DNA in a second PCR reaction using the universal eukaryotic primers NS31 [19] with a GC clamp (NS31-GC) and G101 [20]. The thermo cycling conditions were as described above, with the exception of the primer annealing at 54˚C for 50 s. The primers sequences were as described by Long *et al.* [21].

The PCR-DGGE was carried out using the DCode system (Bio-Rad, CA, USA) and following the manufacturer's instructions. The concentration of acrylamide gel (37.5:1) was 8%; the denaturing gradient was 30% - 50% (100% denaturant contained 7 M urea and 40% (v/v) deionized formamide); and the electrophoresis buffer fluid was 1 × TAE. 25 μL of PCR product and 5 μL of 6 × loading buffer were mixed before adding this to the loading wells using a micro-injector. This was pre-run for 30 min at 70 V and 60˚C, and then electrophoresis was carried out for 15 h at 110 V and 60˚C. The gel was then removed, and immediately stained with SYBR™ Green I (1:10,000) (Invitrogen) for 28 min. Images of the DGGE profile were captured and recorded using the Gel Dox XR gel imaging system (Bio-Rad, USA).

2.5. Statistical Analysis

Quantity One software (Bio-Rad, USA) was used to analyze DGGE banding patterns by measurement of migration distance and intensity of the bands within each lane of a gel. The impacts of invasive plants on the community structure of AM fungi in soils were assessed using the Shannon-Wiener index (*H*), the abundance index (*S*) and the evenness index (*E*). These indexes were calculated as follows: $H = -\sum p_i \ln p_i$ and $E = H/\ln S$, where *S* is the abundance index, indicating the number of bands in the DGGE gel. p_i corresponds to the proportion of the gray level of band i_{th} in the total gray level of the sample. Comparisons of soil physicochemical properties and AM fungal community diversity indexes among sites were performed by means of one-way analysis of variance (ANOVA) and post hoc (least significant difference) tests. The DGGE pattern profiles were compared using pairwise similarity coefficient (*Cs*) analysis. The *Cs* values were calculated as follows: $Cs = 2j/(a+b)$, where *a* is the number of DGGE bands in lane 1, *b* is the number of bands in lane 2, and *j* is the number of common bands. Dendrogram analysis was employed based on the *Cs* value. The relationship between soil AM fungi diversity index and selected soil physio-chemical parameters was assessed based on a Pearson correlation analysis. All other data analyses were conducted using SPSS 16.0 software.

3. Results

3.1. Soil Physicochemical Properties

In the soils invaded by *F. bidentis*, the organic matter and NH_4^+-N increased by 69.8% and NO_3^--N by 68.26% when compared with the native soil (CK). Conversely, the total P and available P decreased significantly ($P <$ 0.05) after invasion by *F. bidentis*, but there was no significant change detected in total N or pH ($P > 0.05$) (**Table 1**).

In soils invaded by the other three Asteraceae, only the soil pH in sites with *M. micrantha* was remarkably higher than that in native soil. Total N and NH_4^+-N were both significantly increased, and enhanced NO_3^--N was also found in the soils invaded by *M. micrantha* and *C. odorata* ($P < 0.05$). However, in soils invaded by *A. adenophora* a significant decrease was observed in available P (81.16%) and similarly in sites invaded by

Table 1. Soil physio-chemical parameters of invaded and non-invaded blocks by *Flaveria bidentis*.

Treatments	pH	Organic matter (g/kg)	Total N (g/kg)	NH_4^+-N (mg/kg)	NO_3^--N (mg/kg)	Total P (g/kg)	Available P (mg/kg)
CK	8.23 ± 0.53a	5.98 ± 0.19b	0.97 ± 0.02a	5.63 ± 1.14b	4.60 ± 0.94b	0.63 ± 0.03a	41.33 ± 3.53a
F. bidentis	8.18 ± 0.06a	7.35 ± 0.42a	0.99 ± 0.01a	9.56 ± 1.11a	7.74 ± 0.90a	0.57 ± 0.03a	6.27 ± 0.65b

The numbers are mean ± standard errors (SE) with 6 repetitions. Data with different letter in the same column are significantly different at $P = 0.05$. The same below.

C. odorata (59.41%). It was concluded that soil properties underwent significant changes in comparison with native soil after invasion by these plants, and that soils invaded by *A. adenophora*, *C. odorata*, and *F. bidentis* appeared to follow similar patterns (**Table 2**).

3.2. Soil Enzyme Activity

One-way analysis of variance indicated that invasion by *F. bidentis* led to increased enzyme activity in soils ($P < 0.05$, **Table 3**). Acidic phosphatase activity increased by 42.42%, neutral phosphatase activity by 58.33%, and alkaline phosphatase activity by 41.44%, respectively.

Enzyme activities in invaded soils sampled in Yunnan Province were significantly different between *M. micrantha*, *A. adenophora*, *C. odorata*, and the native soils (**Table 4**). Urease activity, for example, was significantly higher in soils with *M. micrantha* and *C. odorata* than in native soils or those with *A. adenophora*. Acid phosphatase activity was only lower in soil invaded by *M. micrantha* ($P < 0.05$). Alkaline phosphatase activity showed no significant difference between soils with *A. adenophora*, those with *C. odorata*, and native soil, but was significantly increased in soils with *M. micrantha*. Catalase activity showed the same trend as alkaline phosphatase.

3.3. Community Structure of AM Fungi

The composition of AM fungi communities in *F. bidentis* and native soils were compared using PCR-DGGE. Gel profiles of 18S rDNA fragments of AM fungi are presented in **Figure 1**. The *Cs* between soil invaded by *F. bidentis* and native soil was 63%, and a dendrogram created using hierarchical clustering analysis showed all samples clustered into two main groups, indicating a marked difference caused by plant invasion. **Table 5** shows our calculations of species richness, the Shannon index, and the evenness index of AM fungal communities. The values for the diversity index (*H*) and evenness index (*E*) of AM fungi in invaded soil were significantly higher than those of the native control soil.

Table 2. Soil physio-chemical parameters of invaded and non-invaded blocks in *Mikania micrantha*, *Chromolaena odorata* and *Ageratina adenophora* plots, respectively.

Treatments	pH	Organic matter (g/kg)	Total N (g/kg)	NH_4^+-N (mg/kg)	NO_3^--N (mg/kg)	Total P (g/kg)	Available P (mg/kg)
CK	4.98 ± 0.13b	18.31 ± 0.56b	1.16 ± 0.02c	9.69 ± 1.31b	3.76 ± 0.10b	0.41 ± 0.01a	35.82 ± 0.59a
M. micrantha	6.93 ± 0.06a	13.98 ± 0.73d	1.43 ± 0.02b	12.36 ± 1.63a	4.30 ± 0.31a	0.38 ± 0.03a	35.85 ± 1.78a
C. odorata	4.92 ± 0.13c	16.93 ± 0.52c	1.40 ± 0.08b	14.15 ± 0.55a	4.83 ± 0.67a	0.28 ± 0.02b	6.75 ± 0.77c
A. adenophora	4.91 ± 0.10c	20.82 ± 0.60a	1.63 ± 0.05a	12.84 ± 1.13a	3.86 ± 0.66b	0.41 ± 0.01a	14.54 ± 1.20b

Table 3. Soil enzymatic activities in invaded and non-invaded blocks by *Flaveria bidentis*.

Treatments	Urease (ug/g·h)	Acid phosphatase (mg/g·24h)	Neutral phosphatase (mg/g·24h)	Alkaline phosphatase (mg/g·24h)	Total phosphatase activity (mg/g·24h)	Catalase (mg/g·24h)
CK	5.88 ± 0.71b	0.19 ± 0.05b	0.05 ± 0.01b	0.65 ± 0.21b	0.89 ± 0.23b	9.76 ± 0.08b
F. bidentis	6.76 ± 0.53a	0.33 ± 0.03a	0.12 ± 0.02a	1.11 ± 0.06a	1.56 ± 0.08a	11.06 ± 0.47a

Table 4. Soil enzymatic activities of invaded and non-invaded blocks in *Mikania micrantha*, *Chromolaena odorata* and *Ageratina adenophora* plots, respectively.

Treatments	Urease (ug/g·h)	Acid phosphatase (mg/g·24h)	Neutral phosphatase (mg/g·24h)	Alkaline phosphatase (mg/g·24h)	Total phosphatase activity (mg/g·24h)	Catalase (mg/g·24h)
CK	15.21 ± 1.18c	2.36 ± 0.13b	0.71 ± 0.03c	0.25 ± 0.02b	3.32 ± 0.43b	2.79 ± 1.02b
M. micrantha	19.11 ± 0.36a	2.05 ± 0.18c	0.86 ± 0.02b	0.88 ± 0.10a	3.79 ± 0.26ab	5.44 ± 1.12a
C. odorata	16.91 ± 0.84b	2.87 ± 0.08a	0.61 ± 0.06c	0.26 ± 0.07b	3.74 ± 0.18ab	3.08 ± 0.79b
A. adenophora	14.49 ± 0.95c	2.78 ± 0.12a	1.01 ± 0.09a	0.26 ± 0.06b	4.05 ± 0.16a	3.39 ± 1.30b

The PCR-DGGE profiles revealed a difference in the AM fungal communities between the three invasive Asteraceae plants and their corresponding native soils in the Yunnan ecosystem (**Figure 2**). The *Cs* between soil invaded by *C. odorata* and native soil was 64%, followed by *M. micrantha* (60%), and *A. adenophora* (56%). A dendrogram created using hierarchical clustering analysis showed all samples clustered into two main groups with similarity of 0.56. One group included soils invaded by *M. micrantha* (lanes 4-6), those invaded by *C. odorata*, and native soil (**Figure 2**). Based on the *Cs*, *A. adenophora* had the greatest effect on the AM fungal community in native soil. The highest values for diversity, species richness, and evenness of AM fungi were measured in soils with *A. adenophora* (averages of 2.69, 25.33 and 0.83, respectively, $P < 0.05$), followed by *C. odorata*, but no significant differences were detected between *C. odorata* and the control (**Table 6**).

3.4. Correlation Analysis between AM Fungal Diversity and Environmental Variables

As shown in **Table 7**, correlation analysis of AM fungal community structure and soil physiochemical properties revealed that fungal abundance was significantly positively correlated with urease and catalase activities, but significantly negatively correlated with levels of soil organic carbon and phosphatase activities. The values for AM fungal diversity were significantly positively correlated with soil organic carbon, total N, NH_4^+-N, and

Table 5. Shannon-Wiener index (*H*), richness (*S*) and evenness (*E*) of soil AM fungi in *Flaveria bidentis* invaded plots.

Treatments	*S*	*H*	*E*
CK	27.67 ± 2.08a	2.23 ± 0.08b	0.67 ± 0.01b
F. bidentis	27.67 ± 1.53a	2.52 ± 0.09a	0.76 ± 0.02a

Table 6. Shannon-Wiener index (*H*), richness (*S*) and evenness (*E*) of soil AM fungi in Yunnan.

Treatments	*S*	*H*	*E*
CK	21.33 ± 0.58c	2.38 ± 0.08b	0.78 ± 0.02b
M. micrantha	19.67 ± 1.16bc	2.37 ± 0.11b	0.80 ± 0.02b
C. odorata	22.67 ± 0.58b	2.50 ± 0.02b	0.80 ± 0.01b
A. adenophora	25.33 ± 2.31a	2.69 ± 0.05a	0.83 ± 0.01a

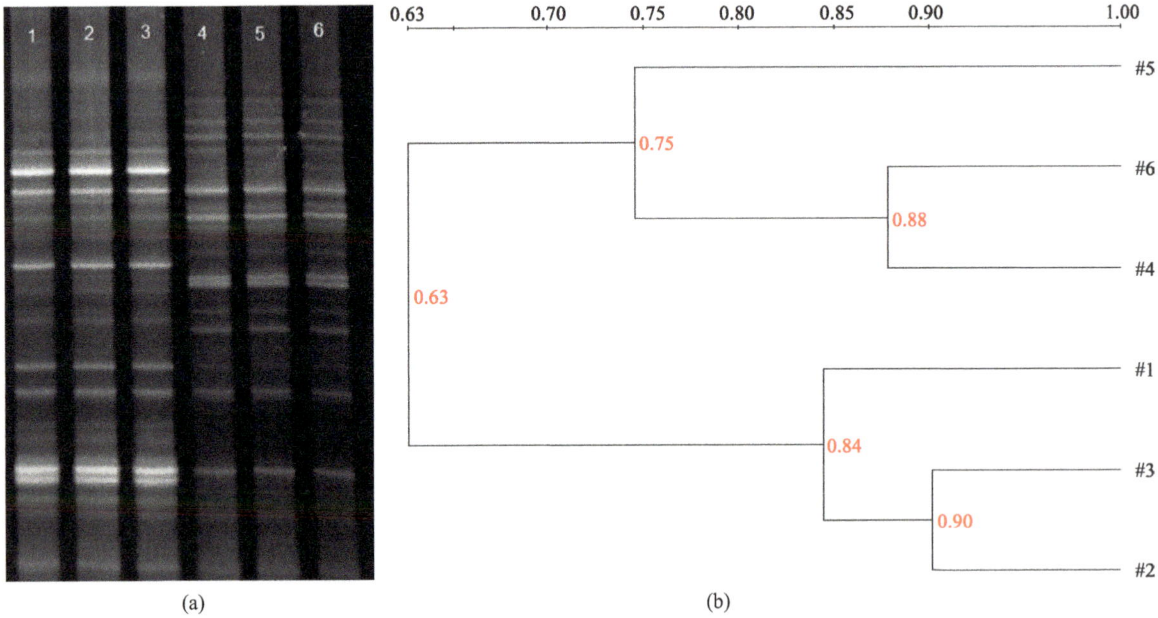

Figure 1. PCR-DGGE fingerprinting (a) and dendrogram using UPGMA (b) according to the *Cs* calculated on the basis of PCR-DGGE profiles of AM fungal communities in the soils invaded by *F. bidentis* (lane 1-3) and native soils (lane 4-6) of AM fungi in Hebei.

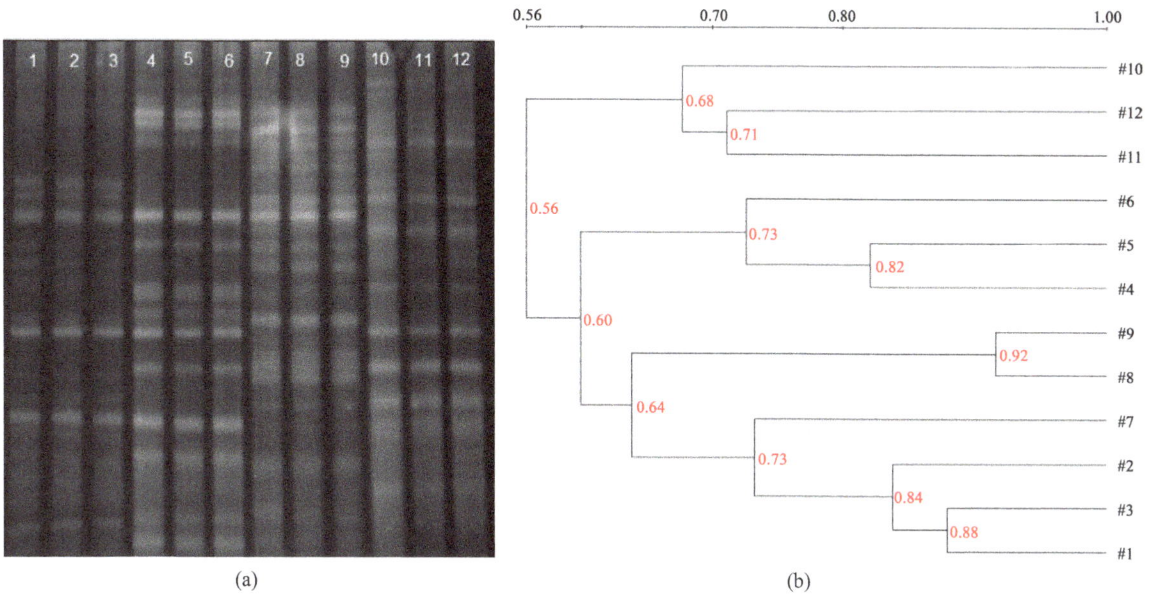

Figure 2. PCR-DGGE fingerprinting (a) and dendrogram using UPGMA (b) according to the *Cs* calculated on the basis of PCR-DGGE profiles of AM fungal communities in the soils invaded by *M. micrantha* (lane 4-6), by *C. odorata* (lane 7-9), by *A. adenophora* (lane 10-12) and native soils (lane 1-3) of AM fungi in Yunnan, respectively.

Table 7. Correlation analysis of soil AM fungi diversity index with soil physio-chemical parameters.

Indexes	pH	Organic matter (g/kg)	Total N (g/kg)	NH_4^+-N (mg/kg)	NO_3^--N (mg/kg)	Total P (g/kg)	Available P (mg/kg)	Urease	Phosphatase	Catalase
S	0.502$^{*1)}$	−0.558*	−0.468	−0.467	0.470*	0.708**	0.231	0.841**	−0.556*	0.659**
H	−0.460	0.526*	0.585*	0.625*	0.103	−0.354	−0.709**	0.227	0.555*	−0.308
E	−0.713**	0.814**	0.822**	0.825**	−0.195	−0.756**	−0.481*	0.736**	0.853**	0.680**

Data with * and ** are indicating significantly correlated at $P = 0.05$ and $P = 0.01$, respectively.

phosphatase activities. The values for the evenness index of the AM fungal community showed a highly significant positive correlation with soil organic carbon, total N, NH_4^+-N, urease, phosphatase, and catalase activities. Conversely, total P and available P were negatively correlated with almost all AM fungal parameters except for abundance.

4. Discussion

Invasion by alien plants can increase or decrease soil fertility [22]. These changes will affect competitive relationships between exotic and native species, and likely promote further invasion by invasive species [23]. Our research showed that apart from *M. micrantha*, the three invasive Asteraceae led to significant changes in soil ecosystems, and the direction of impacts was relatively consistent, suggesting that the soil N and P cycles changed upon plant invasion. Invasion by three Asteraceae, *A. adenophora*, *C. odorata*, and *F. bidentis* increased the N pool, and also lowered soil pH and P availability in the soil.

As the Asteraceae are typical mycorrhizal symbionts, AM fungi may be involved in the influence of these Asteraceae on the N cycle [24]. Plant invasion promoted an increase in soil N, and a closer relationship was found between the AM fungal community structure and ammonium than with nitrate. There is evidence that AM fungi preferentially absorb ammonia rather than other forms of N [25]. The absorption and transfer rate of ammonia in several external hyphae of AM fungi was more than ten times that of nitrates [25] [26]. Typically, nitrates have high mobility in soils, and can quickly reach plant roots through diffusion [27], while ammonia is likely to be absorbed by soil colloids, has poor mobility and is possibly deficient in the rhizosphere. The external hyphae of AM fungi can extend the surface area of root absorption and allow plants to acquire more ammonia.

This suggests that the ecological significance of AM fungi may also be in the exploitation of pulses in ammonia availability in the soil environment [24].

In addition, AM fungi play an important role in the ability of plants to absorb nutrients from soil [7]. They can, for example, efficiently help P absorption by *Spartina alterniflora* [28]. The decreased total P in soil invaded by Asteraceae, and the substantial reduction in available P in soil at sites where the other three Asteraceae were present (not *M. micrantha*), is consistent with the findings of Zhang *et al.* (2010) in relation to *F. bidentis*. Jeffries *et al.* [29] also verified a higher infection rate of AM fungi when there were low P levels in soil. Infection by a large number of AM fungal hyphae expands the surface area of root systems, reduces the distance for nutrient absorption by the roots, and improves spatial effectiveness of P in soil. Hyphae can partially replace the absorptive function of root systems [30], and increase the capacity of plants to absorb P. This might also explain why available P in invaded soils dropped substantially. Jiang *et al.* [31] maintained that *A. adenophora* strongly absorbs P in soils and limits the growth of other plants, thus outcompeting native species. The role of AM fungi in improving absorption of P has also been suggested by studies of other invasive plants, such as spotted knapweed [32]. The impacts of AM fungal infection on Asteraceae root systems and changes in community structure on the P cycle, as well as resultant changes in the competitive patterns of plants may play an important role in the ecosystem function.

AM fungi can directly influence the chemical properties of soil by secreting H^+ and some acidic compounds, and improving nutrient availability [7], which corroborates with the decrease in pH with Asteraceae species studied except for *M. micrantha*. Ezawa *et al.* [33] demonstrated that the invader *Gigaspora margarita* activated both acid and alkaline phosphatases, but P availability in mycorrhizal plants is typically higher than that in non-mycorrhizal plants. In our experiment, significantly increased phosphatase activity in the soil around invasive Asteraceae was remarkably correlated with the diversity of AM fungi, thus affecting the P cycle.

Higher diversity of AM fungi was detected in the soils invaded by *A. adenophora* than in the soils of the other invasive species. It is well acknowledged that AM fungi do not possess absolute specificity, but to some extent display host preferences [34]. Some AM fungi will not colonize the roots of certain plant species, or even germinate in response to the roots of certain plant species [35]. Evidence suggested that the more closely related are two co-occurring plants (e.g., *A. adenophora* and *C. odorata*), the more dissimilar are their AM fungal communities, and this might be due to the belowground niche partitioning of the AM fungal community. Other studies of plant-AM fungal associations reported similar findings [36] [37]. This might help to explain the coexistence of co-occurring species, especially in the presence of interspecific competition [38]. As AM fungi effectively increase resource acquisition, plants (such as *M. micrantha*, *A. adenophora*, and *C. odorata*) that associate with distinct AM fungi may experience enhanced functional complementarity relative to other species [37]. Also, this association might enable the more complete use of available resources and an overall increase in plant productivity [14]. Exploring further why the shifting patterns of AM fungal communities diverge from those of other Asteraceae species constitutes an interesting prospect for additional research.

In contrast with our expectations, the pattern of impact of *M. micrantha* on invaded soils differed somewhat from that of the other three Asteraceae. *Mikania* possesses many of the biological characteristics associated with successful invasive plant species [39]; it has a smothering habit, grows very fast, develops adventitious roots in each vine node as an asexual reproduction strategy, and also produces allelochemicals that mainly leach from stems [40]. It was speculate, therefore, that *M. micrantha* is more reliant on aboveground competition for invasion of new habitats. Owing to this difference in invasion strategies between *M. micrantha* and the other Asteraceae studied, impacts on the properties of invaded soil and the microbial flora also vary; highlighting the importance of the biological characteristics of each invasive species.

5. Conclusion

In summary, with the exception of invasion by *M. micrantha*, Asteraceae invasion leads to significant changes in soil ecosystems in similar pattern, suggesting that N and P cycles in invaded soil change upon plant invasion. Pulses in ammonia and a drop in available P seem well correlated with AM fungal diversity. Moreover, intrafamilial comparisons of the Asteraceae have shown that close relatives tend to host more divergent AM fungi, and that host preferences might favor co-occurring species. Thus, in the area studied, a feedback system in plant-AM fungi-soil nutrients might be one of the mechanisms underlying the success of the invasive Asteraceae, suggesting belowground plant-AM fungal associations in niche partitioning and coexistence of confamilial species.

Acknowledgements

This research was financially supported by the Natural Science Foundation of Tianjin (12JCQNJC09800), Special Fund for Agro-scientific Research in the Public Interest (201103027) and the National Science Foundation of China (31401811).

References

[1] Xie, Y., Li, Z.Y., William, P.G. and Li, D. (2000) Invasive Species in China—An Overview. *Biodiversity and Conservation*, **10**, 1317-1341.

[2] Weber, E., Sun, S. and Li, B. (2008) Invasive Alien Plants in China: Diversity and Ecological Insights. *Biological Invasions*, **10**, 1411-1429. http://dx.doi.org/10.1007/s10530-008-9216-3

[3] John, P.S. and John, M.D. (2011) Why Are Some Plant Genera More Invasive than Others? *PLoS ONE*, **6**, 1-8.

[4] Chen, B.M., Peng, S.L. and Ni, G.Y. (2009) Effects of the Invasive Plant *Mikania micrantha* H.B.K. on Soil Nitrogen Availability through Allelopathy in South China. *Biological Invasions*, **11**, 1291-1299. http://dx.doi.org/10.1007/s10530-008-9336-9

[5] Huangfu, C.H., Zhang, T.R., Chen, D.Q., Wang, N.N. and Yang, D.L. (2011) Residual Effects of Invasive Weed Yellowtop (*Flaveria bidentis*) on Forage Plants for Ecological Restoration. *Allelopathy Journal*, **27**, 55-64.

[6] Huangfu, C.H., Li, H.Y., Chen, X.W., Liu, H.M. and Yang, D.L. (2015) The Effects of Exotic Weed *Flaveria bidentis* with Different Invasion Stages on Soil Bacterial Community Structures. *African Journal of Biotechnology*, **14**, 2636-2643. http://dx.doi.org/10.5897/AJB2015.14658

[7] Smith, S.E., Jakobsen, I., Gronlund, M. and Smith, F.A. (2011) Roles of Arbuscular Mycorrhizas in Plant Phosphorus Nutrition: Interactions between Pathways of Phosphorus Uptake in Arbuscular Mycorrhizal Roots Have Important Implications for Understanding and Manipulating Plant Phosphorus Acquisition. *Plant Physiology*, **156**, 1050-1057. http://dx.doi.org/10.1104/pp.111.174581

[8] Helgason, T., Merryweather, J.W., Denison, J., Wilson, P., Young, J.P.W. and Fitter, A.H. (2002) Selectivity and Functional Diversity in Arbuscular Mycorrhizas of Co-Occurring Fungi and Plants from a Temperate Deciduous Woodland. *Journal of Ecology*, **90**, 371-384. http://dx.doi.org/10.1046/j.1365-2745.2001.00674.x

[9] Moora, M., Öpik, M., Sen, R. and Zobel, M. (2004) Native Arbuscular Mycorrhizal Fungal Communities Differentially Influence the Seedling Performance of Rare and Common Pulsatilla Species. *Functional Ecology*, **18**, 554-562. http://dx.doi.org/10.1111/j.0269-8463.2004.00876.x

[10] Uibopuu, A., Moora, M., Saks, U., Daniell, T., Zobel, M. and Öpik, M. (2009) Differential Effect of Arbuscular Mycorrhizal Fungal Communities from Ecosystems along Management Gradient on the Growth of Forest Understorey Plant Species. *Soil Biology and Biochemistry*, **41**, 2141-2146. http://dx.doi.org/10.1016/j.soilbio.2009.07.026

[11] Zhang, T.R., Huangfu, C.H., Bai, X.M. and Yang, D.L. (2010) Effects of *Flaveria bidentis* Invasion on Soil Nutrient Contents and Enzyme Activities. *Chinese Journal of Ecology*, **29**, 1353-1358.

[12] Pringle, A., Bever, J.D., Gardes, M., Parrent, J.L., Rillig, M.C. and Klironomos, J.N. (2009) Mycorrhizal Symbioses and Plant Invasions. *Annual Review of Ecology and Systematics*, **40**, 699-715. http://dx.doi.org/10.1146/annurev.ecolsys.39.110707.173454

[13] Shah, M.A., Reshi, Z.A. and Khasa, D.P. (2009) Arbuscular Mycorrhizas: Drivers or Passengers of Alien Plant Invasion. *Botanical Review*, **75**, 397-417. http://dx.doi.org/10.1007/s12229-009-9039-7

[14] Montesinos-Navarro, A., Segarra-Moragues, J.G., Valiente-Banuet, A. and Verdú, M. (2012) Plant Facilitation Occurs between Species Differing in Their Associated Arbuscular Mycorrhizal Fungi. *New Phytologist*, **196**, 835-844. http://dx.doi.org/10.1111/j.1469-8137.2012.04290.x

[15] Zhu, H.H., Yao, Q., Sun, X.T. and Hu, Y.L. (2007) Colonization, ALP Activity and Plant Growth Promotion of Native and Exotic Arbuscular Mycorrhizal Fungi at Low pH. *Soil Biology and Biochemistry*, **39**, 942-950. http://dx.doi.org/10.1016/j.soilbio.2006.11.006

[16] Avio, L., Castaldini, M., Fabiani, A., Bedini, S., Sbrana, C., Turrini, A. and Giovannetti, M. (2013) Impact of Nitrogen Fertilization and Soil Tillage on Arbuscular Mycorrhizal Fungal Communities in a Mediterranean Agroecosystem. *Soil Biology and Biochemistry*, **67**, 285-294. http://dx.doi.org/10.1016/j.soilbio.2013.09.005

[17] Hart, M.M., Reader, R.J. and Klironomos, J.N. (2003) Plant Coexistence Mediated by Arbuscular Mycorrhizal Fungi. *Trends in Ecology & Evolution*, **18**, 418-423. http://dx.doi.org/10.1016/S0169-5347(03)00127-7

[18] Hausmann, N.T. and Hawkes, C.V. (2009) Plant Neighborhood Control of Arbuscular Mycorrhizal Community Composition. *New Phytologist*, **183**, 1188-1200. http://dx.doi.org/10.1111/j.1469-8137.2009.02882.x

[19] Kowalchuk, G.A., de Souza, F.A. and van Veen, J.A. (2002) Community Analysis of Arbuscular Mycorrhizal Fungi

Associated with *Ammophila arenaria* in Dutch Coastal Sand Dunes. *Molecular Ecology*, **11**, 571-581. http://dx.doi.org/10.1046/j.0962-1083.2001.01457.x

[20] Cornejo, P., AzcÓn-Aguilar, C., Barea, J.M. and Ferrol, N. (2004) Temporal Temperature Gradient Gel Electrophoresis (TTGE) as a Tool for the Characterization of Arbuscular Mycorrhizal Fungi. *FEMS Microbiology Letters*, **241**, 265-270. http://dx.doi.org/10.1016/j.femsle.2004.10.030

[21] Long, L.K., Yao, K., Guo, J., Yang, R.H., Huang, Y.H. and Zhu, H.H. (2010) Molecular Community Analysis of Arbuscular Mycorrhizal Fungi Associated with Five Selected Plant Species from Heavy Metal Polluted Soils. *European Journal of Soil Biology*, **46**, 288-294. http://dx.doi.org/10.1016/j.ejsobi.2010.06.003

[22] Liao, C.Z., Peng, R.H., Luo, Y.Q., Zhou, X.H., Wu, X.W., Fang, C.M., Chen, J.K. and Li, B. (2008) Altered Ecosystem Carbon and Nitrogen Cycles by Plant Invasion: A Meta-Analysis. *New Phytologist*, **177**, 706-714. http://dx.doi.org/10.1111/j.1469-8137.2007.02290.x

[23] Ehrenfeld, J.G. (2003) Effects of Exotic Plant Invasions on Soil Nutrient Cycling Processes. *Ecosystems*, **6**, 503-523. http://dx.doi.org/10.1007/s10021-002-0151-3

[24] Veresoglou, S.D., Sen, R., Mamolos, A.P. and Veresoglou, D.S. (2011) Plant Species Identity and Arbuscular Mycorrhizal Status Modulate Potential Nitrification Rates in Nitrogen-Limited Grassland Soils. *Journal of Ecology*, **99**, 1339-1349. http://dx.doi.org/10.1111/j.1365-2745.2011.01863.x

[25] Hawkins, H. and George, E. (2001) Reduced ^{15}N-Nitrogen Transport through Arbuscular Mycorrhizal Hyphae to *Triticumestivum* L. Supplied with Ammonium vs Nitrate Nutrition. *Botany*, **87**, 303-311. http://dx.doi.org/10.1006/anbo.2000.1305

[26] Yoko, T. and Katsuya, Y. (2005) Nitrogen Delivery to Maize via Mycorrhizal Hyphae Depends on the Form of N Supplied. *Plant Cell and Environment*, **28**, 1247-1254. http://dx.doi.org/10.1111/j.1365-3040.2005.01360.x

[27] Tinker, P.B.H. and Nye, P.H. (2000) Solute Transport in the Rhizosphere. Oxford University Press, Oxford.

[28] McHugh, J.M. and Dighton, J. (2004) Influence of Mycorrhizal Inoculation, Inundation Period, Salinity and Phosphorus Availability on the Growth of Two Salt Marsh Grasses, *Spartina alterniflora* Lois. And *Spartina cynosuroides* (L.) Roth in Nursery Systems. *Restoration Ecology*, **12**, 533-545. http://dx.doi.org/10.1111/j.1061-2971.2004.03109.x

[29] Jeffries, P., Gianinazzi, S., Sperotto, S., Turnau, K. and Barea, J. (2003) The Contribution of Arbuscular Mycorrhizal Fungi Sustainable Maintenance of Plant Health and Fertility. *Biology and Fertility of Soils*, **37**, 1-16.

[30] Vance, C.P., Uhde-Stone, C. and Allan, D.L. (2003) Phosphorus Acquisition and Use: Critical Adaptations by Plants for Securing a Nonrenewable Resource. *New Phytologist*, **157**, 423-447. http://dx.doi.org/10.1046/j.1469-8137.2003.00695.x

[31] Jiang, Z.L., Liu, W.X., Wan, F.H. and Li, Z.Y. (2008) Effects of *Ageratina adenophora* (Spreng.) Invasion on Soil Nutrient Properties and Their Seasonal Dynamics. *Journal of Agro-Environment Science*, **27**, 267-272.

[32] Zabinski, C.A., Quinn, L. and Callaway, R.M. (2002) Phosphorus Uptake, Not Carbon Transfer, Explains Arbuscular Mycorrhizal Enhancement of *Centaurea maculosa* in the Presence of Native Grassland Species. *Function Ecology*, **16**, 758-765. http://dx.doi.org/10.1046/j.1365-2435.2002.00676.x

[33] Ezawa, T., Saito, M. and Yoshida, T. (1995) Comparison of Phosphatase Localization in the Intraradial Hyphae of Arbuscular Mycorrhizal Fungi, *Glomus* Spp and *Gigaspora* Spp. *Plant and Soil*, **176**, 57-63. http://dx.doi.org/10.1007/BF00017675

[34] Sanders, I.R. (2003) Preference, Specificity and Cheating in the Arbuscular Mycorrhizal Symbiosis. *Trends in Plant Science*, **8**, 143-145. http://dx.doi.org/10.1016/S1360-1385(03)00012-8

[35] Clapp, J.P., Helgason, T., Daniell, T.J. and Young, J.P.W. (2002) Genetic Studies of the Structure and Diversity of Arbuscular Mycorrhizal Fungal Communities. In: van der Heijden, M.G.A. and Sanders, I.R., Eds., *Mycorrhizal Ecology*, Springer-Verlag, Berlin, 201-224.

[36] Veresoglou, S.D. and Rillig, M.C. (2014) Do Closely Related Plants Host Similar Arbuscular Mycorrhizal Fungal Communities? A Meta-Analysis. *Plant and Soil*, **377**, 395-406. http://dx.doi.org/10.1007/s11104-013-2008-2

[37] Reinhart, K.O. and Anacker, B.L. (2014) More Closely Related Plants Have More Distinct Mycorrhizal Communities. *AoB Plants*, **6**, plu051. http://dx.doi.org/10.1093/aobpla/plu051

[38] Aanderud, Z.T. and Bledsoe, C.S. (2009) Preferences for ^{15}N-Ammonium, ^{15}N-Nitrate, and ^{15}N-Glycine Differ among Dominant Exotic and Subordinate Native Grasses from a California Oak Woodland. *Environmental and Experimental Botany*, **65**, 205-209. http://dx.doi.org/10.1016/j.envexpbot.2008.10.003

[39] Yang, Q.H., Feng, H.L. and Ye, W.H. (2003) An Investigation of the Effects of Environmental Factors on the Flowering and Seed Setting of *Mikania micrantha* H.B.K. *Journal of Tropical and Sub-Tropical Botany*, **11**, 123-126.

[40] Zhang, M.X., Ling, B., Kong, C.H., Zhao, H. and Pang, X.F. (2002) Allelopathic Potential of Volatile Oil from *Mikania micrantha*. *Chinese Journal of Applied Ecology*, **13**, 1300-1302.

7

Differences in Root Growth and Permeability in the Grafted Combinations of Dutch Tomato Cultivars (Starbuck and Maxifort) and Japanese Cultivars (Reiyo, Receive, and Magnet)

Tomomi Kakita, Ayami Abe, Takashi Ikeda*

School of Agriculture, Meiji University, Kawasaki, Japan
Email: *agrisys@meiji.ac.jp

Abstract

Grafting is widely established in agriculture and provides practical advantages for vegetable production. We investigated physiological differences between the grafted combinations of Dutch (Starbuck and Maxifort) and Japanese (Reiyo, Receive and Magnet) tomato cultivars. Plants were grown hydroponically until the flowers on the first truss bloomed, and the following parameters were measured: fresh weight of the aerial parts, root surface area, root permeability (by using a pressure chamber), and water potential of exudates (by using an isopiestic psychrometer). The Starbuck/Maxifort combination had higher values of the aerial part weight, root surface area, and root permeability than Reiyo/Receive, whereas Reiyo/Maxifort tended to have higher values of these parameters than Reiyo/Receive and Reiyo/Magnet. Maxifort had a significantly larger root surface area than Receive, but root permeability was not significantly different. InReiyo/Maxifort and Starbuck/Receive, these parameters were not significantly different except for a single comparison of root permeability. Thus, root permeability and root surface area may depend not only on the rootstock but be also affected by scion in grafted plants. Water potential of exudates was similar in most combinations and experiments. This shows that three rootstock cultivars provided similar nutrient concentrations even with different scions.

Keywords

Isopiestic Psychrometer, Pressure Chamber, Root Surface, Water Potential

*Corresponding author.

1. Introduction

Grafting is widely established in agriculture and provides practical advantages for vegetable production by increasing resistance to soilborne pathogens [1]-[4], tolerance to low root-zone temperatures [5]-[7], salinity tolerance [8]-[10], and endogenous hormone production [11]-[13]. In particular, increased salinity tolerance improves fruit yield of grafted tomato [8] [14]. Higher yield of grafted plants has been also observed in watermelon (*Citrullus lanatus*) [15] and in melon (*Cucumismelo*) [16] [17]. Increased plant growth and crop yield upon grafting in melon have been attributed to the increase in water and nutrient uptake due to the vigorous root systems of rootstocks [16]. Therefore, we think grafting would alter plant water relations.

We previously reported that seedlings of the Japanese cultivar Reiyo had higher root permeability to nutrient solution than those of the Dutch cultivar Dundee [18]; this study investigated only non-grafted plants, whereas most tomatoes in commercial production are grafted, including about 75% of the hydroponic tomatoes in the Netherlands [7]. However, the effect of grafting on root growth and permeability in Japanese and Dutch tomatoes has not been addressed. In this study, we used own-rooted Japanese and Dutch tomato cultivars and their grafted combinations to investigate how grafting affects root growth and permeability.

2. Materials and Methods

2.1. Cultivars and Growing Conditions

Two beef tomato (*Solanum lycopersicum* L.) cultivars, the Japanese cultivar Reiyo (Sakata Seed Co., Kanagawa, Japan) and the Dutch cultivar Starbuck (De Ruiter Seeds, Bleiswijk, the Netherlands) were usedas scions. Japanese cultivars Receive and Magnet (Sakata Seed Co.) and the Dutch cultivar Maxifort (*S. lycopersicum* × *S. habrochaites*) (De Ruiter Seeds) were used as rootstocks. Seeds were germinated in a dark incubator (HNM-S10; Koito Industries, Ltd., Kanagawa, Japan) at 25˚C. When root length reached 2 - 3 mm, germinated seeds were transferred into commercial soil (Zen-Noh, Tokyo, Japan) in plug trays. Seedlings were grown in a glasshouse; the roof and side windows were opened when the internal temperature reached 25˚C and the ceiling fan was activated at 30˚C. In winter, a heater was operated when the internal temperature became below 12˚C.

At the 4- to 5-leaf stage, Reiyo, Receive, Starbuck, and Maxifort were transplanted into a nutrient film hydroponic system (NFT) laid out to keep the root systems of each plant separate as described in Furukawa *et al.* [19] and Kakita *et al.* [20] (**Figure 1**). Plants of each cultivar were randomly placed 15 cm apart from each other. The nutrient solution (Otsuka-A; Otsuka Chemical Co., Ltd., Osaka, Japan) was adjusted to electrical conductivity (EC) of 2.2 - 2.4 dS·m^{-1} and pH of 5.5 - 6.5. Plants were pinched at two leaves above the primary flower cluster.

Figure 1. The hydroponic culture system (nutrient film technique, NFT) designed so that tomato roots do not grow over each other. Nutrient solution (EC 2.2 - 2.4 dS·m^{-1}, pH 5.5 - 6.5) flowed along the slope near the plants.

2.2. Grafting

Seedlings at the 3-leaf stage were used. Plants were cut diagonally at 30 degrees about 1 cm above the cotyledons and grafted with tubes (Nasunics Co., Ltd., Tochigi, Japan). The grafted seedlings were kept in the dark for 2 - 4 days and then acclimatized for 1 week to greenhouse conditions in plastic boxes (62 × 42 × 22 cm) with the tops lined with wet newspapers on the sides. The humidity was adjusted by spraying water on the newspapers, opening and closing the top, and using a shading net. During this period, plants were supplied with a nutrient solution with EC of 1.2 $dS \cdot m^{-1}$ and pH of 5.5 - 6.5. At the 4- to 5-leaf stage, grafted plants were transplanted into the NFT system as described above. Plants used for the measurements grew in the NFT system until the flowers on the first truss bloomed. In Experiment 1, we prepared the grafted combinations Reiyo/Receive and Starbuck/Maxifort (scion/rootstock) and cultivated these plants from September to November 2011 (n = 10 and 12, respectively), March to May 2012 (n = 8 and 11), and October to January. 2012-2013 (n = 8 for both combinations). In Experiment 2, we prepared the grafted combinations Reiyo/Maxifort and Starbuck/Receive and cultivated them from November to February 2011-2012 (n = 11 and 8, respectively), April to June 2012 (n = 10 and 7), November to February 2012-2013 (n = 7 and 8), and September to November (n = 6 for both combinations). In Experiment 3, we grafted the Japanese scion Reiyoon to Receive, Magnet, and Starbuck. Plants were cultivated from September to January2013-2014 (n = 6, 5, and 10, respectively) and November to February 2013-2014 (n = 5, 7, and 5).

2.3. Determination of the Fresh Weight of the Aerial Parts, Root Surface Area, Root Permeability, and Exudate Water Potential

After the flowers on the first truss bloomed, plants were cut near the cotyledons (non-grafted plants) or 2 cm above the graft union (grafted plants), and fresh aerial parts were weighed.

The root system of each plant was enclosed (1 cm below the cut) in a silicone rubber ring (38 mm outside diameter; 16 mm inside diameter), and the gaps between the ring and the stem were sealed with putty (EN ISO 4823, type 0 putty; Heraeus Kulzer-Japan, Tokyo, Japan). After the putty had set, the root system was fixed in a pressure chamber [18] [21] filled with nutrient solution (EC 2.4 $dS \cdot m^{-1}$).The nutrient solution was forced into the roots under a pressure of 0.25 MPa (permeability remained constant at 0.2 - 1.0 MPa; our unpublished data).The emerging exudate was discarded for the first 2 - 3 min and then absorbed on pre-weighed cotton for the next 2 min; the difference in cotton weight before and after absorption was considered as the weight of the exudate. The root surface area was measured with a root scanner (WinRHIZO 2007; Regent Instruments Inc., Quebec, Canada) and used as a measure of root growth. Root permeability was calculated according to the following equation:

Permeability = exudate weight (g)/[root surface area (m^2) × extraction time (s) × applied pressure (MPa)]

Water potential was used as a proxy for the total amount of solutes in the exudate. The exudate was collected with a microsyringe (Hamilton Company, Reno, Nevada, USA) and water potential was measured by using an isopiestic psychrometer (Model-3; Isopiestics Co., Lewes, Delaware, USA) according to Boyer [21]. Sucrose solution of known potential and the exudate were placed into the sample chambers and on the thermocouple, respectively.

Student's *t*-test was used to compare plants growing under the same conditions.

3. Results

In a comparison between non-graftedscion cultivars, the aerial part weight of Reiyo was 1.2 times that of Starbuck (**Figure 2(a)**), where as the root surface area, root permeability, and water potential of exudates were not significantly different (**Figures 2(b)-(d)**). In a comparison between rootstock cultivars, Maxifort had a higher weight of the aerial parts and larger root surface area than Receive (**Figure 2(a)**, **Figure 2(b)**), where as the root permeability and exudate water potential were not significantly different (**Figure 2(c)**, **Figure 2(d)**).

In experiment 1, we compared sympatric grafted combinations. The combination of Dutch cultivars, Starbuck/Maxifort, had higher values of the weight of the aerial parts (2.1 to 4.4 times), root surface area (2.3 to 3.6 times), and root permeability (1.1 to 1.7 times) than the combination of Japanese cultivars Reiyo/Receive (**Figures 3(a)-(c)**), whereas the water potential of exudates was not significantly different (**Figure 3(d)**).

In Experiment 2, the allopatric combinations Reiyo/Maxifort and Starbuck/Receive did not differ significantly in the aerial part weight and root surface area (**Figure 4(a)**, **Figure 4(b)**). These combinations also did

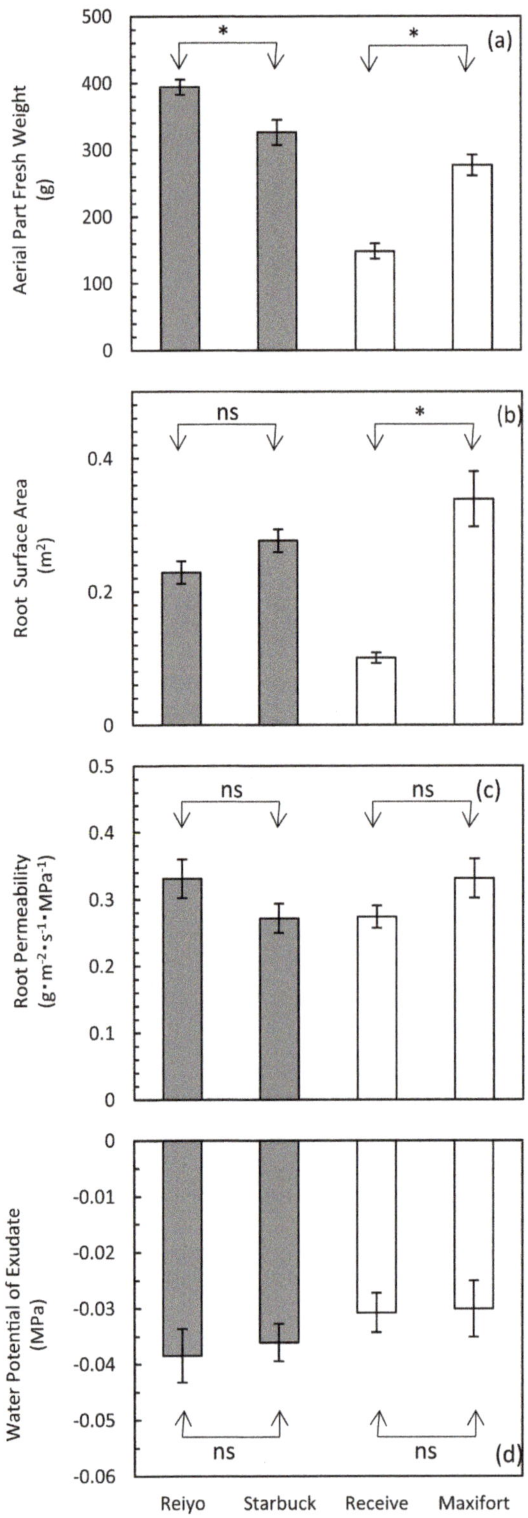

Figure 2. Fresh weight of aerial parts (a), root surface area (b), root permeability (c), and exudate water potential (d) in non-grafted plants. Plants were grown from September to November 2012. Values are means ± SE (n = 6; $^{*}p < 0.05$; ns, not significantby Student's *t*-test).

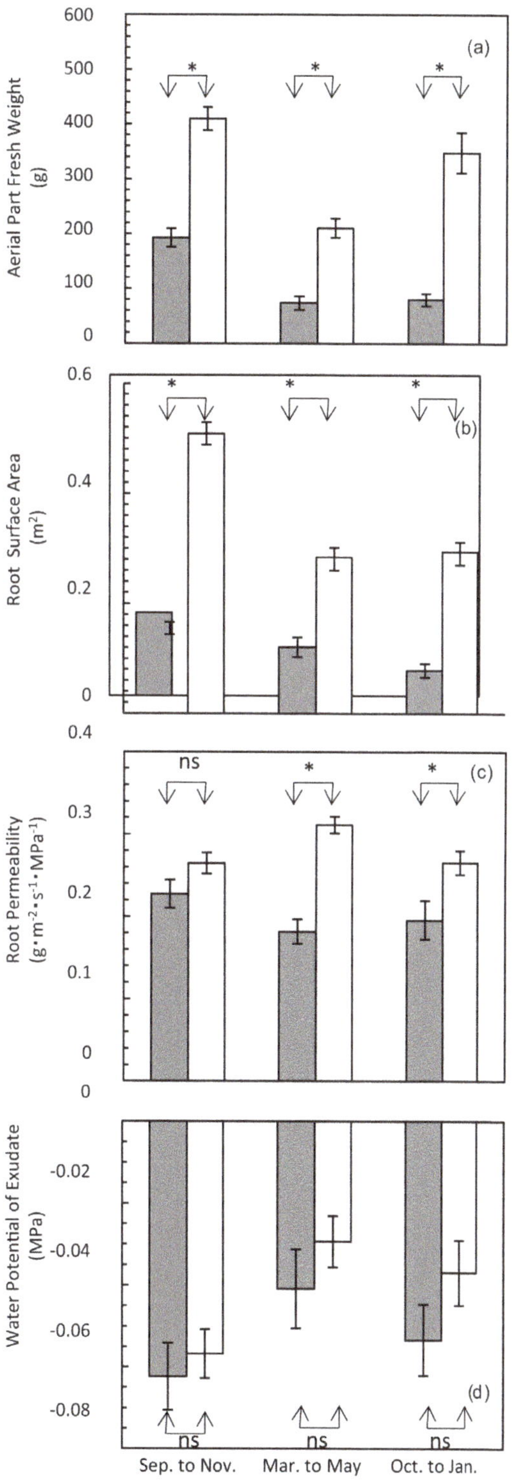

Figure 3. Fresh weight of aerial parts (a), root surface area (b), root permeability (c), and exudate water potential (d) insympatric combinations: Reiyo/Receive (shaded bars) and Starbuck/Maxifort (open bars). Plants were grown from September to November (Reiyo/Receive: n = 10; Starbuck/Maxifort: n = 12), from March to May (Reiyo/Receive: n = 8; Starbuck/Maxifort: n = 11), and from October to January (n = 8 for both combinations). Values are means ± SE ($^{*}p < 0.05$; ns, not significantby Student's *t*-test).

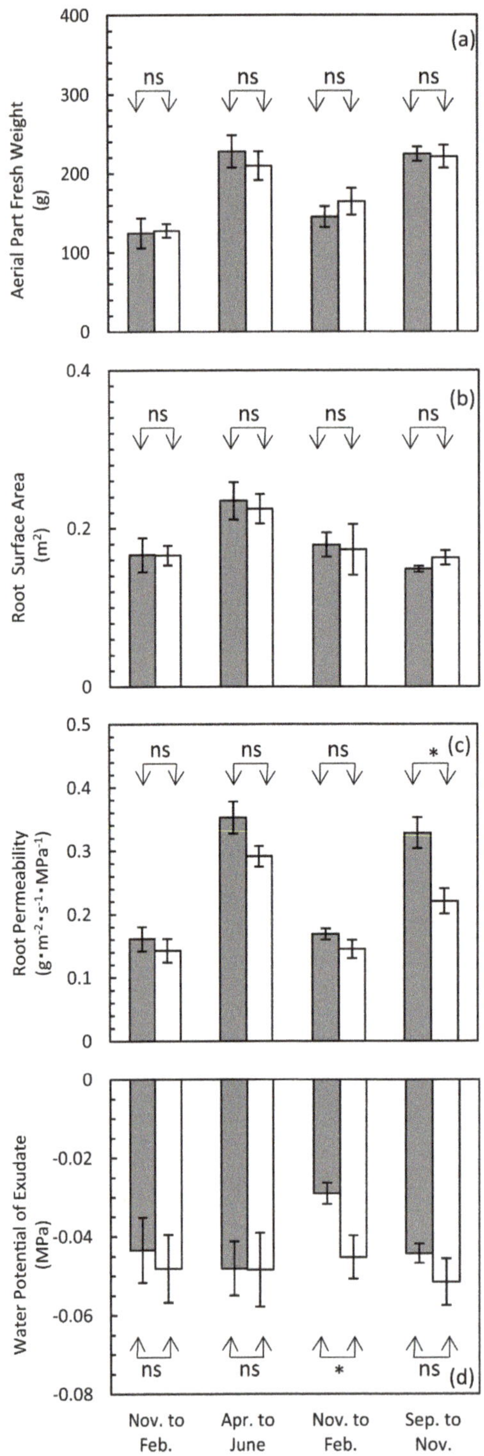

Figure 4. Fresh weight of aerial parts (a), root surface area (b), root permeability (c), and exudate water potential (d) in allopatric combinationsReiyo/Maxifort (shaded bars) and Starbuck/Receive (open bars). Plants were grown from November to February (Reiyo/Maxifort: n = 11; Starbuck/Receive: n = 8), from April to June (Reiyo/Maxifort: n = 10; Starbuck/Receive: n = 7), from November to February (Reiyo/Maxifort: n = 7; Starbuck/Receive: n = 8), and from September to November (n = 6 for both combinations). Values are means ± SE ($^{*}p < 0.05$; ns, not significantby Student's *t*-test).

not differ significantly in root permeability and exudate water potential, except that in 1 out of 4 cultivation periods Reiyo/Maxifort had significantly higher root permeability (**Figure 4(c)**; September to November) and significantly lower exudate water potential (**Figure 4(d)**; November to February) than Starbuck/Receive.

In Experiment 3, the Reiyo/Maxifort combination had a higher root permeability than Reiyo/Receive and Reiyo/Magnet (**Figure 5(c)**). The same tendency was observed for the aerial part fresh weight and root surface area, although no significant difference was observed between the aerial part fresh weight of Reiyo/Maxifort and Reiyo/Receive in 1 out of the 2 cultivation periods (**Figure 5(a)**; November to February) and between the root surface area of Reiyo/Maxifort and Reiyo/Receive (**Figure 5(b)**). In both cultivation periods, exudate water potential tended to be lowest in Reiyo/Receive and highest in Reiyo/Maxifort, but the difference reached statistical significance only for Reiyo/Receive in one of the two cultivation periods (**Figure 5(d)**; November to January).

4. Discussion

Diverse experimental approaches have been used to compare various physiological characteristics in Dutch and Japanese tomatoes, including fruit weight and mesocarp thickness [22], effects of CO_2 concentration on fruit growth, relative source and sink strengths [23], leaf characteristics, and photosynthetic rate [24]. These studies mainly considered the differences between the aerial parts of Dutch and Japanese tomatoes, whereas little is known about the root systems despite their importance for vegetative growth and fruit yield. Most of these studies used non-grafted plants [18] [22]-[25], whereas grafted plants are morphologically and physiologically different.

In this study, we documented the differences in aerial part weight, root surface area, and root permeability in non-grafted Japanese and Dutch tomato cultivars and in their allopatric and sympatric grafted combinations. Collection of exudate from the stem by using root pressure has been used to estimate plant water uptake, but this method is prone to errors if plants actively evaporate [26] [27]. Therefore, to measure root permeability, we chose the pressure chamber technique, which has been previously used in tomato [18] [20] [28]-[30] and rice [31]-[33].

Japanese and Dutch tomato cultivars grown under the same conditions differ in productivity and plant vigor [34]: Japanese cultivars tend to be more vigorous [34]-[36]. This may be explained by differences in water and nutrient uptake. Previous studies mainly investigated the aerial parts of Dutch and Japanese tomatoes, whereas little is known about the root systems. Furukawa *et al.* [18] demonstrated that the aerial part weight and root permeability in seedlings of the Japanese cultivar Reiyo are significantly higher than in those of the Dutch cultivar Dundee. This was also observed between seedlings of Reiyo and those of the Dutch cultivar Starbuck [30]. In this study, root permeability of Reiyo tended to be higher than that of Starbuck, but the difference was not statistically significant (**Figure 2(c)**). Thus, one of the reasons for growth differences between the cultivars observed by Furukawa *et al.* [19] and Ahn *et al.* [36] may be the difference in root permeability. Further studies of other cultivars from both countries at various growing stages are needed.

In this study, in rootstock cultivars, the value of root surface area (**Figure 2(b)**) but not that of root permeability (**Figure 2(c)**) was significantly higher in Maxifort than in Receive. The combination of Dutch cultivars Starbuck/Maxifort had a larger root surface area (**Figure 3(b)**) and higher root permeability (in two out of three cultivation periods; **Figure 3(c)**) than the combination of Japanese cultivars Reiyo/Receive. These differences would result in higher water and nutrient uptake, which would increase the aerial part growth (**Figure 3(a)**). We also found that Reiyo on Maxifort had higher root permeability than Reiyo grafted on two Japanese rootstocks, Receive and Magnet (**Figure 5(c)**), which probably resulted in a larger fresh weight of aerial parts (**Figure 5(a)**). Thus, grafted combinations with Maxifort rootstock had increased root surface area and root permeability. However, root surface area and root permeability did not differ significantly between Reiyo/Maxifort and Starbuck/Receive (**Figure 4(b)**, **Figure 4(c)**). These results do not allow us to conclude that the physiological effects of rootstocks are primarily a function of root permeability. Melon yield depends on the combination of scion and rootstock, and the content of inorganic ions in the leaves is altered by the rootstock [16]. Santa-Cruz *et al.* [37] observed that the same rootstock differently affects growth of scions from different tomato cultivars. Thus, the root surface area and root permeability in grafted plants may be affected by the scion cultivar. To elucidate the reasons for the difference in fresh weight of the aerial part between the sympatric combinations (**Figure 3**) and the absence of such difference between the allopatric combinations (**Figure 4**) in our study, we need to investigate grafting compatibility [16] [38], hydraulic resistance of xylem vessels [39], or other physiological factors.

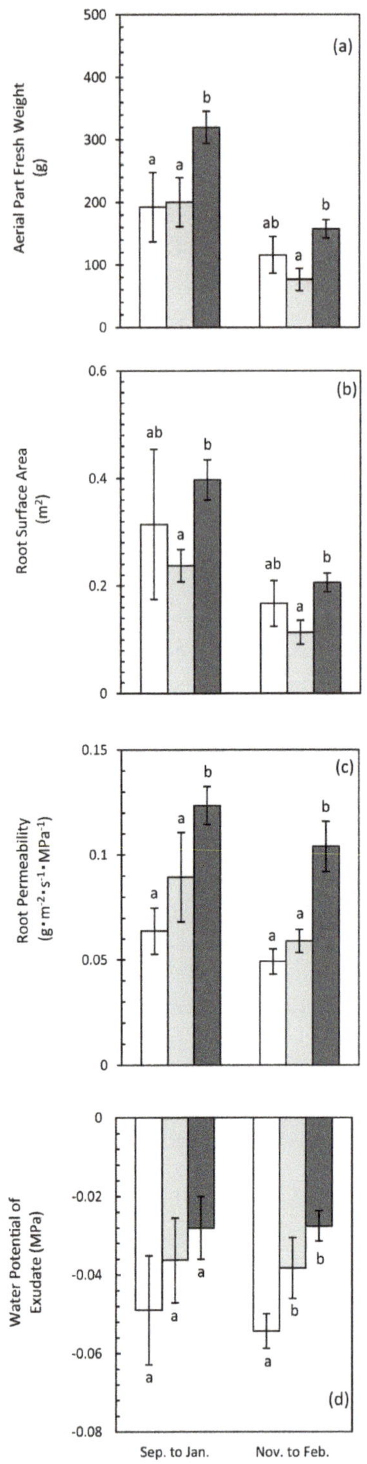

Figure 5. Fresh weight of aerial parts (a), root surface area (b), root permeability (c), and exudate water potential (d) in the combinations Reiyo/Receive (open bars), Reiyo/Magnet (light shaded bars), and Reiyo/Maxifort (darkshaded bars). Plants were grown from September to January (Reiyo/Receive: n = 6; Reiyo/Magnet: n = 5; Reiyo/Maxifort: n = 10) and from November to February (Reiyo/Receive: n = 5; Reiyo/Magnet: n = 7; Reiyo/Maxifort: n = 5). Values are means ± SE. The same letters above bars indicateno significant difference by Student's *t*-testbetween plants grown in same period.

The pressure chamber method requires analysis of many plants in a short time to prevent the impact of differences in plant growth. In each cultivation period, measurements can be conducted for only a few plants. We therefore decided to improve the accuracy by increasing the number of replications of each experiment. Although air temperature, day length, nutrient solution, and temperature during cultivation are expected to affect the absolute values of the parameters measured, this should not affect the validity of our comparisons within the same groups (*i.e.* sympatric or allopatric combinations), because comparisons within the groups were performed at the same time. Because root permeability is affected by the root temperature [33] [40], its variability between growing seasons was probably due to temperature differences during greenhouse cultivation. Notably, the relative differences between different grafted combinations were not affected by ambient temperature.

It is noteworthy that the water potential of exudates was similar in most experiments (except in one out four cultivation periods in **Figure 4(d)** and one out of two cultivation periods in **Figure 5(d)**). Since water potential reflects solute concentration [21], this shows that solute concentration in exudates was similar in grafted plants with different rootstocks, indicating that all rootstock cultivars provided similar nutrient concentrations to the scions. We may need to consider the effect of scion cultivar on growth and rootstock root surface area in future studies.

5. Conclusion

We found that the combination of Dutch cultivars (Starbuck grafted onto Maxifort) had a higher root surface area and root permeability than a combination of Japanese cultivars (Reiyo grafted onto Receive), which may in turn enhance the aerial part growth. Root permeability of the Reiyo/Maxifort combination was higher than that ofReiyo/Receive and Reiyo/Magnet, although the difference between allopatric combinations (Reiyo/Maxifort and Sturbuck/Recieve) was not significant. Further studies are needed to explore the physiological effects of grafting in other cultivars at different growth stages.

References

[1] Ioannou, N. (2001) Integrating Soil Solarization with Grafting on Resistant Rootstocks for Management of Soil-Borne Pathogens of Eggplant. *Journal of Horticultural Science & Biotechnology*, **76**, 396-401.

[2] Lee, J.M. (1994) Cultivation of Grafted Vegetables. I. Current Status, Grafting Methods and Benefits. *HortScience*, **29**, 235-239.

[3] King, S.R., Davis, A.R., Liu, W. and Levi, A. (2008) Grafting for Disease Resistance. *HortScience*, **43**, 1673-1676.

[4] Keinath, A.P. and Hassell, R.L. (2014) Control of Fusarium Wilt of Watermelon by Grafting onto Bottlegourd or Interspecific Hybrid Squash despite Colonization of Rootstocks by Fusarium. *Plant Disease*, **98**, 255-266. http://dx.doi.org/10.1094/PDIS-01-13-0100-RE

[5] Den Nijs, A.P.M. and Smeets, L. (1987) Analysis of Differences in Growth of Cucumber Genotypes under Low Light Conditions in Relation to Night Temperature. *Euphytica*, **36**, 19-32. http://dx.doi.org/10.1007/BF00730643

[6] Tachibana, S. (1982) Comparison of Effects of Root Temperature on the Growth and Mineral Nutrition of Cucumber Cultivars and Figleaf Gourd. *Journal of Japanese Society for Horticultural Science*, **51**, 299-308.

[7] Venema, J.H., Dijk, B.E., Bax, J.M., van Hasselt, P.R. and Elzenga, J.T.M. (2008) Grafting Tomato (*Solanum lycopersicum*) onto the Rootstock of a High-Altitude Accession of *Solanum habrochaites* Improves Suboptimal-Temperature Tolerance. *Environmental &Experimental Botany*, **63**, 359-367. http://dx.doi.org/10.1016/j.envexpbot.2007.12.015

[8] Estañ, M.T., Martinez-Rodriguez, M.M., Perez-Alfocea, F., Flowers, T.J. and Bolarin, M.C. (2005) Grafting Raises the Salt Tolerance of Tomato through Limiting the Transport of Sodium and Chloride to the Shoot. *Journal of Experimental Botany*, **56**, 703-712. http://dx.doi.org/10.1093/jxb/eri027

[9] Martinez-Rodriguez, M.M., Estañ, M.T., Moyano, E., Garcia-Abellan, J.O., Flores, F.B. Campos, J.F., Al-Azzawi, M.J., Flowers, T.J. and Bolarin, M.C. (2008) The Effectiveness of Grafting to Improve Salt Tolerance in Tomato When an "Excluder" Genotype Is Used as Scion. *Environmental & Experimental Botany*, **63**, 392-401. http://dx.doi.org/10.1016/j.envexpbot.2007.12.007

[10] Zekri, M. and Parsons, L.P. (1992) Salinity Tolerance of Citrus Rootstocks: Effects of Salt on Root and Leaf Mineral Concentrations. *Plant &Soil*, **147**, 171-181. http://dx.doi.org/10.1007/BF00029069

[11] Aloni, B., Cohen, R., Karni, L., Aktas, H. and Edelstein, M. (2010) Hormonal Signaling in Rootstock-Scion Interactions. *Scientia Horticulturae*, **127**, 119-126. http://dx.doi.org/10.1016/j.scienta.2010.09.003

[12] Dong, H.H., Niu, Y.H., Li, W.J. and Zhang, D.M. (2008) Effects of Cotton Rootstock on Endogenous Cytokinins and

Abscisic Acid in Xylem Sap and Leaves in Relation to Leaf Senescence. *Journal of Experimental Botany*, **59**, 1295-1304. http://dx.doi.org/10.1093/jxb/ern035

[13] Proebsting, W.M., Hedden, P., Lewis, M.J., Croker, S.J. and Proebsting, L.N. (1992) Gibberellin Concentration and Transport in Genetic Lines of Pea. *Plant Physiology*, **100**, 1354-1360. http://dx.doi.org/10.1104/pp.100.3.1354

[14] Fernández-García, N., Martínez, V., Cerda, A. and Carvajal, M. (2004) Fruit Quality of Grafted Tomato Plants Grown under Saline Conditions. *Journal of Horticultural Science & Biotechnology*, **79**, 995-1001.

[15] Alan, Ö., Özdemir, N. and Gimen, Y. (2007) Effect of Grafting on Watermelon Plant Growth, Yield and Quality. *Journal of Agronomy*, **6**, 362-365. http://dx.doi.org/10.3923/ja.2007.362.365

[16] Ruiz, J.M., Belakbir, A., Lopez-Cantarero, A. and Romero, L. (1997) Leaf-Macronutrient Content and Yield in Grafted Melon Plants: A Model to Evaluate the Influence of Rootstock Genotype. *Scientia Horticulturae*, **71**, 227-234. http://dx.doi.org/10.1016/S0304-4238(97)00106-4

[17] Ruiz, J.M. and Romero, L. (1999) Nitrogen Efficiency and Metabolism in Grafted Melon Plants. *Scientia Horticulturae*, **81**, 113-123. http://dx.doi.org/10.1016/S0304-4238(98)00200-3

[18] Furukawa, T., Ishikawa, K., Onuki, S., Mukai, K. and Ikeda, T. (2010) Permeability of Root Cells of Dutch "Dundee" and Japanese "Reiyo" Varieties of Tomato Seedlings to Hydroponic Solution and Potassium Ions. *Journal of Japanese Society for Horticultural Science*, **79**, 287-293. http://dx.doi.org/10.2503/jjshs1.79.287

[19] Furukawa, T., Ikeda, T. and Tomita, H. (2008) Growth, Quality and Rhizosphere State on Dutch and Japanese Varieties of Tomato Plants at Low Truss and High Planting Density under NFT. *Horticulture Research*, **7**, 524. (In Japanese)

[20] Kakita, T., Mukai, K., Sato, Y., Nakamura, M. and Ikeda, T. (2011) Difference of Root Permeability of Dutch and Japanese Varieties of Single-Truss Tomato Plants. *Proceedings of the Annual Conference of the American Society for Horticultural Science*, Waikoloa, 25-28 September 2011, 116.

[21] Boyer, J.S. (1995) Measuring the Water Status of Plants and Soil. Academic Press, San Diego. http://dx.doi.org/10.1016/b978-012122260-4/50005-6

[22] Oyabu, T., Kato, M. and Sakakibara, M. (2009) Selection of Breeding Materials and Screening Method of Slicing Favorable Tomato Cultivar. *Research Bulletin of Aichi Agricultural Research Center*, **41**, 61-66. (In Japanese)

[23] Matsuda, R., Nakano, A., Ahn, D.H., Suzuki, K., Yasuba, K. and Takaichi, M. (2011) Growth Characteristic and Sink Strength of Fruit at Different CO_2 Concentrations in a Japanese and a Dutch Tomato Cultivar. *Scientia Horticulturae*, **127**, 528-534. http://dx.doi.org/10.1016/j.scienta.2010.12.009

[24] Saito, T., Yong-Gen, Y., Matsuoka, M., Watanabe, S., Matsukura, C. and Fukuda, N. (2011) An Investigation of Differences in Fruit Yield and Components Contributing to Increased Yield in Japanese and Dutch Tomato Cultivars. *Plant Biotechnology*, **28**, 455-461. http://dx.doi.org/10.5511/plantbiotechnology.11.0928a

[25] Matsuda, R., Suzuki, K., Nakano, A., Higashide, T. and Takaichi, M. (2011) Responses of Leaf Photosynthesis and Plant Growth to Altered Source-Sink Balance in a Japanese and a Dutch Tomato Cultivar. *Scientia Horticulturae*, **127**, 1298-1300. http://dx.doi.org/10.1016/j.scienta.2010.12.008

[26] Else, M.A., Hall, K.C., Arnold, G.M., Davies, W.J. and Jackson, M.B. (1995) Export of Abscisic Acid, 1-Aminocyclopropane-1-Carboxylic Acid, Phosphate, and Nitrate from Roots to Shoots of Flooded Tomato Plants. *Plant Physiology*, **107**, 377-384.

[27] Jeschke, W.D. and Pate, J.S. (1991) Cation and Chloride Partitioning through Xylem and Phloem within the Whole Plant of *Ricinus communis* L. under Conditions of Salt Stress. *Journal of Experimental Botany*, **42**, 1105-1116. http://dx.doi.org/10.1093/jxb/42.9.1105

[28] Kanai, S., Moghaieb, R.E., El-Shemy, H.A., Panigrahi, R., Mohapatra, P.K., Ito, J., Nguyen, N.T., Saneoka, H. and Fujita, K. (2011) Potassium Deficiency Affects Water Status and Photosysthetic Rate of the Vegetative Sink in Greenhouse Tomato Prior to Its Effects on Source Activity. *Plant Science*, **180**, 368-374. http://dx.doi.org/10.1016/j.plantsci.2010.10.011

[29] Nonami, H., Fukuyama, T. and Hashimoto, Y. (1998) Presence of Water Chemicals and Water Permeability in Tomato Roots under Salt-Stressed Conditions. *Proceedings of Environmental Control in Biology*, 282-283.

[30] Onuki, S., Mukai, K., Sato, Y., Furukawa, T. and Ikeda, T. (2010) Effect of Mercuric Chloride to Root Water Permeability on Dutch and Japanese Varieties of Tomato Seedlings. *Horticulture Research*, **9**, 369. (In Japanese)

[31] Asanuma, S., Ookawa, T. and Hirasawa, T. (2007) The Resistance to Water Flow of the Osmotic Water Absorption and Transport in Rice Plants for Rapid Determination of Root Resistance to Water Transport. *Japanese Journal of Crop Science*, **76**, 246-247. (In Japanese)

[32] Miyamoto, N., Steudle, H., Hirasawa, T. and Lafitte, R. (2001) Hydraulic Conductivity of Rice Roots. *Journal of Experimental Botany*, **52**, 1835-1486. http://dx.doi.org/10.1093/jexbot/52.362.1835

[33] Murai-Hatano, M., Kuwagata, T., Sakurai, J., Nonami, H., Ahamed, A., Nagasuga, K., Matsunami, T., Fukushi, K.,

Maeshima, M. and Okada, M. (2008) Effect of Low Root Temperature on Hydraulic Conductivity of Rice Plants and the Possible Role of Aquaporins. *Plant Production Science*, **49**, 1294-1305. http://dx.doi.org/10.1093/pcp/pcn104

[34] Ahn, D., Nakano, A., Matsuda, R., Higashide, T., Suzuki, K. and Takaichi, M. (2010) Effect of the Change of Sink Strength by Fruit Thinning on Growth in a Japanese and a Dutch Tomato Cultivar. *Horticulture Research*, **9**, 238. (In Japanese)

[35] Ando, A., Wada, T., Hirai, H. and Abe, K. (2003) Effect of Rockwool-Cube Size and Days of Raising Seedlings on Growth and Root Activity of the Seedlings, Growth after Transplanting and Yield of Tomato Growth on Rockwool. *Horticulture Research*, **2**, 297-301. http://dx.doi.org/10.2503/hrj.2.297

[36] Kobayashi, S. and Kirimura, Y. (1999) Effects of Rockwool Cube Size and Root Limitation on Plant Growth and Fruit Yield of Lower-Truss Tomato by Hydroponics. *Horticulture Research*, **68**, 307. (In Japanese)

[37] Santa-Cruz, A., Martinez-Rodriguez, M.M., Perez-Alfocea, F., Romero-Aranda, R. and Bolarin, M.C. (2002) The Rootstock Effect on the Tomato Salinity Response Depends on the Shoot Genotype. *Plant Science*, **162**, 825-831. http://dx.doi.org/10.1016/S0168-9452(02)00030-4

[38] Oda, M. (2001) Grafting of Vegetable Crops. *Scientific Report of the Graduate School of Agriculture& Biological Science Osaka Prefecture University*, **54**, 49-72.

[39] Oda, M., Maruyama, M. and Mori, G. (2005) Water Transfer at Graft Union of Tomato Plants Grafted onto *Solanum* Rootstocks. *Journal of Japanese Society for Horticultural Science*, **74**, 458-463. http://dx.doi.org/10.2503/jjshs.74.458

[40] Nagasuga, K., Murai-Hatano, M. and Kuwagata, T. (2011) Effect of Root Temperature on Dry Matter Production and Root Water Uptake in Rice Plants. *Plant Production Science*, **14**, 22-29. http://dx.doi.org/10.1626/pps.14.22

8

Phytochemical Screening, Extraction of Essential Oils and Antioxidant Activity of Five Species of Unconventional Vegetables

Marcos Schleiden Sousa Carvalho[1], Maria das Graças Cardoso[2*], Luciane Vilela Resende[1], Marcos de Souza Gomes[3], Luiz Roberto Marques Albuquerque[2], Anni Cristini Silvestri Gomes[2], Thaís Aparecida Sales[2], Karen Caroline Camargo[2], David Lee Nelson[4], Gabriele Mikami Costa[1], Mariana Araújo Espósito[2], Luis Felipe Lima e Silva[1]

[1]Department of Agriculture, Federal University of Lavras, Lavras, Brazil
[2]Department of Chemistry, Federal University of Lavras, Lavras, Brazil
[3]Federal University of Vales de Jequitinhonha e Mucuri, Unaí, Brazil
[4]Federal University of Vales de Jequitinhonha e Mucuri, Diamantina, Brazil
Email: [*]mcardoso@dqi.ufla.br

Abstract

Unconventional vegetables, in general, are plants that have been largely consumed by the population at some point and, because of changes in eating behavior, now present reduced economic and social expression and have lost ground to other vegetables. The objectives of this study were to perform phytochemical screening of the ethanol extracts of *Rumex acetosa* L., *Tropaeolum majus* L., *Solanum muricatum*, *Stachys byzantina* K. Koch and *Solanum betaceum* Cav. and evaluate their antioxidant potentials via the methods involving scavaging of the DPPH free radical and the ABTS radical, phosphomolybdenum and reducing power. In phytochemical screening, five species of unconventional vegetables tested positive for tannins; for sesquiterpene, lactones and other lactones. These tests were positive for *Tropaeolum majus* L. and *Rumex acetosa* L., *Solanum betaceum* Cav. and *Solanum muricatum* tested negative for steroids. Only *Solanum betaceum* Cav. gave positive tests for flavonoids. Among the five plant species studied, *Stachys byzantina* K. Koch presented the greatest antioxidant potential in all the methods evaluated.

Keywords

***Rumex acetosa* L., *Tropaeolum majus* L., *Stachys byzantina* K. Koch, Reducing Power**

[*]Corresponding author.

1. Introduction

Unconventional vegetables are considered to be those that are currently consumed by only a few people, usually in restricted areas or communities. These vegetables have a good taste and nutritional value [1]. Because of globalization and the increasing use of processed foods, cultivation and consumption of unconventional vegetables have decreased in all the regions of Brazil, both urban and rural areas. As a result, the eating habits of Brazilians of all social classes are being modified, and the consumption of food from local and regional sources has been shrinking [2].

Tropical and subtropical countries have the greatest diversity of vascular plant species. However, the number of indigenous fruits and vegetables utilized are negligible. With respect to native vegetables, research, cultivation, use and valorization seem to be even less [3].

Among these vegetables, some stand out in Brazilian cuisine. In southeastern Brazil, more specifically in the state of Minas Gerais, there is *Rumex acetosa* L. (sorrel), which is a herbaceous perennial native to Europe and North Asia (Eurasia), including being considered a "weed" in that continent. It presents sour leaves and is used in raw salads (pure or mixed, providing freshness) or as a vegetable in soups, puree or green sauce and omelets [4].

Another widely used unconventional vegetable is *Tropaeolum majus* (nasturtium). It is an annual, herbaceous plant native to the highlands of Mexico and Peru. It is widely cultivated for ornamental purposes and for consumption in most regions of Brazil. Its flowers and young leaves can be used to prepare raw salads, green pasta, pancakes, pizzas, breads and soups, and it can be cooked with meat, among other uses. The *Stachys byzantina* K. Koch (peixinho; "little fish") species, also known as peixinho-do-jardim, is a herbaceous perennial vegetable native to Turkey, Asia and Caucasus. It is widely cultivated in southern and southeastern Brazil for ornamental purposes in sites with strong sunlight and for consumption as a vegetable. Its leaves can be eaten after cooking and suitable culinary preparation. It has a light texture and fried-fish flavor, hence some of the popular names: peixinho, peixinho-da-horta or lambari-da-horta ("little fish", "little-fish-of-the-garden" or "tetra-of-the-garden") [4]. *Solanum muricatum* (melãozinho; "little melon") is an evergreen shrub with fruits 10 to 15 cm long. It is a plant of the Solanaceae family. It has a sweet flavor and aroma similar to melon (hence the name, despite not being closely related). *Solanum betaceum* (tomate da arvore; "tree tomato") of the Solanaceae family is an erect shrub, sublenhose, perennial, sparsely branched, 2.5 to 4.5 m high and native to Bolivia and Peru. The fruit can be consumed fresh. It has a yellow juicy pulp with a slightly sour flavor and many seeds [4].

Unconventional vegetables are not usually organized in the supply chain itself and do not arouse interest from seed, fertilizer or pesticide companies. However, the work of rescuing these plants from the point of view of increasing their use in food and research represents a cultural, social and economic gain and may stimulate the production and consumption of these foods because of their possible nutritional and pharmaceutical characteristics, in addition to their hardiness in cultivation.

Considering the hardiness to cultivation and desirable nutritional characteristics, the determination of the chemical composition of these species by means of a preliminary survey that can predict the groups of secondary metabolites in the extract is important. Some research has shown that certain compounds present in plant extracts can have antioxidant action, and they are indicated to reduce or prevent the deleterious effects of cellular oxidative stress [5] [6]. There is a growing interest in finding plant species that can act in the fight against free radicals.

The objectives of this study were to evaluate the antioxidant potential using the methods of sequestration of the DPPH free radical or the ABTS radical, phosphomolybdate and reducing power and to perform phytochemical screening of *Rumex acetosa* L. (sorrel), *Tropaeolum majus* L. (capuchin), *Solanum muricatum* (little melon), *Stachys byzantina* K. Koch (little fish) and *Solanum betaceum* Cav. (tree tomato).

2. Material and Methods

2.1. Collection and Extraction of Plant Material

The leaves and fruits of unconventional vegetables (considered to be those that are currently consumed by only a few people, usually in restricted areas or communities) were obtained from the collection of the Department of Agriculture of the Federal University of Lavras. Samples of *Rumex acetosa* L. (sorrel), *Tropaeolum majus* L. (capuchin), *Solanum muricatum* (little melon), *Stachys byzantina* K. Koch (little fish) and *Solanum betaceum* Cav. (tree tomato) were sent to the Organic Chemistry Laboratory–Essential oils of the Department of Chemistry of the same University. The collected material (leaves or fruits) was dried in an oven (35°C) for 72 hours. To 100 mL of ethanol was added 50 g of the plant, and the mixture was heated under reflux for 8 hours. The mixture was filtered through a Buchner funnel, and the ethanol was evaporated on a Büchi B114 rotary evaporator to furnish the plant extract.

2.2. Phytochemical Screening

The presence of the major classes of metabolites was determined according to the method proposed by Matos (1988) [7]. To test for tannins, a few milligrams of extract was dissolved in 3 mL of distilled water and a drop of 1% ferric chloride was added. A change in color or formation of a precipitate indicated a positive reaction. To test for the presence of flavonoids, a few milligrams of extract was dissolved in 10 mL of methanol. If necessary, the mixture was filtered, and five drops of concentrated HCl and a one-cm piece of magnesium tape were added. A pink tint in the solution indicated a positive reaction. For these and all other tests, 1 mg of each extract of leaves was used. Tests were performed for organic acids, reducing sugars, polysaccharides, proteins and amino acids, tannins, catechins, flavonoids, cardiac glycosides, sesquiterpene, lactones and other lactones, azulenes, carotenoids, steroids, depsides, coumarin derivatives, saponins, alkaloids, purines and anthraquinones.

2.3. Essential Oils

To obtain the essential oils, 200 g of edible fruits or leaves was subjected to hydrodistillation for 2 hours, using a modified Clevenger apparatus to furnish the hydrolac [8]. The essential oil was separated from the distillate by centrifugation on a benchtop centrifuge with a horizontal crosspiece (Fanem Model 206 Baby®I BL) at 965.36 g for 5 minutes or by partition with dichloromethane. No essential oil was obtained from any of the plants studied.

2.4. Antioxidant Activity

2.4.1. ABTS Method

The antioxidant activity by the ABTS•+ method [ammonium salt of 2,2'-azinobis (3-ethylbenzenethiazoline-6-sulfonic acid)] was performed according to the method described by Guerreiro *et al.* (2013) [9]. The ABTS radical was formed by the reaction of 7 mM ABTS•+ solution with 2.4 mM potassium persulphate, incubated at 25°C in the dark for 12 - 16 hours. Once formed, the PA radical was diluted with ethanol to obtain the absorbance (from 0.7 to 0.72) at 734 nm. In a dark environment, 50 μL of each of the samples (concentrations 31.25, 62.5, 125, 250 and 500 $\mu g \cdot mL^{-1}$) was transfered to test tubes containing 1950 μL of the ABTS•+ radical. The percentage inhibition of the ABTS•+ radical by the samples was calculated according to the following equation:

$$\% \text{ Effect on radical capture} = \left[\left(A_{CN} - A_{sam}\right)/A_{CN}\right] \times 100$$

where:

A_{CN} = Absorbance of the negative control;

A_{sam} = Absorbance of the sample;

The values were compared with those of the BHT (butyl hydroxy toluene) standard at concentrations of 31.25; 62.5; 125; 250 and 500 $\mu g \cdot mL^{-1}$.

2.4.2. Sequestration of the DPPH Free Radical

Evaluation of antioxidant activity by means of the capture of the DPPH (2,2-diphenyl-1-picrylhydrazyl) free radical was performed according to a modification of the method of Lutz-Lopes *et al.* (2008) [10]. A methanol solution of DPPH was prepared at a concentration of 40 $\mu g \cdot mL^{-1}$. To each test tube was added 2.7 ml of the DPPH solution, followed by 0.3 mL of each extract dissolved in ethanol at concentrations of 31.25; 62.5; 125; 250 and 500 $\mu g \cdot mL^{-1}$. The BHT standard was used as a positive control. The negative control was a solution containing all the reagents except the extracts. After standing for 60 minutes in the dark, absorbance readings were taken at 515 nm (Shimadzu UV-160 1PC). The antioxidant activity (AA%) was calculated from the following equation:

$$AA\% = \left[\left(A_{CN} - A_{sam}\right)/A_{CN}\right] \times 100$$

where:

A_{CN} = Absorbance of the negative control;
A_{sam} = Absorbance of the sample.

2.4.3. Phosphomolybdate

To a test tube containing 1 mL of reagent solution (0.6 M sulfuric acid, 28 mM sodium phosphate and 4 mM ammonium molybdate) was added an aliquot of 0.1 ml of 500 $\mu g \cdot mL^{-1}$ solution of extract in ethanol. The blank contained 1 mL of the reagent solution and 0.1 mL of ethanol. The tubes were capped and incubated in a water bath at 95°C for 60 minutes. After cooling, the test tubes were read at 695 nm. The values for the antioxidant activity of the extracts were compared with a curve obtained with various concentrations of BHT (31.25, 62.5, 125, 250 and 500 $\mu g \cdot mL^{-1}$), and the results were expressed in mg equivalents of BHT per mg of extract [11].

2.4.4. Reducing Power

To a test tube were added a 0.1 mL aliquot of a solution of plant extract with 1 mL of phosphate buffer solution (pH = 7.4) and 1.0 mL of 1% potassium ferricyanide. After incubation at 50°C for 30 minutes, 0.5 mL of 10% trichloroacetic acid, 1.5 mL of distilled water and 0.3 mL of ferric chloride were added, and the absorbance was measured at 700 nm. The antioxidant activity of the extracts were compared with a standard curve for various concentrations of BHT (31.25, 62.5, 125, 250 and 500 $\mu g \cdot mL^{-1}$). Results were expressed in equivalent mg BHT per mg of extract [12].

2.5. Statistical Analysis

Analysis of variance was performed for the ABTS and DPPH tests, followed by the F test, to verify the effect of concentration on the antioxidant activity. For both tests, a completely randomized design (CRD) was used in a 6 × 5 factorial: six samples and five concentrations of extracts (31.25, 62.5, 125, 250 and 500 $\mu g \cdot mL^{-1}$) were analyzed with three replications. Analyses were performed using the Sisvar program [13].

3. Results and Discussion

3.1. Phytochemical Screening

The results of the tests for the classes of compounds in both extracts are presented in **Table 1**.

All five species of vegetables tested positive for tannins. The tannin identification reaction is based on the formation of a green or blue coloration or precipitate. According to Cunha and Batista (2010) [14], the tannins are generally found in different concentrations in vascularized plants. They may be present in many plant organs such as leaf, fruit and bark; they are soluble in water, alcohols and acetone, and exert a protective function in plants.

The test for steroids was positive for *Rumex acetosa L.*, *Tropaeolum majus L.* and *Stachys byzantina* K. Koch. The residue from the *Solanum muricatum* extract was not soluble in chloroform, so it was not possible to perform the test with this extract. The test with the *S. betaceum* extract was negative. The extracts from *Rumex acetosa* L. and *Tropaeolum majus* L. tested positive for sesquiterpene lactones and other lactones,

Tests for flavonoids were negative with the extracts from *Rumex acetosa* L., *Tropaeolum majus* L., *Solanum muricatum* and *Stachys byzantina* K. Koch. A positive result was observed for the extract from *Solanum betaceum* Cav. According to Souza *et al.* (2010) [15], flavonoids have an antioxidant action by reason of their

Table 1. Metabolites whose results were positive in phytochemical screening for ethanol extracts of *Rumex acetosa* L., *Tropaeolum majus* L., *Solanum muricatum*, *Stachys byzantina* K. Koch and *Solanum betaceum* Cav.

Compound	*Rumex acetosa* L.	*Stachys byzantina* K. Koch	*Tropaeolum majus* L.	*Solanum betaceum* Cav.	*Solanum muricatum*
Tannins	+	+	+	+	+
Steroids	+	+	+	-	-
Sesquiterpene, lactones and other lactones	+	-	+	-	-
Flavonoids	-	-	-	+	-

chemical structures and reducing properties. They are phenolic compounds that have a perfect structure for scavenging free radicals and can be more effective as antioxidants than some vitamins.

The antioxidant activity of flavonoids is related to their oxidation power [16]. Suhartonoa *et al.* (2012) [17] studied the antioxidant activity and the presence of flavonoids in some medicinal plants (Kelakai, Kasturi, pasak bumi and ferns) and found that all fractions of the plants tested had a high flavonoid content, unlike four of the five species analyzed in this study. Porto *et al.* (2014) [18] found a high concentration of flavonoids 728.00 mg. 100 g^{-1}) when they analyzed the chemical composition of Jenipapo (*Genipa american* L.), unlike Barreto, Benassi and Mercadante (2009) [19] who obtained a flavonoid concentration of 73.3 mg. 100 g^{-1} for buriti (*Mauritia flexuosa*), 103.8 g/mg. 100 g^{-1} for bacuri (*Platonia insignis* M.), 319.4 mg. 100 g^{-1} for murici (*Byrsonima crassifolia* L.) and 741.2 mg. 100 g^{-1} for pequi (*Caryocar brasilia* C.). According to Porto *et al.* (2014) [18], only the *Caryocar brasilia* C. contained a concentration of flavonoides near that obtained in the previous study, whereas the values for the other plants were well below those recorded in that survey. The remaining phytochemical screening tests with the extracts of these five vegetables were negative.

3.2. Essential Oils

No essential oils were obtained from the five species of non-conventional vegetable surveyed here.

3.3. Antioxidant Activities

The results of evaluation of the antioxidant activity (AA) through the method involving inhibition of the ABTS radical are shown in **Figure 1**. The percent of AA increased with increasing concentration for all the samples. Thus, the highest percentage of antioxidant activity was obtained at a concentration of 500 μg·mL^{-1}.

Among the five species studied, *S. brizantina* presented the greatest activity, and this fact is due to the high concentration of phenolics (tannins) present in the plant. *Solanum muricatum* had the lowest antioxidant potential, and its activities at concentrations of 62.5, 125 and 250 μg·mL^{-1} were statistically equal. Although tannins were found in that extract, these compounds might be present in lower concentrations or complexed that the hydroxyl hydrogen is unable to react with the ABTS radical (**Figure 2**).

Results for the antioxidant activity determined by scavaging of the DPPH free radical are shown in **Figure 3**. The percentages of the antioxidant activity of all the species were similar to those observed with the previous method. There was an increase in the antioxidant activity with increasing concentration of the extract, there by demonstrating a dose-dependent effect (**Figure 4**).

Results of the tests of the extracts from the five vegetables for antioxidant activity performed by means of the

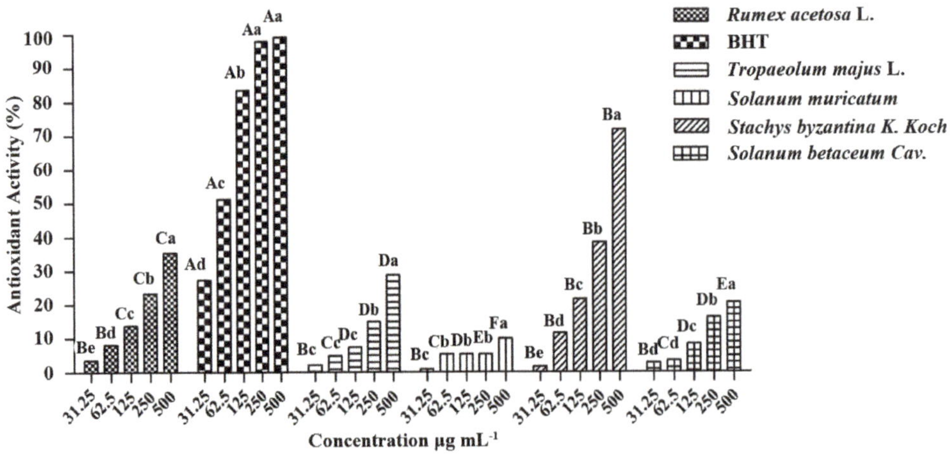

Figure 1. Percentage of the antioxidant activity of plants *Rumex acetosa* L., *Tropaeolum majus* L., *Solanum muricatum*, *Stachys byzantina* K. Koch and *Solanum betaceum* Cav. the method of inhibiting the ABTS radical. The means followed by the same letter: in lower case are the comparison of concentrations in each plant and the upper letter of each concentration compared between different plants do not differ significantly at probability of 5%, by Scott-Knott test. *BHT = Butyl hydroxy toluene.

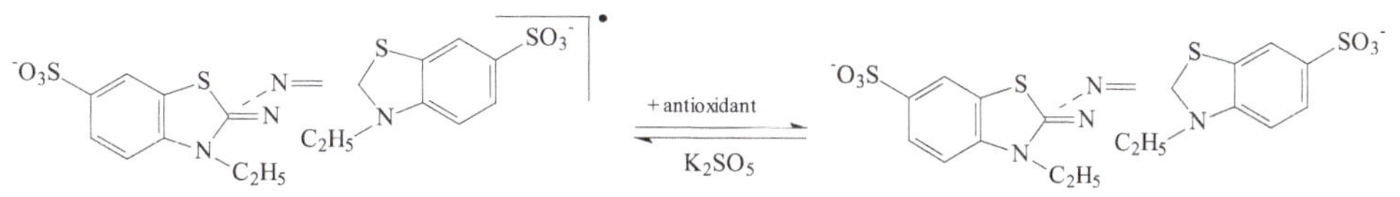

Figure 2. Antioxidant stabilizing ABTS radical and formation of potassium persulfate.

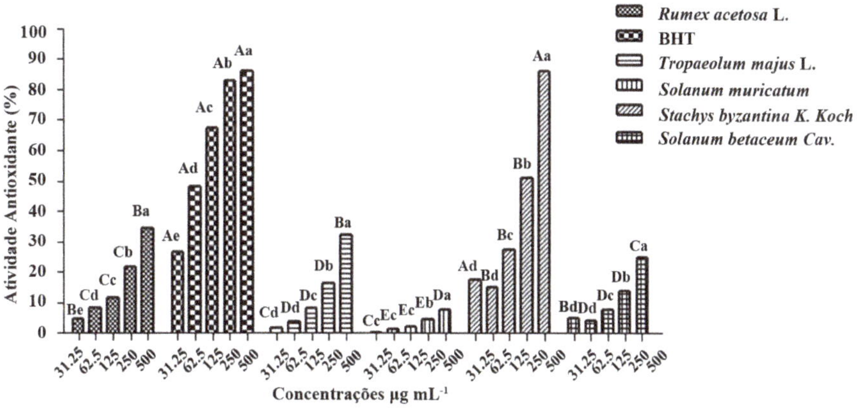

Figure 3. Percentage of the antioxidant activity of plants *Rumex acetosa* L., *Tropaeolum majus* L., *Solanum muricatum*, *Stachys byzantina* K. Koch and *Solanum betaceum* Cav. by DPPH radical inhibition method. The means followed by the same letter: in lower case are the comparison of concentrations in each plant and the upper letter of each concentration compared between different plants do not differ significantly at probability of 5%, by Scott-Knott test. *BHT = Butyl hydroxy toluene.

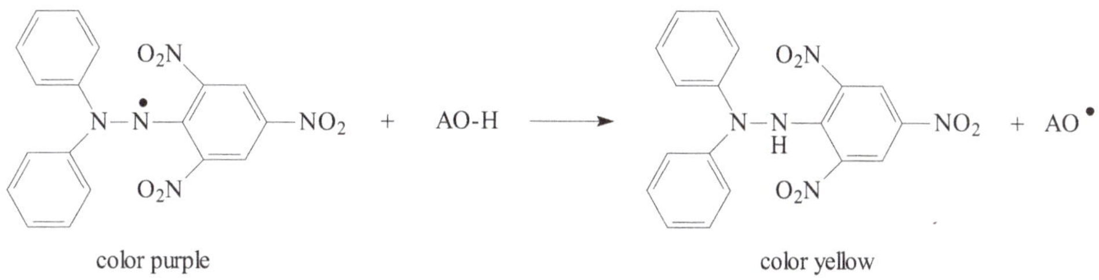

Figure 4. Antioxidant stabilizing DPPH• radical. *AO-H = representation of an antioxidant substance.

formation of complexes with phosphomolibdenum and by determination of the reducing power were expressed in mg equivalent BHT/mg of plant extract (**Table 2**). The extracts from *Tropaeolum majus* L. and *Stachys byzantina* K. Koch presented the highest antioxidant activities as measured by formation of the phosphomolibdenum complex. This method is a simple and inexpensive way to measure the total antioxidant capacity of a complex mixture of compounds containing both lipophilic and hydrophilic substances, such as extracts obtained from plants [20].

The *Stachys byzantina* K. Koch extract presented the highest antioxidant activity by the reducing power method (**Table 2**). This method is based on the reduction of ferricyanide to ferrocyanide ion in the presence of ferric ion (derived from $FeCl_3$) (**Figure 5**) [21].

According Meir *et al.* (1995) [22], the reducing capacity of a compound may serve as an important indicator of its antioxidant potential. Sudha *et al.* (2012) [23] found that the reducing power of the ethanol extracts of ripe and unripe exotic cucumbers in concentrations of 0.190; 0.177 to 1 $mg{\cdot}mL^{-1}$ and 0.921; 0.672 to 5 $mg{\cdot}mL^{-1}$, respectively. The values observed in this study were lower than those found by Sudha *et al.* (2012) [23].

4. Conclusion

The extracts of five species of unconventional vegetables tested positive for tannins, whereas only the extracts from *Rumex acetosa* L. and *Tropaeolum majus* L. tested positive for sesquiterpene lactones and other lactones.

$$Fe^{2+} + [Fe(CN)_6]^{3-} \rightleftharpoons Fe^{3+} + [Fe(CN)_6]^{4-}$$

Figure 5. Reduction of ferricyanide ion to ferrocyanide.

Table 2. Antioxidant activity of *Rumex acetosa* L. (sorrel), *Tropaeolum majus* L. (Capuchinha), *Solanum muricatum* (melaozinho), *Stachys byzantina* K. Koch (peixinho) and *Solanum betaceum* Cav. (tree tomato), by phosphomolybdenum complex and the reducing power methods.

Plants	Phosphomolybdenum complex	Reducing power
	mg BHT Equivalent/mg of sample extract	
Rumex acetosa L. (leaf)	0.11	0.06
Tropaeolum majus L. (leaf)	0.30	0.07
Solanum muricatum (fruit)	0.07	0.00
Stachys byzantina K. Koch (leaf)	0.36	0.28
Solanum betaceum Cav. (fruit)	0.11	0.04

No steroids were found in the tests of the extracts of *Solanum betaceum* Cav. and *Solanum muricatum*, and only the extract from *Solanum betaceum* Cav. tested positive for flavonoids. The results were negative in the remaining tests. The percentage of antioxidant activity determined by sequestering of the DPPH and ABTS radicals increased with increasing concentration, thereby showing a dose-dependent effect for the five species studied. The highest antioxidant activity was observed for *Tropaeolum majus* L. and *Stachys byzantina* K. Koch using the phosphomolibdenum method. Of the five species of vegetables evaluated, *Stachys byzantina* K. Koch presented antioxidant activity by all the methods evaluated. The plants appeared to be devoid of essential oils.

Acknowledgements

The authors thank the Conselho Nacional de Desenvolvimento Científico e Tecnológico (CNPq), the Coordenação de Aperfeiçoamento de Pessoal de Nível Superior (CAPES) and the Fundação de Amparo à Pesquisa do Estado de Minas Gerais (FAPEMIG) for financial support, a scholarship and a PVNS fellowship (D.L. Nelson).

References

[1] Brasil. Ministério da Saúde (2002) Alimentos regionais brasileiros. Brasília, DF, 140 p.

[2] Brasil. Ministério da Agricultura, Pecuária e Abastecimento (2010) Hortaliças não-convencionais: (Tradicionais)/Ministério da Agricultura, Pecuária e Abastecimento. Secretaria de Desenvolvimento Agropecuário e Cooperativismo. MAPA/ACS, Brasília, 52 p.

[3] Kinupp, V.F. (2009) Plantas Alimentícias Não-Convencionais (PANCs): Uma Riqueza Negligenciada. Anais da 61ª Reunião Anual da SBPC - Manaus, AM, 1-4. http://www.sbpcnet.org.br/livro/61ra/mesas_redondas/MR_ValdelyKinupp.pdf

[4] Kinupp, V.F. and Lorenzi, H. (2014) Plantas alimentícias não convencionais (PANC) no Brasil: Guia de identificação, aspectos nutricionais e receitas. Instituto *Plantarum* de estudos da Flora, São Paulo.

[5] Andrade, C.A., Costa, C.K., Bora, K., Miguel, M.D., Miguel, O.G. and Kerber, V.A. (2007) Determinação do conteúdo fenólico e avaliação da atividade antioxidante de *Acacia podalyriifolia* A. Cunn. ex G. Don, *Leguminosae mimosoideae*. *Revista Brasileira de Farmacognosia*, **17**, 231-235. http://dx.doi.org/10.1590/S0102-695X2007000200017

[6] Balestrin, L., Dias, J.F.G., Miguel, O.G., Dallstella, D.S.G. and Miguel, M.D. (2008) Contribuição ao estudo fitoquimico de *Dorstenia multiformis* Miquel (*Moraceae*) com abordagem em atividade antioxidante. *Revista Brasileira de Farmacognosia*, **18**, 230-235. http://dx.doi.org/10.1590/S0102-695X2008000200016

[7] Matos, F.J.A. (1988) Introdução à Fitoquímica Experimental. 2ª Edition, Edições UFC, Fortaleza.

[8] Farmacopeia Brasileira, Brasil. Agência Nacional de Vigilância Sanitária (2010) Farmacopeia brasileira. 5th Edition, Anvisa, Brasília, 1, 198-199.

[9] Guerreiro, A.C., Gago, C.M.L., Miguel, M.G.C. and Antunes, M.D.C. (2013) The Effect of Temperature and Film Covers on the Storage Ability of *Arbutus unedo* L. fresh fruit. *Scientia Horticulturae*, **159**, 96-102.

http://dx.doi.org/10.1016/j.scienta.2013.04.030

[10] Lopes-Lutz, D., Alviano, D.S., Alviano, C.S. and Kolodziejczyk, P.P. (2008) Screening of Chemical Composition, Antimicrobial and Antioxidant Activities of *Artemisia* Essential Oils. *Phytochemistry*, **69**, 1732-1738. http://dx.doi.org/10.1590/S0100-40422007000200026

[11] Oliveira, C.M. (2015) Caracterização química, atividade antibacteriana, antitumoral e ensaios antioxidantes do óleo essencial das folhas e flores de *Callistemon viminalis*. 111 p. Dissertação (mestrado)—Universidade Federal de Lavras, Lavras. 2015.

[12] Kim, D.O., Lee, K.W., Lee, H.J. and Lee, C.Y. (2002) Vitamin C Equivalent Antioxidant Capacity (VCEAC) of Phenolics Phytochemicals. *Journal of Agricultural Food and Chemistry*, **50**, 3713-3717. http://dx.doi.org/10.1021/jf020071c

[13] Ferreira, D.F. (2011) Sisvar: A Computer Statistical Analysis System. *Ciência e Agrotecnologia*, **35**, 1039-1042.

[14] Cunha, A.P. and Batista, M.T. (2010) Taninos. Livro: Farmacognosia e Fitoquímica. Parte III. Capítulo 14. 3ª Edição. Lisboa. Fundação Calouste Gulbenkian. 292-310.

[15] Souza, M.D., Fernandes, R.R. and Pasa, M.C. (2010) Estudo etnobotânico de plantas medicinais na comunidade São Gonçalo Beira Rio, Cuiabá, MT. *Revista Biodiversidade*, **9**, 91-100.

[16] Barreiros, A.L.B.S., David, J.M. and David, J.P. (2006) Estresse oxidativo: Relação entre geração de espécies reativas e defesa do organismo. *Química Nova*, **29**, 113-123. http://dx.doi.org/10.1590/S0100-40422006000100021

[17] Suhartonoa, E., Vianib, E., Rahmadhanb, M.A., Gultomb, I.S., Rakhmanb, M.F. and Indrawardhanac, D. (2012) Total flavonoid and Antioxidant Activity of Some Selected Medicinal Plants in South Kalimantan of Indonesian. *APCBEE Procedia*, **4**, 235-239. http://dx.doi.org/10.1016/j.apcbee.2012.11.039

[18] Porto, R.G.C.L., Cardoso, B.V.S., Barros, N.V.A., Cunha, E.M.F., ARAÚJo, M.A.M. and Moreira-Araújo, R.S.R. (2014) Chemical Composition and Antioxidant Activity of *Genipa Americana* L. (Jenipapo) of the Brazilian Cerrado. *Journal of Agriculture and Environmental Sciences*, **3**, 51-61. http://dx.doi.org/10.15640/jaes

[19] Barreto, G.P.M., Benassi, M.T. and Mercadante, A.Z. (2009) Bioactive Compounds from Several Tropical Fruits and Correlation by Multivariate Analysis to Free Radical Scavenger Activity. *Journal of the Brazilian Chemical Society* (Online), **20**, 1856-1861. http://dx.doi.org/10.1590/S0103-50532009001000013

[20] Prieto, P., Pineda, M. and Aguilar, M. (1999) Spectrophotometric Quantitation of Antioxidant Capacity through the Formation of a Phosphomolybdenum Complex: Specific Application to the Determination of Vitamin E. *Analytical Biochemistry*, **269**, 337-341. http://dx.doi.org/10.1006/abio.1999.4019

[21] Dos Santos, M.H., Batista, B.L, Duarte, S.M.S., Abreu, C.M.P. and Gouvêa, C.M.C.P. (2007) Influência do processamento e da torrefação sobre a atividade antioxidante do café (*Coffea arabica*). *Química Nova* (*Impresso*), **30**, 604-610. http://dx.doi.org/10.1590/S0100-40422007000300020

[22] Meir, S., Kanner, J., Akiri, B. and Philosoph-Hadas, S. (1995) Determination and Involvement of Aqueous Reducing Compounds in Oxidative Defense Systems of Various Senescing Leaves. *Journal of Agricultural Food and Chemistry*, **43**, 1813-1819. http://dx.doi.org/10.1021/jf00055a012

[23] Sudha, G., Priya, M.S., Shree, R.B.I. and Vadivukkarasi, S. (2012) Antioxidant Activity of Ripe and Unripe Pepino Fruit (*Solanum muricatum* Aiton). *Journal of Food Science*, **77**, 1131-1135.

9

Morphological Comparison of Five Varieties of *Colocasia esculenta* (L.) Schott in Anambra State, Southeastern Nigeria

Chinelo A. Ezeabara*, C. U. Okeke, J. E. Amadi, A. I. Izundu, Bibian O. Aziagba, P. T. Egboka, C. D. Udechukwu

Department of Botany, Nnamdi Azikiwe University, Awka, Nigeria
Email: *e.chinelo5@yahoo.com

Abstract

Comparative morphological study of five varieties of *Colocasia esculenta* (L.) Schott present in Anambra State, Nigeria was carried out, in order to furnish plant taxonomists with information which could be of great help in delimitation of the varieties. Significant difference was established at $p < 0.05$. The result showed that adaxial surface of "kochuo" had a purplish dot on the centre, abaxial leaf surface had a conspicuous purplish colour at the point of attachment to the leaf, and the cormels had numerous striking vertical purple stripes on the surface. There was presence of foliaceous (leaf-like) appendages at the veins of the abaxial surface of the leaf of "ogeriobosi". The leaf length ranged from 35.6 ± 7.70 cm ("kochuo") to 49.9 ± 3.55 cm ("ogeriobosi"). Petiole length of "ogeriobosi" was the highest (63.3 ± 3.83 cm), whereas the least was *Colocasia esculenta* var. *antiquorum* (26.67 ± 2.20 cm). The corm length ranged from 4.10 ± 0.10 cm (*Colocasia esculenta* var. *antiquorum*) to 8.60 ± 0.35 cm ("ogeriobosi"), while the cormel length ranged from 3.70 ± 0.96 cm (*Colocasia esculenta* var. *antiquorum*) to 7.03 ± 0.36 cm ("ogeriobosi"). This work has revealed diagnostic and differential morphological characters, which could be useful for identification and description of varieties of *C. esculenta*. In addition, it provided additional information which might be helpful in resolving the on-going controversy in the taxonomy of *Colocasia*, which would, in turn, probably lead to possible delimitation of *C. esculenta*.

Keywords

***Colocasia*, Corms, Cormels, Diagnostic Characters, Foliaceous Appendages, Morphological Characters, Plant Taxonomy**

*Corresponding author.

1. Introduction

Colocasia esculenta (L.) Schott of the family Araceae are popular tuber crops in Southeastern Nigeria. *Colocasia esculenta* and *Xanthosoma sagittifolium* serve as staple food in Southeastern Nigeria, and are commonly known as cocoyam. Various ethnic groups in Nigeria have different names for *C. esculenta*, which attested to its nationwide distribution and use. It is known as ede/akaso/uli in Ibo, guaza in Hausa, koko in Yoruba, mkpon in Efik and ikereburu in Ijo [1]-[4].

The observed varieties of *C. esculenta* showed a lot of variability in morphological characters. This variability is common among crops that have been cultivated for a long time [5]. Currently, *C. esculenta* is considered as the major species with two varieties: *Colocasia esculenta* var. *antiquorum* (L.) (Schott) Hubbard & Rehder and *Colocasia esculenta* var. *esculenta* (L.) Schott [2] [6].

The knowledge of variability of *C. esculenta* is deficient and limited. In addition, there is paucity of literature on its taxonomy, which implies that thorough taxonomic research has not been done on *C. esculenta*. As a result, a morphological study on varieties of *C. esculenta* becomes a necessity; because morphological characters are the strongest tools used in taxonomic classification of plants, and this makes its application very crucial. The objective of this work, therefore, is to investigate the morphology of varieties of *C. esculenta*, with a view to exploring the macro-characters, which are the major criterion for plant classification.

2. Materials and Methods

2.1. Sources of Materials

Cormels of varieties of *C. esculenta* were obtained from the local farmers from the three senatorial districts in Anambra State, namely: Agulu in Anaocha LGA (Anambra North); Umuikwu-Anam in Anambra West LGA (Anambra Central) and Uga in Aguata LGA (Anambra South).

Five varieties were then selected from the three senatorial districts and grown in a common garden at Uga, Anambra State, Nigeria in June, 2013. They include *Colocasia esculenta* var. *antiquorum* (eddoe), *Colocasia esculenta* var. *esculenta* (dasheen), "Kochuo", "Nwine" and "Ogeriobosi" with collection numbers: ACE 34, ACE 35, ACE 36, ACE 37 and ACE 38, respectively. The voucher specimens were authenticated by Prof. C. U. Okeke and C. A. Ezeabara, Plant Taxonomists, and deposited in herbarium of Department of Botany, Nnamdi Azikiwe University, Awka, Anambra State, Nigeria.

2.2. Morphological Study

The petiole and leaf measurements were done in October 2013; hence the plants were four months old when the measurements were done. The corm (main stem) and cormels were measured in December 2013. The meristem of the corm (main stem) were removed until they cannot be removed without force; third cormels from the base of the main stem were randomly chosen and third fully opened active leaves and petioles from the base were measured. The petiole measurement started from the ligule to the base of the leaf. The raw corms and third cormels from the base of the main stem were peeled with a knife. Observations and measurements of the plants parts were done using eye lens, thread and ruler. Photographs of the habit and plant parts were taken with digital camera (Sony DSC-W230, China).

2.3. Statistical Analysis

One-Way-Anova (F-Test) was used to analyze the data at $p < 0.05$. Duncan's Multiple Range Test (DMRT) was then used to separate the means and data were expressed as mean ± standard deviation of triplicate determinations.

3. Results

The leaf of all the varieties was thick, succulent and downward-pointing, with reticulate venation. Three strong midribs arose from the point of attachment to the petiole, the main midrib pointed towards the basal lobe, while other two smaller midribs extended to the two posterior lobes, with one midrib on each lobe. The primary lateral veins originated from the three midribs, while the veinlets emanated from these primary lateral veins. The leaf margin was entire; the leaf blade was sagittate and not shiny, the abaxial and adaxial surfaces of the leaf blade were glabrous with the exception of abaxial surface of "ogeriobosi"; the anterior lobes were twice as large as the posterior lobes, which were round in shape; and the petiole attachment was peltate. Petiole had a wide range of colours, including pale green, dark green, yellowish green and purple (**Figures 1(a)-(j)**). The base of petiole of *C. esculenta* var. *antiquorum* was deep purple (**Figure 1(a)**) when compared with *C. esculenta* var. *esculenta* (**Figure 1(b)**). The leaf of "kochuo" has a prominent distinctive feature, which was presence of a purplish dot on the centre of the adaxial surface (**Figure 1(c)**). The petiole of "nwine" was yellowish green (**Figure 1(d)**, **Figure 1(i)** and **Table 1**), while that of "ogeriobosi" was purple in colour (**Figur 1(e)** and **Table 1**). Abaxial leaf surface of "kochuo" had a conspicuous purplish colour at the point of attachment to the leaf (**Figure 1(h)**). "Ogeriobosi" had foliaceous (leaf- like) appendages at the veins of the abaxial surface of the leaf (**Figure 1(j)**).

C. esculenta var. *antiquorum* and "kochuo" had large corms which were more or less orbicular (approximately circular) in shape, whereas corms of *C. esculenta* var. *esculenta*, "nwine" and "ogeriobosi" were more or less oval. "Ogeriobosi" had the largest corm as well as the largest cylindrical cormels (**Figures 2(a)-(f)**). "Nwine" possessed large corm with slender cormels (**Figure 2(c)** and **Figure 2(f)**). Cormels of *C. esculenta* var. *antiquorum* were small, numerous and round. They were relatively the smallest, whereas those of "ogeriobosi" were the biggest (**Figure 2(a)**, **Figure 2(e)**, **Figure 2(f)** and **Table 1**). The cormels of *C. esculenta* var. *esculenta* were few and more or less cylindrical shaped, while cormels of "kochuo" were numerous with vertical purplish stripes on the surface.

Figure 1. Photomicrographs of habit and abaxial leaf surfaces of five varieties of *Colocasia esculenta*. (a) Habit of *C. esculenta* var. *antiquorum*. (b) Habit of *C. esculenta* var. *esculenta*. (c) Habit of "kochuo". (d) Habit of "nwine". (e) Habit of "ogeriobosi". (f) Abaxial leaf surface of *C. esculenta* var. *antiquorum*. (g) Abaxial leaf surface of *C. esculenta* var. *esculenta*. (h) Abaxial leaf surface of 'kochuo'. (i) Abaxial leaf surface of "nwine". (j) Abaxial leaf surface of "ogeriobosi".

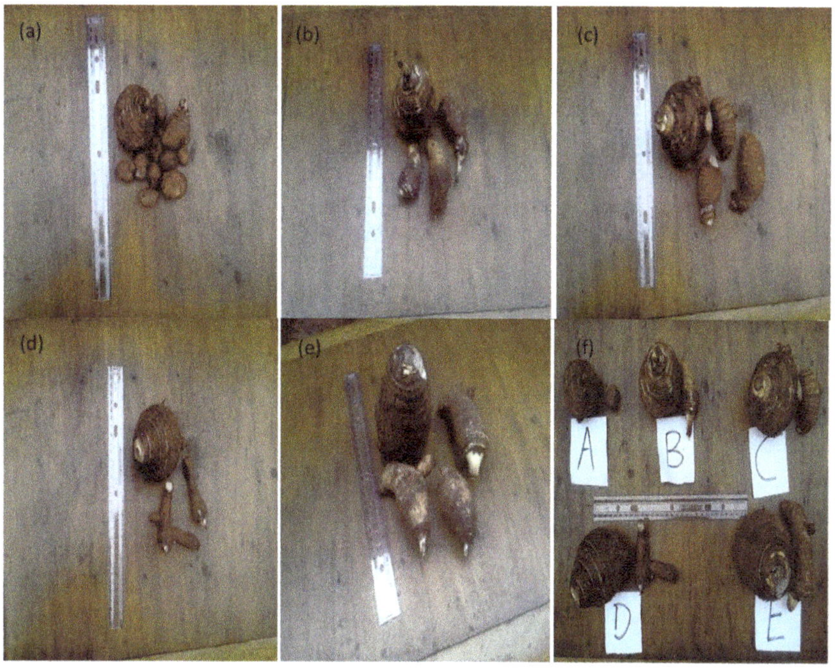

Figure 2. Photomicrographs of five varieties of *Colocasia esculenta* showing the shape, size and texture of the main stem (corm) and cormels. (a) *C. esculenta* var. *antiquorum*. (b) *C. esculenta* var. *esculenta*. (c) "kochuo". (d) "nwine". (e) "ogeriobosi". (f) [(A) *C. esculenta* var. *antiquorum*, (B) *C. esculenta* var. *esculenta*, (C) "kochuo", (D) "nwine", (E) "ogeriobosi"].

Table 1. Morphological characters (cm) of *Colocasia esculenta* var. *antiquorum*, *Colocasia esculenta* var. *esculenta*, "Kochuo", "Nwine" and "Ogeriobosi".

Varieties	Petiole length	Petiole colour	Leaf length	Leaf width	Leaf colour	Corm length	Corm width	Cormel length	Peeled tuber (raw)
Antiq	26.67 ± 2.20^a	Pale green	42.80 ± 5.86^a	36.90 ± 4.25^a	Pale green	4.10 ± 0.10^a	5.23 ± 1.28^a	3.70 ± 0.96^a	Green
Esc	38.07 ± 2.22^b	Pale green	44.80 ± 4.33^b	36.00 ± 2.65^a	Deep green	6.17 ± 0.50^b	4.47 ± 0.72^b	5.00 ± 1.35^b	Green
"Kochuo"	40.33 ± 22.35^c	Green	35.6 ± 7.70^c	34.13 ± 4.80^b	Deep green	4.86 ± 1.95^c	6.33 ± 1.27^c	5.45 ± 3.02^c	Purple
"Nwine"	41.4 ± 5.65^d	Yellowish green	41.4 ± 4.23^d	45.27 ± 3.66^c	Pale green	4.55 ± 0.87^d	5.67 ± 0.59^d	6.63 ± 1.42^d	Pink
"Ogeriobosi"	63.3 ± 3.83^e	Purple	49.9 ± 3.55^e	35.8 ± 3.64^d	Deep green	8.60 ± 0.35^e	5.53 ± 0.76^e	7.03 ± 0.36^e	Milk

Antiq = *Colocasia esculenta* var. *antiquorum*; Esc = *Colocasia esculenta* var. *esculenta*. Data are means ± standard deviation of triplicate determinations. Columns with different subscripts are significantly different at ($p < 0.05$).

There was significant difference among the petiole length of all the varieties at $p < 0.05$. Petiole length of "ogeriobosi" was the highest (63.3 ± 3.83 cm), while the least was *C. esculenta* var. *antiquorum* (26.67 ± 2.20 cm). The petiole colour of *C. esculenta* var. *antiquorum* and *C. esculenta* var. *esculenta* were the same, whereas others differed. There was significant difference among the leaf length of all the varieties at $p < 0.05$. The leaf length of "ogeriobosi" was the highest (49.9 ± 3.55 cm), while the least was "kochuo" (35.6 ± 7.70 cm). There was no significant difference between the leaf width of *C. esculenta* var. *antiquorum* (36.90 ± 4.25 cm) and *C. esculenta* var. *esculenta* (36.00 ± 2.65 cm), whereas the leaf width of "nwine" (45.27 ± 3.66 cm) was the highest. The leaves of *C. esculenta* var. *esculenta*, "kochuo" and "ogeriobosi" were deep green in colour, while those of *C. esculenta* var. *antiquorum* and "nwine" were light green. There was significant difference among the corm lengths of all the varieties. The highest corm length was 8.60 ± 0.35 cm ("ogeriobosi"), and the least was 4.10 ± 0.10 (*C. esculenta* var. *antiquorum*). There was also significant difference among the corms width of all the varieties. "Kochuo" (6.33 ± 1.27 cm) had the greatest corm width, whereas the least was found in *C. esculenta* var. *esculenta* (4.47 ± 0.72 cm). The cormel length of "ogeriobosi" (7.03 ± 0.36 cm) was the highest, while *C. esculenta* var. antiquorum (3.70 ± 0.96 cm) was the least. The colour of the peeled raw tubers of all the varieties varied with *C. esculenta* var. *antiquorum* and *Colocasia esculenta* var. *esculenta* having the same colour (**Table 1**).

4. Discussion

The common leaf features of all the varieties with the exception of presence of foliaceous outgrowth at the abaxial surface of "ogeriobosi" suggested their inter-relatedness. These characters could be regarded as their field characters and probably as the generic characters of *Colocasia*; which some could be applicable in distinction of *Colocasia* from *Xanthosoma*. The appearance of *C. esculenta* and *X. sagittifolium* are sometimes confusing, as a result of their similarity. The leaves of *Xanthosoma* (Tannia) are large, approximately 20 cm in length, 15 cm in width, hastate (saggitate-ovate) in shape, with the anterior lobe twice as large as the posterior lobe, with distinct marginal vein, and round basal lobes [7]. In addition, the petiole attachment of *Xanthosoma* is at the margin of the leaf. Leaves of *X. sagittifolium* are nearly in a rosette in acaulescent plants, or in a distal crown in mature plants; blades horizontal to slight ascending, with the posterior lobes ascending, simple, upper surface dark green with light green primary secondary veins on basal lobes, lower surface light green with dark green venation; petioles light green; basal lobes of the leaf sub-rhomboid obtuse [8]. *Colocasia* can therefore, be easily distinguished from *Xanthosoma* by the point where the petiole is attached to the leaf. In *Colocasia*, the petiole attachment was peltate, whereas for *Xanthosoma*, the petiole attachment is at the margin of the leaf. Moreover, *C. esculenta* can be differentiated from *X. sagittifolium* by shape of the basal lobes, position of the leaf blades, and colour of petiole and leaf. In *Colocasia esculenta*, the basal lobe of leaf was round, leaf blade pointed downward, and leaf colour was dark green with petiole colour ranging from pale green, dark green, yellowish green and purple; while for *X. sagittifolium*, basal lobes is sub-rhomboid obtuse, blades horizontal to slight ascending, with leaf and petiole light green. It has been stated that morphological characteristics are the strongest determinants of the agronomic value and taxonomic classification of plants [9].

Generally, all the varieties of *C. esculenta* looked alike in the field until a closer look was made. Interestingly, however, there were remarkable differences in some parts of the varieties. Presence of foliaceous appendages at the veins of the abaxial surface of "ogeriobosi" leaves could be regarded as a diagnostic character; in addition, the leaf length of "ogeriobosi" (49.9 ± 3.55 cm) was the highest; and deep purple colour of the petiole was the most conspicuous morphological feature differentiating it from others. Diagnostic character for "nwine" was possession of yellowish green petiole, with highest leaf width (45.27 ± 3.66 cm), whereas that of "kochuo" was location of purplish spot in the centre of the adaxial surface of the leaves. In addition, varieties of *C. esculenta* can be delimited on the basis of their leaf, petiole as well as peeled corm and cormel colour. The leaves of *C. esculenta* var. *esculenta*, "kochuo" and "ogeriobosi" were deep green in colour, indicating a close relatedness, whereas those of *C. esculenta* var. *antiquorum* and "nwine" were pale green, also suggesting a close affinity. In addition, variation in the colour of the peeled raw corm and cormel of all the varieties with the exception of *C. esculenta* var. *antiquorum* and *C. esculenta* var. *esculenta* could be beneficial in differentiating them. The use of leaf characters in classification and identification of plants has been extensively reported. It has been documented that the leaf characters, such as arrangement, type, form, duration and venation are widely used in both classification and identification [10]. In *Ulmus* and *Betula*, the species are delimited only on the basis of leaf characters. In *Trifolium*, the species were separated on the basis of stipule morphology. *Dalbergia* species were distinguished on the basis of their leaflet, size, shape and arrangement on the rachis. The leaves are important for identification in palms, *Salix* and *Populus* [11]. Leaf character has been reported to be a critical tool in the hand of taxonomists in the classification and separation of taxa [12].

Moreover, there were some similarities between *C. esculenta* var. *antiquorum* and *C. esculenta* var. *esculenta*, which happened to be the only two taxonomically recognized varieties of *C. esculenta*. The colour of the peeled raw tubers of all the varieties varied, but *C. esculenta* var. *antiquorum* and *C. esculenta* var. *esculenta* had the same colour. There was also no significant difference in the leaf width of *C. esculenta* var. *antiquorum* (36.90 ± 4.25 cm) and *C. esculenta* var. *esculenta* (36.00 ± 2.65 cm), indicating a close affinity between them. However, some morphological characters can be used in differentiating them. The corm and cormel length of *C. esculenta* var. *esculenta* (6.17 ± 0.50 cm; 5.00 ± 1.35 cm) were bigger than those of *C. esculenta* var. *antiquorum* (4.10 ± 0.10 cm; 3.70 ± 0.96 cm), respectively. Besides, *C. esculenta* var. *antiquorum* had large corm which was more or less orbicular (approximately, circular) in shape with numerous small round cormels, whereas the corm of *C. esculenta* var. *esculenta* was large and oval with few more or less cylindrical shaped cormels. The petiole of *C. esculenta* var. *antiquorum* was deep purple beneath, while *C. esculenta* var. *esculenta* was pale green. This suggested that the petiole colour, size, shape and number of corm and cormel of *C. esculenta* var. *antiquorum* and *C. esculenta* var. *esculenta*, therefore, were the most prominent morphological characters which distinguished them from each other. Some of these characters have extensively been used to differentiate between *C. esculenta* var. *antiquorum* and *C. esculenta* var. *esculenta* by various workers [2] [6] [13] [14].

There was also significant difference in the corm length, corm width and cormel length of all the varieties of *C. esculenta*. This indicated that these characters could be useful in distinguishing the varieties. Long slender cormel of "nwine" could be used as a ready distinctive character, as well as the largest corm and cormel size of "ogeriobosi". It has been specified that underground parts, such as roots and tubers, are of some taxonomic value in plants [10]. The tubers are helpful in the taxonomy of *Dioscorea* and Cyperaceae. Some vegetative characters that play a major role in plant taxonomy and in deducing phylogeny include growth habit, phonological characters, underground organs, stems, leaves, petiole and stipules. Three sesame species, namely *Sesamum alatum*, *S. radiatum* and *S. indicum* were differentiated on the basis of their vegetative and the pod characteristics [15]. In addition, such characteristics, because of their high taxonomic importance, could be used in constructing a taxonomic key for the purpose of easy and quick identification of the three sesame species irrespective of their growth environment.

5. Conclusions

This work provided information on the morphology of these five varieties of *C. esculenta* present in Anambra State, Nigeria; which was previously lacking. Secondly, the overwhelming evidence from this study suggested close relatedness between *C. esculenta* var. *antiquorum* and *C. esculenta* var. *esculenta*, and provided diagnostic characters for "kochuo", "nwine" and "ogeriobosi".

Conspicuous diagnostic characters observed in "kochuo", were presence of purplish colour at the point of attachment to the leaf of the abaxial leaf surface, as well as presence of a purplish dot on the centre of the adaxial surface of the leaf, and numerous vertical purplish stripes on the surface of the cormels. For "ogeriobosi", the prominent diagnostic features include: presence of foliaceous appendages at the veins of the abaxial surface of the leaves, as well as the purplish colour of the petiole; while yellowish green petiole colour of "nwine" was only the striking diagnostic character. The differential characters include, large corm of *C. esculenta* var. *antiquorum*, which was more or less orbicular (approximately circular) in shape with numerous small round cormels; large and oval corm of *C. esculenta* var. *esculenta* with few more or less cylindrical shaped cormels; slender cormels of "nwine"; and possession of the largest oval corm of "ogeriobosi".

Plants are generally grouped by their relationship to one another based on their similarities and differences, which is based on the characters they possess. This study, therefore, supplied additional morphological information which might be helpful in resolving the on-going controversy in the taxonomy of *Colocasia*, which would, in turn, probably lead to possible delimitation of *C. esculenta*.

References

[1] Aiyeloja, A.A. and Bello, O.A. (2006) Ethnobotanical Potentials of Common Herbs in Nigeria: A Case Study of Enugu State. *Educational Research and Review*, **1**, 16-22.

[2] Nyananyo, B.L. (2006) Plants from the Niger Delta. Onyoma Research Publications, Port Harcourt, 403 p.

[3] Okujagu, T.F. (2008) Medicinal Plants of Nigeria; South East Zone, Vol. 1. Lisida Consulting, Lagos, 204 p.

[4] United States Department of Agriculture (2013) Taxon: *Colocasia esculenta* (L.) Schott. National Germplasm Resources Laboratory, Beltsville, 5 p.

[5] Gomez-Beloz, A. and Rivero, T. (2006) Ure (*Colocasia esculenta* Araceae): An Edible Aroid of the Warao. *Ethnobotany Research and Applications*, **4**, 103-111.

[6] Brooks, F. (2001) Crop Profile for Taro in American Samoa. ASCC Land Grant Program, Pago, 15 p.

[7] Manner, H.I. (2011) Farm and Forestry Production and Marketing Profile for Tannia (*Xanthosoma* spp.). In: Elevitch C.R., Ed., *Specialty Crops for Pacific Island Agroforestry*, Permanent Agriculture Resources (PAR), Holualoa, 1-16.

[8] Acevedo-Rodriguez, P. and Strong, M.T. (2005) Monocots and Gymnosperms of Puerto Rico and Virgin Islands. Contribution from the United National Herbarium, Volume 52, 1-415.

[9] Cholastova, T. and Knotova, D. (2012) Using Morphological and Microsatellite (SSR) Markers to Assess the Genetic Diversity in Alfalfa (*Medicago sativa* L.). *World Academy of Science, Engineering and Technology*, **69**, 856-862.

[10] Sharma, O.P. (1993) Plant Taxonomy. Tata McGraw-Hill Publishing Company Limited, New Delhi, 482 p.

[11] Singh, G. (2004) Plant Systematics. 3rd Edition. Oxford IDH Publishing Co Pvt. Ltd., New Delhi, 598 p.

[12] Illoh, H.C., Oladipo, O.T. and Adedeji, A.A. (2011) Comparative Systematic Foliar Morphological and Anatomical Studies of Three *Cleome* (Linn.) Species in Nigeria. *Nigerian Journal of Botany*, **24**, 17-42.

[13] Udealor, A. and Ezulike, T.O. (2011) Evaluation of Different Crop Species for Compatibility with Cocoyam Minisett. *Nigerian Journal of Plant Protection*, **25**, 122-129.

[14] Ugwuaja, F.N. and Chiejina, N.V. (2011) Preliminary Investigation of the Causes of Leaf Rot of Cocoyam in Nsukka Zone of Nigeria. *Nigerian Journal of Plant Protection*, **25**, 203-213.

[15] Alege, G.O., Akinyele, B.O., Ayodele, S.M. and Ogbode, A.V. (2011) Taxonomic Importance of the Vegetative and Pod Characteristics in Three Nigerian Species of Sesame. *African Journal of Plant Science*, **5**, 213-217.

10

Determining Nodulation Regulatory (*Rj*) Genes of Myanmar Soybean Cultivars and Their Symbiotic Effectiveness with *Bradyrhizobium japonicum* USDA110

Aung Zaw Htwe[1*], Yuchi Saeki[2], Kyi Moe[3], Takeo Yamakawa[4]

[1]Laboratory of Plant Nutrition, Graduate School of Bioresource and Bioenvironmental Sciences, Faculty of Agriculture, Kyushu University, Fukuoka, Japan
[2]Department of Biochemistry and Applied Biosciences, Faculty of Agriculture, Miyazaki University, Miyazaki, Japan
[3]Department of Agronomy, Yezin Agricultural University, Naypyitaw, Myanmar
[4]Laboratory of Plant Nutrition, Division of Molecular Biosciences, Department of Bioresource and Bioenvironmental Sciences, Faculty of Agriculture, Kyushu University, Fukuoka, Japan
Email: [*]aungzawhtwe333@gmail.com

Abstract

Soybean (*Glycine max* L.) plays an essential role in human nutrition as a protein source, and in plant nutrition as a N source. The rate of N fixation varies depending on the cultivars and compatibility between the inoculated *Rhizobium* strain and the host cultivar. Characterizing the nodulation regulatory (*Rj*) genes is necessary to determine the compatibility of cultivars and *Rhizobium* strains. *Rj* genes were previously identified based on inoculation tests and PCR analyses. The six cultivars Yezin-3, Yezin-7, Yezin-11, Shan Seine (Local), Madaya (Local), and Hinthada (Local) were identified as harboring the Rj_4 gene. Two cultivars, Yezin-6 and Yezin-8, were classified as *non-Rj*-gene harboring. Two other cultivars, Yezin-9 and Yezin-10, were identified as Rj_3- and Rj_2Rj_3-gene harboring, respectively. Ours is the first report on Rj_3- and Rj_2Rj_3-gene harboring cultivars in Myanmar. We evaluated Myanmar soybean cultivars for symbiotic effectiveness, relying on the standard strain *Bradyrhizobium japonicum* USDA110. In our first experiment, the soybean cultivar Yezin-11 (Rj_4) showed the highest N fixing potential. Based on their potential for fixing N and nodulation, the top six soybean cultivars were Yezin-11 (Rj_4), Yezin-9 (Rj_3), Yezin-6 (*non-Rj*), Yezin-8 (*non-Rj*), Yezin-3 (Rj_4) and Yezin-10 (Rj_2Rj_3). These cultivars were selected for a second experiment, which revealed that the N fixation, nodulation, and plant growth of Yezin-11 (Rj_4)

[*]Corresponding author.

were superior to the other cultivars. We conclude that Yezin-11 (*Rj_4*) is the most efficient cultivar for nodulation and N fixation when inoculated with *B. japonicum* USDA110.

Keywords

B. japonicum **USDA110, Inoculation Test, PCR Analysis, Nodulation Regulatory Genes (*Rj* Gene), Symbiotic Effectiveness**

1. Introduction

Myanmar is an agricultural country, as agriculture is the backbone of its economy. Legumes are the second largest crops in Myanmar, following rice (*Oryza sativa* L.), in terms of cultivated hectares. Due to the relatively low cost of cultivation and increasing demand for domestic consumption and export, the total cultivated area of pulses has increased from 0.73 million hectares in 1988-89 to 4.4 million hectares in 2011-12 [1]. Soybean (*Glycine max* L.) is an important cash crop and the second largest cultivated crop after rice in Myanmar [2]. Soybean is also one of the most efficient leguminous crops in terms of fixing N [3] [4].

Nodule formation by a cultivar is often dependent on a specific *Rhizobium* strain [5], which may be attributed to nodulation regulatory genes called *Rj* genes. Different soybean cultivars possess different nodulation *Rj* genes. In soybean, the alleles *Rj*(s) and *rj*(s) are dependent on their compatibility with *Bradyrhizobium* and *Ensifer*/*Sinorhizobium* species [6]. Some nodulation *Rj* genes are found in nature, while others resulted from artificially induced mutations [6]. Williams and Lynch [7] found a non-nodulating soybean line, the rj_1 genotype, which resulted from a cross between the cultivars Lincoln and Richard. The *Rj* genes Rj_2, Rj_3 and Rj_4 inhibit the formation of functional nodules by certain *Bradyrhizobium* strains [8]-[11]. The regulatory gene *Rfg*1 restricts nodulation by the fast-growing strain *Sinorhizobium fredii* USDA257 [12].

Ishizuka *et al.* [13] [14] tested the compatibility and preference of the *Rj*-genotype with specific *Bradyrhizobium* strains. *Bradyrhizobium* strains are classified into nodulation Types A, B, and C based on their compatibility with *Rj* cultivars. Type A strains are capable of forming nodules on all *Rj* genotype cultivars. Type B strains cannot form nodules on the Rj_2Rj_3-gene harboring cultivars. Type C strains are inhibited from nodule formation by Rj_4 genotype cultivars. When different *Rj*-gene harboring cultivars are planted in the same field, non-*Rj*, Rj_4 and Rj_2Rj_3 cultivars selectively form nodules with the Types A, B and C strains, respectively. Many scientists reported that the indigenous *Bradyrhizobium* strains in the soil exhibited preferences for nodulation on compatible *Rj* genotypes [15]-[18]. Recently, Soe *et al.* [19] identified Myanmar soybean cultivars with *non-Rj* and Rj_4 genotypes. Soe *et al.* [20] [21] pointed out that the cultivars Yezin-6, harboring the *non-Rj* gene, and Yezin-3, harboring the Rj_4 gene, had enhanced nodulation and N fixation when inoculated with *B. japonicum* USDA110 and indigenous strains. Yamakawa *et al.* [22] [23] stated that $Rj_2Rj_3Rj_4$ conferred improved nodulation when inoculated with *B. japonicum* USDA110. *Bradyrhizobium japonicum* USDA110, which is a Type A strain, could form functional nodules in all dominant *Rj* genotypes. Therefore, *B. japonicum* USDA110 is used in many countries as an inoculant to increase soybean yield.

The *Rj* genes are identified by an inoculation test, which uses strains that restrict nodulation on specific *Rj* genotype soybean cultivars. As an accelerated method for molecularly identifying *Rj* genes, Yang *et al.* [24], Tang *et al.* [25] and Hayashi *et al.* [26] used cloning to identify the genes Rj_2, *Rfg*1 and Rj_4. Yang *et al.* [24] classified the *Rj*2 and *Rfg*1 genes that encoded a member of the Toll-interleukin receptor/nucleotide-binding site/leucine-rich repeat (TIRNBS- LRR) class of plant resistance (R) proteins involved in host resistance to microbial pathogens through an effector-triggered immune (ETI) response. Recently, Hayashi *et al.* [26] described the molecular identification of the Rj_4 gene based on map-based cloning of several mapping populations. They identified the Rj_4 genes that encoded a thaumatin-like protein (TLP) belonging to the pathogenesis-related (PR) protein family 5, which was involved in inhibition of nodulation with specific *Rhizobia* strains. Cloning of the Rj_3 gene has not been reported, so its identification is only based on inoculation test results.

In Myanmar, many researchers have been focusing on selecting strains to increase soybean N fixation. Recently, the Department of Agricultural Research (DAR) has developed improved soybean varieties, such as Yezin-9, Yezin-10, and Yezin-11. However, *Rj* genes have not been identified in some of the released cultivars. To recommend the most efficient N-fixing cultivars, it is necessary to evaluate symbiotic effectiveness with inoculated strains and identify the nodulation *Rj* genes. In the past, *Rj* genes were identified based on inoculation tests. Therefore, our goal in this study was to identify *Rj* genes of Myanmar soybean cultivars based on inoculation tests and multiplex PCR analysis and to screen the cultivars for N-fixing efficiency by using the standard strain *B. japonicum* USDA 110.

2. Materials and Methods

2.1. Origin of Soybean Varieties

Ten soybean varieties (Shan Seine [local], Hinthada [local], Madaya [local], Yezin-3, Yezin-6, Yezin-7, Yezin-8, Yezin-9, Yezin-10, Yezin-11) were collected from the Food Legume Section, Department of Agricultural Research, Yezin, Myanmar. These varieties were grown in the glasshouse of the Plant Nutrition Laboratory, Kyushu University, Japan from July to November 2013 to obtain genetically pure and viable seeds. The focus was to study the ability of cultivars to adapt to weather in Japan. Shan Seine [local], Hinthada [local], Madaya [local] were widely grown in Shan State, Ayeyawaddy Region, Mandalay and Sagaing Regions, respectively. Yezin cultivars used in this experiment were mainly grown in Yezin, Mandalya Region and Shan State, and recommended for farmers to improve soybean production. Flower color, days to maturity and origin of these varieties are shown in **Table 1**.

2.2. Determination of Nodulation Regulatory Genes by Inoculation Test

The *Rj* genotypes of 10 soybean cultivars, including three reference cultivars D51 (Rj_3), CNS (Rj_2Rj_3) and Hill (Rj_4), were investigated to estimate their compatibility with native bradyrhizobia. These varieties were inoculated with the three bradyrhizobial strains *B. japonicum* Is-1, *B. elkanii* USDA33 and *B. japonicum* Is-34 [13]. The strains Is-1, USDA33 and Is-34 failed to produce nodules on the roots of soybean cultivars harboring the Rj_2Rj_3, Rj_3 and Rj_4 genes, respectively [10] [27].

The seeds were sterilized by soaking them in 2.5% sodium hypochlorite solution for 5 min, rinsing five times with 10 mL of 99.5% ethanol, and washing five times with sterilized half-strength modified Hoagland Nutrient (MHN) solution [28]. Five surface-sterilized seeds were sown in pots filled with 1 L of vermiculite and 0.6 L of N-free MHN solution. The strains mentioned above were cultured in A1E liquid media [29] and incubated on a rotary shaker at 30°C for 7 days. Inoculant was prepared by diluting 1 mL of liquid bacterial culture with 99 mL of sterilized MHN solution to obtain a bacterial suspension of about 10^7 cells·mL^{-1}. Seeds were inoculated with the bacterial suspension at 5 mL per seed.

Table 1. Origin of Myanmar soybean varieties.

Variety	Flower color	Days to maturity	Origin
Yezin-3	Violet	115	DAR, Yezin, MR
Yezin-6	Violet	115	DAR, Yezin, MR
Yezin-7	Violet	100	DAR, Yezin, MR
Yezin-8	Violet	115	DAR, Yezin, MR
Yezin-9	White	120	DAR, Yezin, MR
Yezin-10	White	120	DAR, Yezin, MR
Yezin-11	Violet	110	DAR, Yezin, MR
Shan Seine	Violet	120	Shan State
Hinthada	Violet	120	Hinthada, AR
Madaya	Violet	130	Madaya, MR

DAR: Department of Agricultural Research; AR: Ayeyawaddy Region; MR: Mandalay Region.

The plants were cultivated in an environmentally-controlled room (25˚C and 75% Relative Humidity) under natural light for 4 weeks. Control pots were used to check for contamination by non-relevant strains and inoculated strains used in this experiment. Watering was done weekly with autoclaved deionized water. After 1 month, the formation of effective nodules was checked to identify nodulation types of all isolates being tested. This experiment was conducted three times, from January to June 2015.

2.3. Determination of Nodulation Regulatory Genes by PCR Analysis

Multiplex PCR analysis was used to identify *Rj* genes and confirm the Rj_2 and Rj_4 alleles, but it could not detect the Rj_3 allele. For DNA extraction, the plants were cultivated in a growth chamber (28˚C for 16 hours for the light condition and 23˚C for 8 hours for the dark condition). Genomic DNA for PCR templates was extracted from the leaves of seedlings using Takara Bio, following the manufacturer's instructions. Primers were designated from sequence information in reports identifying the Rj_2 and Rj_4 genes [24]-[26]. The primers are described in **Table 2**. The PCR reaction consisted of a pre-run at 94˚C for 5 min, denaturation at 94˚C for 30 s, annealing at 65˚C for 30 s, and extension at 72˚C for 30 s for the first 10 cycles, with a decrease in annealing temperature of 1˚C per cycle. The remaining 20 cycles were repeated at the same temperatures for denaturing and annealing, annealing at 55˚C, and extension at 72˚C for 30 s, followed by the final extension at 72˚C for 10 min and preservation at 4˚C. The reaction producer of PCR analysis of *Rj* genes was innovated by Dr. Yuchi Saeki (Professor, Department of Biochemistry and Applied Biosciences, Miyazaki University). Photos of PCR products were taken after agarose gel electrophoresis (3% agarose gel in 1x TAE buffer) to check the band placement and identify the *Rj* genes.

2.4. Evaluation of Symbiotic Effectiveness of Myanmar Soybean Cultivars

The seeds were surface-sterilized as described above. Six surface-sterilized seeds were sown in pots filled with 1 L of vermiculite and 0.6 L of N-free MHN solution. *Bradyrhizobium japonicum* USDA110 was cultured in A1E liquid media and incubated on a rotary shaker at 30˚C for 7 days. Inoculant was prepared as described above. Seeds were inoculated with the bacterial suspension at 5 mL per seed. The cultivation conditions were the same as described above. Three plants were chosen from each pot for data collection in the first experiment. Six plants were taken from three different pots for the second experiment.

For the acetylene reduction assay (ARA), the soybean plants with intact nodules were placed in 100-mL conical flasks, sealed with a serum stopper and injected with 12 mL of acetylene (C_2H_2) gas to replace air with acetylene.

Table 2. Primers sets and amplification of Multiplex PCR analysis.

Primer sets	Primer sequences	Annealing temperature (˚C)	Amplification site (bp)
Rj_2 specific primers			
K452E-F	(5'-GCTTCAATAGATATGACTTGACAG-3')	58.5	161
R490I-R	(5'-AATCAAGTCATGCATTGTAACTA-3')	57.8	
rj_2 specific primers			
K452-F	(5'-GCTTCAATAGATATGACTTGACAA-3')	58.9	161
R490-R	(5'-ATCAAGTCATGCATTGTAACTC-3')	58.2	
Rj_4 specific primers			
T107A-F	(5'-TTGGAGGAAACGCCG-3')	62.2	324
202-203AY-R	(5'-AATCATGAGAAGAACAAGTATAAGC-3')	58.5	
rj_4 specific primers			
T107-F	(5'-TTGGAGGAAACGCCA-3')	59.9	318
202-203del-R	(5'-AATCATGAGAAGAACAAGTATGGA-3')	60.4	

The nitrogenase activity, in terms of ethylene (C_2H_4) concentration of the plants, was measured using a flame ionization gas chromatograph (GC-14A, Shimadzu, Kyoto, Japan) at 5 and 65 min after injecting with C_2H_2 gas as described by Soe *et al.* [20]. After completing the assay, nodules were counted by removing them from the roots. Shoots, roots, and nodules were collected separately and oven dried at 70°C for 24 hours to record their dry weights. STATISTIX 8 was used for data analysis (Analytical Software, Tallahassee, FL, USA). Means were compared by using Tukey's HSD test at $P < 0.05$.

3. Results

3.1. Nodulation Regulatory (*Rj*) Genes of Myanmar Soybean Cultivars Identified by an Inoculation Test

Identifying the *Rj* genes of soybean varieties is important to determine their host specificity and compatibility with specific bradyrhizobia. We evaluated the *Rj* genotypes of soybean cultivars from Myanmar to identify their nodulation *Rj* genes and estimate their compatibility with strains to be inoculated. Among the tested cultivars, six (Shan Seine [local], Hinthada [local], Madaya [local], Yezin-3, Yezin-7 and Yezin-11) were identified as harboring the Rj_4-gene. Only two cultivars, Yezin-6, Yezin-8, were classified as *non-Rj*-gene harboring cultivars. Yezin-9 and Yezin-10 were identified as Rj_3- and Rj_2Rj_3-gene harboring cultivars, respectively. The results of inoculation testing are shown in **Table 3**.

3.2. Nodulation *Rj* Genes of Myanmar Soybean Cultivars Identified by PCR Analysis

Among the tested cultivars, six (Shan Seine [local], Hinthada [local], Madaya [local], Yezin-3, Yezin-7 and Yezin-11) harbored Rj_4 gene alleles. Yezin-6, Yezin-8 and Yezin-9 did not harbor Rj_4 or Rj_2 genes, although Yezin-6 and Yezin-8 harbored the recessive alleles rj_4 and rj_2, and Yezin-9 and Yezion-10 harbored the recessive allele rj_4. We found the Rj_2 gene in Yezin-10. The results from the inoculation test and PCR analysis are shown in **Figure 1**.

3.3. Symbiotic Effectiveness of USDA 110 on Myanmar Soybean Cultivars

The number of nodules produced was significantly different when inoculated with *B. japonicum* USDA 110 (**Table 4**). The numbers of nodules ranged from 5 to 13 per plant. The most nodules were obtained from Yezin-10 (Rj_2Rj_3), but it was not significantly different from other cultivars, except for Madaya Local (Rj_4), which

Table 3. Nodulation regulatory genes (*Rj* genes) of cultivars.

Cultivar	Nodule No. plant^{-1} on inoculated strains			*Rj* gene
	USDA 33	Is-1	Is-34	
Shan Seine	Medium	Medium	None	Rj_4
Madaya	Medium	Medium	Low	Rj_4
Hinthada	Medium	High	None	Rj_4
Yezin-3	Medium	High	Low	Rj_4
Yezin-6	Medium	Medium	Medium	*non-Rj*
Yezin-7	Medium	High	None	Rj_4
Yezin-8	Medium	Medium	Medium	*non-Rj*
Yezin-9	Low	Medium	Medium	Rj_3
Yezin-10	None	None	High	Rj_2Rj_3
Yezin-11	Medium	Medium	None	Rj_4

High = 10 - 15 nodules plant^{-1}; Medium = 4 - 9 nodules plant^{-1}; Low = 1 - 3 nodules plant^{-1}; None = No nodulation. This division was based on Htwe *et al.* [43].

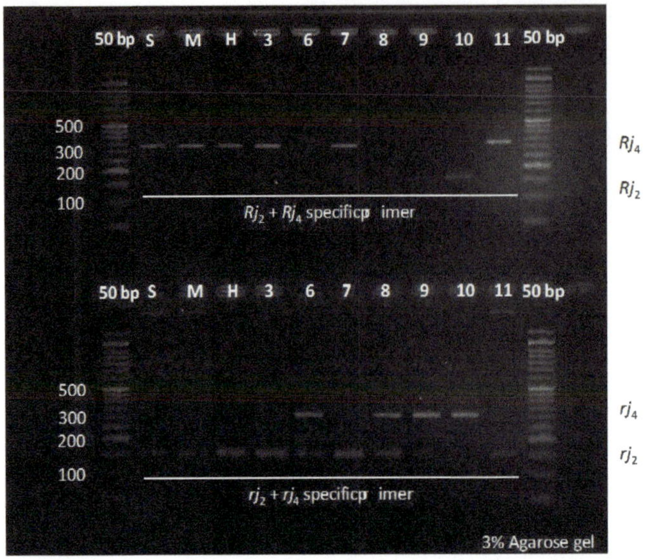

Figure 1. Expression of *rj* or *Rj* gene alleles amplified by multiplex PCR with Rj_2 and Rj_4, and rj_2 and rj_4 specific primers. S: Shan Seine; M: Madaya; H: Hinthada; 3: Yezin-3; 6: Yezin-6; 7: Yezin-7; 8: Yezin-8; 9: Yezin-9; 10: Yezin-10; 11: Yezin-11.

Table 4. Effect of *B. japonicum* USDA 110 strain on acetylene reduction activity, nodulation and plant growth of Myanmar soybean cultivars at 28 DAS.

Cultivars	NN (No. plant^{-1})	NDW (mg·plant^{-1})	SDW (g·plant^{-1})	RDW (g·plant^{-1})	ARA (μmol C_2H_4 h^{-1} plant^{-1})
Yezin-3	10.67 ab	15.70 ab	0.22 ab	0.09 cde	0.49 ab
Yezin-6	9.67 ab	15.40 ab	0.23 ab	0.13 ab	0.52 ab
Yezin-7	10.67 ab	12.60 ab	0.16 bc	0.08 de	0.35 ab
Yezin-8	9.00 ab	19.20 a	0.23 ab	0.11 bcd	0.51 ab
Yezin-9	5.33 ab	11.40 ab	0.21 ab	0.12 abc	0.58 ab
Yezin-10	13.00 a	10.00 ab	0.20 ab	0.11 bcd	0.43 ab
Yezin-11	12.00 ab	23.80 a	0.29 a	0.15 a	0.65 a
Shane Seine	8.00 ab	11.80 ab	0.14 bc	0.08 de	0.11 ab
Hinthada	10.67 ab	16.30 ab	0.27 a	0.12 abc	0.41 ab
Madaya	5.00 b	2.80 b	0.09 c	0.05 e	0.08 b

Mean values in each column followed by the same letters are not significantly different at $P > 0.05$ (Tukey's test). NN: nodule number; NDW: nodule dry weight; SDW: shoot dry weight; RDW: root dry weight; ARA: acetylene reduction activity. Yezin-6 was used as control. Nodule number, nodule dry weight and ARA value for control were zero. Shoot and root dry weight of control was 0.20 and 0.12 g·plant^{-1}, respectively.

produced the fewest nodules. Nodule dry weights also differed significantly among the cultivars. The nodule dry weights of Yezin-11 (Rj_4) and Yezin-8 (*non-Rj*) were greater than weights of the other cultivars but, with the exception of Madaya Local (Rj_4), these differences were not statistically significant.

Shoot dry weights were significantly different between some of the cultivars. The highest shoot biomass (0.29 g·plant^{-1}) was obtained from Yezin-11 (Rj_4), but this did not differ significantly from that of Yezin-3, Yezin-6, Yezin-8, Yezin-9, Yezin-10, or Hinthada local cultivars. Significantly greater root dry weight was recorded for Yezin-11 (Rj_4), but it was not significantly different from Yezin-6, Yezin-9, or Hinthada local cultivars. The nitrogenase activities varied significant among soybean cultivars when inoculated with *B. japonicum* USDA110 (**Table 4**). The highest ARA values were obtained from Yezin-11 (Rj_4), at 0.65 μmol·h^{-1}·plant^{-1}, but this did not differ significantly from that of the other cultivars, except for the Madaya local cultivar with the lowest nitrogenase activity. These results indicated that nodulation, N fixation, and plant growth of soybean cultivars differed when inoculated with *B. japonicum* USDA110. Higher N-fixing cultivars, in terms of ARA per plant, were Yezin-11 (*Rj*$_4$), Yezin-9 (*Rj*$_3$), Yezin-6 (*non-Rj*), Yezin-8 (*non-Rj*), Yezin-3 (*Rj*$_4$), and Yezin-10 (*Rj*$_2$*Rj*$_3$). These top six cultivars, with higher N fixing potential, were selected for the next experiment.

3.4. Symbiotic Effectiveness of USDA 110 on Selected Soybean Cultivars

The number of nodules, ranging from 10 to 17.67 per plant, differed significantly among cultivars when inoculated with *B. japonicum* USDA 110 (**Figure 2(a)**). We observed a significant increase in the number of nodules in Yezin-11, followed by Yezin-10. Nodule dry weight was also significantly different among treatments, due to inoculation with *B. japonicum* USDA 110 (**Figure 2(b)**). The highest nodule dry weight was observed for Yezin-11, but it was not statistically different from that of other cultivars, except for Yezin-10 with the lowest nodule dry weight. Shoot dry weights varied significantly among cultivars, ranging from 0.45 to 0.63 g·plant^{-1} (**Figure 3(a)**). The highest shoot dry weight was obtained for Yezin-11 (*Rj*$_4$), but it was not statistically different from that of other cultivars, except for Yezin-6 (*non-Rj*). Root dry weight differed significantly among the cultivars (**Figure 3(b)**). The highest root dry weight was obtained from Yezin-11 (*Rj*$_4$). There were significant differences between the cultivars in nitrogenase activity, which ranged from 0.36 to 1.49 μmol·h^{-1}·plant^{-1} (**Figure 4**). The highest ARA value was obtained from Yezin-11 (*Rj*$_4$). Although the lowest ARA value was obtained from Yezin-8 (*non-Rj*), it did not differ statistically from Yezin-3 (*Rj*$_4$), Yezin-6 (*non-Rj*), Yezin-9 (*Rj*$_3$), or Yezin-10 (*Rj*$_2$*Rj*$_3$). These results indicated that Yezin-11 (*Rj*$_4$) was the most efficient cultivar, with the most nodules, the highest nodule, shoot, and root dry weights, and the greatest nitrogenase activity.

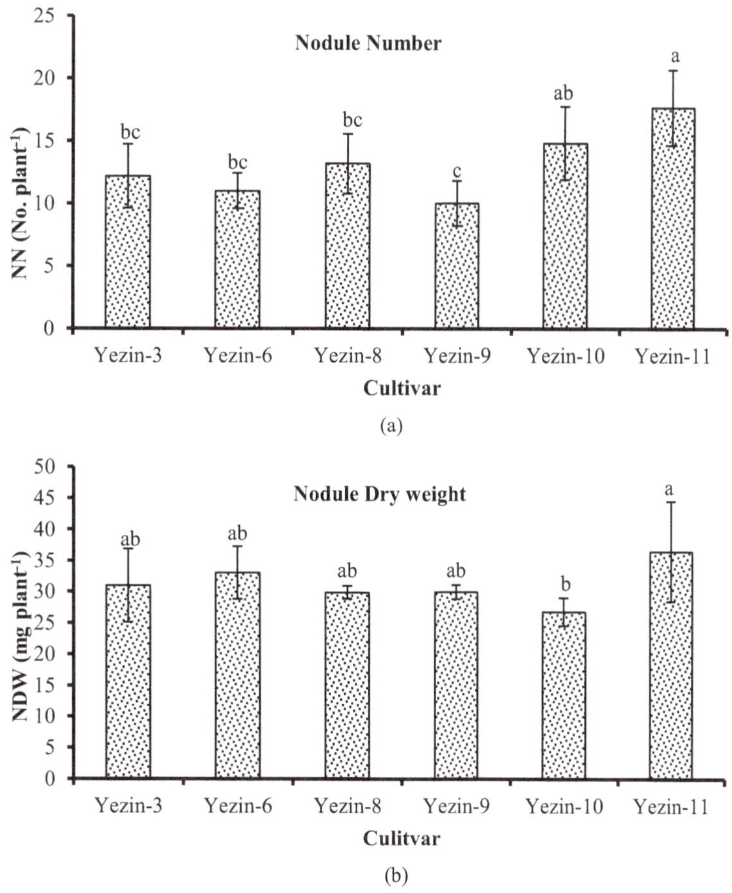

Figure 2. Effect of *B. japonicum* USDA 110 strain on (a) Nodule number and (b) Nodule dry weight of selected Myanmar soybean cultivars at 28 DAS. Mean values followed by same letters are not significantly different at $P > 0.05$ (Tukey's test). Yezin-6 and Yezin-8 were used as control treatments. Nodule number and nodule dry weight for controls were zero.

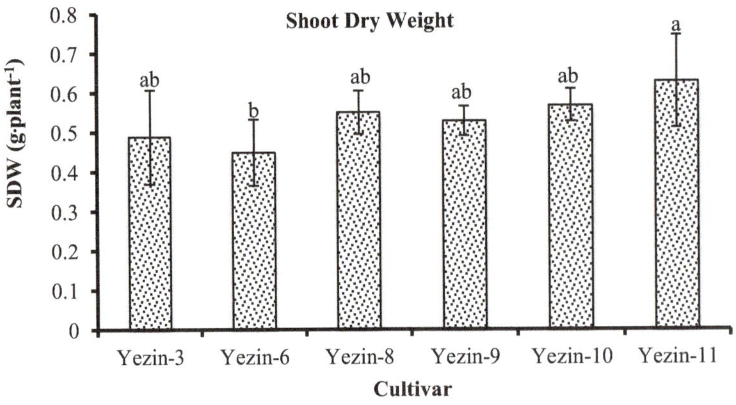

(a)

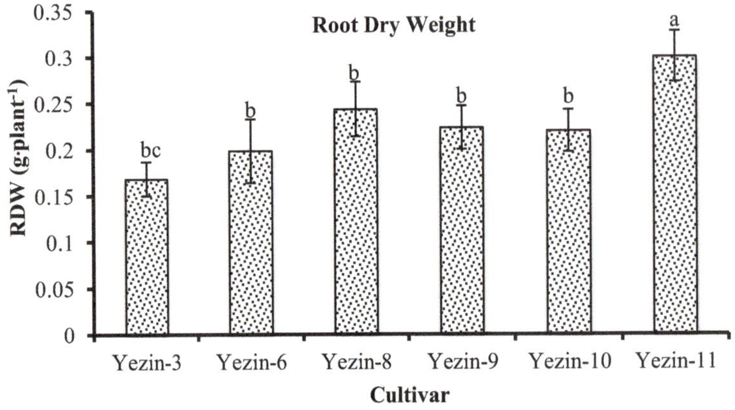

(b)

Figure 3. Effect of *B. japonicum* USDA 110 strain on (a) Shoot dry weight (SDW) and (b) Root dry weight of selected Myanmar soybean cultivars at 28 DAS. Mean values followed by same letters are not significantly different at $P > 0.05$ (Tukey's test). Yezin-6 and Yezin-8 was used as control treatments. Shoot and root dry weight of controls were 0.37 and 0.19 g·plant^{-1}, and 0.46 and 0.22 g·plant^{-1}, respectively.

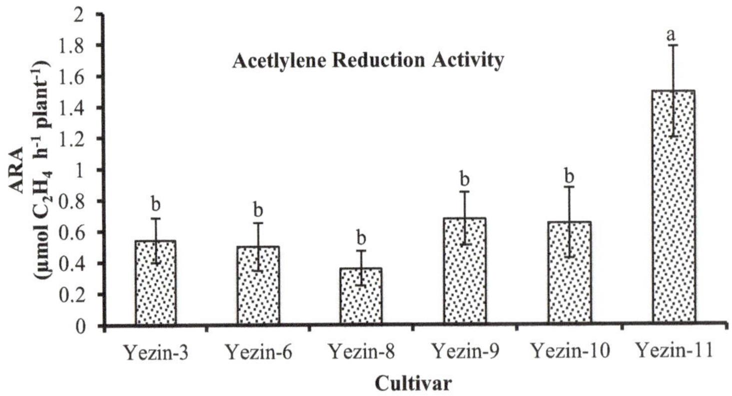

Figure 4. Effect of *B. japonicum* USDA 110 strain on Acetylene reduction activity (ARA) of Selected Myanmar soybean cultivars at 28 DAS. Mean values followed by same letters are not significantly different at $P > 0.05$ (Tukey's test). Yezin-6 and Yezin-8 were used as control treatments. ARAs for each control were zero.

4. Discussion

Symbiotic N fixation of soybean could provide 65 to more than 160 kg fixed N·ha^{-1} [30], accounting for 40 to 70% of the total N requirement. This symbiosis is highly specific, as a particular species or strain of *Rhizobia* could induce a symbiotic association with only a specific leguminous species or cultivars [31]. This specificity involves molecular recognition between the host and the bacterium, through exchange of compound signals that induce nodulation and N fixation [32] [33]. A *Rhizobium* strain that is effective on one legume might not be highly effective on other legumes [34], as the host legume has a dominant role in determining the nodule forming strain. Saeki *et al.* [17] stated that *Rj* genotypes of soybean cultivars have the ability to affect both compatibility and preference for nodulation between the host cultivar and soybean *Rhizobia*.

Compatibility between *Rj* soybean genotypes and soybean-nodulating bradyrhizobia must be considered in selecting the best varieties and strains for soybean cultivation. Therefore, nodulation *Rj* genes collected from soybean cultivars in Myanmar were evaluated to determine compatibility and preference between strains and cultivars. The *Rj* genes could be identified by inoculating with specific strains of *Bradyrhizobium*, such as *B. japonicum* Is-1, *B. elkanii* USDA33 and *B. japonicum* Is-34 [13]. The strains Is-1, USDA33, and Is-34 failed to form nodules on the roots of soybean cultivars harboring the Rj_2Rj_3, Rj_3 and Rj_4 genes, respectively [10] [27]. In this study, Yezin-6 and Yezin-8 formed nodules with all inoculated strains. Therefore, we identified them as non-*Rj*-gene harboring cultivars. The soybean cultivars Shan Seine (local), Hinthada (local), Yezin-7 and Yezin-11 were restricted in nodule formation when inoculated with *B. japonicum* Is-34. Therefore, we classified them as Rj_4 genotype cultivars. In the inoculation test, a few nodules were produced on the roots of Yezin-3 and Madaya (local) when inoculated with Is-34, but we assumed that they harbored the Rj_4 gene. One or two nodules were formed on the roots of Yezin-9 and D51 when inoculated with USDA33. Yamakawa *et al.* [22] stated that a few effective nodules and numerous ineffective nodules were produced by D51 when inoculated with USDA33. This agrees with our findings. Therefore, we assumed that Yezin-9 was an Rj_3 gene-harboring cultivar. Soe *et al.* [19] identified the *Rj* genes of some Myanmar cultivars. They reported that Hinthada, Southern Shan local, Northern Shan local, Yezin-3, and Yezin-11 cultivars harbored the Rj_4-gene, whereas Shan Sein, Shan Wha, Yezin-6, Yezin-8, and Yezin-14 cultivars did not harbor *Rj* genes.

In this study, we performed PCR analysis to confirm the *Rj* genotypes. Although we could not identify Rj_3, the analysis was very useful in identifying the Rj_4 and Rj_2 alleles. The cultivars Shan Seine (local), Hinthada (local), Madaya (local), Yezin-3, Yezin-7 and Yezin-11 harbored Rj_4 alleles. Yezin-6, Yezin-8, and Yezin-9 cultivars did not harbor other dominant *Rj* alleles, although they harbored the recessive alleles rj_2 and rj_4. Only Yezin-9 was identified as an Rj_2-gene harboring cultivar. Based on PCR analysis, we confirmed that Yezin-3 and Madaya (local) were Rj_4-genotype soybean cultivars, although they could also form a few nodules with their nodulation restricting strain Is-34. Contrary to Soe *et al.* [19], the *Rj* genotypes of Hinthada (local), Yezin-3, Yezin-6, Yezin-8 and Yezin-11 were the same and the results for Shan Seine (local) differed from those reported by Soe *et al.* [19]. In this study, Shan Seine (local) was identified as an Rj_4-gene harboring cultivar, according to the inoculation test and PCR analysis results. Therefore, PCR analysis for *Rj* gene determination was deemed necessary if the inoculation test results were uncertain.

According to the inoculation test and PCR results, the cultivars Shan Seine (local), Hinthada (local), Madaya (local), Yezin-3, Yezin-7 and Yezin-11 were identified as Rj_4-genotype cultivars. Yezin-6 and Yezin-8 were identified as non-*Rj* genotype cultivars. Yezin-9 and Yezin-10 were identified as harboring the Rj_3 and Rj_2Rj_3 genes, respectively. In Myanmar, the Rj_4 genotype cultivars were the most widely grown cultivars, accounting for 60% of the tested cultivars. Devine and Kuykendall [35] reported the Rj_4 genotype cultivars as most frequently found in Southeast Asia, but not common in North Asia. Devine and Breithaupt [36] also reported that 60% of the cultivars from Southeast Asia and 71.2% of those from Myanmar harbored Rj_4 genes.

In Myanmar, DAR has developed improved soybean cultivars to replace local cultivars. However, some Myanmar farmers have continued to grow local soybean cultivars, such as Shan Seine (local), Madaya (local) and Hinthada (local). Proper matching of soybean cultivars and *Rhizobia* strains optimizes performance through enhanced N fixation. In this study, we screened improved and local cultivars using the strain *B. japonicum* USDA110, as several studies have reported significant increases in growth, yield, nodulation, and N fixation of Myanmar soybean cultivars due to inoculation with this strain [20] [37] [38].

We found that local cultivars such as Madaya (local) (Rj_4), Shan Seine (local) (Rj_4) and Yezin-7 (Rj_4) showed lower nitrogenase activity, nodulation and plant growth. Hinthada (local) produced an increased number of nodules and nodule dry weight, but its N fixation was lower than other improved cultivars such as Yezin-11 and Yezin-3, despite having the Rj_4 genes in common. Soe *et al.* [20] also reported that improved Yezin cultivars were more efficient for N fixation compared with local cultivars. We also discovered that the improved Yezin cultivars Yezin-11 (Rj_4), Yezin-9 (Rj_3), Yezin-6 (*non-Rj*), Yezin-8 (*non-Rj*), Yezin-3 (Rj_4), and Yezin-10 (Rj_2Rj_3) were superior in N fixing capacity. These top six cultivars, showing higher nitrogenase activity, were selected for the second screening experiment. Selection of cultivars was based on nitrogenase activity and nodulation. Wani *et al.* [39] stated that the number of nodules or nitrogenase activity were genotypically variable within grain legume species. A legume plant with effective nodules could meet not only its own N requirements, but also enrich soil N content, thus improving soil fertility and sustainability [40].

In results from the second screening experiment, Yezin-11 (Rj_4) was the most efficient cultivar, with the most nodules, and highest nodule, shoot and root dry weights, and the greatest nitrogenase activity. When we compared the N fixing rates, in terms of ARA per plant inoculated with *B. japonicum* USDA 110, Yezin-9 (Rj_3) and Yezin-10 (Rj_2Rj_3) were most efficient in N fixation, though they did not differ significantly from the N fixation of Yezin-6 (*non-Rj*), Yezin-8 (*non-Rj*) and Yezin-3 (Rj_4). This might be related to the *Rj* genes, which can affect both preference and compatibility for nodulation between the host cultivar and soybean *Rhizobia* [13] [14] [17]. Yamakawa *et al.* [22] [23] stated that $Rj_2Rj_3Rj_4$-gene harboring cultivars had improved N fixation compared with the *non-Rj*, Rj_2, Rj_2Rj_3 and Rj_4 soybean cultivars when inoculated with *B. japonicum* USDA110. In both experiments, the Rj_4-gene harboring Yezin-11 cultivars showed higher N fixation capacity, followed by the Rj_3-gene harboring Yezin-9 cultivars. The N fixation activity and the ratio of N fixed from the atmosphere to the total N accumulation in plants vary significantly within soybean cultivars [41] [42].

5. Conclusion

Most Myanmar soybean cultivars were identified as harboring the Rj_4 gene. A few cultivars were classified as *non-Rj*, Rj_2Rj_3 and Rj_3 gene harboring. This was the first report of Rj_2Rj_3 and Rj_3 genotype soybean cultivars in Myanmar. We evaluated the N fixation and nodulation of Myanmar soybean cultivars by using the standard strain *B. japonicum* USDA 110. In both experiments, Yezin-11 (Rj_4) was the most efficient for N fixation and nodulation. It appeared that Yezin-11 (Rj_4) was more compatible with *B. japonicum* USDA110 based on the results from both experiments. Yezin-11 had about two to three times higher symbiotic N fixation capacity than the other soybean cultivars. Our study provides useful information for breeders seeking to produce cultivars efficiently at N fixation. Further study is needed on the effectiveness of different *Rj* genotypes with indigenous bradyrhizobia to increase soybean productivity via symbiotic N fixation.

Acknowledgements

This work was supported by Ministry of Education, Culture, Sports, Sciences and Technology of Japan. We are grateful to Dr. Htun Shwe (Researcher, Food Legume Section, Department of Agricultural Research) for providing soybean seeds.

References

[1] MOAI (2013) Myanmar Agriculture in Brief. Ministry of Agriculture and Irrigation, Naypyitaw.

[2] CSO (2006) Myanmar Agricultural Statistics (1992-1993 to 2004-2005). Central Statistical Organization, Ministry of National Planning and Economic Development, Yangon.

[3] Unkivich, M.J. and Pate, J.S. (2000) An Appraisal of Recent Field Measurements of Symbiotic N_2 Fixation by Annual Legumes. *Field Crop Research*, **65**, 211-228. http://dx.doi.org/10.1016/S0378-4290(99)00088-X

[4] van Kessel, C. and Hartley, C. (2000) Agricultural Management of Grain Legumes: Has It Led to An Increased in Nitrogen Fixation? *Field Crop Research*, **65**, 165-181. http://dx.doi.org/10.1016/S0378-4290(99)00085-4

[5] Van, K., Kim, M.Y. and Lee, S.H. (2007) Genomic of Root Nodulation. *Genomic-Assisted Crop Improvement*, **2**, 435-452. http://dx.doi.org/10.1007/978-1-4020-6297-1_16

[6] Hayashi, M., Saeki, Y., Haga, M., Harada, K., Kouchi, H. and Umehara, Y. (2012) A Review: *Rj* (*rj*) Genes Involved in Nitrogen-Fixing Root Nodule Formation in Soybean. *Breeding Science*, **61**, 544-553. http://dx.doi.org/10.1270/jsbbs.61.544

[7] Williams, L.F. and Lynch, D.L. (1954) Inheritance of a Non-Nodulating Character in the Soybean. *Agronomy Journal*, **46**, 28-29. http://dx.doi.org/10.2134/agronj1954.00021962004600010008x

[8] Caldwell, B.E. (1966) Inheritance of a Strain-Specific Ineffective Nodulation in Soybeans. *Crop Science*, **6**,427-428. http://dx.doi.org/10.2135/cropsci1966.0011183X000600050010x

[9] Caldwell, B.E., Hinson, K. and Johnson, H.W. (1966) A Strain-Specific Ineffective Nodulation Reaction in the Soybean *Glycine max* L. Merrill. *Crop Science*, **6**, 495-496. http://dx.doi.org/10.2135/cropsci1966.0011183X000600050033x

[10] Vest, G. (1970) Rj_3-A Gene Conditioning Ineffective Nodulation. *Crop Science*, **10**, 34-35. http://dx.doi.org/10.2135/cropsci1970.0011183X001000010013x

[11] Vest, G. and Caldwell, B.E. (1972) Rj_4—A Gene Conditioning Ineffective Nodulation in Soybean. *Crop Science*, **12**, 692-693. http://dx.doi.org/10.2135/cropsci1972.0011183X001200050042x

[12] Trese, A.T. (1995) A Single Dominant Gene in McCall Soybean Prevents Effective Nodulation with *Rhizobium fredii* USDA257. *Euphytica*, **81**, 279-282. http://dx.doi.org/10.1007/BF00025618

[13] Ishizuka, J., Suemasu, Y. and Mizogami, K. (1991) Preference of *Rj*-Soybean Cultivars for *Bradyrhizobium japonicum* for Nodulation. *Soil Science and Plant Nutrition*, **37**, 15-21. http://dx.doi.org/10.1080/00380768.1991.10415005

[14] Ishizuka, J., Yokoyama, A. and Suemasu, Y. (1991) Relationship between Serotypes of *Bradyrhizobium japonicum* and Their Compatibility with *Rj*-Cultivars for Nodulation. *Soil Science and Plant Nutrition*, **37**, 23-30. http://dx.doi.org/10.1080/00380768.1991.10415006

[15] Minami, M., Yamakawa, T., Yamamoto, A., Akaso, S. and Saeki, Y. (2009) Estimation of Nodulation Tendency among *Rj*-Genotype Soybeans Using Bradyrhizobial Community Isolated from an Andosol. *Soil Science and Plant Nutrition*, **55**, 65-72. http://dx.doi.org/10.1111/j.1747-0765.2008.00333.x

[16] Saeki, Y., Akagi, I., Takaki, H. and Nagatomo, Y. (2000) Diversity of Indigenous *Bradyrhizobium* Strains Isolated from Three Different *Rj*-Soybean Cultivars in Terms of Randomly Amplified Polymorphic DNA and Intrinsic Antibiotic Resistance. *Soil Science and Plant Nutrition*, **46**, 917-926. http://dx.doi.org/10.1080/00380768.2000.10409157

[17] Saeki, Y., Kaneko, A., Hara, T., Suzuki, K., Yamakawa, T., Nguyen, M.T., Nagatomo, Y. and Akao, S. (2005) Phylogenetic Analysis of Soybean Nodulating Rhizobia Isolated from Alkaline Soils in Vietnam. *Soil Science and Plant Nutrition*, **51**, 1043-1052. http://dx.doi.org/10.1111/j.1747-0765.2005.tb00143.x

[18] Saeki, Y., Minami, M., Yamamoto, A. and Akao, S. (2008) Estimation of the Bacterial Community Diversity of Soybean-Nodulating Bradyrhizobia Isolated from *Rj*-Genotype Soybeans. *Soil Science and Plant Nutrition*, **54**, 718-724. http://dx.doi.org/10.1111/j.1747-0765.2008.00300.x

[19] Soe, K.M., Yamakawa, T., Hashimoto, S. and Sarr, P. (2013) Phylogenetic Diversity of Indigenous Soybean Bradyrhizobia from Different Agro-Climatic Regions in Myanmar. *Science Asia*, **39**, 574-583. http://dx.doi.org/10.2306/scienceasia1513-1874.2013.39.574

[20] Soe, K.M. and Yamakawa, T. (2013) Evaluation of Effective Myanmar *Bradyrhizobium* Strains Isolated from Myanmar Soybean and Effects of Coinoculation with *Streptomyces griseoflavus* P4 for N Fixation. *Soil Science and Plant Nutrition*, **59**, 361-370. http://dx.doi.org/10.1080/00380768.2013.794437

[21] Soe, K.M. and Yamakawa, T. (2013) Low-Density Co-Inoculation of Myanmar *Bradyrhizobium yuyanmingense* MAS34 and *Streptomyces griseoflavus* P4 to Enhance Symbiosis and Seed Yield in Soybean Varieties. *American Journal of Plant Science*, **4**, 1879-1892. http://dx.doi.org/10.4236/ajps.2013.49231

[22] Yamakawa, T., Eriguchi, M., Hussain, A.K.M.A. and Ishizuka, J. (1999) Soybean Preference for *Bradyrhizobium japonicum* for Nodulation: Nodulation by $Rj_2Rj_3Rj_4$-Genotypes Isolated from the Progenies of a Cross between Soybean cv. ICA-2 (Rj_2Rj_3) and Hill (Rj_4). *Soil Science and Plant Nutrition*, **45**, 461-469. http://dx.doi.org/10.1080/00380768.1999.10409360

[23] Yamakawa, T., Hussain, A.K.M.A. and Ishizuka, J. (2003) Soybean Preference for *Bradyrhizobium japonicum* for Nodulation Occupation of Serogroup USDA110 in Nodules of Soybean Plants Harboring Various *Rj*-Genes Grown in a Field. *Soil Science and Plant Nutrition*, **49**, 835-841. http://dx.doi.org/10.1080/00380768.2003.10410345

[24] Yang, S., Tang, F., Gao, M., Krishnan, H.B. and Zhu, H. (2010) R Gene-Controlled Host Specificity in the Legume-Rhizobia Symbiosis. *Proceedings of the National Academy of Sciences of the United States of America*, **107**, 18735-18740. http://dx.doi.org/10.1073/pnas.1011957107

[25] Tang, F., Yang, S., Liu, J., Gao, M. and Zhu, H. (2014) Fine Mapping of the Rj_4 Locus, a Gene Controlling Nodulation Specificity in Soybean. *Molecular Breeding*, **33**, 691-700. http://dx.doi.org/10.1007/s11032-013-9985-y

[26] Hayashi, M., Shiro, S., Kanamori, H., Mori-Hosokawa, S., Sasaki-Yamagata, H., Sayama, T., Nishioka, M., Takahashi, M., Ishimoto, M., Kataoyose, Y., Kaga, A., Harada, K., Kouchi, H., Saeki, Y. and Umehara, Y. (2014) A Thaumatin-Like Protein, Rj_4, Controls Nodule Symbiotic Specificity in Soybean. *Plant Cell Physiology*, **55**, 1679-1689. http://dx.doi.org/10.1093/pcp/pcu099

[27] Ishizuka, J., Kim, S.D., Hussain, A.K.M.A. and Yamakawa, T. (1993) Soybean Preference for *Bradyrhizobium japonicum* for Nodulation: Isolation of Rj_2Rj_4-Lines from the Cross of Soybean cvs. IAC-2 (Rj_2) and Hill (Rj_4). *Soil Science*

and Plant Nutrition, **39**, 79-86. http://dx.doi.org/10.1080/00380768.1993.10416977

[28] Nakano, Y., Yamakawa, T., Ikeda, M. and Ishizuka, J. (1997) Nodulation of *Rj*-Soybean Varieties with *Rhizobium fredii* USDA193 under Limited Supply of Nutrients. *Soil Science and Plant Nutrition*, **43**, 929-932. http://dx.doi.org/10.1080/00380768.1997.10414659

[29] Kuykendall, L.D. (1979) Transfer of R Factors to and between Genetically Marked Sublines of *Rhizobium japonicum*. *Applied Environmental Microbiology*, **37**, 862-866.

[30] Klubeck, B.P., Hendrickson, L.L., Zablotowicz, R.M., Skwara, J.E., Varsa, E.C., Smith, S., Isleib, T.G., Maya, J., Valdes, M., Dazzo, F.B., Todd, R.L. and Walgenback, D.D. (1988) Competitiveness of Selected *Bradyrhizobium japonicum* Strains in Midwestern USA Soils. *Soil Science Society American Journal*, **52**, 662-666. http://dx.doi.org/10.1080/00380768.1997.10414659

[31] Somasegaran, P. and Hoben, H.J. (1994) Handbook for Rhizobia: Methods in Legumes-Rhizobium Technology. Springer-Verlag Inc, New York, 450. http://dx.doi.org/10.1007/978-1-4613-8375-8

[32] Denarie, J., Debelle, E. and Rosenberg, C. (1992) Signalling and Host Range Variation in Nodulation. *Annual Review Microbiology*, **46**, 446-453. http://dx.doi.org/10.1146/annurev.mi.46.100192.002433

[33] Perret, X., Staechelin, C. and Brouthton, W.J. (2000) Molecular Basis of Symbiotic Promiscuity. *Microbiology and Molecular Biology Reviews*, **64**, 180-201. http://dx.doi.org/10.1128/MMBR.64.1.180-201.2000

[34] Boonkerd, N. and Singleton, P. (2002) Production of *Rhizobium* Biofertilizer. In: Kannaiyan, S., Ed., *Biotechnology of Biofertilizers*, Narosa Publishing House, New Delhi, 122-128.

[35] Devine, T.E. and Kuykendall, L.D. (1996) Host Genetic Control of Symbiosis in Soybean (*Glycine max* L.). *Plant Soil*, **186**, 173-187. http://dx.doi.org/10.1007/BF00035072

[36] Devine, T.E. and Breithaupt, B.H. (1981) Frequencies of Nodulation Response Alleles, Rj_2 and Rj_4 in Soybean Plant Introductions and Breeding Lines. USDA Technical Bulletin No. 2618.

[37] Soe, K.M., Bhromsiri, A., Karladee, D. and Yamakawa, T. (2012) Effects of *Endophytic actinomycetes* and *Bradyrhizobium japonicum* Strains on Growth, Nodulation, Nitrogen Fixation and Seed Weight of Different Soybean Varieties. *Soil Science and Plant Nutrition*, **58**, 319-325. http://dx.doi.org/10.1080/00380768.2012.682044

[38] Aung, T.T., Tittaburt, P., Boonkerd, N. and Herridge, D. (2013) Co-Inoculation Effects of *Bradrhizobium japonicum* and *Azospirillum* sp. on Competitive Nodulation and Rhizosphere Eubacterial Community Structures of Soybean under Rhizobia-Established Soil Conditions. *African Journal of Biotechnology*, **12**, 2850-2862.

[39] Wani, S.P., Rupela, O.P. and Lee, K.K. (1995) Sustainable Agriculture in a Semiarid Tropic through Biological Nitrogen Fixation in Grain Legume. *Plant Soil*, **174**, 29-49. http://dx.doi.org/10.1007/BF00032240

[40] Thabgaraju, M. and Werner, D. (2004) Indigenous Strains of Rhizobia and Their Performance in Specific Regions of India. In: Werner, D., Ed., *Biological Resources and Migration*, Springer-Verlag, Berlin, 174. http://dx.doi.org/10.1007/978-3-662-06083-4_17

[41] Hungria, M. and Bohrer, T.R.J. (2000) Variability of Nodulation and Dinitrogen Fixation Capacity among Soybean Cultivars. *Biology and Fertility of Soils*, **31**, 45-52. http://dx.doi.org/10.1007/s003740050622

[42] Nohara, T., Nakayama, N., Nakamura, T., Takahashi, M., Maruyama, S., Arihara, J. and Shimada, S. (2006) Cultivar Differences of Nitrogen Fixation Capacity and Its Contribution to Nitrogen Accumulation in Soybean Grown in the Field with a High Soil Nitrate Level. *Japanese Journal of Crop Science*, **75**, 350-359. http://dx.doi.org/10.1626/jcs.75.350

[43] Htwe, A.Z., Yamakawa, T., Sarr, P.S. and Sakata, T. (2015) Diversity and Distribution of Soybean-Nodulation Bradyrhizobia Isolated from Major Soybean-Growing Regions in Myanmar. *African Journal of Microbiology Research*, **9**, 2183-2196. http://www.academicjournals.org/journal/AJMR/article-full-text-pdf/C58126E55962

11

Cloning and Characterization of δ-Guaiene Synthase Genes Encoding a Sesquiterpene Cyclase from *Aquilaria microcarpa* Cell Cultures

Fumiya Kurosaki, Syun Hirohashi, Takahiro Katoh, Futoshi Taura, Jung-Bum Lee

Laboratory of Medicinal Bio-Resources, Graduate School of Medicine and Pharmaceutical Sciences for Research, University of Toyama, Sugitani, Toyama, Japan

Email: kurosaki@pha.u-toyama.ac.jp

Abstract

Three cDNA clones encoding δ-guaiene synthase, a sesquiterpene cyclase, were isolated from tissue cultures of *Aquilaria microcarpa*, and data mining analysis of the orthologous genes suggested that 10 and 9 amino acid residues of N- and C-terminal ends of the translated products of these clones remained undefined. The recombinant enzyme proteins, to which the putative missing N- and C-terminal amino acid sequences (MSSAKLGSAS and ALLRHAIEI, respectively) were ligated, exhibited the catalytic activities of sesquiterpene biosynthesis. Among these three δ-guaiene synthases, two isoforms were capable of liberating α-guaiene, δ-guaiene, β-elemene plus α-humulene as a minor product, while remaining one isoenzyme generated α-, δ-guaiene and β-elemene but not α-humulene. Although the enzyme protein solely lacking in the N-terminal 10 amino acid residues was capable of synthesizing the sesquiterpenoids, the protein without 9 amino acids at C-terminal did not exhibit the catalytic activity. These results suggest that two types of δ-guaiene synthase; α-, δ-guaiene, β-elemene-producing type, and α-, δ-guaiene, β-elemene plus α-humulene-producing type; concomitantly occur in *A. microcarpa* cell cultures, and several amino acid residues at C-terminal of the synthase protein are essential to exhibit the catalytic activities as the sesquiterpene cyclase.

Keywords

Methyl Jasmonate, Yeast Extract, Sesquiterpene, δ-Guaiene Synthase, *Aquilaria microcarpa*

1. Introduction

It has been widely known [1] that *Aquilaria*, and *Gyrinops* species sometimes form dark resinous heartwood, called agarwood, in the trunk of the plants in response to mechanical wounding or microbial infection. Agarwood produces a variety of sesquiterpene compounds with aroma, and, therefore, it has been used as the scent, perfume and traditional medicines [1]. These highly valuable tissues are not formed under healthy conditions, and artificial transformation of these trees to agarwood has been very difficult. Okudera and Ito, however, reported [2] that biosynthetic activities of several sesquiterpenoids, such as δ-guaiene, α-guaiene and α-humulene, are induced in cell cultures of *A. crassna* by the treatment with methyl jasmonate or yeast extract. Guaianolide sesquiterpene compounds are thought to be synthesized via two cyclization reactions [3]: i) C1-to-C10 cyclization generating germacrene-like macrocyclic intermediate; and ii) C2-to-C6 cyclization forming guaiane-skeleton (**Figure 1**). A part of the macrocyclic intermediate is converted to α-humulene or germacrene A without the second cyclization. It has been also demonstrated [4] [5] that β-elemene is an artificially formed compound by the pyrolysis of germecrene A. Recently, several cDNA clones encoding δ-guaiene synthase, a sesquiterpene cyclase, have been isolated from *A. crassna* and *A. sinensis* [3] [6]. Catalytic specificities of the translated proteins of these genes have been extensively studied, and it has been shown that the enzymes liberate multiple sesquiterpene compounds, such as δ-guaiene, α-guaiene, β-elemene and α-humulene, from farnesyl diphosphate as the substrate. However, only very little is known about the molecular basis of the enzymatic reaction of the synthase generating multiple products.

We have recently shown [5] [7] that treatment of cell cultures of *A. microcarpa* with methyl jasmonate or yeast extract results in the marked transcriptional activation of δ-guaiene synthase gene, and a cDNA clone (*GS*-1, GenBank accession No. KF800046) encoding the enzyme protein has been isolated [5]. It has been also demonstrated that *GS*-1 translate catalyzes the formation of δ-guaiene, α-guaiene and β-elemene, but not α-humulene [5]. In the present experiments, we have isolated three additional clones of the synthase gene although several amino acid residues at both N- and C-terminal ends of the translated proteins remain undefined. In order to understand the inducible biosynthetic activities of *A. microcarpa* liberating multiple sesquiterpene compounds, we have ligated putative missing amino acid sequences to N- and C-terminals of the recombinant proteins, and examined the catalytic properties of these 'deficiencies supplied' δ-guaiene synthases.

2. Materials and Methods

2.1. Cloning of δ-Guaiene Synthase Genes from *A. microcarpa* Cell Culture

Seeds of *A. microcarpa* were generous gift from Professor Shigetoshi Kadota, Institute of Natural Medicine, University of Toyama, and they were germinated and grown in a greenhouse at Experimental Station for Medi-

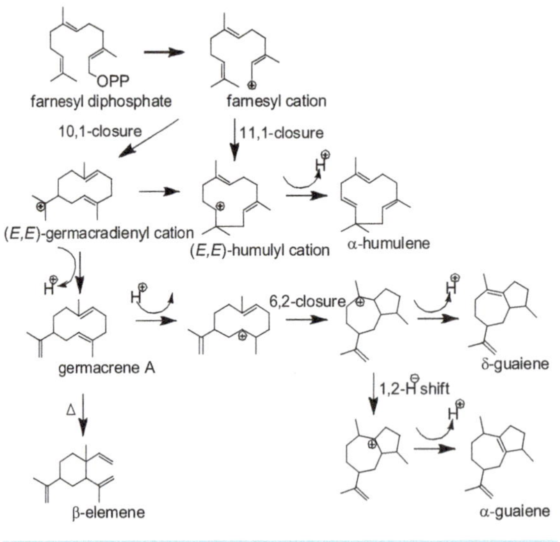

Figure 1. Predicted biosynthetic pathway of sesquiterpene compounds of *Aquilaria* plants.

cinal Plant Research of our university. Sterilized young leaf tissues of the plant were placed on Murashige and Skoog's agar medium [8] supplemented with 3% (w/v) sucrose and 3 μM of 2,4-dichlorophenoxyacetic acid, and callus formation was observed in the leaf veins after 4 weeks. The dedifferentiated cells were then incubated on the medium in the presence of 3 μM of 2,4-dichlorophenoxyacetic acid and 3 μM of N^6-benzyladenine at 26˚C under darkness. After establishment of the callus culture, the cells were transferred onto a fresh medium every 3 weeks as described previously in detail [9] [10]. Total RNA was isolated from yeast extract-treated cultured cells (approximately 100 mg) with RNeasy Plant Mini Kit (Qiagen, Hilden, Germany) [5] [9], and cDNA templates were generated by reverse transcription reaction using Transcriptor First Strand cDNA Synthesis Kit (Roche, Basel, Switzerland). The cDNA fragments presumably encoding the translatable region of *δ*-guaiene synthase were isolated by PCR amplification employing the primer pair designated from the nucleotide sequence of the homologous gene previously isolated [5] (5'-ATG TCT TCG GCA AAA CTA GGT TCT GCC TCC-3' as the forward and 5'-TCA GAT TTC AAT AGC ATG ACG CAA CAA GGC-3' as the reverse primer). The amplified cDNAs were subcloned into the pGEM-T Easy vector (Promega, Fitchburg, WI, USA), and the nucleotide sequences were determined on both strands using the dye-terminator method on a PRISM 3130 Genetic Analyzer (Applied Biosystems, Waltham, MA, USA).

2.2. Preparation and Purification of Recombinant *δ*-Guaiene Synthase Proteins

The modified *δ*-guaiene synthase genes, *GS*-2, 3 and 4, in which undefined sequences of the translatable regions were ligated, were amplified using a common primer pair (5'-ATG TCT TCG GCA AAA CTA GGT TCT GCC TCC-3' as the forward and 5'-TCA GAT TTC AAT AGC ATG ACG CAA CAA GGC-3' as the reverse primer, 1644 mer as the product). Amplification of *GS-2ΔNΔC*, the undefined regions-omitted gene, was performed with 5'-GAA GAT GTT AGC CGC CGA GAT GCC-3' as the common forward primer and 5'-TCA GGC AAT TCT ATC CTT GGT CAC-3' as the gene specific reverse primer employing pGEM-T Easy vector harboring *GS*-2 as the template. *GS-3ΔNΔC* and *GS-4ΔNΔC* were similarly amplified by the combination of the common forward primer and the gene specific reverse primer (5'-TCA GGC AAT TCT ACC CTT GGT CAC-3' for *GS-3ΔNΔC*, and 5'-TCA GGC AAT TCT ATC TTT GGT CAC-3' for *GS-4ΔNΔC*, respectively). *GS-4ΔN* and *GS-4ΔC* were also amplified by the appropriate combinations of the common primer and the specific primers (5'-GAA GAT GTT AGC CGC CGA GAT GCC-3' and 5'-TCA GAT TTC AAT AGC ATG ACG CAA CAA GGC-3' for *GS-4ΔN* while 5'-ATG TCT TCG GCA AAA CTA GGT TCT GCC TCC-3' and 5'-TCA GGC AAT TCT ATC TTT GGT CAC-3' for *GS-4ΔC*) using *GS*-4 subcloned into pGEM-T Easy vector as the template.

Over expression of *δ*-guaiene synthase genes in *Escherichia coli* was performed using the *E. coli* Expression System with Gateway Technology (Invitrogen, Waltham, MA, USA). Desired regions of the synthase genes amplified by PCR were successively subcloned into pENTR and pDEST15 (Invitrogen, Waltham, MA, USA), and then, *E. coli* BL21 cells were transformed with the constructed expression vectors. The cell cultures were grown in LB medium overnight at 25˚C, and, after transfer into the fresh medium, isopropyl *β*-D-1-thiogalactopyranoside (IPTG, final concentration 0.1 mM) was added to the cultures at an optical density of 0.4 - 0.5 at 600 nm. The cultures were further incubated at 25˚C for 5 h. *δ*-Guaiene synthase proteins were recovered as the glutathione-S-transferase (GST) tagged form, and were purified by an affinity chromatography on a Glutathione-Sepharose 4B column (GE Healthcare Life Sciences, Little Chalfont, UK) according to the instruction manual. Protein concentrations were determined by the method of Bradford [11], and, if necessary, the purified enzyme solutions were appropriately concentrated by Amicon Ultra-15 (Merck Millipore, Billerica, MA, USA). The samples were, then, subjected to sodium dodecyl sulfate polyacrylamide gel electrophoresis (SDS-PAGE, 10% gel) according to the method of Laemmli [12], and the separated proteins were stained with Coomassie brilliant blue.

2.3. Catalytic Activities of Recombinant *δ*-Guaiene Synthase Proteins

In order to examine sesquiterpene biosynthetic activities of the recombinant *δ*-guaiene synthases, the purified enzyme proteins were incubated with farnesyl diphosphate as the substrate, and the reaction products were analyzed by GC-MS. The assay mixtures (200 μl) contained 50 mM Tris-HCl (pH 7.0), 20 mM $MgCl_2$, 10 mM dithiothreitol, 46 μM farnesyl diphosphate (Sigma-Aldrich, St. Louis, MO, USA), 10% (v/v) glycerol, purified GST-*δ*-guaiene synthase proteins (approximately 10 μg) and limonene (0.2 μg, nacalai tesque, Kyoto, Japan) as an internal standard. In a control experiment, the enzyme proteins were heat denatured (95˚C, 10 min) prior to the assay. The mixtures were incubated at 25˚C for 4 h, and the reaction products were extracted using head-space solid phase microextraction assembly with a 100 μm polydimethylsiloxane fiber (Supelco, Bellefonte, PA, USA). The fiber was exposed to the sample headspace for 30 min, and, after adsorption of the reaction products, sesquiterpene compounds were analyzed by GC-MS (Shimadzu, GCMS-QP5000; Agilent, J&W DB-1ms 0.32 mm × 30 m column) in the splitless mode. The injection port and the interface temperature were adjusted at 250˚C, and the mass range was set from m/z 40 to 400. The carrier flow (He) was at 1.4 ml/min. The oven temperature was started at 50˚C for 2 min and was increased to 170˚C at the rate of 4˚C/min, then raised to 300˚C at 30˚C/min. The reaction products were separated under the above conditions to calculate retention indices, and the identification of the compounds was based on the comparison of reported retention indices [13] [14] and mass spectra with those in the NIST02 and Wiley 7 MS databases.

3. Results

3.1. δ-Guaiene Synthase Genes of *A. microcarpa* Cell Culture

It has been demonstrated [3] [6] that sesuqiterpene biosynthetic genes encoding δ-guaiene synthase in *Aquilaria* plants are usually organized as the multigene family. To our knowledge, seven orthologues of the synthase genes have been isolated from *Aquilaria* plants (*A. crassna*, *A. sinensis* and *A. microcarpa*), and the translated products of these genes are found to be highly conserved. All of the synthase proteins consist of 547 amino acid residues and the sequences also show very high homology (98% - 99% identity) [3] [5] [6]. In order to isolate δ-guaiene synthase genes other than *GS*-1 from *A. microcarpa* cell cultures, we employed the primer pair corresponding to the N- and C-terminal ends of the translatable region of *GS*-1 (30 nucleotides each, including stop codon for C-terminal). Three new clones presumably encoding δ-guaiene synthase were isolated (*GS-2ΔNΔC*, *GS-3ΔNΔC* and *GS-4ΔNΔC*; GenBank accession No. KT283579, KT283580 and KT283581, respectively), and the putative amino acid sequences of the translated products showed high homology with *GS*-1 (**Figure 2**). Two specific motifs, the double arginine residue (RRX8W) at the N-terminal end and the aspartate rich motif (DDXXD) in the reaction center of the protein [15] [16], were observed in these three clones, and replacement of amino acid residues toward *GS*-1 was found at only three positions in *GS*-2, four in *GS*-3 and eight in *GS*-4, respectively (**Figure 2**).

Since nucleotide sequences corresponding to N- and C-terminal ends of *GS*-1 were employed as primers for PCR amplification, 10 and 9 amino acid residues of the terminals remained to be defined. A number of attempts were made to elucidate the undefined translatable and untranslatable regions of these three δ-guaiene synthase genes, and the rapid amplification of cDNA ends method have been systematically performed to obtain the DNA fragments. However, these experiments have not succeeded possibly because only very few nucleotides are replaced in the newly isolated three clones.

3.2. Preparation and Purification of Recombinant δ-Guaiene Synthase Proteins

Among seven δ-guaiene synthases reported from *Aquilaria* [3] [5] [6], six enzyme proteins share the common N-terminal sequence MSSAKLGSAS while remaining one is MSSAKLGSTS (**Table 1**). On the other hand, C-terminal sequence of all of the seven proteins is identical (ALLRHAIEI). Therefore, we ligated the most probable "missing" amino acid sequences to N- and C-terminal ends, and catalytic properties of these deficiencies-supplied enzymes (*GS*-2, *GS*-3 and *GS*-4) were examined together with the terminals-lacking translates of *GS-2ΔNΔC, GS-3ΔNΔC* and *GS-4ΔNΔC*. The recombinant enzymes were obtained as GST-tagged proteins, and were purified by an affinity chromatography on a Glutathione-Sepharose 4B column. In SDS-PAGE analysis (**Figure 3**), the protein bands corresponding to the position of GST-fused *GS*-2, 3 and 4 in IPTG-treated cell extracts (approximately 91.5 kDa) were shown to be sufficiently purified. The recombinant enzymes lacking in the terminal structures were also prepared as GST-tagged forms and purified in a similar manner.

3.3. Catalytic Activities of Recombinant δ-Guaiene Synthase Proteins

The recombinant δ-guaiene synthases of *A. microcarpa* were incubated with farnesyl diphosphate, and the reaction products were analyzed by GC-MS. We have previously reported [5] that translate of *GS*-1 liberates δ-guaiene, α-guaiene and germacrene A as the reaction products. We have also demonstrated [5] that germa-

```
GS-1        MSSAKLGSASEDVSRRDANYHPTVWGDFFLTHSSNFLENNHSILEKHEELKQEVRNLLVVETSDLPSKIQLTDKIIRLGV 80
GS-2ΔNΔC    ----------EDVSRRDADYHPTVWGDFFLTHSSNFLENNHSILEKHEELKQEVRNLLVVETSDLPSKIQLTDKIIRLGV 70
GS-3ΔNΔC    ----------EDVSRRDANYHPTVWGDFFLTHSSNFLENNDNILEKHEELKQEVTNLLVVETSDLPSKIQLTDEIIRLGV 70
GS-4ΔNΔC    ----------EDVSRRDANYHPTVWGDFFLTHSSNFLENNDNILEKHEELKQEVRNLLVVETSDLPSKIQLTDKIIRLGV 70

GS-1        GYHFEMEIKAQLEKLHDHQLHLNFDLLTTSVWFRLLRGHGFSISSDVFKRFKNTKGEFETEDARTLWCLYEATHLRVDGE 160
GS-2ΔNΔC    GYHFEMEIKAQLEKLHDHQLHLNFDLLTTSVWFRLLRGHGFSISSDVFKRFKNTKGEFETEDARTSWCLYEATHLRVDGE 150
GS-3ΔNΔC    GYHFEMEIKAQLEKLHDHQLHLNFDLLTTSVWFRLLRGHGFSISSDVFKRFKNTKGEFETEDARTLWCLYEATHLRVDGE 150
GS-4ΔNΔC    GYHFEMEIKAQLEKLHDHQLHLNFDLLTTSVWFRLLRGHGFSISSDVFKRFKNTKGEFETEDARTLWCLYEATHLRVDGE 150

GS-1        DILEEAIQFSRKKLEALLPELSFPLNECVRDALHIPYHRNVQRLAARQYIPQYDAELTKIESLSLFAKIDFNMLQALHQS 240
GS-2ΔNΔC    DILEEAIQFSRKKLEALLPELSFPLNECVRDALHIPYHRNVQRLAARQYISQYDAELTKIESLSLFAKIDFNMLQALHQS 230
GS-3ΔNΔC    DILEEAIQFSRKKLEALLPELSFPLNECVRDALHIPYHRNVQRLAARQYIPQYDAELTKIESLSLFAKIDFNMLQALHQS 230
GS-4ΔNΔC    DILEEAIQFSRKKLEALLPELSFPLNECVRDALHIPYHRNVQRLAARQYIPQYDAELTKIESLSLFAKIDFNMLQALHQS 230

GS-1        ELREASRWWKEFDFPSKLPYARDRIAEGYYWMMGAHFEPKFSLSRKFLNRIIGITSLIDDTYDVYGTLEEVTLFTEAVER 320
GS-2ΔNΔC    ELREASRWWKEFDFPSKLPYARDRIAEGYYWMMGAHFEPKFSLSRKFLNRIIGITSLIDDTYDVYGTLEEVTLFTEAVER 310
GS-3ΔNΔC    ELREASRWWKEFDFPSKLPYARDRIAEGYYWMMGAHFEPKFSLSRKFLNRIIGITSLIDDTYDVYGTLEEVTLFTEAVER 310
GS-4ΔNΔC    ELREASRWWKEFDFPSKLPYARDRIAEGYYWMMGAHFEPKFSLSRKFLNRIIGITSLIDDTYDVYGTLEEVTLFTKAVER 310

GS-1        WDIEAVKDIPKYMQVIYTGMLGIFEDFKDNLINARGKDYCIDYAIEVFKEIVRSYQREAEYFHTGYVPSYDEYMENSIIS 400
GS-2ΔNΔC    WDIEAVKDIPKYMQVIYTGMLGIFEDFKDNLINARGKDYCIDYAIEVFKEIVRSYQREAEYFHTGYVPSYDEYMENSIIS 390
GS-3ΔNΔC    WDIEAVKDIPKYMQVIYTGMLGIFEDFKDNLINARGKDYCIDYAIEVFKEIVRSYQREAEYFHTGYVPSYDEYMENSIIS 390
GS-4ΔNΔC    WDIEAVQDIPKYMQVIYTGMLGIFEDFKDNLINARGKDYCIDYAIEVFKEIVRSYQREAEYFHTGYVPSYDEYMENSIIS 350

GS-1        GGYKMFIILMLIGRAEFELKETLDWASTIPEMVKASSLIARYIDDLQTYKAEEERGETVSAVRCYMREYGVSEEEACKKM 480
GS-2ΔNΔC    GGYKMFIILMLIGRAEFELKETLDWASTIPEMVKASSLIARYIDDLQTYKAEEERGETVSAVRCYMREYGVSEEEACKKM 470
GS-3ΔNΔC    GGYKMFIILMLIGRAEFELKETLDWASTIPEMVKASSLIARYIDDLQTYKAEEERGETVSAVRCYMREYGVSEEEACKKM 470
GS-4ΔNΔC    GGYKMFIILMLIGRGEFELKETLDWASTIPEMVKASSLIARYIDDLQTYKAEEKRGETVSAVRCYMREYGVSEEEACKKM 470

GS-1        REMIEIEWKRLNKTTLEADEISSSVVIPSLNFTRVLEVMYDKGDGYSDSQGVTKDRIAALLRHAIEI 547
GS-2ΔNΔC    REMIEIEWKRLNKTTLEADEISSSVVIPSLNFTRVLEVMYDKGDGYSDSQGVTKDRIA--------- 528
GS-3ΔNΔC    REMIEIEWKRLNKTTLEADEISSSVVIPSLNFTRVLEVMYDKGDGYSDSQGVTKGRIA--------- 528
GS-4ΔNΔC    REMIEIEWKKLNKTTLEANEISSSVVIPSLNFTRVLEVMYDKGDGYSDSQGVTKDRIA--------- 528
```

Figure 2. Alignment of amino acid sequence of δ-guaiene synthases from *A. microcarpa*. Non-conserved amino acids were shaded, and GenBank accession numbers are as follows; *GS*-1, KF800046; *GS-2ΔNΔC*, KT283579; *GS-3ΔNΔC*, KT283580; *GS-4ΔNΔC*, KT283581.

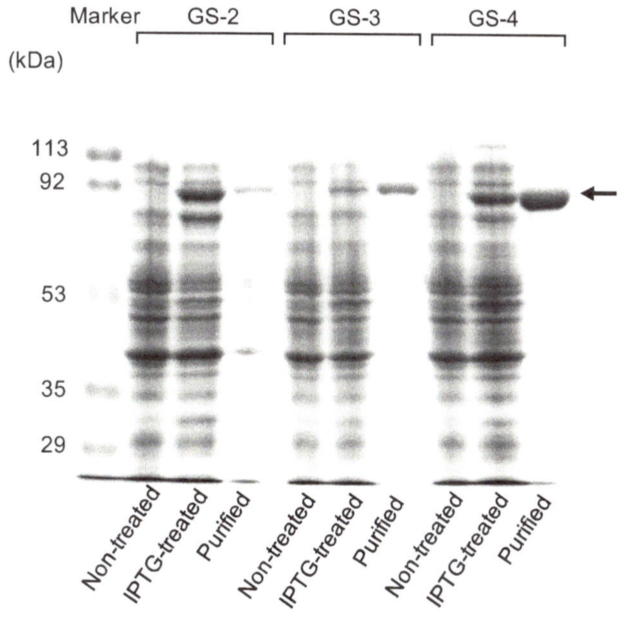

Figure 3. SDS-PAGE analysis of production and purification of δ-guaiene synthase proteins. *GS*-2, *GS*-3 and *GS*-4 were overexpressed in *E. coli* cells, and the translated products were obtained as GST-fusion proteins. The crude extracts prepared from non-treated control, IPTG-treated cells and the purified proteins were analyzed by SDS-PAGE, and the expected position of GST-fused δ-guaiene synthase (approximately 91.5 kDa) was indicated by an arrow.

Table 1. Amino acid sequences of N- and C-terminal ends of δ-guaiene synthases isolated from *Aquilaria* plants.

	N-terminal end	
A. microcarpa GS-1	MSSAKLGSAS	10
A. sinensis-1	MSSAKLGSAS	10
A. sinensis-2	MSSAKLGSAS	10
A. sinensis-3	MSSAKLGSAS	10
A. crassna-1	MSSAKLGSAS	10
A. crassna-2	MSSAKLGSAS	10
A. crassna-3	MSSAKLGSAS	10
	C-terminal end	
A. microcarpa GS-1	ALLRHAIEI	547
A. sinensis-1	ALLRHAIEI	547
A. sinensis-2	ALLRHAIEI	547
A. sinensis-3	ALLRHAIEI	547
A. crassna-1	ALLRHAIEI	547
A. crassna-2	ALLRHAIEI	547
A. crassna-3	ALLRHAIEI	547

GenBank accession numbers are as follows; *A. sinensis*-1, JQ712682; *A. sinensis*-2, JQ712683; *A. sinensis*-3, JQ712684; *A. crassna*-1, GU083697; *A. crassna*-2, GU083698; *A. crassna*-3, GU083699.

crene A is easily converted to β-elemene by pyrolysis, and, therefore, under the usual GC-MS conditions, together with α- and δ-guaiene, β-elemene was observed as the apparent product of the enzymatic reaction instead of germacrene A. As shown in **Figure 4(a)**, *GS*-2 produced at least three sesquiterpene compounds, and peak numbers 1, 2 and 4 were assigned to be β-elemene, α-guaiene and δ-guaiene by the retention indices and mass spectra (**Figure 5**) in the data base, respectively [13] [14]. In sharp contrast, detectable amount of the reaction products was not observed when the assay was carried out with the terminals sequences-omitted form, *GS-2ΔNΔC* (**Figure 4(a)**). Heat denatured enzyme proteins also did not show the catalytic activity (data not shown). *GS*-3 was also capable of producing β-elemene, α-guaiene and δ-guaiene as the major products, however, unlike in *GS*-1 [5] and *GS*-2, this protein was found to liberate α-humulene (peak 3) as an additional minor product (**Figure 4(b)** and **Figure 5**). As well as *GS-2ΔNΔC*, *GS-3ΔNΔC* did not exhibit the sesuqiterpene biosynthetic activity. *GS*-4 showed the catalytic specificity essentially similar to that of *GS*-3, and this enzyme produced β-elemene, α-guaiene, δ-guaiene and low amount of α-humulene (**Figure 4(c)** and **Figure 5**). As was in other δ-guaiene synthases, *GS-4ΔNΔC* was shown to be an inactive protein. In **Figure 4(a)**, it appeared that a minor peak could be observed at the retention time similar to α-humulene (peak 3), and, therefore, we carefully examine the possible formation of this sesquiterpene compound in *GS*-2-catalyzed reaction. However, based on the mass spectra analysis, it was revealed that the apparent minor peak was due to the formation of unknown compound but not α-humulene.

In order to elucidate whether or not both N- and C-terminal structures of δ-guaiene synthase are essential to exhibit the catalytic activity, 10 amino acid residues at N-terminal or 9 amino acids at C-terminal of the enzyme were selectively omitted, and the catalytic activities of these proteins were examined. In this set of the experiments, *GS*-4 which showed the highest recovery in the protein purification processes was selected, and N-terminal-lacking and C-terminal-lacking enzyme proteins (*GS*-4*Δ*N and *GS*-4*Δ*C) was prepared as the GST-fused form. As well as *GS*-4, N-terminal lacking enzyme *GS*-4*Δ*N exhibited the catalytic activity, if somewhat reduced, and the peaks of the sesquiterpene compounds were clearly detected (**Figure 6**). In sharp contrast, *GS*-4*Δ*C apparently did not show the enzymatic activity, and no peak of the reaction product was observed as far as tested. These results strongly suggest that 9 amino acid residues at C-terminal of δ-guaiene synthase proteins should be an essential structure to function as the sesquiterpene cyclase.

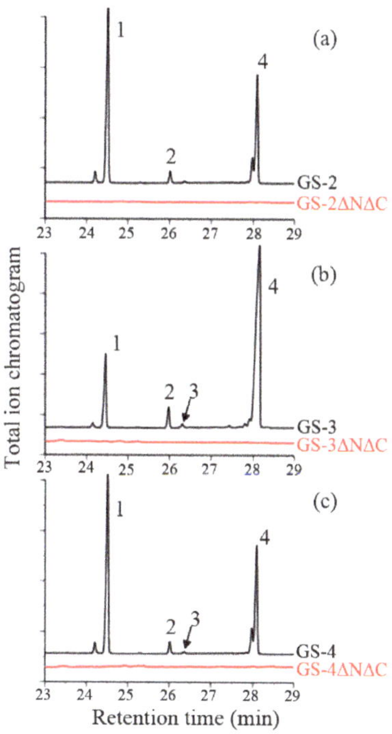

Figure 4. Total ion chromatograms of the reaction products formed by the incubation of recombinant GST-δ-guaiene synthase proteins with farnesyl diphosphate. (a) *GS*-2 and *GS*-2*ΔNΔC*; (b) *GS*-3 and *GS*-3*ΔNΔC*; (c) *GS*-4 and *GS*-4*ΔNΔC*.

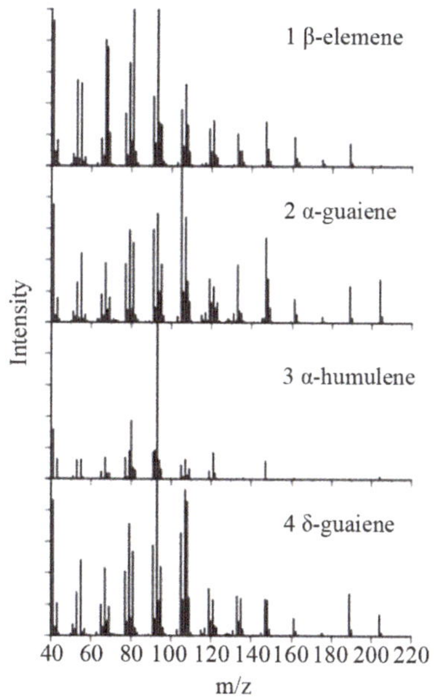

Figure 5. Mass spectra of the compounds corresponding to the peaks shown in **Figure 4**.

4. Discussion

As shown in **Figure 1**, α- and δ-guaiene are thought to be synthesized from germacrene A as an intermediate. On the other hand, α-humulene biosynthesis shares the common pathway with guaienes-forming processes only in the early stage. Kumeta and Itoh [3] isolated three δ-guaiene synthase genes from *A. crassna*, and showed that the translated proteins of all of these genes generate δ-guaiene, α-guaiene plus α-humulene as the minor product.

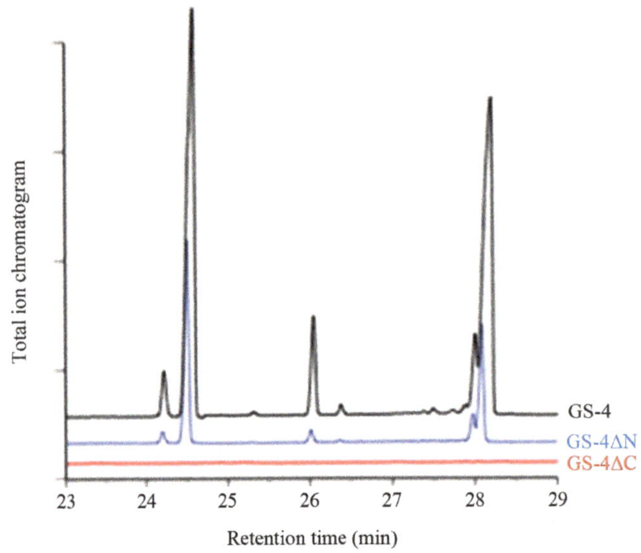

Figure 6. Total ion chromatograms of the reaction products liberated by recombinant δ-guaiene synthases; *GS*-4, *GS*-4*Δ*N and *GS*-4*Δ*C.

In contrast, Xu *et al.* have recently reported [6] that three homologous genes of δ-guaiene synthase were isolated from *A. sinensis*, and the translated proteins liberate δ-guaiene, α-guaiene and β-elemene but not α-humulene. Therefore, it is assumed that two classes of δ-guaiene synthase would occur in *Aquilaria* plants; δ-guaiene, α-guaiene and β-elemene producing group (*A. sinensis*), and δ-guaiene, α-guaiene and α-humulene producing group (*A. crassna*). We have previously reported [5] that *GS*-1, the translated product of a δ-guaiene synthase gene isolated from *A. microcarpa* cell culture, is capable of producing δ-guaiene, α-guaiene and β-elemene, however, the formation of α-humulene by *GS*-1 was not observed. In the present experiments, we have isolated three additional cDNA clones encoding the partial structures of δ-guaiene synthase (*GS-2ΔNΔC*, *GS-3ΔNΔC* and *GS-4ΔNΔC*) from *A. microcarpa* (**Figure 2**). We have also shown that the modified proteins, in which probable missing amino acid residues at N- and C-terminals were supplied (*GS*-2, *GS*-3 and *GS*-4), showed unique catalytic properties (**Figure 4** and **Figure 5**). The reaction products of *GS*-2 were α-guaiene, δ-guaiene and β-elemene, and liberation of α-humulene was not observed. Therefore, catalytic specificity of *GS*-2 was found to be similar to that of *GS*-1 isolated from *A. microcarpa* [5] and three δ-guaiene synthases from *A. sinensis* [6]. In contrast, *GS*-3 and *GS*-4 produced α- and δ-guaiene, β-elemene plus α-humulene, and, therefore, characteristics of the enzyme reaction of *GS*-3 and *GS*-4 should resemble the synthases from *A. crassna* [3]. These observations imply that two classes of δ-guaiene synthase, α-humulene producing and non-producing types, would concomitantly occur in *A. microcarpa* cells. In order to clarify the molecular bases of these unique specificities of δ-guaiene synthases in *Aquilaria* plants, preparation and characterization of several mutant proteins of the synthase are in progress in our laboratory in which the non-conserved amino acids presented in **Figure 2** are appropriately replaced.

Although recombinant *GS*-2, 3 and 4 proteins showed the biosynthetic activity of the sesquiterpene compounds, N- and C-terminals-deleted proteins, *GS-2ΔNΔC*, *GS-3ΔNΔC* and *GS-4ΔNΔC*, did not catalyze the terpenoids production (**Figure 4**). However, *GS*-4*Δ*N, deleted solely N-terminal end, clearly showed the enzymatic activity (**Figure 6**) while *GS*-4*Δ*C without C-terminal was found to lose the catalytic function. These observations strongly suggest that several amino acid residues at C-terminal end of *GS*-4 protein are essential to exhibit the terpene cyclase activity. We are attempting to construct the 3D model of the enzyme protein employing the suitable related protein(s) as the template for the elucidation of the biochemical functions of C-terminal amino acid sequence of δ-guaiene synthase.

Acknowledgements

This work was supported in part by Grant-in-Aid (No. 15K07990 to FK) from the Ministry of Education, Culture, Sports, Science and Technology of Japan, and by funding from Takahashi Industrial and Economic Research Foundation to FK.

References

[1] Ng, L.T., Chang, Y.S. and Kadir, A.A. (1997) A Review on Agar (gaharu) Producing *Aquilaria* Species. *Journal of Tropical Forest Products*, **2**, 272-285.

[2] Okudera, Y. and Ito, M. (2009) Production of Agarwood Fragrant Constituents in *Aquilaria* Calli and Suspension Cultures. *Plant Biotechnology*, **26**, 307-315. http://dx.doi.org/10.5511/plantbiotechnology.26.307

[3] Kumeta, Y. and Ito, M. (2010) Characterization of *δ*-Guaiene Synthases from Cultured Cells of *Aquilaria* Responsible for the Formation of the Sesquiterpenes in Agarwood. *Plant Physiology*, **154**, 1998-2007. http://dx.doi.org/10.1104/pp.110.161828

[4] Kraker, J.W., Franssen, M.C.R., Groot, A., Koenig, W.A. and Bouwmeester, H.J. (1998) (+)-Germacrene A Biosynthesis. The Committed Step in the Biosynthesis of Bitter Sesquiterpene Lactones in Chicory. *Plant Physiology*, **117**, 1381-1392. http://dx.doi.org/10.1104/pp.117.4.1381

[5] Lee, J.B., Hirohashi, S., Yamamura, Y., Taura, F. and Kurosaki, F. (2014) Induction, Cloning and Functional Expression of a Sesquiterpene Biosynthetic Enzyme, *δ*-Guaiene Synthase, of *Aquilaria microcarpa* Cell Cultures. *Natural Product Communications*, **9**, 1231-1236.

[6] Xu, Y., Zhang, Z., Wang, M., Wei, J., Chen, H., Gao, Z., Sui, C., Luo, H., Zhang, X., Yang, Y., Meng, H. and Li, W. (2013) Identification of Genes Related to Agarwood Formation: Transcriptome Analysis of Healthy and Wounded Tssues of *Aquilaria sinensis*. *BMC Genomics*, **14**, 227-242. http://dx.doi.org/10.1186/1471-2164-14-227

[7] Kenmotsu, Y., Ogita, S., Katoh, Y., Yamamura, Y., Takao, Y., Tatsuo, Y., Fujino, H., Kadota, S. and Kurosaki, F. (2011) Methyl Jasmonate-Induced Enhancement of Expression Activity of *Am-FaPS*-1, a Putative Farnesyl Diphosphate Synthase Gene from *Aquilaria microcarpa*. *Journal of Natural Medicines*, **65**, 194-197. http://dx.doi.org/10.1007/s11418-010-0451-4

[8] Murashige, T. and Skoog, F. (1962) A Revised Medium for Rapid Growth and Bioassays with Tobacco Tissue Culture. *Physiologia Plantarum*, **15**, 473-479. http://dx.doi.org/10.1111/j.1399-3054.1962.tb08052.x

[9] Kenmotsu, Y., Asano, K., Yamamura, Y. and Kurosaki, F. (2013) Cloning and Expression of Putative Rac/Rop GTPase Genes, *Am-rac1* and *Am-rac2*, Involved in Methyl Jasmonate-Induced Transcriptional Activation of Farnesyl Diphosphate Synthase in Cell Cultures of *Aquilaria microcarpa*. *Plant Molecular Biology Reporter*, **31**, 539-546. http://dx.doi.org/10.1007/s11105-012-0529-0

[10] Kurosaki, F. and Taura, F. (2015) Transcriptional Activation of Sesquiterpene Biosynthetic Enzyme *δ*-Guaiene Synthase Gene in Cell Cultures of *Aquiralia microcarpa* Overexpressing *cam*1 and *rac*2 Encoding Calmodulin and Rac GTPase. *Plant Gene*, **2**, 25-28. http://dx.doi.org/10.1016/j.plgene.2015.03.004

[11] Bradford, M.M. (1976) A Rapid and Sensitive Method for Quantitation of Microgram Quantities of Protein Utilizing the Principle of Protein-Dye Binding. *Analytical Biochemistry*, **72**, 248-254. http://dx.doi.org/10.1016/0003-2697(76)90527-3

[12] Laemmli, U.K. (1970) Cleavage of Structural Proteins during the Assembly of the Head of Bacteriophage T4. *Nature*, **227**, 680-685. http://dx.doi.org/10.1038/227680a0

[13] Babushok, V.I., Linstrom, P.J. and Zenkevich, I.G. (2011) Retention Indices for Frequently Reported Compounds of Plant Essential Oils. *Journal of Physical and Chemical Reference Data*, **40**, Article ID: 043101. http://dx.doi.org/10.1063/1.3653552

[14] Bos, R., Woerdenbag, H.J., Hendriks, H., Smit, H.F., Wikstroem, H.V. and Scheffer, J.J. (1997) Composition of the Essential Oil from Roots and Rhizomes of *Valeriana wallichii* DC. *Flavour and Fragrance Journal*, **12**, 123-131. http://dx.doi.org/10.1002/(SICI)1099-1026(199703)12:2<123::AID-FFJ613>3.0.CO;2-4

[15] Bohlmann, J., Meyer-Gauen, G. and Croteau, R. (1998) Plant Terpenoid Synthases: Molecular Biology and Phylogenetic Analysis. *Proceedings of the National Academy of Sciences of the United States of America*, **95**, 4126-4133. http://dx.doi.org/10.1073/pnas.95.8.4126

[16] Whittington, D.A., Wise, M.L., Urbansky, M., Coates, R.M., Croteau, R.B. and Christianson, D.W. (2002) Bornyl Diphosphate Synthase: Structure and Strategy for Carbocation Manipulation by a Terpenoid Cyclase. *Proceedings of the National Academy of Sciences of the United States of America*, **99**, 15375-15380. http://dx.doi.org/10.1073/pnas.232591099

12

Photosynthetic Light Utilization Efficiency, Water Relations and Leaf Growth of C_3 and CAM Tropical Orchids under Natural Conditions

Shawn Tay[1], Jie He[1*], Tim Wing Yam[2]

[1]Natural Sciences and Science Education Academic Group, National Institute of Education, Nanyang Technological University, Singapore
[2]Singapore Botanic Gardens, National Parks Board, Singapore
Email: *jie.he@nie.edu.sg

Abstract

Native orchid species of Singapore in their natural conditions experience stress from high irradiance, high temperatures and periods of extended low rainfall, which impact orchid plant physiology and lead to reduced growth and productivity. In this study, it was found that there was a reduction in photochemical efficiency of photosystem II (PSII) in 6 native orchid species under high light (HL) and *Bulbophyllum membranaceum* under low light (LL). There was chronic photoinhibition in these 6 orchid species over a period of 3 months after transplanting onto the tree trunks without watering and fertilization, especially in *Coelogynes mayeriana* and *Bulbophyllum membranaceum* under both HL and LL. This chronic photoinhibition caused by sustained period of water deficit in their natural conditions was later reversed by natural re-watering conditions from higher rainfall. These results indicate that water deficit has a greater impact on photosynthetic light utilization efficiency than excess light. The present study also showed that after natural re-watering, relative water content (RWC) of leaves and pseudobulbs generally increased. During the natural re-watering, total leaf area also gradually increased and reached maximum expansion after 7 weeks under both HL and LL, with some exceptions due to leaf abscission or decline in total leaf area, possibly a strategy for water conservation.

Keywords

Orchid, CAM, C_3, Photosynthetic Light Utilization Efficiency, Water Relations and Leaf Growth

*Corresponding author.

1. Introduction

Singapore has a tropical rainforest climate with no distinctive seasons, temperatures ranging from 22°C to 35°C, high relative humidity averages ranging from 60% to 100%, and annual rainfall about 2300 mm, with November to January being the wettest months. Considerable efforts have been made in the past decade to re-introduce native orchid species back into Singapore as part of conservation [1]-[3]. Many of these re-introduced species are epiphytes adapted to relatively dry habitats, native to low-elevation forests, and are dependent upon their microclimate. Crucial to these epiphytic orchids is the requirement for water economy where high evapotranspiration combined with scarce water supply challenges plant water balance [4]. Epiphytic orchids grow well under optimal environmental conditions and propagate naturally. However, when they experience stress from high irradiance, high temperatures and periods of extended low rainfall, these factors significantly impact their physiology and lead to reduced growth and productivity.

High irradiance and high temperature cause a reduction in photosynthetic quantum yield, loss in leaf chlorophyll, and a greater susceptibility to photoinhibition in orchids [5]-[7]. The reduction in photosynthesis can be attributed to dynamic photoinhibition, photooxidation, photobleaching of chlorophyll and chloroplast damage leading to deletrious effects of the leaf photosynthetic apparatus [7]-[9]. Furthermore, orchid leaves exposed to drastic increase in photon flux densities experienced a drop in CO_2 gas exchange and leaf chlorosis [10].

Epiphytic orchids experiencing stress from water deficit show significant decrease in relative water content (RWC) in leaves and pseudobulbs, reduction in photosynthetic light utilization, and greater susceptibility to photoinhibition caused by high irradiance [11]. In other plants, periodic drought stress also leads to a closure of stomata [12], through the reduction of shoot water potential directly or reduced turgor [13]. As a result, CO_2 depletes in the intercellular spaces and at the chloroplast level, limiting photosynthesis. Plants grown under low irradiance may also experience a reduction in the photosynthetic rate, resulting in a decline in growth rate [5] [14]. Therefore it is necessary to strike a good balance in maximising photon energy utilization of leaves to increase photosynthetic activity whilst minimising photoinhibition [6].

Orchids are also able to tolerate and acclimatise to stressful conditions, dynamically achieved by the physiological and morphological adaptations. In some orchid species, carbon fixation through Crassulacean acid metabolism [15] provides a means of water conservation during times of water deficit, while preventing loss of carbon [16]-[19]. Thick-leaved orchids, such as *Bulbophyllum vaginatum*, *Dendrobium leonis* and *Phalaenopsis cornu-cervi*, have features that are characteristic of plants that fix carbon primarily through Crassulacean acid metoblism [15], while thin-leaved orchids such as *Coelogynes rochussenii*, *Coelogynes mayeriana* and *Bulbophyllum membranaceum* are plants that fix carbon primarily through C_3 pathway [20]-[22].

There is little information about the photosynthetic light utilization efficiency and water relations of tropical orchid species in their natural environments, especially with regards to their responses to varying environmental conditions and the stress that comes with it. The information would be useful for improving the approach towards conservation of orchid species in their natural environments. Therefore, to complement the conservation efforts in Singapore, this study aims to better understand the photosynthetic light utilization efficiency, water relations and leaf growth of six selected C3 and CAM native tropical orchids under natural conditions. The results benefit sustainable conservation of these orchid species under natural conditions.

2. Materials and Methods

2.1. Plant Cultivation under Natural Conditions

Mature plants of *Coelogynes rochussenii*, *Coelogynes mayeriana*, *Bulbophyllum vaginatum*, *Bulbophyllum membranaceum*, *Dendrobium leonis* and *Phalaenopsis cornu-cervi* planted by the National Parks Board (NParks) Singapore in National Institute of Eduction (NIE), Singapore, were used in this study. These epiphytic orchid plants were planted on the trunks of rain trees, *Albizia saman* (syn. *Samanea saman*), under natural high light (HL) and low light (LL) conditions on 12 January 2015. LL was achieved by natural shading provided by the trees and positioning the plants out of direct sunlight. These orchids were neither watered nor fertilized after planting, and depended upon rainfall as the only source of water. The maximal PPFD ranged from 90 to 200 $\mu mol \cdot m^{-2} \cdot s^{-1}$ for LL and from 400 to 1300 $\mu mol \cdot m^{-2} \cdot s^{-1}$ for HL, with an ambient day temperature of about 30°C to 35°C during the photoperiod. 6 plants of each species for each light condition were planted on the trees.

2.2. Measurement of Photosynthetic Photon Flux Density (PPFD)

PPFD was measured using a photosynthetically available radiation quantum sensor and reading unit (Skye Instruments Ltd, Llandrindod, UK). The stabilized PPFD, within a range of 0 to 1999 $\mu mol \cdot m^{-2} \cdot s^{-1}$, was measured from 6 different positions above the leaves for HL and LL respectively, and an average was calculated from the 6 values measured.

2.3. Measurement of Chl Fluorescence F_v/F_m Ratio

Diurnal changes of F_v/F_m ratios were taken every two hours from 0800 to 1600 h with the Plant Efficiency Analyser, PEA (Hansatech Instruments Ltd, England) on sunny and cloudy days, while weekly F_v/F_m ratios were taken at the time corresponding to the maximal midday PPFD. 6 samples were taken from each species under each light condition. The leaves were pre-darkened with clips for 15 min prior to measurements. Dark adapted leaves were placed under the light pipe to measure F_o, initial chlorophyll (Chl) fluorescence during the very early onset of illumination with excitation light. F_m, maximum Chl fluorescence was determined by 0.8 seconds of saturated pulse (>6000 $\mu mol \cdot m^{-2} \cdot s^{-1}$). The variable fluorescence yield F_v, was determined by $F_m - F_o$. The efficiency of excitation energy captured by open PSII reaction centres in dark adapted leaves was estimated by the fluorescence F_v/F_m ratio.

2.4. Measurement of Chl Reading by Hand Held Chlorophyll Meter (SPAD-502)

Leaf Chl content was read by using a hand-held chlorophyll meter (SPAD-502, Minolta, Japan), which measures the absorbances of the leaf at wavelengths 645 nm and 663 nm. Using these two absorbances, the meter calculates a numerical SPAD value which is proportional to the amount of Chl present in the leaf, and expresses in terms of SPAD units. 6 samples were taken from each species under each light condition.

2.5. Measurement of Midday RWC

Square cuts (1 cm by 1 cm) were made out of the leaves and 5 mm thick slices were made out of the pseudobulbs. These sample were first weighed with an analytical balance to determine fresh weight (FW), followed by immersion in water in the dark for 3 hours (for pseudobulb of C. mayeriana and leaf of *B. membranaceum*) or 24 hours (for the rest of the samples) prior to measurement of their saturated weight (SW). The samples were then dried in the oven at 80˚C for at least 72 hours to obtain their dry weight. RWC was expressed as RWC = (FW − DW)/(SW − DW) × 100%. 10 samples each for leaves and for pseudobulbs were obtained from each species.

2.6. Measurement of Total Leaf Area

Young, budding leaves from 6 plants of each orchid species, cultivated under the respective light conditions (HL or LL), were selected and tracked over 16 weeks. The individual leaf area was measured every week by capturing the image of selected leaves against a measured scale and then using Image J software to calculate the leaf area from the image according the method outlined by Schneider *et al.* [23]. The total leaf area was calculated as a sum of the individual leaf areas. Total leaf area is one of the parameters for leaf growth.

2.7. Statistical Analysis

One-way ANOVA was used to test for significant differences among different growth irradiances, using Tukey's multiple comparison tests to discriminate the means (IBM SPSS Statistics for Macintosh, Version 22.0, 2013).

3. Results

3.1. Photosynthetic Light Utilization Efficiency

Diurnal PPFD was much higher for the plants grown under HL than LL on a sunny day at 1000 h, 1200 h and 1400 h (**Figure 1(a)** and **Figure 1(b)**) and the diurnal F_v/F_m ratio at these timings were also correspondingly lower in HL than LL (**Figure 1(c)** and **Figure 1(d)**), with the biggest difference in *B. vaginatum* and *B. membranaceum*. This suggests a reduction in the efficiency of open PSII reaction centres of the orchids under exposure

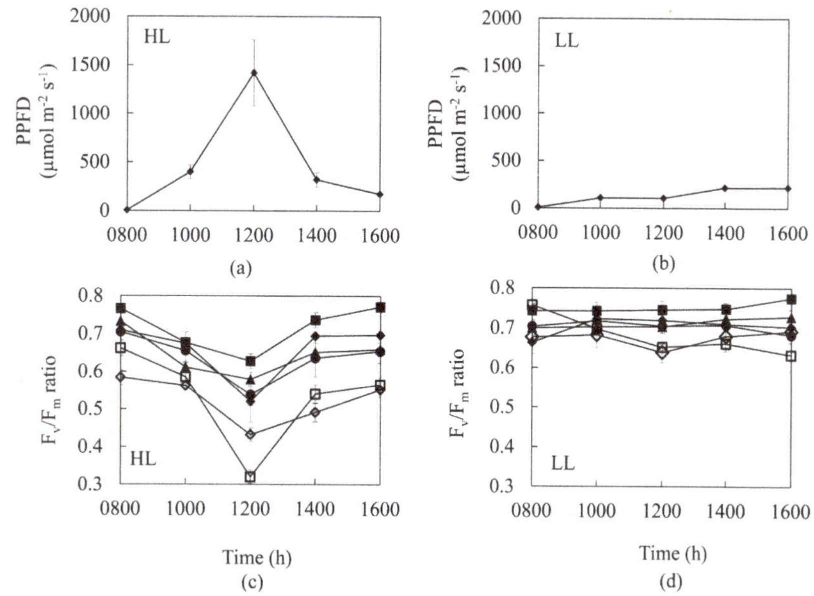

Figure 1. Diurnal changes of PPFD and F_v/F_m ratio on sunny days for *C. rochussenii* (▲), *C. mayeriana* (■), *B. vaginatum* (◇), *B. membranaceum* (□), *D. leonis* (●) and *P. cornu-cervi* (◆) grown under HL and LL.

to HL (PPFDs of 400 to 1300 μmol·m^{-2}·s^{-1}) towards midday, particularly more severe in *B. vaginatum* and *B. membranaceum*, among the 6 orchid species examined. A reduction in the efficiency of open PSII reaction centres was also shown in *B. membranaceum* under exposure to LL (with a maximal PPFD of 200 μmol·m^{-2}·s^{-1}) during midday. F_v/F_m ratio at 1600 h of all plants grown under HL except *C. mayeriana*, and *B. membranaceum* grown under LL did not recover to the same levels as at 0800 h, suggesting that these plants were suffering from chronic photoinhibition.

However, there was evidence of recovery from this chronic photoinhibition as shown in the difference in F_v/F_m ratios at midday, between 1 and 3 months of growing under HL and LL. Significant increase in the midday F_v/F_m ratios from 1 to 3 months of cultivation were observed in all species under both HL (**Figure 2(a)**) and LL (**Figure 2(b)**) except for *C. mayeriana* under LL (One-way ANOVA, $p < 0.05$, $n = 6$). Between species, *B. vaginatum* and *B. membranaceum* after 1 month under HL show significantly lower F_v/F_m ratios as compared with the other species, but recovered to levels comparable to *D. leonis* and *P. cornu-cervi* after 3 months under HL (Tukey's multiple comparison test, $p < 0.05$, $n = 6$). No significant difference in F_v/F_m ratio was found between species after 1 month and 3 months under LL. Further evidence of recovery from this chronic photoinhibition was provided in the data collected over 16 weeks for the changes to midday maximal PPFD and F_v/F_m ratio. Over the 16 weeks of cultivation under HL, midday maximal PPFD exceeded 800 μmol·m^{-2}·s^{-1} only in weeks 2 to 7 and week 13 (**Figure 3(a)**) while little fluctuation in midday PPFD (<200 μmol·m^{-2}·s^{-1}) was seen in LL (**Figure 3(b)**). Based on rainfall data of the micro-climate of the area where the orchids were cultivated, weeks 2 to 7 coincided with a period of low rainfall (**Figure 3(a)**), while the drop in midday maximal PPFD (< 800 μmol·m^{-2}·s^{-1}) from week 8 onwards, with the exception of week 13, coincided with an increase in weekly total rainfall. Corresponding with the fluctuations in midday maximal PPFD under HL and the weekly total rainfall, the F_v/F_m ratio decreased from weeks 2 to 8 for all 6 orchid species, under both HL (**Figure 3(c)** and **Figure 3(e)**) and LL (**Figure 3(d)** and **Figure 3(f)**), and started increasing back to original levels or higher thereafter. The decrease in F_v/F_m ratio was greater in HL than LL, for *B. vaginatum*, *B. membranaceum* and *P. cornu-cervi*, *C. rochussenii*, *C. mayeriana* and *D. leonis*. Therefore, the period of higher weekly total rainfall from week 8 onwards provided re-watering conditions, and this corresponded with plants showing recovery from chronic photoinhibition, where the F_v/F_m ratio that was decreasing below 0.700 in the first 8 weeks started to increase back to levels above 0.700 after week 8.

In contrast to the changes in F_v/F_m ratio over the 16 weeks of cultivation, difference between Chl content under HL and LL were lesser, with the exception of *B. membranaceum* which showed a higher Chl content under

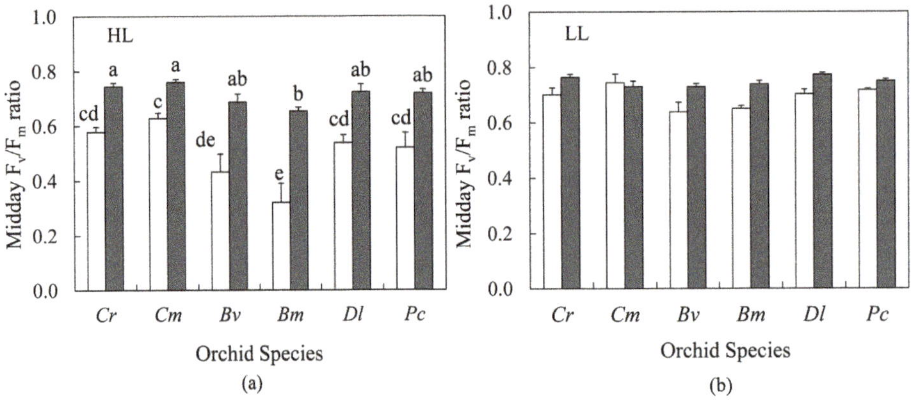

Figure 2. F_v/F_m ratio at midday maximal PPFD after 1 month (open bars) and 3 months (solid bars) of cultivation under HL and LL (*Cr*: *C. rochussenii*, *Cm*: *C. mayeriana*, *Bv*: *B. vaginatum*, *Bm*: *B. membranaceum*, *Dl*: *D. leonis*, *Pc*: *P. cornu-cervi*). Significant difference observed between the F_v/F_m ratio after 1 and 3 months, under HL and LL, in all species except for *Cm* under LL (One-way ANOVA, $p < 0.05$, $n = 6$). Different letters above bars indicate significant difference in F_v/F_m ratio between species after 1 month ((a), (b)) and 3 months ((c), (d), (e)) under HL (Tukey's multiple comparison, $p < 0.05$, $n = 6$). No significant difference found between species after 1 month and 3 months under LL.

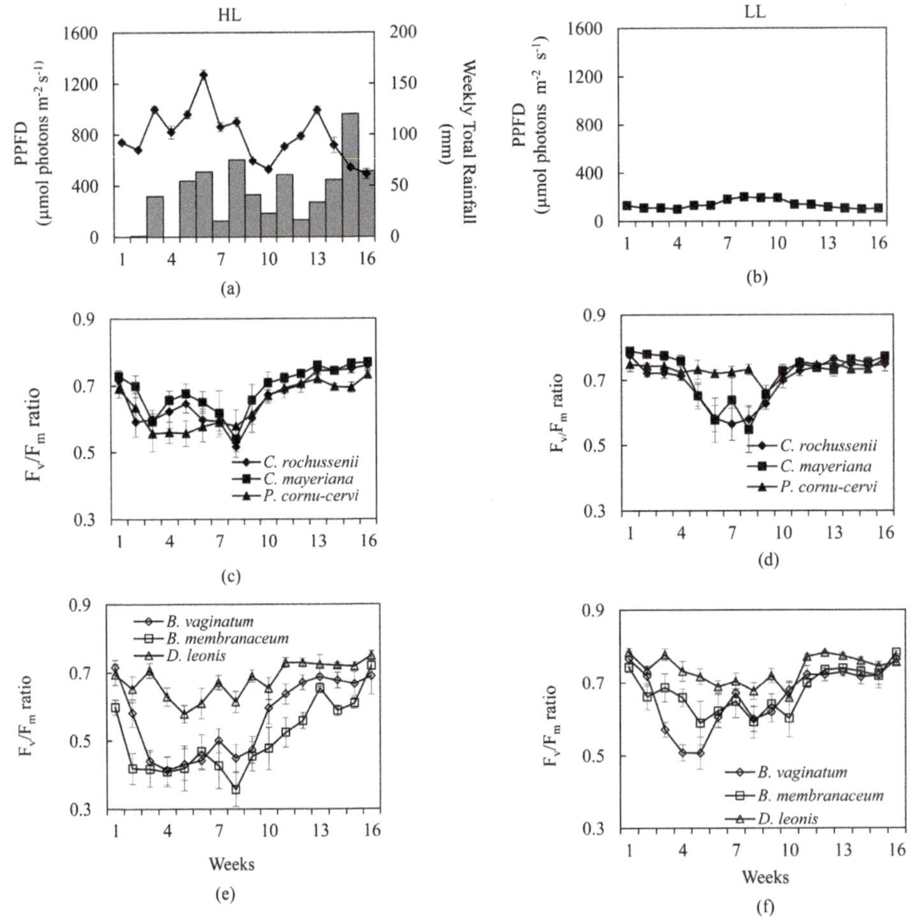

Figure 3. Changes of midday PPFD and F_v/F_m ratios of different orchid species over 16 weeks under HL and LL. Rainfall data adapted with permission from Meteorological Services Singapore (Historical Daily Records, 2015).

LL (**Figure 4(d)**) than HL (**Figure 4(c)**). Chl content fluctuateed little between 20 to 40 SPAD units for *C. rochussenii*, *C. mayeriana* and *P. cornu-cervi* (**Figure 4(a)** and **Figure 4(b)**), whereas Chl content showed greater fluctuations between 30 to 70 SPAD units for *B. vaginatum* and *D. leonis* (**Figure 4(c)** and **Figure 4(d)**).

3.2. Water Relations

To further understand the significance of the natural re-watering conditions provided by the higher weekly total rainfall from week 8 onwards, the water status of the plants was examined. RWC of leaves (**Figure 5(a)** and **Figure 5(b)**) and pseudobulbs (**Figure 5(c)** and **Figure 5(d)**) increased significantly in all 6 orchid species, between 1 to 4 months of cultivation under HL and LL, with the exception of the pseudobulbs of *B. vaginatum* and *B. membranaceum* under HL (One-way ANOVA, $p < 0.05$, $n = 10$). Between species, leaves of *B. vaginatum, B. membranaceum, D. leonis and P. cornu-cervi* after 1 month under HL and LL had significantly lower RWC compared to C. *rochussenii* and *C. mayeriana* (Tukey's multiple comparison test, $p < 0.05$, $n = 10$). However, after 4 months, RWC in the leaves of *P. cornu-cervi* under HL and LL and *B. membranaceum* under LL increased back to levels comparable with *C. rochussenii* and *C. mayeriana*. After 1 and 4 months under HL and LL, pseudobulbs of *C. rochussenii*, *B. vaginatum* and *B. membranaceum* remained significantly lower than that of *C. mayeriana*.

3.3. Leaf Growth

The difference in leaf growth under HL and LL was also examined through the changes in total leaf area for all 6 species. Total leaf area generally increased for the first 7 weeks for all 6 orchid species, with the exception of *D. leonis* cultivated under HL (**Figure 6(e)**), where there was a sharp drop in total leaf area during weeks 5 to 7. The sharp drop was due to a loss of the selected leaves, and this could be further attributed to the sustained low rainfall period coupled with HL from weeks 2 to 9. Total leaf area continued to be sustained at around the same level from week 7 onwards for the rest of the plants, with the exception of *C. rochussenii* (**Figure 6(a)**) and *B. membranaceum* (**Figure 6(d)**) cultivated under LL, and *C. mayeriana* (**Figure 6(b)**) and *P. cornu-cervi* (**Figure 6(f)**) cultivated under HL. For the plants with total leaf area remaining unchanged from week 7 onwards, it could be attributed to the leaves reaching their maximum expanded area.

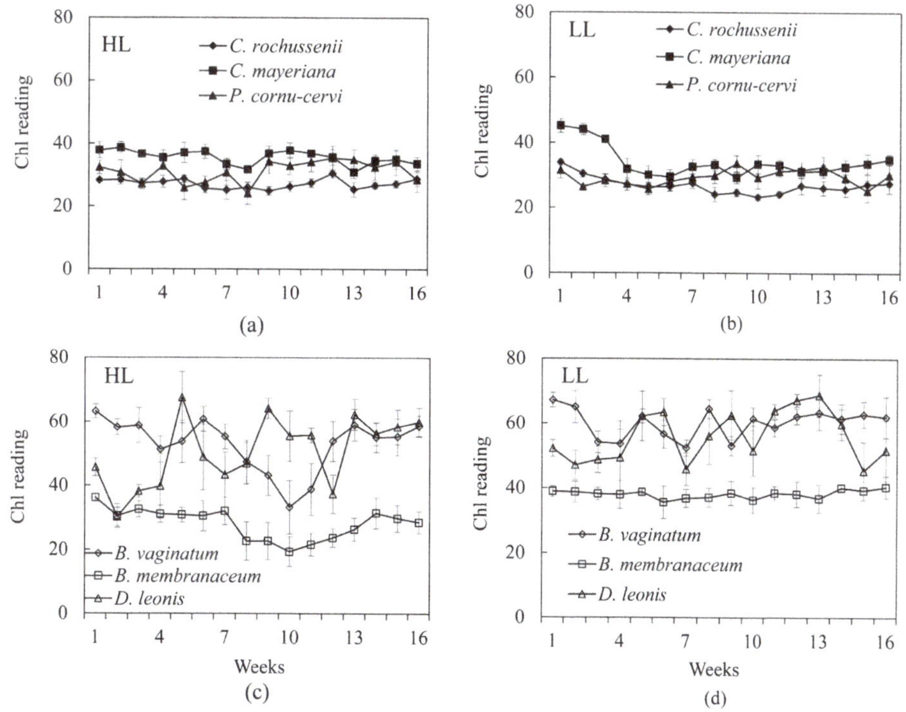

Figure 4. Changes of Chl reading of different orchid species over 16 weeks under HL and LL.

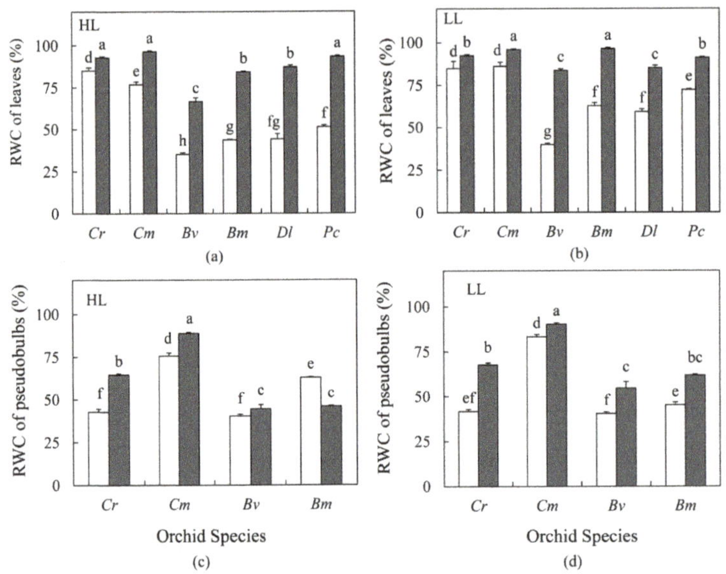

Figure 5. RWC of leaves and pseudobulbs after 1 month (open bars) and 4 months (solid bars) of cultivation under HL and LL (*Cr*: *C. rochussenii*, *Cm*: *C. mayeriana*, *Bv*: *B. vaginatum*, *Bm*: *B. membranaceum*, *Dl*: *D. leonis*, *Pc*: *P. cornu-cervi*). Significant difference was observed between 1 and 4 months under HL and LL, in all species except for *Bv* pseudobulb under HL (One-way ANOVA, $p < 0.05$, $n = 10$). Different letters above bars indicate significant difference in RWC between species after 1 and 4 months under HL ((a), (b), (c)) and LL ((d), (e), (f)) (Tukey's multiple comparison, $p < 0.05$, $n = 10$).

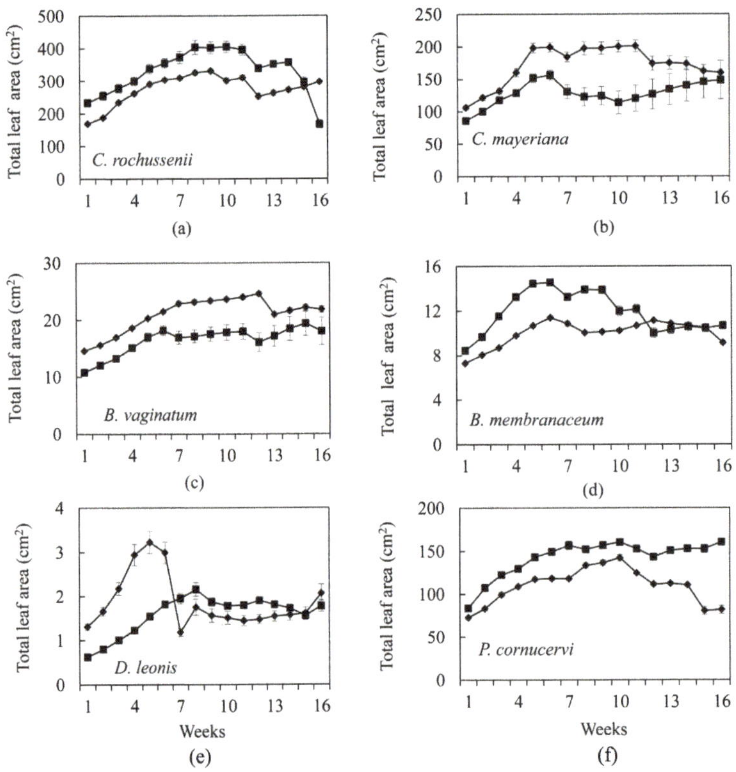

Figure 6. Total leaf area of different orchid species over 16 weeks under HL (◆) and LL (■).

4. Discussion

Some 70% of the native orchid species of Singapore are mostly vascular epiphytes [1]. Vascular epiphytes are dependent upon their micro-climate in photosynthetic light utilization and water economy [4]. In a review by Ort [24], it was reported that light intensities often exceed photosynthetic capacity in plants, while stomatal conductance limits CO_2 uptake into leaves, rendering even moderate irradiances in excess of photosynthetic capacity. Plants grown under high irradiance may lead to photoinhibition of the leaves, and this will result in a reduction in growth and productivity. Both C_3 and CAM orchids have been reported to show stress from exposure to high irradiances under natural conditions, such as decrease in photosynthetic light utilization, photoinhibition, photooxidative damage, and leaf chlorosis [5] [10] [11] [25]-[29]. Our present study reports a reduction in the efficiency of open PSII reaction centres of all the 6 orchid species under exposure to HL and C_3 orchid *B. membranaceum* under exposure to LL, towards midday (**Figure 1**), indicating photoinhibition. This photoinhibition is particularly more severe in *B. vaginatum* and *B. membranaceum* grown under HL. All plants under HL were also not able to fully recover from the photoinhibition over the first month (**Figure 2(a)**, open bars), indicating chronic photoinhibition.

However, photoinhibition of PSII *in vivo* provides photoprotection and is an acclimation strategy for plants under stress from excessive irradiances [30]. Orchids have protective stategies to overcome the stress from high irradiances and to help balance maximising photon energy utilization while minimising photoinhibition [6]. Previous studies reported higher tolerance to high PPFD and faster recovery from photoinbition in C_3 *Oncidium* orchid as compared to CAM *Phalaenopsis* orchid [29], greater photoprotection in leaves of C_3 orchid *C. rochussenii* under HL compared to LL [28], and recovery from photoinhibition in C_3 orchid *Gramatophyllum speciosum* in HL [27]. The results of our study seem to contradict these cases because the diurnal F_v/F_m ratios decline sharply towards 1200 h in all the orchids exposed to HL (**Figure 1(c)**) as compared to LL (**Figure 1(d)**). All 6 species also showed chronic photoinhibition after 1 month under HL, but recovered after 3 months (**Figure 2**). However, in all previous studies, water deficit was not severe and this could be the reason why the C_3 orchids could better utilize photosynthetic light energy as compared to the CAM orchids.

Ort [24] argued that at high irradiances, water status take physiological precedence over maximizing photosynthesis. It has been reported that in thin-leaved, C_3 epiphytic orchid *Catasetum viridiflavum*, and other epiphytes, re-watered plants show recovery from photoinhibition despite high PPFD, indicating that water deficit has a greater impact on photoinhibition than high light [31] [32]. Furthermore, strong photoinhibition and reduced photosynthetic capacity was observed in epiphytic CAM orchid hybrid, *Cattleya forbesii* Lindl. × *Laelia tenebrosa* Rolfe., following sustained water deficit for 42 days [11]. It has also been reported that under water deficit stress, *C. rochussenii* under HL had a reduced photosynthetic light utilization than plants under LL, and water deficit stress took precedence over light stress in reducing the Chl reading of *C. rochussenii*. [28]. Drought stress coupled with high light occurs when water supply is insufficient to sustain plant growth [27] [33]. Our present study showed that a sustained period of water deficit caused a decline in photosynthetic light utilization in all 6 orchid species, followed by a period of natural re-watering conditions that enabled recovery from this chronic photoinhibition (**Figure 3**). The changes in photosynthetic light utilization were less likely to be attributed to changes in Chl content since there was little difference in Chl content between HL and LL (**Figure 4**), with the exception of *B. membranaceum* which showed a higher Chl content under LL than HL. Therefore, the results suggest that water deficit has a greater impact on photosynthetic light utilization efficiency than HL, and that *B. vaginatum* and *B. membranaceum* thrive better in LL, well-watered conditions.

Water deficit that drives closure of stomata result in reduction in CO_2 uptake and subsequent down-regulation of photosynthetic efficiency [34], which is alleviated in epiphytic orchids by carbon fixation through CAM, while achieving water economy [4] [35] [36]. Epiphytic orchids have also been reported to lose water from the leaf slowly, yet maintain high values of RWC [37]. However, another study reported a reduction in RWC and osmotic potential during a period of water shortage in some tropical epiphytes [38] [39]. Moreover, studies had shown that pseudobulbs helped to reduce the loss of leaf water content during periods of water deficit stress by storing water in the pseudobulbs and then supplying this water to the leaves, thereby providing orchids with the ability to survive [11] [40]-[43]. A previous study had also reported that *C. rochussenii* grown under LL had a significantly higher leaf RWC than those under HL [28]. Our present study showed that after a period of natural re-watering, RWC of leaves and pseudobulbs increased significantly in all 6 orchid species, with the exception of the pseudobulbs of *B. vaginatum* and *B. membranaceum* under HL and *C. mayeriana* under LL (**Figure 5**).

Large pseudobulbs in C_3 orchids *C. rochussenii* and *C. mayeriana* could have stored water and helped maintain leaf water content so as to alleviate photoinhibition. The small size and thin leaves of *B. membranaceum* could have disadvantaged it in its ability to store water in the pseudobulbs to overcome periods of water deficit and for the leaves to overcome the stress from high irradiance, which explains the severity of photoinhibition. On the other hand, in both *D. leonis* and *P. cornu-cervi*, having fleshy leaves and the ability to fix carbon through CAM could be significant strategies in response to photoinhibition. Therefore, both of these species are more adaptable to both HL and LL conditions, and could have a wider range of tolerable irradiance.

Water deficits lead to reduction in leaf area, leaf abscission, and dieback of individual plant parts. However, reduction in leaf area and leaf abscission strongly reduced water loss, and epiphytic orchids that encounter shortages of water may stop growing when water is in short supply, only to resume growth when water supply is restored [31] [44]. Our study showed a general increase in total leaf area for the 6 species for the first 7 weeks under HL and LL and maintained onwards as they had reached maximum leaf expansion. However, the drop in total leaf area in some species, such as *D. leonis*, was due to leaf abscission and decline in total leaf area. This could be a strategy for water conservation through reducing the total number of leaves, and therefore number of stomata, as well as reducing the surface area, leading to lesser water loss by evapotranspiration.

5. Conclusion

The results suggest that water deficit has a greater impact on photosynthetic light utilization efficiency than excess light. Over a period of water deficit, 6 orchid species experienced a decrease in F_v/F_m ratio and RWC of leaves and pseudobulbs, which could both be reversed by natural re-watering conditions from higher rainfall. Over 16 weeks, total leaf area increased and leaves reached maximum expansion after 7 weeks under both HL and LL, with some exceptions due to leaf abscission or decline in total leaf area, possibly a strategy for water conservation. Future work would be required to fully understand the adaptation strategies in response to stress from HL and water deficit that may occur in the natural environment. This will aid in improving the approach of re-introduction of native orchid species in Singapore.

Acknowledgements

This project was carried out in National Institute of Education (NIE), Nanyang Technologial University (NTU), Singapore. We would like to thank NIE, NTU for providing the space for orchid cultivation and also the facilities required for this study, as well as the Singapore Botanic Gardens and the National Parks Board for providing the plant materials. We would like to also thank Miss Choong Tsui Wei for her valuable inputs and assistance in optimizing the methods of this study and analyzing the results.

References

[1] Yam, T.W. (2013) Native Orchids of Singapore: Diversity, Identification and Conservation. National Parks Board, Singapore.

[2] Yam, T.W., *et al.* (2011) Conservation and Reintroduction of Native Orchids of Singapore—The Next Phase. *European Journal of Environmental Sciences*, **1**, 38-47.

[3] Yam, T.W. and Thame, A. (2005) Conservation and Reintroduction of the Native Orchids of Singapore. *Selbyana*, **26**, 75-80.

[4] Benzing, D. (1998) Vulnerabilities of Tropical Forests to Climate Change: The Significance of Resident Epiphytes. *Climatic Change*, **39**, 519-540. http://dx.doi.org/10.1023/A:1005312307709

[5] He, J., Khoo, G.H. and Hew, C.S. (1998) Susceptibility of CAM Dendrobium Leaves and Flowers to High Light and High Temperature under Natural Tropical Conditions. *Environmental and Experimental Botany*, **40**, 255-264. http://dx.doi.org/10.1016/S0098-8472(98)00042-2

[6] He, J. and Teo, L.C.D. (2007) Susceptibility of Green Leaves and Green Flower Petals of CAM Orchid Dendrobium cv. Burana Jade to High Irradiance under Natural Tropical Conditions. *Photosynthetica*, **45**, 214-221. http://dx.doi.org/10.1007/s11099-007-0035-z

[7] Lichtenthaler, H.K., Buschmann, C., Döll, M., Fietz, H.-J., Bach, T., Kozel, U., *et al.* (1981) Photosynthetic Activity, Chloroplast Ultrastructure, and Leaf Characteristics of High-Light and Low-Light Plants and of Sun and Shade Leaves. *Photosynthesis Research*, **2**, 115-141. http://dx.doi.org/10.1007/BF00028752

[8] Chow, W.S. (1994) Photoprotection and Photoinhibitory Damage. In: Bittar, E.E. and Barber, J. Eds., *Advances in*

Molecular and Cell Biology, Elsevier, London, 151-196. http://dx.doi.org/10.1016/s1569-2558(08)60397-5

[9] Osmond, C.B. (1994) What Is Photoinhibition? Some Insights from Comparisons of Shade and Sun Plants. Bios Scientific Publication, Oxford.

[10] Johnson, S.R. (1993) Photosynthesis and Aspects of Phenology of the Rapidly Dispersing Orchid *Oeceoclades maculata*. Lindleyana, **8**, 69-72.

[11] Stancato, G.C., Mazzafera, P. and Buckeridge, M.S. (2001) Effect of a Drought Period on the Mobilisation of Non-Structural Carbohydrates, Photosynthetic Efficiency and Water Status in an Epiphytic Orchid. *Plant Physiology and Biochemistry*, **39**, 1009-1016. http://dx.doi.org/10.1016/s0981-9428(01)01321-3

[12] Pinheiro, C. and Chaves, M.M. (2011) Photosynthesis and Drought: Can We Make Metabolic Connections from Available Data? *Journal of Experimental Botany*, **62**, 869-882.

[13] Turner, N.C. and Jones, M.M. (1980) Turgor Maintenance by Osmotic Adjustment: A Review and Evaluation. In: Turner, N.C. and Kramer, P.J., Eds., *Adaptation of Plants to Water and High Temperature Stress*, Willey & Sons, New York, 87-103.

[14] Lüttge, U., Ball, E., Fetene, M. and Medina, E. (1991) Flexibility of Crassulacean Acid Metabolism in Kalanchoe Pinnata (Lam.) Pers. I. Response to Irradiance and Supply of Nitrogen and Water. *Journal of Plant Physiology*, **137**, 259-267. http://dx.doi.org/10.1016/S0176-1617(11)80129-X

[15] Cameron, K.M., Chase, M.W., Whitten, W.M., Kores, P.J., Jarrell, D.C., Albert, V.A., *et al.* (1999) A Phylogenetic Analysis of the Orchidaceae: Evidence from rbcL Nucleotide Sequences. *American Journal of Botany*, **86**, 208-224. http://dx.doi.org/10.2307/2656938

[16] Adams III, W.W. and Osmond, C.B. (1988) Internal CO_2 Supply during Photosynthesis of Sun and Shade Grown CAM Plants in Relation to Photo Inhibition. *Plant Physiology*, **86**, 117-123. http://dx.doi.org/10.1104/pp.86.1.117

[17] Benzing, D.H. (1989) The Evolution of Epiphytism. In: Lüttge, U., Ed., *Vascular Plants as Epiphytes: Evolution and Ecophysiology*, Springer-Verlag, Berlin, 15-41. http://dx.doi.org/10.1007/978-3-642-74465-5_2

[18] Silvera, K., Santiago, L.S., Cushman, J.C. and Winter, K. (2009) Crassulacean Acid Metabolism and Epiphytism Linked to Adaptive Radiations in the Orchidaceae. *Plant Physiology*, **149**, 1838-1847. http://dx.doi.org/10.1104/pp.108.132555

[19] Silvera, K., Santiago, L.S., Cushman, J.C. and Winter, K. (2010) The Incidence of Crassulacean Acid Metabolism in Orchidaceae Derived from Carbon Isotope Ratios: A Checklist of the Flora of Panama and Costa Rica. *Botanical Journal of the Linnean Society*, **163**, 194-222. http://dx.doi.org/10.1111/j.1095-8339.2010.01058.x

[20] Arditti, J. and Woolhuse, W.H. (1980) Aspects of the Physiology of Orchids. In: Woolhouse, H., Ed., *Advances in Botanical Research*, Academic Press, London, 421-655. http://dx.doi.org/10.1016/s0065-2296(08)60091-9

[21] Hew, C.S. and Yong, J.W.H. (2004) The Physiology of Tropical Orchids in Relation to the Industry. World Scientific Publishing Co. Pte. Ltd., Singapore.

[22] Neales, T.F. and Hew, C.S. (1975) Two Types of Carbon Fixation in Tropical Orchids. *Planta*, **123**, 303-306. http://dx.doi.org/10.1007/BF00390710

[23] Schneider, C.A., Rasband, W.S. and Eliceiri, K.W. (2012) NIH Image to Image J: 25 Years of Image Analysis. *Nature Methods*, **9**, 671-675. http://dx.doi.org/10.1038/nmeth.2089

[24] Ort, D.R. (2001) When There Is Too Much Light. *Plant Physiology*, **125**, 29-32. http://dx.doi.org/10.1104/pp.125.1.29

[25] Ding, T.H., Ong, H.T. and Yong, H.C. (1980) Factors Affecting Flower Development and Production of Golden Shower (*Oncidium goldiana*). *Proceedings of the 3rd ASEAN Orchid Congress*, Kuala Lumper, 22-26 August, 65-78.

[26] He, J., Norhafis, H. and Qin, L. (2013) Responses of Green Leaves and Green Pseudobulbs of CAM Orchid *Cattleya laeliocattleya* Aloha Case to Drought Stress. *Journal of Botany*, **2013**, 1-9.

[27] He, J., Yong, T.Z. and Yam, T.W. (2014) Orchid Conservation in Singapore under Natural Conditions: Responses of *Grammatophyllum speciosum* to Growth Irradiances. *Plant Science International*, **1**, 11-23. http://dx.doi.org/10.12735/psi.v1n1p11

[28] Tan, S.H.G. (2012) Photosynthetic Characteristics of Green Leaves and Green Pseudobulbs of *Coelogyne rochussenii* Orchid and Their Responses to Different Growth Irradiance and Drought Stress. National Institute of Education, Nanyang Technological University, Singapore.

[29] Wang, S.Z. (2007) Effects of Growth Irradiance and Virus Infection on the Physiology of C3 Oncidium and CAM Phalaenopsis Orchids under Natural Tropical Conditions. National Institute of Education, Nanyang Technological University, Singapore.

[30] Anderson, J.M., Park, Y.I. and Chow, W.S. (1997) Photoinactivation and Photoprotection of Photosystem II in Nature. *Physiologia Plantarum*, **100**, 214-223. http://dx.doi.org/10.1111/j.1399-3054.1997.tb04777.x

[31] Zotz, G. and Tyree, M. (1996) Water Stress in the Epiphytic Orchid, *Dimerandra emarginata* (G. Meyer) Hoehne. *Oecologia*, **107**, 151-159. http://dx.doi.org/10.1007/BF00327898

[32] Zotz, G. and Winter, K. (1994) Annual Carbon Balance and Nitrogen-Use Efficiency in Tropical C3 and CAM Epiphytes. *New Phytologist*, **126**, 481-492. http://dx.doi.org/10.1111/j.1469-8137.1994.tb04245.x

[33] Ashraf, M. (2010) Inducing Drought Tolerance in Plants: Recent Advances. *Biotechnology Advances*, **28**, 169-183. http://dx.doi.org/10.1016/j.biotechadv.2009.11.005

[34] Cornic, G. and Fresneau, C. (2002) Photosynthetic Carbon Reduction and Carbon Oxidation Cycles Are the Main Electron Sinks for Photosystem II Activity during a Mild Drought. *Annals of Botany*, **89**, 887-894. http://dx.doi.org/10.1093/aob/mcf064

[35] Haag-Kerwer, A., Franco, A.C. and Luttge, U. (1992) The Effect of Temperature and Light on Gas Exchange and Acid Accumulation in the C3-CAM Plant *Clusia minor* L. *Journal of Experimental Botany*, **43**, 345-352. http://dx.doi.org/10.1093/aob/mcf064

[36] Rodrigues, M.A., *et al.* (2013) Spatial Patterns of Photosynthesis in Thin- and Thick-Leaved Epiphytic Orchids: Unravelling C3-CAM Plasticity in an Organ-Compartmented Way. *Annals of Botany*, **112**, 17-29. http://dx.doi.org/10.1093/aob/mct090

[37] Goh, C.J. and Kluge, M. (1989) Gas Exchange and Water Relations in Epiphytic Orchids. In: Lüttge, U., Ed., *Vascular Plants as Epiphytes, Evolution and Ecophysiology*, Springer-Verlag, Heidelberg, 139-166. htt0.1007 p://dx.doi.org/1/978-3-642-74465-5_6

[38] Sinclair, R. (1983) Water Relations of Tropical Epiphytes: I. Relationships between Stomatal Resistance, Relative Water Content and the Components of Water Potential. *Journal of Experimental Botany*, **34**, 1652-1663. http://dx.doi.org/10.1093/jxb/34.12.1652

[39] Sinclair, R. (1983) Water Relations of Tropical Epiphytes: II. Performance during Droughting. *Journal of Experimental Botany*, **34**, 1664-1675. http://dx.doi.org/10.1093/jxb/34.12.1664

[40] Chiang, Y.-L. and Chen, Y.-R. (1968) Observations on *Pleione formosana* Hayata. *Taiwania*, **14**, 271-301.

[41] Ertelt, J.B. (1992) Horticultural Aspects of Growing and Displaying a Wide Variety of Epiphytes. *Selbyana*, **13**, 95-98.

[42] Stern, W.L. and Morris, M.W. (1992) Vegetative Anatomy of Stanhopea (Orchidaceae) with Special Reference to Pseudobulb Water-Storage Cells. *Lindleyana*, **7**, 34-53.

[43] Zheng, X.N., *et al.* (1992) Response of Cymbidium Sinense to Drought Stress. *The Journal of Horticultural Science & Biotechnology*, **67**, 295-300.

[44] Yam, T.W. (2011) Conservation and Reintroduction of the Native Epiphytic Orchids of Singapore—A Physiological and Developmental Biology Perspective. In: *Proceedings of the International Conference on Biological Science*, Faculty of Biology Universitas Gadjah Mada, Yogyakarta, 2-9.

13

Design and Test of the Semi-Automatic Test-Bed with Inclined Belt of Garlic Transplanting Machine

Zhaolei Zhang[1], Aijun Geng[2,3*], Jianning Yang[2], Ruxin Li[2], Jialin Hou[2]

[1]Shandong Changlin Deutz-Fahr Machinery Ltd., Linyi, China
[2]Mechanical and Electrical Engineering College, Shandong Agriculture University, Tai'an, China
[3]Shandong Provincial Key Laboratory of Horticultural Machineries and Equipments, Tai'an, China
Email: zhangzhaolei218@126.com, *zhilan@yeah.net, 990169562@qq.com, rxli@sdau.edu.cn

Abstract

In order to verify the feasibility of semi-automatic garlic planter with inclined belt program and determine its reasonable operating parameters, the semi-automatic test-bed with inclined belt of garlic transplanting machine was designed, and the garlic box experiments were conducted. The angle of inclined belt on the test-bed and its running speed were adjustable. Single factor test results showed that the program of the semi-automatic garlic planter with inclined belt was feasible, and the angle of inclined belt and the test bed running speed affected the indicators. Orthogonal experiment results showed that the angle of inclined belt was the main factor affecting the test indicators. It is also found that the best angle was 30 degrees, while the most reasonable running speed was 0.75 Km/h.

Keywords

Garlic, Planter, Inclined Belt, Test-Bed

1. Introduction

In order to meet the agricultural requirements of planting garlic [1] [2], garlic is mainly planted manually in our country at present [3]. With low efficiency, each person can only plant about 230 m^2 per day, and the working conditions are very poor. As the labor cost rises year by year, the garlic planting cost greatly increases [4]. So garlic planting has become the key restricting the developing of garlic industry [5], and the farmers are crying

*Corresponding author.

for the mechanization of garlic planting [6].

There are two problems of garlic planting machine [7] in our country: the first is that mechanization cannot solve the problem of turning around the garlic perfectly by various means of gravity, image identification [8] [9]; the second is that the bud is not sure to erect upward after being planted into seed ditch [10]. In order to solve the two problems, a new way is proposed in the paper-using the semi-automatic test-bed with inclined belt of garlic transplanting machine. Before planting, we put the garlic seed in the garlic box made by degradable material [11] [12], which is then put on the belt to be transported into seed ditch. In order to verify the new way of planting, the semi-automatic test-bed with inclined belt of garlic transplanting machine was designed and tested to provide data support for the theory.

2. Test-Bed and Major Parameter

2.1. The Structure and Working Principles

As showed in **Figure 1**, the semi-automatic test-bed with inclined belt of garlic transplanting machine includes the institution of planting garlic box, the institution of support garlic box, the rod of angle change, transmission systems, the deflector, the rack, the motor and the motor control system. The institution of support garlic box prevents the garlic box from slipping and turning over on the belt, and guides the garlic box into seed ditch. The rod of angle adjusting could adjust the angle between the belt and earth, to obtain the consequences of changes in different angle; The deflector is an institution of transition between the belt and earth, parallel to belt, to guide the garlic box into seed ditch stably at the end of the belt.

The speed of the test-bed was controlled by IPC. When IPC motor is started, the garlic box with garlic seed can be put on the belt manually, which will be transported to the deflector with guidance of the institution of support. The deflector then drops the garlic box into seed ditch stably and completes the progress of planting.

2.2. Major Parameter

The parameter was shown in **Table 1**.

3. Transmission Systems and Motion Analysis of Garlic Box

3.1. Transmission Systems Design

3.1.1. The Structure of Transmission Systems Design

Figure 2 is the sketch of transmission systems, the arrows point out the positions of power transmission. The middle shaft makes sure that the transmission point of running shaft and transport shaft move in opposite direc-

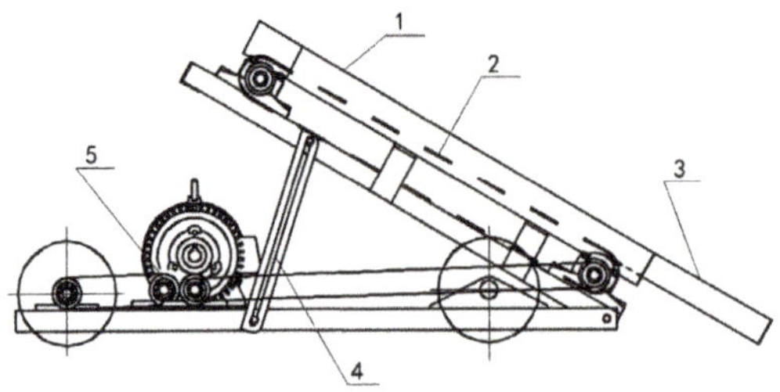

Figure 1. Text-bed. 1: The institution of support garlic box; 2: The institution of planting garlic box; 3: The deflector; 4: The rod of angle change; 5: Transmission systems.

Table 1. Test-bed parameter.

Parameter	
Power	1 kW
Rows	3
Belt angle	0° - 40°
Speed	0 - 1.8 km/h

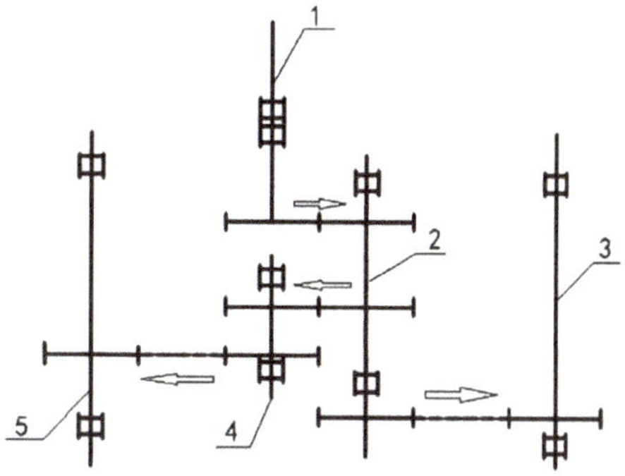

Figure 2. Sketch of transmission systems. 1: power output shaft; 2: middle shaft I; 3: transport shaft; 4: middle shaft II; 5: running shaft.

tion. The transmission uses triple retarding mechanism in running structure and two-grade reducing mechanism in transport structure to improve the motor speed [13].

3.1.2. Transmission Ratio

The suitable transmission ratio guarantees the garlic box to fall end to end in seed ditch. This ratio can be fixed according to the test-bed's running speed and the transport speed of belt.

v—the test-bed's running speed.

v_1—the transport speed of belt.

t—the time from the front end of garlic box N touching ditch to the front end of garlic box F touching seed ditch.

l—the length of garlic box.

From **Figure 3** it can be seen that in the time of t the garlic box transport the length is l in test-bed. In order to make sure the garlic box F and N end in end in seed ditch, in the time of t the garlic box transport the length relative to earth is

$$l - l\cos\beta \tag{1}$$

From the relationship between time and speed of the garlic box on belt, it can be obtained:

$$\left(v - v_{1x}\right)\cdot t = l - l\cos\beta \tag{2}$$

From **Figure 3** it is known that

$$t = \frac{l}{v_1} \quad v_{1x} = v_1 \cdot \cos\beta \tag{3}$$

From Formulas (2) and (3), the result is

$$v = v_1 \tag{4}$$

According to Formula (4), the transmission ratio can be worked out from the walking wheel's and belt wheel's diameter.

3.2. Motion Analysis of Garlic Box

The motion of the garlic box on belt and the garlic box front end touching the seed ditch are analyzed to research the tail of garlic box N's speed and the front end of garlic box F's speed. In this way the length between two garlic box end to end can be worked out.

When the garlic box is transported on the belt, the speed is v. when the garlic box front end touches the seed ditch, the speed of garlic box tail can be got from **Figure 4** [14]

$$v_a = v_e + v_r \tag{5}$$

In **Figure 4**, v_a—the absolute speed of the garlic box tail when the garlic box front end touches the seed ditch. Number and point change with the transport of garlic box.

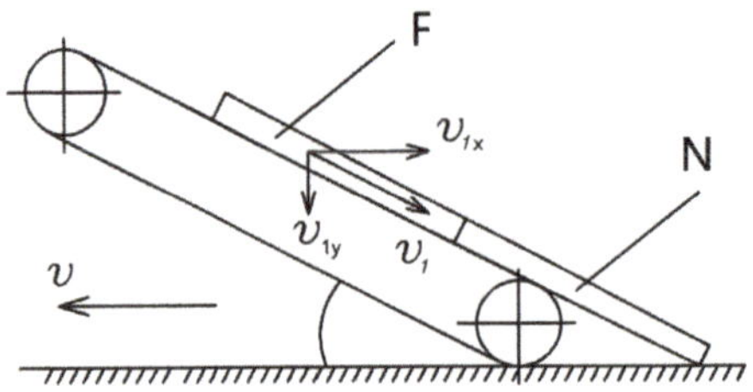

Figure 3. Sketch of garlic box motion.

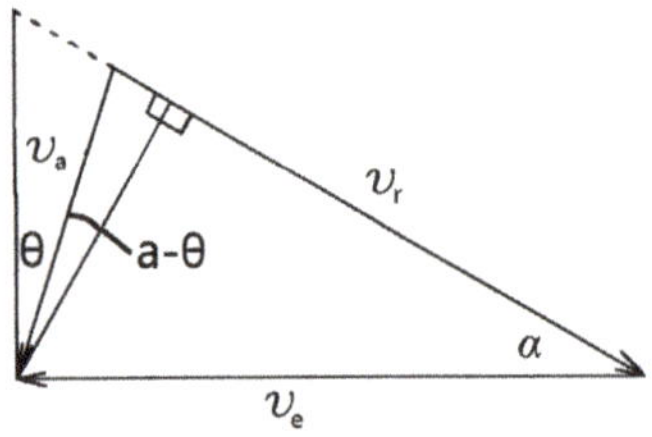

Figure 4. The garlic speed synthesis.

v_e—the convected speed, the number is same as v.
v_r—the relative speed, fixed orientation, number is change with v_a.
θ—the angle between the garlic box and earth.
α—the belt angle.
From **Figure 4** and Formula (5), it can be obtained

$$v_r = v\cos\alpha + v\sin\alpha\tan(\alpha-\theta) \tag{6}$$

Figure 5 the relationship between the relative speed and θ from Formula (6).

From **Figure 5**, v_r is increasing with the angle decreasing between garlic box and earth. when $\theta = 15°$, $v_r = v$. There are three steps in the transport of the garlic boxes: first, before the front end of garlic box N touches the seed ditch, there is no relative motion between the two garlic boxes; the second, when the front end of garlic box N touches the seed ditch to $v_r = v$, in this step $v_r < v$, so the garlic box F will extrude the garlic box N; the third, from $v_r = v$ to all of the garlic box N touches the seed ditch, in this step $v_r > v$, the a space will appear

According to the above analysis, the result is that the length between two garlic boxes end to end will appear, when the garlic boxes all touch the seed ditch.

t_1 is the time the garlic box transport from state b to state c. so

$$t_1 = \frac{l-x}{v} \tag{7}$$

Because in the time t_1, $v_r > v$, the two garlic box F and N will separate, t_2 is the time the garlic box transport from state b to the head of garlic box F touch earth

$$t_2 = \frac{p}{v} \tag{8}$$

In **Figure 6**, it is known that the transport length of tail garlic F is p,

$$p = \frac{l\sin\theta}{\sin\alpha},\quad x = \frac{l\sin(\alpha-\theta)}{\sin\alpha} \tag{9}$$

So the length between two garlic boxes end to end is d,

$$d = (t_1 - t_2)v = p + x - l = \frac{\sin(\alpha-\theta)+\sin\theta-\sin\alpha}{\sin\alpha}l \tag{10}$$

From Formulas (6) and (10) it can be worked out that the relationship between belt angle and length, **Figure 6**. Through **Figure 6** we can know the length in any belt angle.

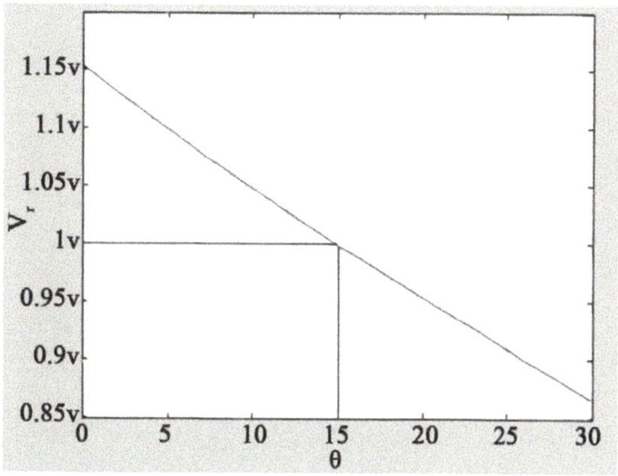

Figure 5. Speed changed curve.

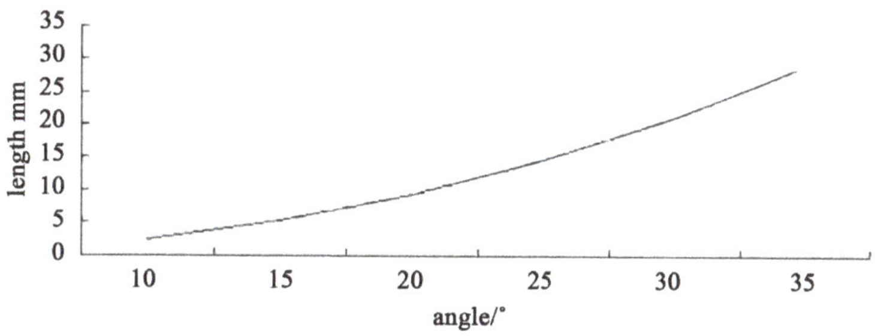

Figure 6. Relationship between belt angle and length.

4. Test of Test-Bed

Through test the feasibility of semi-automatic garlic planter with inclined belt program is to be verified and its reasonable operating parameters are to be determined. Field experiment is conducted to make sure that it will never appear slip in every belt angle and at every transport speed after the front end of garlic box touches seed ditch. **Figure 7** is test-bed.

4.1. Single Factor Experiment

Experimental factor: the belt angle, the running speed [15].

Experimental index: the length between two garlic boxes end to end.

Figure 8 is the state of garlic box in seed ditch, the distance between two garlic box is fixed.

The angle of repose is 43° which is the garlic box on belt [16], the belt angle is 0° - 38°. One person can make sure the biggest speed of planting machine is 0.82 km/h for 3 rows normally.

Figure 9 and **Figure 10** show the result.

Through the single factor experiment it can be proved that the semi-automatic garlic planter with inclined belt is feasible; Experimental factors have influence on index.

4.2. Orthogonal Test

The belt angle and test-bed running speed both have influence on index, so belt angle and test-bed speed are selected to take the orthogonal test [17].

4.2.1. Experiment Design

In test we choose orthogonal experimental design of $L_9(3^4)$ [18] [19], each set of test is repeated for 5 times. The experiment arrangement is shown in **Table 2**.

4.2.2. Result Analysis

The results are shown in **Table 3**.

Figure 7. Test-bed.

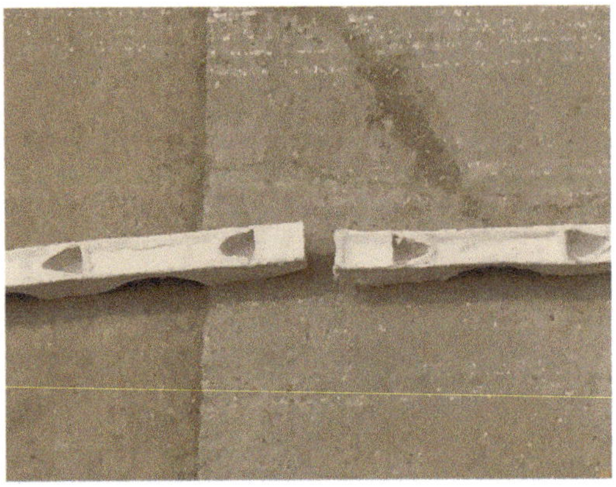

Figure 8. The state of garlic box in seed ditch.

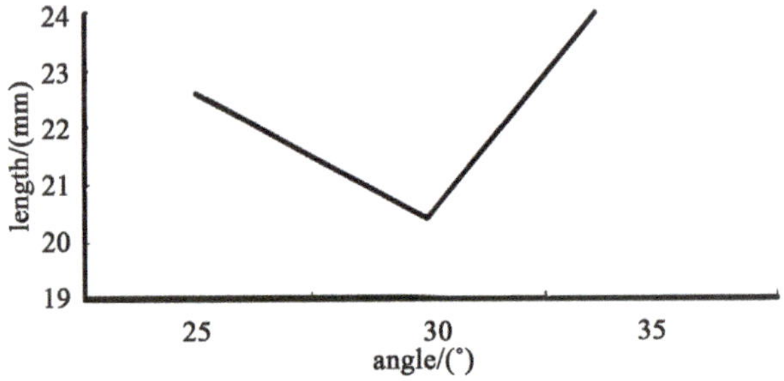

Figure 9. Belt angle single factor experiment.

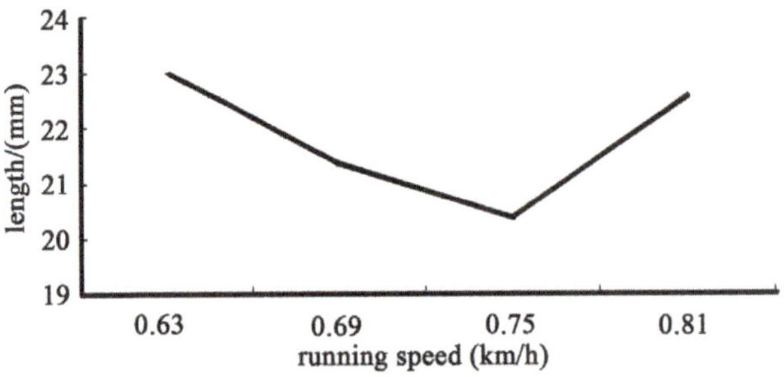

Figure 10. Running speed single factor experiment.

From **Table 2** we can see the orthogonal experiment shows interaction, no empty row, having repetition test. I work out the DEVSQ, sum of squared error and DOF [20] [21]. **Table 4** shows the variance analysis [22], **Table 5** shows the binary collocation.

In **Table 4**, $F_A = 4.0295 > F_{0.05} = 3.23$, but $F_A = 4.0295 < F_{0.01} = 5.18$, so factor A has the greatest effect on experimental index, and the reliability is 95%, while the influence factors of B and interaction of A × B on experimental index are less. Because $F_A > F_B$, so the order of each factor which influence the index is A, B.

Because A × B significantly affect the index, binary collocation of factor A, B (**Table 5**) is used to make sure the best combination. In the test, the smaller data is better. From **Table 5** the best combination is A_2B_3.

Table 2. Orthogonal experimental design of $L_9(3^4)$.

number	A (angle)	B (speed)	A × B	
1	1 (25)	1 (0.63)	1	1
2	1	2 (0.69)	2	2
3	1	3 (0.75)	3	3
4	2 (30)	1	2	3
5	2	2	3	1
6	2	3	1	2
7	3 (35)	1	3	2
8	3	2	1	3
9	3	3	2	1

Table 3. Test index.

The length between two garlic boxes/mm				
20	25	25	29	25
24	25	20	21	25
24	24	24	21	20
18	23	26	26	22
24	20	19	19	25
26	18	17	21	20
35	26	23	21	24
21	26	22	26	25
24	22	27	21	33

Table 4. Table of variance analysis.

factor	S	f	S/f	F ratio	significant	F_α
A	0.903	2	0.4515	4.0295	※	F0.01 = 5.18
B	0.301	2	0.1505	1.3432		F0.05 = 3.23
E1	0.098	4	0.0245			F0.10 = 2.44
E2	4.384	36	0.1218			F0.25 = 1.44
E	4.482	40	0.11205			
total	5.686	8				

Table 5. Binary collocation of factor A, B.

	A_1	A_2	A_3
B_1	24.8	23	25.8
B_2	23	21.4	24
B_3	22.6	20.4	25.4

5. Conclusions

1) The test-bed can plant the garlic box stably; the semi-automatic garlic planter with inclined belt is feasible.

2) The single factor experiment shows that belt angle and test-bed running speed have effect on length between two garlic boxes.

3) Through the orthogonal test, the best test combination is fixed—the belt angle is 30° and the test-bed running speed is 0.75 km/h.

4) We should cut down 20 mm at two ends of garlic box to make sure the length between the caves remain unchanged.

References

[1] Jin, C.Q., Yuan, W.S., Wu, C.Y. and Zhang, M. (2008) Experimental Study on Effects of the Bulbil Direction on Garlic Growth. *Transactions of the Chinese Society of Agricultural Engineering*, **24**, 155-157.

[2] Gao, R.X. (2010) Technology of Garlic Planting. Farmer's Friend, **12**, 20-21.

[3] Wang, X.H. (2005) Experts Answer for You. China Agriculture Press, Beijing, 46-55.

[4] Xu, P.W. (2004) The Situation of Garlic Production Abroad. Modern Vegetable, **7**, 40.

[5] Shang, S.Q., Lin, Y.X. and Yu, Y. (2011) Garlic Production Machinery Is Essential. *Xinjiang Agriculture*, **1**, 59-60.

[6] Garg, I.K., Dixit, A. (2003) Development and Evaluation of Manually-Operated Garlic Planter. *Agricultural Mechanization in Asia, Africa and Latin America*, **34**, 19-22.

[7] Choi, D.-K. (2003) Garlic Upright Positioning and Planting Device: USA, US6619219B2.

[8] Bai, Y.-C., Liu, Y.-D. and Di, M.-L. (2007) Discussing on the Present Situation and Development Trends of Garlic Transplanting Machine. *Journal of Agricultural Mechanization Research*, **11**, 247-248.

[9] Jian, S.-C. and Liu, Y.-D. (2009) Experimental Research on the Garlic Clove Automatic Orientation Control Device of Garlic Planting Machine. *Agricultural Equipment & Vehicle Engineering*, **10**, 28-29, 37.

[10] Guo, Y. and Zhang, Z.L. (2009) Garlic Planter Research Status. *Agriculture Machinery Research*, **31**, 222.

[11] Shandong Agriculture University (2013) Garlic Box. Chinese Patent No. 201220560362.2.

[12] Shandong Agriculture University (2013) Garlic Box Planting Machine. Chinese Patent No. 201320452758.X.

[13] Wei, H.A., Zhang, J.L. and Yang, X.P. (2014) Improved Design and Test of 4UFD-1400 Type Potato Combine Harvester. *Transactions of the Chinese Society of Agricultural Engineering*, **30**, 12-17.

[14] Wang, D. (2009) Rational Mechanics (I). China Machine Press, Beijing, 172-180.

[15] Tian, S.B., Yang, J.F. and Wang, R.L. (2014) Optimization Experiment of Operating Parameters on Vibration Sorting-Clip Device for Vegetable Grafting Machine. *Transactions of the Chinese Society of Agricultural Engineering*, **30**, 9-16.

[16] Zhou, Z.E. (1994) Physical Properties of Agricultural Materials. China Agriculture Press, Beijing.

[17] He, Y.E. and Yang, X.W. (1986) Agricultural Machinery Test Design. China Machine Press, Beijing.

[18] Zhang, G.Z., Luo, X.W. and Zang, Y. (2013) Experiment of Sucking Precision of Sucking Plate with Group Holes on Rice Pneumatic Metering Device. *Transactions of the Chinese Society of Agricultural Engineering*, **29**, 13-20.

[19] Yuan, Z.F. (2000) Test Design and Analysis. China Machine Press, Beijing.

[20] Zhuang, C.Q. (2005) Basis of Applied Mathematical Statistics. South China University of Technology Press, Guangzhou, 222-305.

[21] Cheng, S.H. (2005) Probability and Statistics. China Agriculture Press, Beijing, 165-179.

[22] Wang, D.P., Bai, W.D. and Li, Z.X. (2005) Soaking Technology for Raising Starch Extraction Ratio from Mung Bean during Production of Thin Noodles from Mung Bean Starch. *Transactions of the Chinese Society of Agricultural Engineering*, **21**, 151-156.

14

Nursery Fertilizer Application Increases Rice Growth and Yield in Rainfed Lowlands with or without Post-Transplanting Crop Stress

Chhay Ros[1], Peter F. White[2], Richard W. Bell[3*]

[1]Department of Foreign Affairs and Trade, Australian Embassy, Phnom Penh, Cambodia
[2]Department of Agriculture and Food WA, Baron-Hay Court, S. Perth, Australia
[3]School of Veterinary and Life Sciences, Murdoch University, Murdoch, Australia
Email: [*]R.Bell@murdoch.edu.au

Abstract

In rainfed lowlands, the beneficial effects on rice (*Oryza sativa*) yield from fertilizing the seedling nursery may depend on the occurrence of water and nutrient stress after transplanting into the mainfield. The aim of the study was to test the effect of seedling age and nursery fertilisation on rice yield when plants experienced nutrient and water stress regimes after transplanting. The first experiment, conducted during the wet season in a low-fertility rainfed lowland field in south-east Cambodia, examined the effect of three levels of nitrogen (N), phosphorus (P) and cow manure (CM) application to the nursery and three ages of seedlings at transplanting (4, 6 and 9 weeks after sowing: WAS) on the growth of a traditional photoperiod-sensitive rice cultivar. Increasing fertilizer application to the nursery increased shoot dry weight and grain yield at each age of seedling transplanted. Shoot dry weight and grain yield from seedlings transplanted 6 WAS were 20% and 10% higher than those from the seedlings transplanted at the other ages. Seedlings transplanted at 4 WAS, which experienced a temporary submergence for 3 days after transplanting, recorded 5% greater seedling mortality from the unfertilized than fertilized nursery soil. The taller seedlings transplanted at 6 WAS experienced less mortality despite more severe submergence. The second experiment was conducted at seven rainfed lowland sites across Cambodia to test the benefit of N alone and N+P application to the nursery on rice yields in farmers' fields. In four out of seven trials, grain yield increases averaged 13% with N+P applied to the nursery whereas N alone increased grain yield at two sites and depressed yield at two. Overall, fertilizer application in the nursery produced more vigorous seedlings with higher nutrient concentration which survived submer-

[*]Corresponding author.

gence and grew better with or without drought or nutrient stress after transplanting, leading to increased yields of rice. The beneficial effect of nursery fertilizer was additional to any obtained by fertilizer applied to the main field at or after transplanting.

Keywords

Nutrient Uptake, Seedling Age, Seedling Vigor, Submergence

1. Introduction

About 15% of the rainfed lowland rice fields in South and Southeast Asia can experience both drought and submergence during the growing season but in Cambodia up to 57% of paddy fields experience both these stresses [1]. Large areas of rainfed lowlands in the lower Mekong basin also have low soil fertility on account of the prevalence of sandy surface horizons [2]. Hence, rice crops in this region may encounter multiple stresses after transplanting. The extent to which nursery fertiliser application to increase seedling vigour can alleviate these stresses and enhance rice production in the main field has not been examined. In a prior glasshouse study, we reported that increased seedling vigour imposed by root pruning, leaf clipping or seedling submergence treatments did not enhance post-transplanting growth under water or nutrient stress [3]. However, the effect of seedling vigour on tolerance of rice in the field to post-transplanting stresses that are common in the low fertility, submergence- and drought-prone rainfed lowlands has not been assessed.

In order to produce more vigorous rice seedlings, farmers in Cambodia apply fertilizers, especially cow manure, to the nursery [4]. Furthermore, rice grain yield in pots was increased 10% by applying N and P to the nursery and this effect could not be replaced by increasing fertilizer application after transplanting [5]. An increase in grain yield by nursery fertilizer application may be related to increased nutrient content or to increased vigor of the seedlings that are transplanted. Either of these improvements in seedling growth may enhance post-transplanting tolerance to low main field soil nutrient status or to crop stress (e.g. drought, submergence).

In the rainfed lowlands, the age at which seedlings are transplanted can also be varied, depending on the seasonal rainfall [6]. The age of seedlings at transplanting in Cambodia ranges from as young as three weeks to as old as three months. Seedlings of advanced age may behave differently in their response to nursery fertilizer application and to post-transplanting stress. Adhikari *et al.* reported that transplanting 40-day-old rice seedlings increased rice yield and effective tiller number in the monsoon season compared to 20-day-old rice seedlings on a loamy, moderately fertile soil in Nepal [7]. By contrast, a number of recent studies, spurred by the interest in the system for rice intensification (SRI), have found increased post-transplanting rice yield from seedlings as young as 10 days compared to those 20 - 40 days old [8].

The present experiment was designed to examine the effect of applying fertilizer to the nursery on the grain yield of transplanted rice under water or nutrient stress after transplanting seedlings of various ages. The hypothesis was that the benefits from fertilizing the nursery were related to larger and more vigorous seedlings better able to withstand submergence, drought or nutrient stress.

2. Materials and Methods

2.1. Experiment 1

The experiment was conducted on a farmer's field 20 km south-west of Phnom Penh in the wet season. Soil in the mainfield was a sandy-loam with pH (H_2O) of 6.3, 0.23% organic carbon, 0.026% total nitrogen, 1 mg Olsen-extractable P/kg and and cation exchange capacity of 1.4 cmol/kg. The soil is classified as a Plinthustalf and belongs to the locally known Prateah Lang Soil Group [9].

Treatments were applied separately to the nursery and the main field. The nursery was laid out in a randomized complete block design with four replications and three treatments; nil (NF0), 20 t of cow manure per ha (NF1), and 20 t of cow manure per ha plus 60 kg of N per ha and 45 kg of P per ha (NF2), giving a total of 12 plots. Nursery plots 7 × 6 m were separated with bunds. The plots were subdivided into three unbunded subplots representing three ages of seedlings at transplanting. Seedlings were transplanted 4, 6 or 9 weeks after sowing

(WAS). All nursery plots received supplemental fertilizers consisting of H_3BO_3 (0.5 kg B/ha), $MgCl_2$ (1 kg Mg/ha), and KCl (30 kg K/ha).

Treatments were also imposed on plants in the main field after transplanting. Two levels of NPK (nil; 60 kg N/ha as urea, 40 kg P /ha as triple super phosphate, 30 kg K/ha as potassium chloride) and two water supply regimes (drainage of free-standing water and retention of free-standing water) were applied to the main field. The combination of main field and nursery treatments resulted in 36 treatments and four replicates of each. The main field had a complete randomized block design and plot size was 4 × 4 m of which 9 m^2 in the center of the plot was harvested. Main field plots were separated by bunds.

For the nursery treatments, cow manure, urea and triple superphosphate and supplemental fertilizers were applied a day before sowing and incorporated into the soil by manual hoeing. Seeds of rice cv. Chhmar Laeat, a medium maturing photoperiod-sensitive cultivar, were pre-germinated by soaking overnight then incubated for 24 h. Pre-germinated seeds were broadcast in nurseries at a rate of 50 g of dry seed per m^2. The seeds were protected from birds by covering the whole nursery with netting. Nursery plots were maintained wet without standing water during the first week after sowing. Thereafter, water was maintained 15 - 20 cm deep by irrigation or drainage as necessary.

Seedlings in appropriate treatments were pulled and bundled one day before each time of transplanting. Only seedlings in the center of the plot were used for transplanting. Seedling bundles were then, according to normal farmers' practice, stored overnight with roots standing in water in nursery plots.

Half of the N (urea) and all the P (triple superphosphate) and K (KCl) was broadcast and then incorporated into the main field plots by hoeing a day before transplanting. The rest of N was broadcast at panicle initiation (PI). Three seedlings were transplanted per hill. Spacing between hills was 20 by 20 cm. A week after transplanting, missing or dead hills were counted in each plot to determine plant mortality and then replanted with spare seedlings from the same nursery plot.

Water treatments were imposed immediately after transplanting. The adequate water level was imposed by maintaining free-standing water (0 - 30 cm depth) until week 18 (**Figure 1**). Irrigation water was supplied or rainwater drained as necessary. The drained treatment had repeated drainage of free-standing water from the plots after rainfall events. Plots were kept free of standing water except for periods up to week 10, when heavy rain caused flooding and plots were unable to be fully drained.

Immediately before pulling at 4, 6 or 9 WAS, seedling height, leaf number and tiller number were measured in the nursery and root samples were taken from nurseries by carefully digging up the roots and washing them free from soil. A further 15 seedlings were sampled for dry matter determination. Shoot and roots were separated and samples were dried (70°C), chopped, ground and then sub-sampled for analysis of N, P and K.

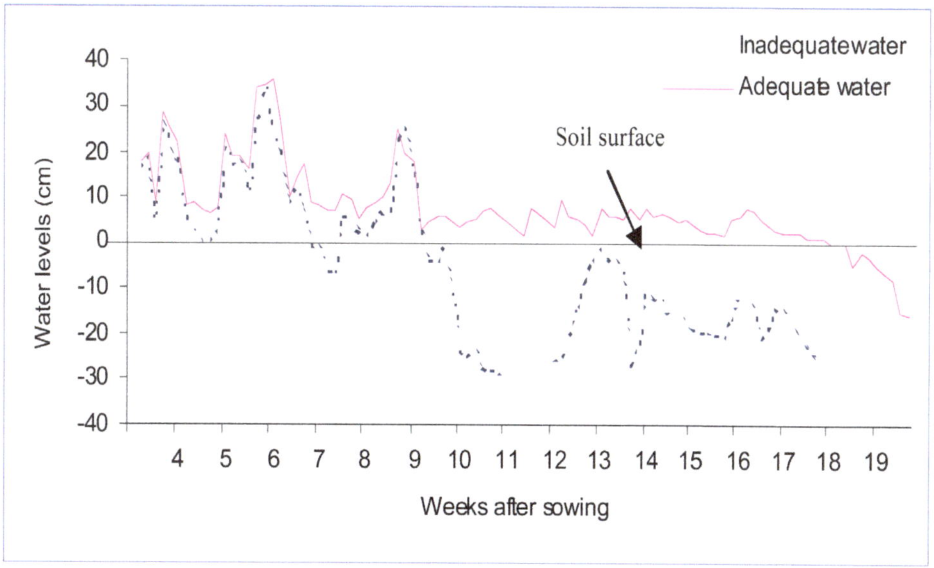

Figure 1. Perched water table level in soil of the experiment site in water adequate and inadequate treatments. Experiment 1.

In the main field, plant height and leaf number and tiller number were recorded at PI, and panicle number, grain number per panicle, straw dry matter and grain yield were recorded in the main field at maturity. Three plant samples consisting of four adjacent hills were randomly taken from outside the harvest area. Plant samples were prepared by oven drying and milling for nutrient analysis.

After transplanting, water level in the main field was measured daily in each replicate block from adequate and drained water regime plots using piezometers. The shallow piezometer penetrated only 20 cm below the ground surface and measured the depth of water perched above the plough layer. The deep piezometer penetrated 70 cm below the ground surface and measured the depth of the ground water table below the plough layer.

The experiment was laid out in a randomized complete block design and analyzed as a 4-way analysis of variance using IRRISTAT software [10]. None of the four way interactions, and few of the three way interactions were significant.

2.2. Experiment 2

Experimental sites were selected in farmers' rainfed lowland fields in Kampong Chhnang, Kampong Speu, Kandal, Prey Veng (2 sites), Pursat and Siem Reap Provinces of Cambodia. The soil groups were classified as Bakan at Kampong Chhnang, Kandal and one site at Prey Veng (Prey Veng 2), Prey Khmer at Siem Reap, and Prateah Lang at other experiments [9]. Seedlings were raised in nurseries with three treatments (nil; 60 kg N/ha as urea; 60 kg N/ha and 17 kg P/ha as di-ammonium phosphate) and then transplanted into main field plots with low (30:4.4:8.2 kg NPK/ha) or medium levels of compound NPK fertilizer (60:8.8:16.4 kg NPK/ha). Each treatment combination was replicated five times. All P and K was applied to main fields at transplanting as triple superphosphate and KCl. Nitrogen as urea was split between transplanting (40%), 30 days after transplanting (30%) and PI (30%). Dry matter of seedlings before transplanting was determined in nursery plots. In the main field, shoot dry matter was determined at panicle initiation and grain yield at maturity. All these assessments were based on sampling from three 2 × 2 m quadrats per plot

The experiment was analyzed as a 2-way analysis of variance with five replicates for each site [10].

3. Results

3.1. Experiment 1

3.1.1. Seedling Growth before Transplanting

At 4 weeks after sowing (WAS), application of cow manure (CM) to the nursery increased seedling height, leaf number, and root dry matter by 8%, 25% and 40%, respectively (**Table 1**). It also increased P and K concentration in seedling shoots (**Table 1**), but had no effect on tiller number and seedling shoot dry matter or N concentration. Adding both CM and inorganic fertilizer further increased plant height and leaf number by about 28%, and root and shoot dry matter by 60% and 13%, respectively, but had no effect on N and P concentrations compared with the CM treatment (**Table 1**).

At 6 WAS, CM continued to increase leaf number, and root dry matter: it also strongly increased shoot and root dry matter but had no significant effect on seedling height (**Table 1**). Cow manure increased N, P and K concentration in shoots (**Table 1**) but P and K increased to a lesser degree than for plants at 4 WAS. Combined application of CM, N and P to the nursery further increased seedling height, shoot dry matter and root dry matter compared to CM alone, but did not increase shoot N, P or K concentration or leaf number (**Table 1**).

At 9 WAS, the effects of CM alone, or combined CM, N and P application on seedling height, leaf number, and shoot and root dry matter were similar to those at 6 WAS (**Table 1**). By contrast, to the seedlings at 4 and 6 WAS, however, tiller number was increased with both CM alone, and combined CM, N and P application (**Table 1**).

With increasing age in the nursery, seedling height, root and shoot dry matter increased but N, P and K concentrations in the shoots decreased (**Table 1**).

3.1.2. Plant Growth after Transplanting

Heavy rain which occurred continuously for about 4 days after transplanting seedlings aged 4 and 6 WAS, respectively caused temporary submergence of the shorter plants (**Figure 1** and **Figure 2**). Transplants which had received fertilizer (CM with or without NP) in the nursery recovered better from transplanting shock and

Table 1. Effect of nursery fertilizer application on plant height, leaf and tiller number, shoot and root dry matter and seedling nitrogen (N), phosphorus (P) and potassium (K) concentrations at 4, 6, and 9 weeks after sowing (WAS). Values are means of four replicates.

Nursery fertilizer	Plant height (cm)	Leaf number/plant	Tiller number/plant	Shoot dry matter (mg/plant)	Root dry matter (mg/plant)	N (%)	P (%)	K (%)
				At 4 WAS				
NF0	36	3.3	1	102	33	1.32	0.15	1.68
NF1	39	4.4	1	106	53	1.43	0.35	2.53
NF2	54	4.7	1	254	61	1.54	0.37	2.66
LSD	3**	0.4**	ns	18**	6**	ns	0.05**	0.38**
				At 6 WAS				
NF0	46	3.8	1	189	75	0.73	0.12	1.53
NF1	52	5.0	1	273	100	0.91	0.27	1.97
NF2	82	5.7	1	497	184	0.96	0.30	2.06
LSD:	8**	0.8**	ns	25**	16**	0.11**	0.04**	0.12**
				At 9 WAS				
NF0	51	4.1	1.8	1519	942	0.58	0.11	1.37
NF1	64	5.4	3.2	2258	1253	0.85	0.26	1.46
NF2	90	5.5	4.7	3961	2979	0.86	0.30	1.60
LSD:	18**	0.6**	1.3**	41**	23**	0.18*	0.07*	ns

** = significant at $P < 0.01$; * = significant at $P < 0.05$; NF0 = control; NF1 = 20 t cow manure/ha; NF2 = 20 t cow manure/ha, 60 kg N/ha and 45 kg P/ha.

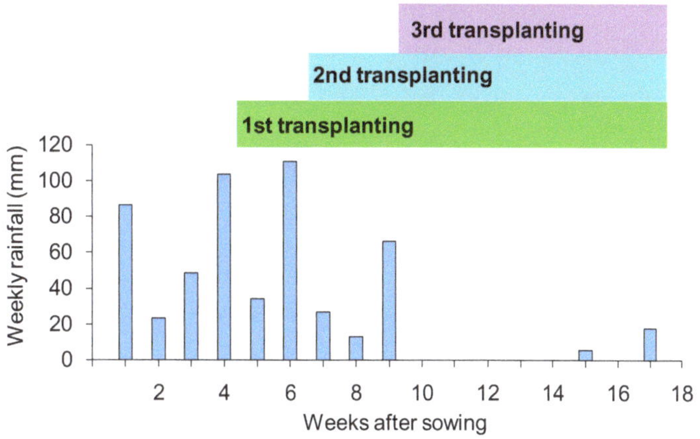

Figure 2. Weekly rainfall distribution during the experiment in relation to transplanting and harvest times. Data was collected during the wet season from Pochentong, Phnom Penh, 7 km north of the experimental site. Experiment 1.

submergence than those which had not received fertilizer. The percentage mortality of the plants 7 days after transplanting from the fertilised nursery was less than 1% (**Table 2**). Mortality was greatest among seedlings transplanted at 4 WAS even though the depth of flooding was shallower than at 6 WAS (**Figure 2**). Replanting of missing hills was carried out a week after each transplanting so that subsequent growth responses were not confounded by differences in plant populations among plots.

In drained plots, heavy rainfall resulted in standing water for most of the first 9 weeks including the first 5 - 3 weeks of post-transplanting growth for the 4 and 6 week-old seedlings, respectively (**Figure 1** and **Figure 2**). By contrast, for the 9 week-old seedlings water level was below the soil surface from the time of transplanting.

Table 2. The effect of main field fertilizer application and two water regimes on tiller and leaf number, transplant mortality and shoot dry matter of rice cv. Chhmar Laeat transplanted from nurseries supplied with varying rates of fertilizer and transplanted at 4, 6 and 9 weeks after sowing (WAS). Values are means of four replicates.

Nursery fertilizer	Tiller number[B] per hill			Leaf number[B] per hill			Transplant mortality[A] (%)			Shoot dry matter[B] (g/hill)		
	4 WAS	6 WAS	9 WAS	4 WAS	6 WAS	9 WAS	4 WAS	6 WAS	9 WAS	4 WAS	6 WAS	9 WAS
Without additional NPK with drained water												
NF0	7.0	7.4	2.9	30	31	11	8.1	3.9	0.2	4.25	4.99	2.50
NF1	7.3	7.2	3.4	32	31	13	5.0	0.7	0.1	6.16	6.03	3.41
NF2	6.7	7.6	3.1	31	27	10	0.9	0.3	0.1	6.90	6.35	3.49
Mean	7.0	7.4	3.1	31	30	11	4.7	1.6	0.1	5.77	5.79	3.13
With additional NPK (60-40-30) *with drained water*												
NF0	8.2	8.8	4.0	37	37	14	7.5	1.7	0.1	8.84	8.21	3.46
NF1	8.2	8.6	3.5	37	38	13	3.4	0.4	0.1	11.14	9.34	4.33
NF2	7.4	8.7	3.2	33	37	11	0.8	0.1	0.1	11.87	10.09	5.88
Mean	7.9	8.7	3.6	36	37	13	3.9	0.7	0.1	10.62	9.21	4.56
Without additional NPK with adequate water supply												
NF0	6.3	7.4	3.5	26	31	13	8.2	1.2	0.1	6.45	5.09	3.35
NF1	6.8	6.8	3.5	31	30	12	2.0	0.3	0.1	7.37	5.91	3.77
NF2	7.0	7.6	3.1	32	31	10	0.7	0.2	0.1	8.81	6.78	4.91
Mean	6.7	7.3	3.4	30	31	12	3.6	0.6	0.1	7.54	5.93	4.01
With additional NPK (60-40-30) *with adequate water supply*												
NF0	8.9	8.0	3.6	36	34	13	7.9	1.3	0.3	9.30	7.57	3.78
NF1	8.2	8.8	4.3	36	38	13	2.9	0.5	0.1	10.63	7.81	4.62
NF2	7.1	9.1	3.7	32	35	12	0.7	0.2	0.1	12.78	12.97	5.65
Mean	8.1	8.6	3.9	35	36	13	3.8	0.7	0.2	10.90	9.45	4.68
LSD:[C]												
Nursery fertilizer (N)	ns			ns			0.05**			0.10**		
Age of seedling (A)	0.4**			1.89**			0.05**			0.10**		
Main field fertilizer (F)	0.3**			1.89**			0.02**			0.08**		
Water regime (W)	ns			ns			0.02**			0.08**		
N × A	ns			ns			0.26**			0.18**		
N × F	ns			ns			0.21**			0.14**		
N × W	ns			ns			0.21**			0.14**		
A × F	ns			2.68*			0.21**			0.14**		
A × W	ns			ns			0.21**			0.14**		
F × W	ns			ns			0.17**			0.11**		
CV (%)	15			18			22			3.7		

** = significant at 1% level; * = significant at 5% level; ns = not significant. NF0 = control; NF1 = 20 t/ha cow manure; NF2 = 20 t/ha cow manure, 60 kg/ha N and 45 kg/ha P. [A]measured one week after transplanting. [B]Sampled at panicle initiation. [C]The four-way and most of the three way interactions were not significant.

3.1.3. Transplanting to Panicle Initiation

At PI, shoot dry matter of the plants transplanted at 4 WAS was the highest, because these plants had grown for the longest period in the main field. Delaying the time of transplanting by 2 or 5 weeks reduced shoot dry matter at PI by 15% and 50%, respectively, regardless of water supply or fertilizer application to the main field (**Table 3**).

Table 3. Effect of main field fertilizer application and two water regimes on nutrient concentration in the shoot of rice cv. Chhmar Laeat transplanted from nurseries supplied with varying rates of fertilizers and transplanted at 4, 6 and 9 weeks after sowing (WAS). Values are means of four replicates. Shoots sampled at panicle initiation.

Nursery fertilizer	Shoot N conc. (%)			Shoot P conc. (%)			Shoot K conc. (%)		
	4 WAS	6 WAS	9 WAS	4 WAS	6 WAS	9 AS	4 WAS	6 WAS	9 WAS
	Without additional NPK with drained water								
NF0	0.89	1.00	1.38	0.12	0.12	0.11	0.12	0.13	0.13
NF1	1.04	1.06	1.46	0.13	0.13	0.13	0.13	0.13	0.13
NF2	1.14	1.02	1.58	0.13	0.14	0.28	0.14	0.15	0.15
Mean	1.02	1.03	1.48	0.13	0.13	0.17	0.13	0.14	0.14
	With additional NPK with drained water								
NF0	1.34	1.21	1.60	0.20	0.21	0.23	0.14	0.13	0.14
NF1	1.33	1.31	1.74	0.21	0.21	0.27	0.13	0.17	0.14
NF2	1.41	1.53	1.84	0.25	0.25	0.30	0.16	0.16	0.08
Mean	1.36	1.35	1.73	0.22	0.22	0.27	0.14	0.15	0.12
	Without additional NPK with adequate water supply								
NF0	0.79	0.75	1.38	0.11	0.12	0.15	0.13	0.14	0.14
NF1	0.81	0.85	1.43	0.12	0.17	0.17	0.13	0.13	0.13
NF2	0.83	0.82	1.48	0.14	0.18	0.19	0.15	0.15	0.15
Mean	0.81	0.81	1.43	0.12	0.16	0.17	0.13	0.14	0.14
	With additional NPK with adequate water supply								
NF0	1.33	1.41	2.15	0.13	0.24	0.31	0.15	0.15	0.17
NF1	1.33	1.43	2.05	0.16	0.25	0.34	0.14	0.15	0.16
NF2	1.49	1.65	2.10	0.24	0.29	0.37	0.15	0.17	0.15
Mean	1.38	1.49	2.10	0.18	0.26	0.34	0.15	0.16	0.16
LSD:[C]									
Nursery fertilizer (N)		0.1*			0.02**			0.13**	
Age of seedling (A)		0.1**			0.02**			0.13**	
Main field fertilizer (F)		0.08**			0.01**			0.11**	
Water regime (W)		ns			0.01*			ns	
N × A		ns			ns			ns	
N × F		ns			ns			ns	
N × W		ns			ns			ns	
A × F		ns			0.02**			0.18**	
A × W		0.1*			0.02**			ns	
F × W		0.1**			ns			ns	
CV (%)		12			13			14	

** = significant at 1% level; * = significant at 5% level; ns = not significant. NF0 = control; NF1 = 20 t/ha cow manure; NF2 = 20 t/ha cow manure, 60 kg/ha N and 45 kg/ha P. [C]The four-way and three way interactions were not significant.

However when CM alone was added to the nursery, shoot dry matter of plants from seedlings transplanted at 4, 6 and 9 weeks was higher than when none was applied to the nursery (**Table 2**). When CM, N and P were applied to the nursery, shoot dry matter of seedlings transplanted at 4, 6 and 9 WAS was even higher than the CM nursery treatment (**Table 2**).

Cow manure application to the nursery had no effect on N concentration in shoots at PI but when combined with N and P application to the nursery, N concentration in shoots increased by 12% and 15% when seedlings were transplanted at 4 and 6 WAS, respectively. When seedlings were transplanted at 4 and 9 WAS, P concentration in shoots at PI increased by 20% and 10%, respectively, due to CM was applied to the nursery (**Table 3**). Furthermore, the combined CM and N and P fertilizer applications significantly increased the P concentration in shoots of plants at PI for each age at which seedlings were transplanted (**Table 3**).

Nursery fertilizer application increased shoot dry matter with and without main field fertilizer but the greater effects were obtained with NPK application in the main field (**Table 2**). Adequate water regime had the greatest positive impact on shoot dry matter when seedlings were transplanted at 4 WAS.

3.1.4. Panicle Initiation to Harvest

At maturity, nursery fertilizer had a positive effect on panicle number but had no effect on panicle length, 100-grain weight or grain fertility (data not shown). The CM application to the nursery increased grain yield by 5%, 12% and 4% when seedlings were transplanted at 4, 6 and 9 WAS, respectively. The application of CM, N and P to the nursery had similar effects on grain yield as the effect of CM application alone (**Table 4**). Straw yield increased 12% by applying CM to the nursery when seedlings were transplanted at 4 WAS but not with older transplants (**Table 4**).

Applying CM to the nursery increased P concentration in straw by 150%, 80% and 100 % when seedlings were transplanted at 4, 6 and 9 WAS, respectively (**Table 5**). However, the CM application to the nursery increased N concentration in straw by 15% only when seedlings were transplanted at 9 WAS (**Table 5**). Combined application of CM and N and P, increased N and P concentrations in straw by up to 35% and 100%, respectively, compared to nutrient concentrations in shoots of plants which received CM alone.

Panicle number and grain and straw yields were all lower when NPK fertilizer was not applied to the main field compared to when it was. However, the effect of nursery fertilizer application on straw yield and grain yield of plants grown without additional NPK applied to the main field was similar to that of plants which received NPK in the main field for all ages of transplanting. Grain yield increased by 10% and 5% with CM application in the nursery when seedlings were transplanted at 4 and 9 WAS (**Table 4**). Cow manure alone, in this case, increased straw yield only when seedlings were transplanted at 4 WAS but the magnitude of increase was two times higher than with NPK in the main field (**Table 4**).

With repeated drainage of free-standing water, the impact of nursery fertilizer application (either CM applied alone or in combination with N and P), on shoot dry matter and grain yield were similar to the effects seen under the adequate water regime (**Table 3** and **Table 4**). With drained water, omitting NPK from the main field decreased grain yield by 0.6 - 1.0 $t{\cdot}ha^{-1}$. Drained water had no effect on grain yield of seedlings transplanted at 4 WAS, but depressed that of seedlings transplanted at 6 and 9 WAS. By contrast, the drainage of free-standing water had minimal effect on straw yield, or on shoot dry matter. The depression in grain yield with drainage of freee-standing water at 6 and 9 WAS was most clearly related to the decline in shoot N concentration, and was also reflected in decreased grain N concentration (data not shown). Furthermore, fertilizer application to the nursery had similar effect on the N and P concentration in shoots and straw as when adequate water was supplied.

Straw and grain yield of seedlings transplanted at 6 WAS were 10% and 15% higher than those of plants grown from seedlings aged 4 and 9 WAS, respectively (**Table 4**). Nitrogen and P concentration in shoot (**Table 3**), and K concentration in straw and in grain of plants grown from seedlings aged 9 WAS were greater than those in plants grown from seedlings aged 4 and 6 WAS (**Table 5**, grain data not shown).

3.2. Experiment 2

Nursery fertilizer increased dry matter of seedlings at transplanting only at Kampong Chhnang and decreased it at the Kandal (**Table 6**). After transplanting, shoot dry matter at PI was generally unaffected by the main field fertilizer or nursery fertilizer treatment except at Pursat and Kampong Chhnang. However, final grain yield was

Table 4. The effect of main field fertilizer application and two water regimes on straw and grain yields of rice cv. Chhmar Laeat transplanted from nurseries supplied with varying rates of fertilizer and transplanted at 4, 6 and 9 weeks after sowing (WAS). Values are means of four replicates.

Nursery fertilizer	Panicle number per m^2			Straw yield (t/ha)			Grain fertility (%)			Grain yield (kg/ha)		
	4 WAS	6 WAS	9 WAS	4 WAS	6 WAS	9 WAS	4 WAS	6 WAS	9 WAS	4 WAS	6 WAS	9 WAS
Without additional NPK with drained water												
NF0	323	350	309	1.19	1.72	1.58	88	90	74	1205	1292	1192
NF1	365	383	342	1.45	1.91	1.60	90	91	77	1309	1455	1287
NF2	373	393	354	1.62	1.99	2.06	91	89	80	1349	1535	1339
Mean	354	375	335	1.42	1.88	1.75	90	90	77	1288	1427	1273
With additional NPK (60-40-30) *with drained water*												
NF0	377	496	467	2.68	3.44	2.83	89	86	80	2001	2338	1811
NF1	394	517	486	2.73	3.42	2.96	91	86	83	2243	2461	1871
NF2	410	544	496	3.07	3.56	2.96	92	87	84	2394	2584	2125
Mean	394	519	483	2.83	3.47	2.92	91	86	82	2213	2461	1936
Without additional NPK with adequate water supply												
NF0	338	377	338	1.70	2.22	2.01	86	83	79	1242	1579	1292
NF1	367	388	363	2.20	2.32	2.04	87	88	81	1355	1604	1338
NF2	375	400	398	2.19	2.42	2.05	87	89	82	1397	1622	1351
Mean	360	388	366	2.03	2.32	2.03	87	87	81	1331	1602	1327
With additional NPK (60-40-30) *with adequate water supply*												
NF0	442	515	473	2.70	3.29	3.16	85	87	84	2168	2479	2136
NF1	459	523	502	3.03	3.23	3.20	87	89	85	2277	2771	2215
NF2	481	531	506	3.20	3.47	3.22	90	89	84	2284	2831	2300
Mean	461	523	494	2.98	3.33	3.19	87	88	84	2243	2694	2217
LSD:[C]												
Nursery fertilizer (N)		20^{**}			0.11^{**}			3.0^{**}			43^{**}	
Age of seedling (A)		20^{**}			0.11^{**}			3.0^{**}			43^{**}	
Main field fertilizer (F)		16^{**}			0.09^{**}			2.46^{**}			35^{**}	
Water regime (W)		16^{**}			0.09^{**}			2.46^{**}			35^{**}	
N × A		ns			ns			5.20^{*}			ns	
N × F		ns			ns			ns			61^{**}	
N × W		ns			ns			ns			ns	
A × F		28^{**}			ns			4.26^{*}			61^{**}	
A × W		ns			ns			ns			61^{**}	
F × W		ns			0.13^{**}			ns			50^{*}	
CV (%)		11.6			10.8			6.9			5.7	

** = significant at 1% level; * = significant at 5% level; ns = not significant. NF0 = control; NF1 = 20 t/ha cow manure; NF2 = 20 t/ha cow manure, 60 kg/ha N and 45 kg/ha P. [C]The four-way and most of the three way interactions were not significant.

increased at these sites and additionally at Kampong Speu and Siem Reap sites (**Table 6**). At Pursat and Siem Reap only, N+P applied to nursery decreased days to flowering by 1 - 2 days.

At 4 sites where N+P applied to the nursery increased grain yield, the average increase was 13%. By contrast, N alone increased grain yield at two sites, and decreased it at 2 sites. Responses of straw dry matter at maturity were similar to those for grain yield (data not shown).

Table 5. Effect of mainfield fertilizer application and two water regimes on nutrient concentration in the straw of rice cv. Chhmar Laeat transplanted from nurseries supplied with varying rates of fertilizers and transplanted at 4, 6 and 9 weeks after sowing (WAS). Values are means of four replicates.

Nursery fertilizer	Straw N conc. (%)			Straw P conc. (%)			Straw K conc. (%)		
	4 WAS	6 WAS	9 WAS	4 WAS	6 WAS	9 WAS	4 WAS	6 WAS	9 WAS
Without additional NPK with drained water									
NF0	0.34	0.38	0.32	0.02	0.02	0.02	0.67	0.98	1.07
NF1	0.39	0.39	0.38	0.04	0.03	0.02	0.63	0.89	1.16
NF2	0.40	0.41	0.44	0.05	0.03	0.03	0.80	1.15	1.25
Mean	0.38	0.39	0.38	0.04	0.03	0.02	0.70	1.01	1.16
With additional NPK (60-40-30) *with drained water*									
NF0	0.38	0.40	0.31	0.09	0.12	0.05	0.54	0.79	1.22
NF1	0.47	0.41	0.37	0.12	0.12	0.09	0.53	1.12	1.40
NF2	0.58	0.48	0.49	0.14	0.15	0.12	0.87	1.38	2.15
Mean	0.48	0.43	0.39	0.12	0.13	0.09	0.64	1.10	1.59
Without additional NPK with adequate water supply									
NF0	0.31	0.35	0.35	0.03	0.02	0.03	0.63	0.82	0.56
NF1	0.35	0.37	0.36	0.05	0.03	0.03	0.67	1.17	0.82
NF2	0.39	0.35	0.39	0.07	0.04	0.04	0.71	1.38	1.14
Mean	0.35	0.35	0.37	0.05	0.03	0.03	0.67	1.12	0.84
With additional NPK (60-40-30) *with adequate water supply*									
NF0	0.42	0.35	0.42	0.02	0.05	0.05	0.54	1.00	1.21
NF1	0.43	0.36	0.49	0.05	0.09	0.10	0.65	1.20	1.45
NF2	0.50	0.59	0.51	0.10	0.12	0.13	0.99	1.47	1.56
Mean	0.45	0.43	0.47	0.05	0.09	0.10	0.72	1.22	1.41
LSD:									
Nursery fertilizer (N)		0.04**			0.01**			0.16**	
Age of seedling (A)		ns			ns			0.16**	
Main field fertilizer (F)		0.03**			0.01*			0.13**	
Water regime (W)		ns			ns			ns	
N × A		ns			ns			ns	
N × F		0.05*			0.02*			ns	
N × W		ns			ns			ns	
A × F		ns			0.02**			0.23**	
A × W		ns			0.02*			0.23*	
F × W		ns			0.01**			ns	
CV (%)		15			34			28	

** = significant at 1% level; * = significant at 5% level; ns = not significant. NF0 = control; NF1 = 20 t/ha cow manure; NF2 = 20 t/ha cow manure, 60 kg/ha N and 45 kg/ha P. [C]The four-way and three way interactions were not significant.

Table 6. Effect of nursery fertilizer application on seedling dry matter (kg DM/m^2) at transplanting and on the response to main field fertilizer (− or +) of shoot DM (t/ha) at panicle initiation (PI) and grain yield (t/ha) at maturity of rice at 7 sites in Cambodia. Values are means of five replicates. Experiment 2.

Nursery	Kandal		Prey Veng 1		Prey Veng 2		Pursat		K. Chhnang		K. Speu		Siem Reap	
fertilizer	+	−	+	−	+	−	+	−	+	−	+	−	+	−
							Seedling DM							
Nil		0.61		1.03		1.02		0.38		0.58		0.75		0.49
N		0.53		0.98		1.14		0.36		0.69		0.74		0.41
NP		0.57		1.09		1.10		0.40		0.70		0.80		0.44
P[A] nursery		0.05		ns		ns		ns		0.09		ns		ns
							Shoot DM at PI							
Nil	2.66	2.58	-	-	1.06	0.89	0.92	0.82	1.43	1.38	2.39	2.28	1.49	1.12
N	2.66	2.49	-	-	1.00	1.01	0.96	0.80	1.49	1.29	2.57	2.29	1.39	1.54
NP	2.86	2.71	-	-	1.05	0.96	1.04	1.06	1.17	1.65	2.49	2.55	1.41	1.42
P[A] main field		ns				ns		ns		ns		ns		ns
nursery		ns				ns		0.02		0.03		ns		ns
							Grain yield							
Nil	6.2	5.7	3.1	2.7	1.4	1.4	2.5	2.3	3.3	2.9	3.3	3.3	2.9	2.9
N	5.6	5.2	2.3	2.2	1.4	1.3	2.6	2.3	3.1	2.8	3.6	3.6	3.7	3.0
NP	6.0	5.9	3.3	2.2	1.3	1.1	3.0	2.6	3.5	3.1	3.3	3.3	3.7	3.0
P[A] main field		ns		ns		ns		0.02		0.01		ns		0.03
nursery		0.01		ns		ns		0.01		0.07		0.1		0.05

[A]Probability of significant effects of nursery fertilizer or main field fertilizer treatments from F test.

4. Discussion

Applying fertilizer to the nursery increased final grain yield of rice under field conditions in low fertility rainfed lowland soils. The present results extend the glasshouse results reported earlier [5] by demonstrating that nursery fertilizer addition can increase yield in the field in both fertilized and unfertilized main fields. Furthermore, and importantly, the effects of low nutrient supply in the nursery were not reversed by applying higher rates of fertilizer after transplanting or by transplanting older seedlings. Similarly, Shalaby *et al.* [11], Rajagopalan and Krishnarajan [12] and Panda *et al.* [13] found that nutrient application to nurseries improved rice grain yield while Adhikari *et al.* [7] found no effect of nursery N fertiliser application on post-transplanting growth or yield of rice on a moderately fertile loamy soil in the mid-hills of Nepal.

The responses in yield to nursery fertilizer application seen in Experiment 1 are relevant to a large proportion of rice area of Cambodia because this experiment is conducted on the Prateah Lang soil which accounts for 35% of the rice soils of Cambodia [9]. Low fertility sandy lowland rainfed soils occur in northeast Thailand and Laos [2], and the present results have relevance to those rice-growing areas as well. As the lack of response to nursery fertiliser application in Nepal shows, the present results may not be relevant where there is a higher inherent fertility of the nursery soils [7]. The fact that the responses to nursery fertilizer application occurred regardless of water regime, and with or without basal fertilizer applied at transplanting, further supports the notion that this finding may have wide practical significance for rice cultivation in rainfed lowlands. Indeed when the effects of fertilizer application on the nursery were tested in a wider range of soils and environments in Experiment 2, grain yield increases were obtained at 4 out 7 sites with N+P, and none of them showed a negative effect of N+P applied to the nursery. At responsive sites the average relative yield increase was 13%. By contrast, N alone had mixed effects, producing negative effects on yield as often as positive effects. The variable response to N alone probably relates to the P status of the nursery and main field. Addition of N alone may accentuate a P deficiency in nursery seedlings or when they are transplanted into the main field. Certainly P deficiency is very widespread in the rainfed lowland rice soils on which the present studies were conducted [9] [14] [15].

Farmers in Cambodia supply up to two-thirds of their cow manure to their nursery [16], resulting in cow manure addition to nurseries at rates 5 - 10 times higher than in main fields [4]. The present results, suggest that farmers' decisions to apply cow manure preferentially on their nurseries are soundly based. However, the present study shows that cow manure alone, at rates greater than most farmers use, will not achieve maximum growth of seedlings in the nursery. Moreover, farmers' old nurseries still contain very low levels of N, P and K and, in their unfertilised state, remain deficient in several elements for seedling growth [4]. At least N+P need to be applied to nursery soils to increase vigor of transplanted seedlings, and although there may also be a case for K and S addition on some soils [15] this was not investigated in the present study.

Part of the increase in grain yield due to fertilizer application to the nursery can be ascribed to the early vigor of the seedlings in the nursery as indicated by greater plant height, increased leaf number, increased tiller number in 9 week-old seedlings, and greater shoot and root dry matter of the fertilised 4, 6 and 9 WAS seedlings. Seedlings with increased vigour may recover faster following transplanting in the rainfed lowlands.

In this experiment, up to 8% of unfertilised transplanted seedlings died after a 7-day period of submergence. Longer periods of submergence are not infrequent in the rainfed lowlands of Cambodia [6], and may therefore have more severe effects on plant survival than recorded here. Plant height (tallness) is one of the important characteristics for evaluating the submergence tolerance of rice by plant breeders. The taller the plant is, the greater the submergence tolerance capacity is [17]. Tall plants may, therefore, cope better with submergence when it occurs during the post-transplanting recovery growth in large areas of the rainfed lowlands, especially in Cambodia but also in Laos and Thailand [1]. When the leaves are completely submerged, oxygen transport to roots fails, causing impairment of root function including respiration, nutrient uptake and nutrient selectivity [18]. Thus, the simple trait of greater height, which was increased by fertilizer applied to the nursery, may be sufficient to increase survival of seedlings when temporary submergence occurs after transplanting. However, in those places where flash-flooding is common such as in Cambodia, tall rice cultivars are likely to be better adapted than short ones [19].

A strong relationship between nutrient (N and P) content of seedlings and nutrient content of shoots at PI and in the grain (**Figure 3**) supports the previous findings that nutrient content in the seedling at transplanting relates

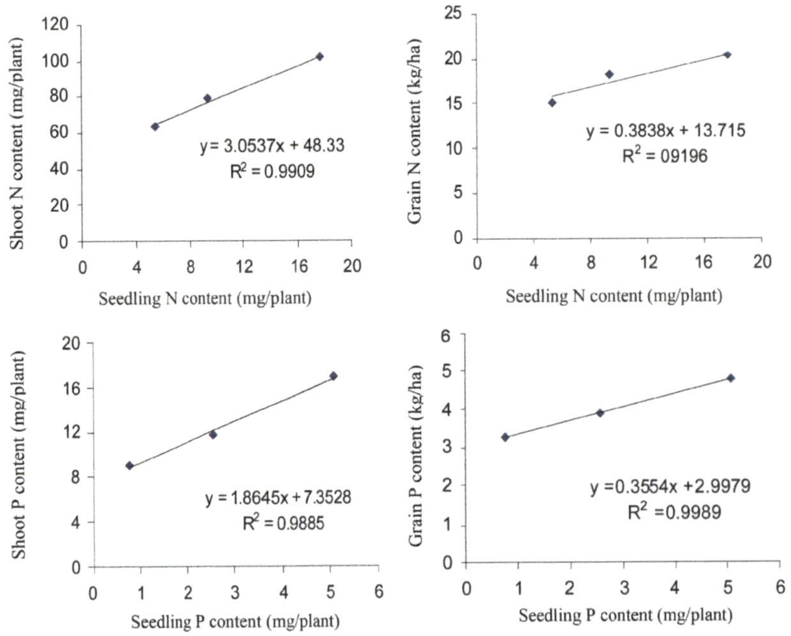

Figure 3. Relationship between N and P content of seedlings at pulling and N and P content of shoots at panicle initiation and of grain at maturity. Values are means for nursery fertiliser treatments, pooled across main field treatments. Experiment 1.

to the plant's subsequent growth and final grain yield [5] [18]. However, the increased post-transplanting shoot growth, and eventually grain yield was not simply the direct consequence of the increased nutrient content in shoots of transplanted seedlings. At PI, shoots contained 6 - 12 times as much as N and 3 - 8 times as much P as transplanted seedlings. Grain contained over a thirty-fold more N and P than transplanted seedlings. The nutrient content in the seedlings at transplanting was only a small proportion of final nutrient content at maturity. Hence the extra nutrient content of fertilised seedlings at transplanting must have acted to accelerate the recovery of seedlings after transplanting, and increase the size of the root and shoot system after transplanting so that plants make more effective use of soil nutrient supply, and tolerate stress factors like nutrient deficiency, drought or submergence. Until clearer definition of the mechanism for the fertilizer stimulation of post-transplanting growth is achieved, it seems reasonable to categorise the response to nursery fertilizer application as a seedling vigor effect. Indeed, in our previous study, a range of treatments (root pruning, leaf clipping, temporary submergence), that depressed seedling vigour impaired post transplanting growth [3]. Decreases in seedling vigour can be induced through nutrient as well as non-nutrient treatments, but the common effect appears to be suppressed recovery after transplanting, with persistent effects including lower grain yield.

Various levels of vigor in the seedlings were created in this experiment by using seedlings with different ages. Six week-old seedlings were more vigorous than younger seedlings in terms of dry matter and stem stiffness. They better withstood submergence soon after transplanting than the 4 WAS seedlings (**Table 3**; **Figure 1**). The ability of the plant to withstand complete submergence increases with age [20]. The poorer growth of the 4 WAS seedlings compared to the 6 WAS seedlings even though submergence was more severe in the latter case might be related to plant age. The risk of submergence is particularly pertinent to the promotion in the rainfed lowlands of rice productions systems such as SRI which transplant seedlings at only 8 - 12 days age [21] [22]. While there is evidence of greater tillering and grain yield from transplanting the very young seedlings used in SRI, this is likely to be a risky strategy for the submergence-prone rainfed lowlands.

Larger but older seedlings (9 WAS) produced less grain yield than the younger seedlings (4 and 6 WAS). The depression in yield may be due to the shortening of the growth period in the main field caused by delaying transplanting (**Figure 2**) [23] [24] although the lower nutrient concentrations and greater leaf chlorosis of these seedlings at transplanting may be partly responsible. Under the 21 days shortened growth period, plants transplanted at 9 WAS were apparently less efficient in nutrient uptake even when the main field was fertilised as reflected in lower shoot dry weight at PI and straw nutrient concentrations at maturity. This also supports the previous findings by many workers who have suggested that seedlings as young as 3 - 4 weeks old produced better growth and higher grain yields [24] [25] and contained higher nutrient concentration than aged seedlings [22]. In a recent review [26], optimal seedling age for transplanted rice yield from a number of studies ranged from 25 to 45 days.

Water regimes in the main field had minimal effects on growth of plants in this study apart from those related to submergence which was discussed above. Flooding of all plots after transplanting of seedlings at 4 and 6 WAS made it difficult to impose a clear water stress treatment in the main field. Indeed in the early stages, the drained water plots had adequate standing water levels and the water adequate plots became severely flooded. Water regime, therefore, had little or no effect on plant dry matter before PI except for 4 and 9 week-old seedlings transplanted into main fields without NPK. Water stress periods occurring at or soon after transplanting may be less limiting to post-transplanting growth of fertilised seedlings with the capacity to establish a larger root system. By contrast, the inhibition of plant growth by water drainage after PI was evident in the main field not fertilised with NPK, but not in those fertilised with NPK. Response of rice yield to transplanting the older seedlings (40 vs. 20 days) in Nepal, was accentuated in a drought year compared to the following year with non-limiting water availability [7]. Further study of the effect of an earlier and more severe water stress on the response of rice to nursery fertilizer is warranted since rainfed lowland rice often experiences such periods of water stress.

In summary, the results suggest that supplying N and P fertilizers in both the nursery and the main field, and transplanting at 6 WAS, offer good prospects for increased yields of rice in the low fertility, submergence- and drought-prone rainfed lowland soils that are common in South and Southeast Asia. The hypothesis that nursery fertilizer application increases the stress tolerance of seedlings post-transplanting was only partially supported since it increased plant survival due to temporary submergence. Nutrient stress in the main fields induced by not supplying NPK lessened rather than strengthened the responsiveness of plants to nursery fertilizer application. This suggests that the benefit of fertilizer application to the nursery is not due to increased nutrient content of the transplanted seedlings *per se*, but rather an indirect response to increased seedling vigor.

Acknowledgements

The authors are grateful to the Australian International Agency for Development for a scholarship to the senior author and for funding support provided through the International Rice Research Institute.

References

[1] Wade, L.J., Fukai, S., Samson, B.K., Ali, A. and Mazid, M.A. (1999) Rainfed Lowland Rice: Physical Environment and Cultivar Requirements. *Field Crops Research*, **64**, 3-12. http://dx.doi.org/10.1016/S0378-4290(99)00047-7

[2] Bell, R.W. and Seng, V. (2004) Rainfed Lowland Rice-Growing Soils of Cambodia, Laos, and Northeast Thailand. In: Seng, V., Craswell, E., Fukai, S. and Fischer, K., Eds., *Water in Agriculture. ACIAR Proceedings*, 116, 161-173.

[3] Ros, C., Bell, R.W. and White, P.F. (2003) Seedling Vigour and the Early Growth of Transplanted Rice (*Oryza sativa*). *Plant and Soil*, **252**, 325-337. http://dx.doi.org/10.1023/A:1024736104668

[4] Ros, C., White, P.F. and Bell, R.W. (1998) Field Survey on Nursery and Mainfield Fertilizer Management. *Cambodian Journal of Agriculture*, **1**, 22-33.

[5] Ros, C., Bell, R.W. and White, P.F. (1997) Nursery Applications of N and P Increase Rice (*Oryza sativa*) Yield. *Kasetsart Journal: Natural Science*, **31**, 96-105.

[6] Javier, E. (1997) Rice Ecosystems and Varieties. In: Nesbitt, H. J., Ed., *Rice Production in Cambodia*, International Rice Research Institute, Los Banos, 39-82.

[7] Adhikari, B.B., Mehera, B. and Haefele, S. (2013) Impact of Rice Nursery Nutrient Management, Seeding Density and Seedling Age on Yield and Yield Attributes. *American Journal Plant Sciences*, **4**, 146-155. http://dx.doi.org/10.4236/ajps.2013.412a3017

[8] Sarwa, N., Maqsood, M., Wajid, S.A. and Anwar-ul-Haq, M. (2011) Impact of Nursery Seeding Density, Nitrogen, and Seedling Age on Yield and Yield Attributes of Fine Rice. *Chilean Journal of Agricultural Science*, **71**, 343-349. http://dx.doi.org/10.4067/S0718-58392011000300001

[9] White, P.F., Oberthür, T. and Pheav, S. (1997) The Soils Used for Rice Production in Cambodia. A Manual for their Identification and Management. International Rice Research Institute, Los Baños, 70.

[10] Gomez, K.A., Bartolome, V.I., Calinga, R.T. and Cosico, A.B. (1994) IRRISTAT User's Manual. Version 3. International Rice Research Institute, Los Banos.

[11] Shalaby, Y.Y., Kobbia, T.E. and Omar, M.A. (1969) Rice Yield as Influenced by N-Fertilization in Nursery. *Journal of Soil Science of the United Arab Republic*, **9**, 185-199.

[12] Rajagopalan, S. and Krishnarajan, J. (1987) Effect of Nursery Management Technique on Phosphorus Nutrition of Rice in Main Field. *Oryza*, **24**, 199-209.

[13] Panda, M.M., Reddy, M.D. and Sharma, A.R. (1991) Yield Performance of Rainfed Lowland Rice as Affected by Nursery Fertilization under Conditions of Intermediate Deepwater (15-50 cm) and Flash Floods. *Plant and Soil*, **132**, 65-71. http://dx.doi.org/10.1007/BF00011013

[14] White, P.F., Nesbitt, H.J., Ros, C., Seng, V. and Pheav, S. (1995) Granulation and Formulation of Local Rock Phosphate Deposits: a Potential Boon for Cambodia. *Agro-Chemicals News in Brief, ESCAP/FAO/UNIDO*, **18**, 26-31.

[15] Seng, V., Ros, C., Bell, R.W., White, P.F. and Hin, S. (2001) Nutrient Requirements for Lowland Rice in Cambodia. In: Fukai, S. and Basnayake, J, Eds., *Increased Lowland Rice Production in the Mekong Region, Proceedings of an International Workshop*, Vientiane, 30 October-1 November 2000, ACIAR, Canberra, 169-178.

[16] Lando, R.P. and Mak, S. (1994) Rainfed Lowland Rice in Cambodia: A Baseline Survey. *IRRI Research Paper Series*, No. 152, International Rice Research Institute, Los Banos.

[17] Adkins, S.W., Shiraishi, T. and McComb, J.A. (1990) Submergence Tolerance of Rice—A New Glasshouse Method for the Experimental Submergence of Plants. *Physiologia Plantarum*, **80**, 642-646. http://dx.doi.org/10.1111/j.1399-3054.1990.tb05691.x

[18] Matsuo, T., Kumazawa, K., Ishii, R., Ishihara, K. and Hirata, H. (1995) Science of the Rice Plant. Vol 2. Physiology. Food and Agricultural Policy Research Center, Tokyo.

[19] Ismail, A.M., Johnson, D.E., Ella, E.S., Vergara, G.V. and Baltazar, A.M. (2012) Adaptation to Flooding during Emergence and Seedling Growth in rice And Weeds, and Implications for Crop Establishment. *AoB Plants*, **2012**, pls019. http://dx.doi.org/10.1093/aobpla/pls019

[20] Palada, M.C. and Vergara, B.S. (1972) Environmental Effects on the Resistance of Rice Seedlings to Complete Sub-

mergence. *Crop Science*, **12**, 209-212. http://dx.doi.org/10.2135/cropsci1972.0011183X001200020018x

[21] Randriamiharisoa, R., Barison, J. and Uphoff, N. (2006) Soil Biological Contributions to the System of Rice Intensification. In: Uphoff, N., Ball, A.S., Fernades, E., Herren, H., Huson, O., Ling, M., Palm, C., Pretty, J., Sanchaez, P., Sanginga, N. and Thies, J., Eds., *Biological Approaches to Sustainable Soil Systems*, CRC Press, Taylor and Francis Group, Boca Raton, 409-424.

[22] Mandal, B.K. and Ray, P.K. (1984) Effect of Age of Seedlings and Nitrogen Levels on the Growth and Yield of Mahsuri. *Indian Journal of Plant Physiology*, **27**, 303-306.

[23] Basnayake, J., Fukai, S., Sipaseuth, Schiller, J.M. and Monthathip, C. (2006) Advances in Agronomic Research in the Lowland Rice Environments of Laos. In: Schiller, J.M., Chanphengsay, M.B., Linquist, B. and AppaRao, S., Eds., *Rice in Laos*, International Rice Research Institute, Los Banos, 349-369.

[24] Singh, D.N. and Bhattacharya, A. (1975) Effect of Age of Seedling at Transplanting on Growth and Yield of Rice Varieties (*Oryza sativa*). *Experimental Agriculture*, **11**, 65-74. http://dx.doi.org/10.1017/S0014479700006244

[25] Patel, C.L., Patel, R.B., Patel, Z.G. and Patel, H.C. (1983) Response of Rice (Var. Mahsuri) to Age of Seedlings at Transplanting, Methods of Transplanting and Depth of Planting. *GAU Research Journal*, **8**, 55-57.

[26] Rajendran, K. and Ganesa raja, V. (2014) Effect of Age of Seedlings on Growth and Yield of Rice. *Indian Journal of Advances in Plant Research*, **1**, 62-66.

15

Early Detection of *Cercospora* Species in Soybean Plants: Immunologic and Molecular Methods

María Gabriela Latorre Rapela[1], María Cristina Lurá[1], Iván Marcipar[2]

[1]Cátedra de Microbiología General, Facultad de Bioquímica y Ciencias Biológicas, Universidad Nacional del Litoral, Santa Fe, Argentina
[2]Cátedra de Tecnología Inmunológica, Facultad de Bioquímica y Ciencias Biológicas, Universidad Nacional del Litoral, Santa Fe, Argentina
Email: latorrerapela@gmail.com

Abstract

Late-cycle diseases (LCD) cause a significant deterioration in quality and reduce yields in soybean crops. In Argentina, in particular, leaf blight and purple seed stain, caused by the agent *Cercospora kikuchii*, and frog eye spot, caused by *C. sojina*, are the prevailing sources of diseases. The early, rapid and accurate detection of these phytopathogens becomes essential, and would contribute to preserving both the environment and the health of humans and animals by preventing the wasteful or improper use of chemicals such as pesticides. In order to detect *Cercospora* species in soybean plants at an early stage, immunochemical and molecular techniques were developed in this work. Strains from the NITE Biological Resource Center collection (Japan): *Cercospora kikuchii* NBRC 6711 and *Cercospora sojina* NBRC 6715 and regional isolates of *C. kikuchii* were used. To develop Dot-Blot and PCR techniques, experiments with plants undergoing different treatments were carried out: those experimentally inoculated with these fungi, those treated with sterile water and healthy plants as well. Both techniques allowed the detection, at early stages, of *Cercospora* species involved in two of the most frequent LCD in the country, when the cercosporin concentration produced by the fungus was higher than 3.93 ± 0.39 nmol·cyl^{-1} ±SD. The sensitivity between both techniques was very different. While Dot-Blot allowed the detection of the disease 4 days after inoculation, PCR detected it after 4 hours, even without visible symptoms of the disease.

Keywords

***Cercospora kikuchii*, *Cercospora sojina*, Dot-Blot, PCR, Late-Cycle Diseases**

1. Introduction

Late-cycle diseases (LCD) are very frequent in soybean crops, quality deterioration and yield decrease being the ordinary consequences. One of the main problems associated with these diseases is that visible symptoms appear only at intermediate or advanced stages, close to the harvest time [1].

Among the causative agents of LCD, *Cercospora* genus species are a major problem. In Argentina, in particular, leaf blight and purple seed stain, whose agent is *Cercospora kikuchii*, and frog eye spot, caused by *C. sojina*, are common diseases that reduce the plant nutritional value and lead to severe crop damage [1] [2]. One of the main pathogenicity factors, shared by many species of this genus, has been demonstrated to be a red toxin called cercosporin. Its production is regulated by CFP (Cercosporin Facilitator Protein), which is specific of the genus and coded by the gene *cfp* (cercosporin facilitator protein) [3]. CFP also exports cercosporin out of the fungus [4], which suggests that the gene plays a fundamental role in toxin production regulation. Therefore, the detection of the *cfp* gene in an infected plant would indicate the presence of a fungus capable of producing cercosporin and consequently, causing the disease.

Plants highly affected lose a large amount of foliage, resulting in an anticipated maturity with an inadequate filling of pods and a reduced germination of seeds. These pathologies can be controlled by many strategies such as the application of fertilizers, fungicides and herbicides, combined with the use of cultivars of greater genetic potential, the massive diffusion of transgenic materials in soybean and handling practices like direct sowing [5].

In general, the diagnosis of fungal phytopathogens is based on the observation of signs and symptoms of the disease which, in the case under study, are visible only when the disease is at an advanced stage. Hence, the early, rapid and accurate detection of these phytopathogens becomes essential, and would contribute to preserving both the environment and the health of humans and animals by preventing the wasteful or improper use of chemicals such as pesticides [6].

As a result of advances in biotechnology, new procedures, such as immunochemical and molecular techniques, are today available to complement or replace the traditional laboratory procedures, so as to accelerate the achievement of results and enable an earlier diagnosis of diseases [7]. Diseases caused by microorganisms can be diagnosed by identifying a unique characteristic of the organism, such as a protein or its genetic material.

Among the various immunochemical techniques, Dot-ELISA or Dot-Blot appear as a more versatile technique applicable to the diagnosis of a large number of samples [8] [9]. It is a rapid ELISA that uses as solid support nitrocellulose paper, characterized by an excellent capacity to adsorb proteins [10]. The (synthetic or recombinant) antigen is usually passively adsorbed (blotting) to the support dropwise ("dot"). This immunoassay is economic and easy to implement since it generates an insoluble, colored precipitate which allows the reaction to be followed by a visual reading [11].

PCR, on the other hand, is a technique with great advantages in fungi identification. It is a rapid method, since the process of extraction and amplification of DNA and electrophoresis can be performed in 24 h, which substantially reduces the diagnosis time as compared to the conventional methods-direct examination of colonies and culture [12]. Its high sensitivity allows detecting low concentrations of the DNA to be amplified in almost any type of sample and with high specificity, which is determined by the sequence of oligonucleotides used and the DNA annealing conditions [13] [14]. Though having all these advantages, a specific laboratory infrastructure, expensive equipment and trained personnel are required, all of which are not always available.

In order to provide tools for the detection of *Cercospora* species in soybean plants at an early stage in laboratories with high or low level of complexity, both an immunochemical (Dot-Blot) technique requiring low level of complexity and a molecular (PCR) technique demanding higher complexity laboratories were developed in this work.

2. Materials and Methods

2.1. Fungi Used

The following strains, belonging to NITE Biological Resource Center (Japan) collection, were used in the experience: *Cercospora kikuchii* NBRC 6711 (CK6711) and *Cercospora sojina* NBRC 6715 (CS6715), as well as regional isolates of *C. kikuchii*, which produced cercosporin in high (CK32, 148.51 ± 3.98 nmol·cyl^{-1} ±SD) and low (CK15, 3.93 ± 0.39 nmol·cyl^{-1} ±SD) concentrations [15], available at the strain collection of the General Microbiology chair at the Facultad de Bioquímica y Ciencias Biológicas of Universidad Nacional del Litoral

(FBCB-UNL). A *Penicillium* sp. isolate was also included for genus specificity study [16].

2.2. Plants Used

Plants at phenological stage V3, experimentally inoculated with CK6711, CS6715, CK15 and CK32, as described in Latorre Rapela *et al.* [17], were used in the experiment. Healthy plants (negative control) and plants treated with sterile water were also used.

In order to determine how early the fungus could be detected, different sampling times (4 and 24 h, and 2, 4 and 6 days) were set, since in previous tests, lesions had been visible to the naked eye in plants infected with *Cercospora* 5 days after inoculation.

2.3. Dot-Blot Technique

2.3.1. Obtaining Recombinant CFP Protein and Anti-CFP Antibodies

The amino acid sequence of CFP has 607 amino acids [3], the antigenic region being located at the C-terminal part and entirely coded in the last exon of the gene (2628 bp··· >3150 bp) (Genbank accession number: AF091042).

Specific primers -"Forward" (FP: 5-gtcctgcccgatgctca-3) and "Reverse" (RP: 5- tcacactgctttgcccac-3)-were designed using the DNA Start Sequence Analysis Software program. These ones were selected between the set of primers suggested by the program taking into account the highest score to amplify nearly all the exon coding region. Primers were synthetized by Genbiotech SRL (Argentina). The selected fragment was ligated firstly to the plasmid vector pGEM-TEasy (Promega®, EE.UU) and then to the expression vector pET-32a (Novagene, EE.UU). Plasmids ligated to PCR amplicon were transformed into *Escherichia coli* BL21 (DE3).

The recombinant protein obtained was purified in an IDA-Sepharose (Novagen, U.S.A) column, over which Ni^{2+} was immobilized.

The specific anti-CFP antibodies were obtained from a rabbit immunized with such protein. Immunization was carried out at the Centro de Experimentaciones Biológicas y Bioterio of Facultad de Ciencias Veterinarias-Universidad Nacional del Litoral (FCV-UNL).

2.3.2. CFP Protein Extraction

The extraction of the CFP protein was carried out according to the protocol proposed by Rollins *et al.* [18], weighing 100 mg of powdered fungal mycelium corresponding to CK6711 and CS6715 (positive controls) and *Penicillium* sp. (negative control) and 100 mg of plant tissue infected with CK32. One ml of extraction buffer was added to each tube and incubated overnight at 4˚C in the dark. After centrifugation at maximum speed (13,000 rpm) for 10 min, the supernatant were transferred to respective Eppendorf tubes and stored at −20˚C until processing.

Total protein determination was performed following Bradford technique [19]. The same methodology was applied to plants treated with water and to the uninoculated. Extractions were carried out with samples taken at 4 and 24 h and 2, 4 and 6 days after inoculations.

2.3.3. Adsorption of Nonspecific Antibodies from Rabbit Serum

For adsorption of nonspecific antibodies, nitrocellulose membrane strips were cut, placed into an Eppendorf tube and embedded with protein extract of the fungus CK15. They were then removed and let dry at room temperature. When dry enough, the membranes were placed into another Eppendorf tube with 1 ml of rabbit serum, stirring for 30 min at room temperature [20].

The procedure was repeated at least 5 times. Finally the adsorbed serum was collected in another Eppendorf tube, fractionated and kept at −20˚C until use.

The presence of antibodies specific for the native protein of CK6711 in the adsorbed serum was determined by Western blot.

2.3.4. Development of Dot-Blot Technique and Determination of Cut-Off Line

The following procedure was applied to every extract obtained: 1 μl of the pure soluble extract, followed by decreasing amounts (1/2, 1/8, 1/32 and 1/64) of this same extract was placed on a nitrocellulose membrane, so as to fix the antigen (CFP protein) on the membrane in decreasing amounts. Each extract was allowed to dry at room temperature and blocked with PBS (Phosphate Buffered Saline-Tris)-skim milk 5% for 60 min. Afterwards, three 5 min washing cycles with PBS-T were performed. It was then incubated for 1 h in a vial with 1ml of rabbit serum adsorbed (1:500) on 1% PBS-milk, subjected to three new washes and incubated again for 1 h with 1 ml of anti-IgG total rabbit serum conjugated with peroxidase (diluted 1/500 with 1% PBS). After a new three times-washing step, the membranes were treated with a solution of PBS, hydrogen peroxide 0.4% and 3%, 3'-diaminobenzidine (Sigma Chemical Co.) for 20 min. The reaction was stopped by washing with water.

All incubation steps were carried out at moderate stirring. All experiments were carried out in duplicate.

This method was designed to facilitate visual reading while decreasing the subjectivity of observations. The applied principle is the semi quantification of the amount of CFP protein present in the plant extracts, while dots were loaded using decreasing amounts of the extract. As a result, more dots are recognized, more amount of protein is present in the sample. Notwithstanding, a weakness of Dot-Blot method for diagnostic purpose is that different users may interpret a weak signal as positive or negative. Then if the result is expressed based only in the last blot with the minimal signal, the variability of the informed result for a defined sampled might be high between users. To obtain a more accurate visual lecture, the signals for the five dots corresponding to a single sample were registered as high or low and were integrated in a score. Every signal obtained on the membrane was considered to have a numerical basis of 4, 2 or 1, depending on the strength of the signal (very strong, low or light, or not visible, respectively). Then, these signals were integrated in a score by multiplying them together. Indeed, different criteria were tested to determine the best way to integrate the signals and the product showed the best discrimination between positives and negatives samples (data not shown).

Once the score was obtained, a cut-off line was defined through the ROC (Receiver Operating Characteristic) curve with the MedCalc program, in order to obtain the best discrimination between reactive and nonreactive samples. For this purpose, plants with symptoms of *Cercospora* disease and healthy plants were used. CK6711 was also included as a positive control. The score was considered less than or equal to 2 for healthy plants and greater than or equal to 4 for infected plants, so that the score value equal to or less than 2 was considered as a marker of lack of CFP in the leaf tissue.

2.4. PCR Technique

Extraction of Total DNA from the Samples and *cfp* Gen Identification

The extraction of total DNA from fungus and plant tissue was carried out according to Lee and Taylor [21] protocol. DNA quality was evaluated by electrophoresis in agarose gel [22].

In a first test, primers FP and RP, previously designed to amplify the coding sequence fragment were used, evaluating different reaction conditions with the DNA of the CK6711 fungus. This technique was carried out with DNA of plants inoculated with the regional isolates CK32 and CK15, using DNA of CK6711 and CS6715 as positive controls.

Then, new primers were designed in order to increase specificity with plant tissues. For this purpose, the alignment and computer analysis of the sequence corresponding to the last exon of the gene, deposited in the GenBank (Accession number AF091042), was made. The program DNA start Sequence Analysis Software was applied, selecting those primers with the highest score (CFP-1: 5-tgaggtcgcagagcaagcat-3 and CFP -2: 5-cgcgtcgtcccattagttttct-3), which were then sent for synthesis to Genbiotech SRL.

For each DNA from fungus and plant tissue obtained, amplification reactions were conducted in a total volume of 50 µl, with 20 ng of DNA and 5 µl of reaction buffer 1× (InbioHighway, Argentina), 1.5 mM $MgCl_2$, 0.2 mM of each dNTPs (Inbio Highway, Argentina), 0.2 µM of each oligonucleotide and 2.5 U of Taq DNA polymerase (Inbio Highway, Argentina). Mineral oil (15 µl) was added at the end of the reaction.

Amplification was carried out as follows: one 3 min denaturalization cycle at 95°C; 35 cycles comprising 1 min at 58°C, 1 min at 72°C and 1 min at 95°C, with a final extension cycle of 5 min at 72°C.

Amplification products were separated by electrophoresis on horizontal submerged agarose gels (1.5% w/v) with TBE 0.5× (Tris-borate 0.089 M, EDTA 0.002 M) [22]. Runtime took 180 min, with a constant voltage of 100 V. The gel staining was performed with ethidium bromide and 100-bp DNA Ladder (Inbio Highway, Argentina) was used as MW marker.

All experiments were carried out in duplicate.

The band profile obtained was photographed with Gel Doc XR System (BIORAD-Life Science Cat. # 170 - 8170) using Quantity One Software.

This test was applied to DNA of the plants inoculated with the regional isolates CK32 and CK15, using DNA of CK6711 and CS6715 as positive controls.

3. Results

The homology between *cfp* sequence and *Cercospora* genus species was corroborated (using BLAST program) when seeking for identity in GenBank database. No homology was found with gene sequences of other fungal species indexed so far.

Figure 1 shows the schematic representation of *cfp* gene regions which are found in the last exon coding CFP protein, its C terminal region being expressed for the immunochemical test (451 bp) and sequence of 264 bp being used for PCR.

3.1. Dot-Blot Technique

The results of Dot-Blot technique are shown in **Figure 2**, where signs of different intensity can be seen, whether they correspond to healthy (lines 9 - 15) or infected plants (lines 1 - 8).

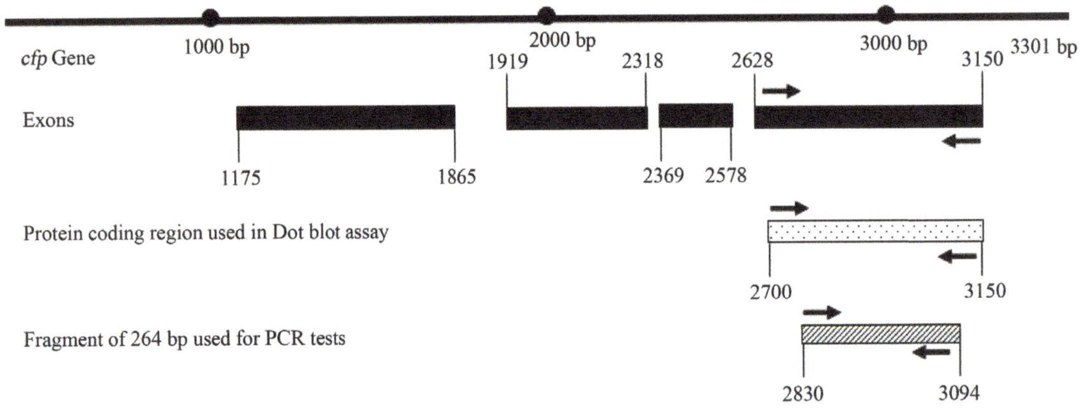

Figure 1. Schematic representation of the *cfp* gene regions used.

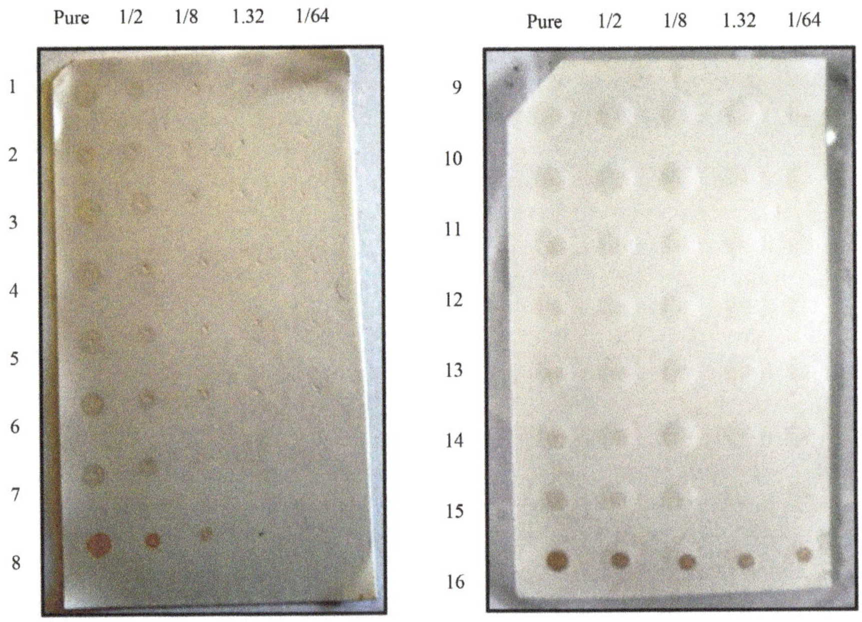

Figure 2. Dot-Blot of soluble extracts from plants and positive control. Lines 1 - 8: plants with symptoms of the disease; lines 9 - 15: healthy plants; line 16: *Cercospora kikuchii* NBRC 6711 (positive control).

The score of each sample is shown in **Table 1**.

According to the criterion adopted, the extracts of the reference strains CK6711 and CS6715 showed score values higher than 4, whereas *Penicillium* sp. extract (negative control) had a value less than 2.

Four days after inoculation, every plant extract inoculated with regional isolate CK32 showed a score 32, which indicated the presence of CFP protein in the leaf tissue. Plants collected 4 and 24 h and 2 days post-inoculation showed values less than 2, as well as plants treated with sterile water and uninoculated ones.

Score values between 128 and 512 appeared for the positive control, whereas it was 1 for the negative control. As an example, Dot-Blot images applied to extracts of plants infected with CK6711 at different times post-inoculation are shown (**Figure 3**). Similar results were obtained with extracts of plants infected with CS6715.

3.2. PCR Technique

When separating the amplification products obtained with oligonucleotides FP and RP, the fungus corresponding to the reference strain could be seen to give a clear band at 451 bp, approximately; however, both infected and healthy plants showed nonspecific bands.

Use of oligonucleotides CFP-1 and CFP-2 allowed the amplification of an inner portion of the *cfp* fragment (about 264 bp). This fragment became visible 4 h after plant inoculation with the reference strains CK6711 and CS6715 and the regional isolate CK32. This signal intensified with time.

No amplification was detected for any of the sampling times in uninoculated plants and plants treated with sterile water. The same response was observed in samples obtained from plants inoculated with CK15, a producer of low concentration of cercosporin.

As an example, **Figure 4** shows one of the results obtained with PCR technique.

Table 1. Score value determination for extracts of plants and positive control.

Sample[a]	Dot-Blot intensity and score calculation	Score	Sample[a]	Dot-Blot intensity and score calculation	Score
1	2 × 2 × 2 × 1 × 1	8	9	1 × 1 × 1 × 1 × 1	1
2	2 × 2 × 2 × 1 × 1	8	10	2 × 1 × 1 × 1 × 1	2
3	2 × 2 × 2 × 1 × 1	8	11	2 × 1 × 1 × 1 × 1	2
4	2 × 2 × 2 × 1 × 1	8	12	1 × 1 × 1 × 1 × 1	1
5	2 × 2 × 2 × 1 × 1	8	13	1 × 1 × 1 × 1 × 1	1
6	2 × 2 × 2 × 1 × 1	8	14	2 × 1 × 1 × 1 × 1	2
7	2 × 2 × 2 × 1 × 1	8	15	2 × 1 × 1 × 1 × 1	2
8	4 × 4 × 2 × 1 × 1	32	16	4 × 4 × 4 × 4 × 4	1024

[a]Samples 1 - 8: plants with symptoms of the disease; samples 9 - 15: plants without symptoms; Sample 16: *Cercospora kikuchii* NBRC 6711.

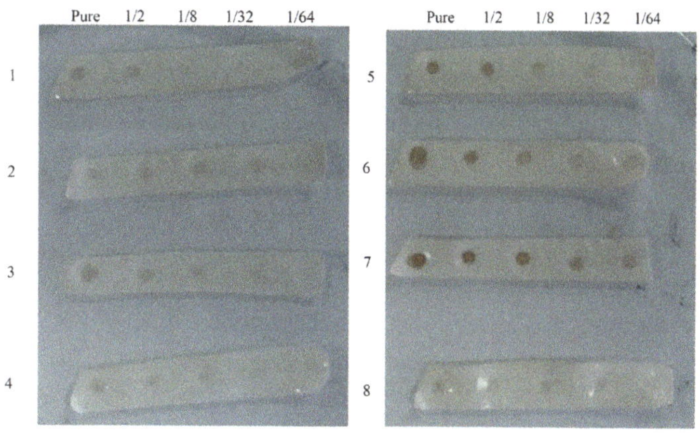

Figure 3. Images of Dot-Blot applied to extracts from plants infected with *Cercospora kikuchii* NBRC 6711 at different times post-inoculation. 1: plant inoculated with water; 2: uninoculated plant; 3, 4, 5 and 6: plants inoculated with *Cercospora kikuchii* NBRC 6711 and collected at 4 h, 24 h, 4 days and 6 days, respectively, 7: positive control (extract of *C. kikuchii* NBRC 6711); 8: negative control (extract of *Penicillium* sp.)

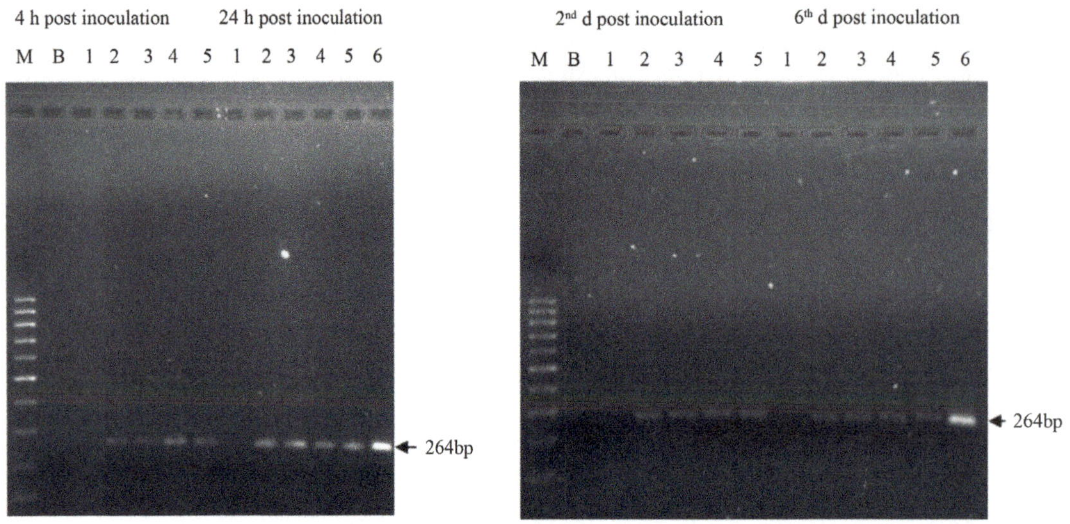

Figure 4. Amplification of *cfp* fragment by PCR in soybeans, at different times post-inoculation. M: marker 100 bp DNA ladder (InbioHighway, Argentina); B: reagent blank; 1: plant treated with water; 2 and 3: plants inoculated with CK32; 4 and 5: plants inoculated with *Cercospora sojina* NBRC 6715; 6: positive control (*Cercospora kikuchii* NBRC 6711).

4. Discussion

Accurate and early identification of the involved pathogen, as well as disease diagnosis, are crucial requisites for a thorough understanding of plant diseases and appropriate decision-making on the effective measures to be taken for crop management.

Traditional methods for phytopathogen identification in the laboratory have turned out to be slow and vague, which has led to the search for specific, sensitive and fast detection techniques [23].

Evaluating the use of an immunochemical technique such as Dot-Blot or Dot-ELISA was the aim of this work. For good phytopathogen detection, a nitrocellulose membrane was used as an adsorption surface, since it allows the immobilization of more physicochemical characteristics of the antigens than ELISA plate [9].

Also, following Lartey *et al.* [20], it was decided to eliminate the nonspecific antibodies present in rabbit serum by adsorbing them with the extract of a fungus which does not produce CFP protein. With this purpose, the regional isolate CK15, which was a cercosporine low producer, was used in our experiment. When it grew on PDA (potato dextrose agar), no red pigment diffused to the medium, causing small chlorotic lesions on the leaf tissue of the soybean plant on which it was inoculated.

The adsorption allowed the elimination of nonspecific reactivities, and the incorporation of a score facilitated interpretation of results, thus resulting in an easy, fast and economic technique that could be used in the field as a screening method [24].

Dot-ELISA can be completed in less than 3 h, allows the simultaneous processing of several samples and, as the interpretation of results is visual, the end point can be easily determined. It is one of the most versatile tests to detect either antigen or antibody and is adaptable to both laboratory and field conditions [8].

This technique has been used in Colombia to detect the mycotoxin beauvericin [25]. It has also been incorporated to the fast detection of different etiologic agents (parasites, bacteria and viruses) responsible for human diseases [8] [26]-[31].

Dot-Blot technique allowed the CFP protein detection 4 days after experimental inoculation, even when no disease symptoms were visible.

Besides the immunochemical technique, a molecular technique (PCR) was developed for the detection of *Cercospora* species attacking soybean.

In 1999, Xu *et al.* [32] developed a PCR technique which allowed detection of *Sporisorium reiliana* in maize, concluding that it was important to carry out epidemiological and etiological studies. This kind of methodology was also applied by other authors for detection of *Phytophtora capsici* in chile plants with wilt symptoms, using oligonucleotides directed against specific sequences of cutinase gene [13]. Pavon Moreno *et al.* [33], on the other hand, were able to detect *Alternaria* spp. in fruit and vegetable products using oligonucleotids specific for the gene Alt a 1 coding for the main allergen of *A. alternate*.

Lartey *et al.* [34] presented a real-time PCR protocol for the rapid detection of *C. beticola* in sugar beet tissues, but, unlike this paper, they amplified a fragment corresponding to the *Cercospora* actin gene.

Results showed that oligonucleotides CFP-1 and CFP-2, designed to amplify an inner portion of *cfp* fragment, were highly specific of *Cercospora* species, so they could determine its presence from both DNA extractions of the fungus mycelium developed "*in vitro*" and plants inoculated with the fungus, even when no disease symptoms were observed. The gen was detected 4 h after experimental inoculation. It is important to point out that when PCR was conducted with plants inoculated with the regional isolate CK15, the 264 bp band corresponding to *cfp* gene was not detected. It should be remembered that this fungus produced very low concentrations of cercosporin (3.93 ± 0.39 nmol·cyl^{-1} ±SD) and hence, small lesions, which could be due to a similar fact reported by Daub *et al.* [35]. They observed that *C. kikuchii* mutants exhibiting a *cfp* gene disruption produced a significant reduction both in cercosporin production and in soybean symptom development.

So far, these methodologies for *Cercospora* species detection have not been developed somewhere in the world. Chanda *et al.* [36] have carried on a quantitative PCR protocol for the detection of *Cercospora kikuchii* in soybean leaves using specific oligonucleotids for the CTB6 gene of *C. kikuchii*. These oligonucleotids are different from the ones described in this investigation and do not determine infections with different *Cercospora* species.

It is important to point out that in this paper, in both techniques, the two strains used yield coincident results, which is because such species produce cercosporin and its transport protein [37].

As a conclusion, the above described techniques allowed detection, at early stages, of the *Cercospora* species involved in two of the most frequent LCD in Argentina, when the cercosporin concentration produced by the fungus was higher than 3.93 ± 0.39 nmol·cyl^{-1} ±SD.

However, sensitivity between both techniques was very different, since PCR always allowed detection at early stages of infection, without visible symptoms of the disease. The usefulness of both techniques should be confirmed by further validation with field samples. It has to be considered that this study was performed with the development of an experimental model to infect soy plants with *Cercospora*. This methodology has allowed reproducing infection symptoms in healthy plant tissues and determining how many hours and days after the inoculation the injuries were visible to the naked eye. This model allowed assessing the previously optimized techniques with high reliability. Unfortunately, a reliable assessment of the method in field crop would be difficult because there is no "gold standard" that allows to know, at early stages, the infection caused by this fungal genus and to determine the time of the crop infection in the field. Then an experimental infection in field crop would be necessary to validate this method.

Acknowledgements

This work was funded by grants from Universidad Nacional del Litoral, Argentina, CAID 2009. The authors would like to thank Eng. Roxana Maumary for her contribution with plant collection and the experimental pattern of soybean plants infected by *Cercospora* species.

References

[1] Couretot, L. and Ivancovich, A. (2010) Enfermedades de final de ciclo de la soja. In: *Taller de Diagnóstico y manejo de Enfermedades de Soja*, Secretaría de Agricultura, Ganadería, Pesca y Alimentos, Argentina, 5-10. http://www.minagri.gob.ar/site/agricultura/roya_soja/07=informes/01_Informes/_archivo/090000-Informes%20campa%C3%B1as%202009-2010/000000-Taller%20de%20Diagn%C3%B3stico%20y%20Manejo%20de%20Enfermedades%20de%20Soja%20PNRS.pdf?PHPSESSID=d2f50273a1d4ccb6df37161bfdc1b76f.

[2] Carmona, M.A., Formento, N.A. and Scandiani, M.M. (2010) Manual mancha ojo de rana. Horizonte A, Buenos Aires.

[3] Callahan, T.M., Rose, M.S., Meade, M.J., Ehrenshaft, M. and Upchurch, R.G. (1999) CFP, the Putative Cercosporin Transporter of *Cercospora kikuchii*, Is Required for Wild Type Cercosporin Production, Resistance, and Virulence on Soybean. *Molecular Plant-Microbe Interactions*, **12**, 901-910. http://dx.doi.org/10.1094/MPMI.1999.12.10.901

[4] Daub, M.E. and Chung, K.R. (2007) Cercosporin: A Photoactivated Toxin in Plant Disease. American Phytopathological Society netFeatures. http://dx.doi.org/10.1094/APSnetFeature/2007-0207

[5] Carmona, M. and Reis, E.M. (2009) Critério: sistema de pontuação para aplicação de fungicidas para as doenças de

final de ciclo na cultura da soja. In: Reis, E.M., Ed., *Critérios indicadores do momento para aplicação de fungicidas visando ao controle de doenças em soja e trigo*, Aldeia Norte Editora, Brasil, 54-65.

[6] Pupiro, M.L. and Malagón Rodríguez, L. (2003) Técnicas del diagnóstico de las enfermedades en las plantas. Trabajo de patología vegetal. Universidad Agraria de La Habana, La Habana.

[7] Macias Figueroa, O.L., Delgado Calvo, Y., Peña Marrero, E., Betancourt, R.L. and Barreto, R.E. (2006) Técnicas para el diagnóstico y determinación de variabilidad genética de fitopatógenos. Universidad de Matanzas "Camilo Cienfuegos", Matanzas.

[8] Pappas, M.G., Hajkowski, R. and Hockmeyer, W. (1983) Dot Enzyme-Linked Immunosorbent Assay (DOT-ELISA) a Microtechnique for the Rapid Diagnosis of Visceral Leishmaniasis. *Journal of Immunological Methods*, **64**, 205-214. http://dx.doi.org/10.1016/0022-1759(83)90399-X

[9] Ferrer Gual, R.M. (2008) Caracterización molecular y desarrollo de métodos de diagnóstico del género *Fabavirus*. Evaluación del BTH como método control. Tesis doctoral. Universidad Politécnica de Valencia, Valencia.

[10] Towbin, A., Staehelin, T. and Gordon, J. (1979) Electrophoretic Transfer of Proteins from Polyacrylamide Gels to Nitrocellulose Sheets: Procedure and Some Applications. *Proceedings of the National Academy of Sciences of the United States of America*, **76**, 4350-4354. http://dx.doi.org/10.1073/pnas.76.9.4350

[11] Agudelo-Flórez, P. and Palacio, L.G. (2009) Dot blot para el diagnóstico de la cisticercosis porcina. *Scientific and Technical Review of the Office International des Epizooties*, **28**, 1077-1084.

[12] Cabezas Vergara, C., Parcha, C., Pineda, J. and Reyes Arenas, E. (2006) Estandarización de una técnica de amplificación genómica (PCR) para el diagnóstico de Dermatofitos. *Ciencia & Trabajo*, **8**, 167-171.

[13] Rico Guerrero, L., Medina Ramos, S., Muñoz Sanchez, C.I., Guevara Olvera, L., Guevara González, R.G., Guerrero Aguilar, B.Z., Torres Pacheco, I., Rodriguez Guerra, R. and González Chavira, M.M. (2004) Detección de *Phytophthora capsici* Leonian en plantas de chile (*Capsicum annuum* L.) mediante PCR. *Revista Mexicana de Fitopatología*, **22**, 1-6.

[14] Saavedra Rodríguez, C.O., Gómez González, S.J. and Ángel Díaz, J.E. (2004) Detección de secuencias específicas de ADN de *Spongospora subterranea* en suelo y tubérculos de papa. *Revista Colombiana de Biotecnología*, **6**, 14-22.

[15] Lurá, M.C., Latorre Rapela, M.G., Vaccari, M.C., Maumary, R., Soldano, A., Mattio, M. and González, A.M. (2011) Genetic Diversity of *Cercospora kikuchii* Isolates from Soybean Cultured in Argentina as Revealed by Molecular Markers and Cercosporin Production. *Mycopathologia*, **171**, 361-371. http://dx.doi.org/10.1007/s11046-010-9362-x

[16] Moreno, N. and Agudelo-Flórez, P. (2010) Aplicación de las pruebas de PCR convencional simple y múltiples para la identificación de aislamientos de *Leptospira* spp. en Colombia. *Revista Peruana de Medicina Experimental y Salud Pública*, **27**, 548-556. http://dx.doi.org/10.1590/S1726-46342010000400009

[17] Latorre Rapela, M.G., Maumary, R., Marcipar, I. and Lurá, M.C. (2013) Modelo experimental de infección de plantas de soja por especies de *Cercospora*. *Revista FABICIB*, **17**, 66-73

[18] Rollins, J.A., Ehrenshaft, M. and Upchurch, R.G. (1993) Effects of Light- and Altered-Cercosporin Phenotypes on Gene Expression in *Cercospora kikuchii*. *Canadian Journal Microbiology*, **39**, 118-124. http://dx.doi.org/10.1139/m93-017

[19] Bradford, M.M. (1976) A Rapid and Sensitive Method for the Quantitation of Microgram Quantities of Protein Utilizing the Principle of Protein-Dye Binding. *Analytical Biochemistry*, **72**, 248-254. http://dx.doi.org/10.1016/0003-2697(76)90527-3

[20] Lartey, R.T., Caesar-TonThat, T.C., Caesar, A.J., Shelver, W.L., Sol, N.I. and Bergman, J.W. (2005) Safflower: A New Host of *Cercospora beticola*. *Plant Disease*, **89**, 797-801. http://dx.doi.org/10.1094/PD-89-0797

[21] Lee, S.B. and Taylor, J.W. (1992) Phylogeny of Five Fungus-Like Protoctistan *Phytophthora* Species, Inferred from the Internal Transcribed Spacers of Ribosomal DNA. *Molecular Biology and Evolution*, **9**, 636-653.

[22] Sambrook, J., Fritsch, E.F. and Maniatis, T. (1989) Molecular Cloning: A Laboratory Manual. Cold Spring Harbor Laboratory Press, New York.

[23] Ward, E., Foster, S.J., Fraaije, B.A. and McCartney, H.A. (2004) Plant Pathogen Diagnostics: Immunological and Nucleic Acid-Based Approaches. *Annals of Applied Biology*, **145**, 1-16. http://dx.doi.org/10.1111/j.1744-7348.2004.tb00354.x

[24] Ndao, M. (2009) Diagnosis of Parasitic Diseases: Old and New Approaches. *Interdisciplinary Perspectives on Infectious Diseases*, **2009**, Article ID: 278246. http://dx.doi.org/10.1155/2009/278246

[25] Arboleda, V.J.W., Delgado, B.F. and Valencia, J.A. (2004) Detección de beauvericina en el hongo entomopatógeno *Beauveria bassiana* mediante el uso de anticuerpos policlonales. *Revista Colombiana de Entomología*, **30**, 125-130.

[26] Shaheen, H.I., Kamal, K.A., Farid, Z., Mansour, N., Boctor, F.N. and Woody, J.N. (1989) Dot-Enzyme-Linked Immunosorbent Assay (Dot-ELISA) for the Rapid Diagnosis of Human Fascioliasis. *Journal of Parasitology*, **75**, 549-552.

http://dx.doi.org/10.2307/3282904

[27] Swarna, S.R. and Parija, S.C. (2008) Dot-ELISA for Evaluation of Hydatid Cyst Wall, Protoscoleces and Hydatid Cyst Fluid Antigens in the Serodiagnosis of Cystic Echinococcosis. *Revista do Instituto de Medicina Tropical de São Paulo*, **50**, 233-236. http://dx.doi.org/10.1590/S0036-46652008000400009

[28] Chen, Y., Xu, F., Fan, X., Luo, H., Ge, S., Zheng, Q., Xia, N, Chen, H., Guan, Y. and Zhang, J. (2008) Evaluation of a Rapid Test for Detection of H5N1 Avian Influenza Virus. *Journal of Virological Methods*, **154**, 213-215. http://dx.doi.org/10.1016/j.jviromet.2008.08.013

[29] Roldán, W.H. and Espinoza, Y.A. (2009) Evaluation of an Enzyme-Linked Immunoelectrotransfer Blot Test for the Confirmatory Serodiagnosis of Human Toxocariasis. *Memorias do Instituto Oswaldo Cruz*, **104**, 411-418. http://dx.doi.org/10.1590/S0074-02762009000300003

[30] Belo, E.F., Farhat, C.K. and Gaspari, E.N. (2010) Comparison of Dot-ELISA and Standard ELISA for Detection of *Neisseria meningitidis* Outer Membrane Complex-Specific Antibodies. *Brazilian Journal of Infectious Diseases*, **14**, 35-40.

[31] Piña, R.R., Gutiérrez, A.H., Gilman, R.H., Rueda, D., Sifuentes, C., Flores, M., Sheen, P., Rodriguez, S., García, H.H. and Zimic, M. (2011) A Dot-ELISA Using a Partially Purified Cathepsin-L-Like Protein Fraction from *Taenia solium* Cysticerci, for the Diagnosis of Human Neurocysticercosis. *Annals of Tropical Medicine and Parasitology*, **105**, 311-318. http://dx.doi.org/10.1179/136485911X12987676649782

[32] Xu, M.L., Melchinger, A.E. and Lübberstedt, T. (1999) Species-Specific Detection of the Maize Pathogens *Sporisorium reiliana* and *Ustilago maydis* by Dot Blot Hybridization and PCR-Based Assays. *Plant Disease*, **83**, 390-395. http://dx.doi.org/10.1094/PDIS.1999.83.4.390

[33] Pavón Moreno, M.A., García Lacarra, T. and Martín de Santos, R. (2008) Utilización del gen *Alt a1* para la detección de *Alternaria* spp. en productos hortofrutícolas mediante una técnica de PCR. *Revista Complutense de Ciencias Veterinarias* (*RCCV*), **2**, 105-113.

[34] Lartey, R., Weiland, J., Caesar, T. and Bucklin Comiskey, S. (2003) A PCR Protocol for Rapid Detection of *Cercospora beticola* in Infected Sugar Beet Tissues. *Journal of Sugar Beet Research*, **40**, 1-10. http://dx.doi.org/10.5274/jsbr.40.1.1

[35] Daub, M.E., Herrero, S. and Chung, K.R. (2005) Photoactivated Perylenequinone Toxins in Fungal Pathogenesis of Plants. *FEMS Microbiology Letters*, **252**, 197-206. http://dx.doi.org/10.1016/j.femsle.2005.08.033

[36] Chanda, A.K., Ward, N.A., Robertson, C.L., Chen, Z.-Y. and Schneider, R.W. (2014) Development of a Quantitative Polymerase Chain Reaction Detection Protocol for *Cercospora kikuchii* in Soybean Leaves and Its Use for Documenting Latent Infection as Affected by Fungicide Applications. *Phytopathology*, **104**, 1118-1124. http://dx.doi.org/10.1094/PHYTO-07-13-0200-R

[37] Agarwal, V.K. and Sinclair, J.B. (1997) Principles of Seed Pathology. 2nd Edition, CRC Press/Lewis Publishers, Boca Ratón.

16

Phelipanche aegyptiaca Management with Glyphosate in Potato

Mustapha Haidar*, Elie Shdeed

American University of Beirut, Beirut, Lebanon
Email: *mhaidar@aub.edu.lb

Abstract

Two years field and greenhouse studies were carried out to evaluate the efficacy of sub-lethal doses of glyphosate (Round up[R]), ammonia gas, phosphoric acid and sulfuric acid against *Phelipanche aegyptiaca* in potato. Results showed that sequential application of sub-lethal doses of glyphosate at all tested rates significantly reduced *P. aegyptiaca* shoot number and shoot dry weight. While, the use of ammonia gas, phosphoric acid and sulfuric acid had no significant effect on the total level of *P. aegyptiaca* infection as compared to the control. The best results considering both *P. aegyptiaca* control and selectivity in potato were obtained by sequential application of sub-lethal doses of glyphosate at 60 and 80 g·ai·ha^{-1}. Sequential application of glyphosate at 60 g·ai·ha^{-1} reduced *P. aegyptiaca* infection by 100% after 100 days after potato emergence (DAPE). Except for sequential application of glyphosate at 60 and 80 g·ai·ha^{-1}, all tested rates enhanced the maturity rate of potato plants and decreased the number of marketable potato tubers.

Keywords

Ammonia Gas, Glyphosate, *Phelipanche aegyptiaca*, Phosphoric Acid, Potato, Sulfuric Acid

1. Introduction

Potato (*Solanum tuberosum*) is considered one of the most important strategic crops in the Mediterranean region. In Lebanon, the Beq'aa and Akkar provinces are the main potato producing areas in the country, with about 68% and 19% of the total production, respectively [1]. Potato is susceptible to several pests among which *P. aegyptiaca*. This parasitic weed parasitizes summer, spring and autumn planted potatoes across Lebanon and the Mediterranean region.

Phelipanche aegyptiaca Forsk (Branched broomrape) is an aggressive root holoparasite that infects roots of

*Corresponding author.

various dicotyledous plants. This parasite is an invasive insidious higher plant in Lebanon and the Mediterranean region; it causes severe yield losses in Solanaceae and Fabaceae crops [2]-[4]. Severe reduction in potato yield quantity and quality were observed due to high levels of *P. aegyptiaca* field infestations in Lebanon [4] [5]. Management of *P. aegyptiaca* remains challenging for potato producers. With variable degrees of success, several control methods have been tested against *Phelipanche* spp. infection including physical, cultural, chemical, soil solarization, trap and catch crops, resistant varieties, synthetic germination stimulants and soil amendments [3] [6]-[13].

One of the most promising methods for controlling *Phelipanche* spp. is the use of sub-lethal doses of glyphosate on crops that show tolerance to glyphosate [3] [7] [8]. The plan is to spray sub-lethal doses of glyphosate on the potato leaves (POE) early in the season, so that the glyphosate would move through the crop phloem to underground *Phelipanche* spp. attachments on the crop roots and inhibit *Phelipanche* spp. growth prior to its above ground shoot emergence. Sub-lethal doses of glyphosate were found to be effective against *Phelipanche* spp. in broad bean [14], sunflower [15], tomato [16], carrot and celery [8] and vetch [17]. Thus, the use of sub-lethal doses of glyphosate as potential and economical method to control *Phelipanche* spp. has become increasingly important. To our knowledge, the use of sub-lethal doses of glyphosate against *P. aegyptiaca* has not been confirmed in potato, most likely because of its limited selectivity, rate, and timing of application.

Inorganic compounds have also been shown to reduce *Phelipanche* spp. growth. Many researchers showed that high nitrogen fertilizers (NH_4NO_3) and sulfur powder reduced *Phelipanche* spp. growth and development [10] [18] [19]. However, our literature search revealed no published work on the effect of ammonia gas, phosphoric acid and sulfuric acid against *P. aegyptiaca*. Acidifying irrigation water may aid in creating optimum soil pH for various crop growth [20] by mitigating a small zone from where the crop roots can better obtain nutrients. A pH range of 5.5 to 6.5 is considered ideal for most crops including potato [20], but may not be for *Phelipanche* spp. because the parasite is commonly distributed in alkaline soils [3]. Most soils in the Beq'aa plain are calcareous in nature with a pH ranging between 7.5 and 8.5 and are thus amenable to *P. aegyptiaca* infestation.

The main objectives of the present work were to examine: a) *P. aegyptiaca* control using ammonia gas, phosphoric acid, sulfuric acid or sub-lethal doses of glyphosate; b) optimal rates of glyphosate application for *P. aegyptiaca* control; and c) response of potato to sub-lethal doses of glyphosate.

2. Materials and Methods

2.1. Greenhouse Experiment

A single greenhouse experiment was done at the greenhouse area of the Faculty of Agricultural Sciences at the Beirut Coastal area, during September 2009 and April 2010. *Phelipanche aegyptiaca* seeds were collected in 2008 from various potato fields in the Beq'aa plain and stored at room temperature. Potato tubers (cultivar Spunta) were spread on paper sheets and kept moist in the lab at room temperature three weeks prior to planting. Boxes were filled with a mixture of potting soil, perlite and peat moss at a rate of 1:1:1, inoculated with 100 mg of *P. aegyptiaca* seeds, and irrigated with water for a period of two weeks prior to planting potato to allow for conditioning of *P. aegyptiaca* seeds. Two tubers were planted in plastic netted boxes 30 × 40 × 30 cm.

Methyl bromide was injected in the boxes prior to planting potato tubers. Methyl bromide was selected as another 100% free *P. aegyptiaca* control (comparison with inorganic chemicals). Four moist boxes containing *P. aegyptiaca* seeds were placed in an area of 10 m^2 covered with nylon sheet and fumigated with methyl bromide at a rate of 900 g per 10 m^2. Sheets were removed 24 hrs after fumigation and boxes placed back in the greenhouse. Potato tubers were planted 20 days after fumigation.

Glyphosate (Round up^R, 48%) was sprayed on potato leaves (POE) at a rate of 125, 135, and 150 $g \cdot ai \cdot ha^{-1}$ (no surfactant was added) in a spray volume of 1000 $L \cdot ha^{-1}$. Spraying rates were selected based on preliminary studies performed in our laboratory. Each rate was tested for single and sequential applications at 20, 40, and 60 DAPE (20, 20 + 40 and 20 + 40 + 60 DAPE). Potato plants were 12 - 18 cm tall at 20 DAPE. Spraying dates were selected based on previous experience with the species as well as on various other considerations. The underground stage of development of *P. aegyptiaca* in the soil is unknown [6], and seeds of the parasite in the soil keep on germinating continuously. Additionally, the best control strategy is to eliminate the parasite before it emerges above the soil [21]. Sulfuric acid or phosphoric acid was mixed with water to have a final solution of pH 2. Diluted solutions of sulfuric or phosphoric acid were then applied at a volume of one liter per box (irrigated) every other week starting at the time of potato planting.

2.2. Field Experiments General Procedure

During the 2009 and 2010 spring seasons, field experiments were carried out at the Advancing Research Enabling Communities (AREC) of the American University of Beirut. The AREC is located in the Northern Central Beqa'a plain of Lebanon. A naturally infested field with *P. aegyptiaca* was used for both years (same field) but at different locations. The soil is a silty clay loam with a pH of 7.41, Electrical Conductivity of 0.24 ms/m and 2.4% organic matter. In both years, all plots were tilled twice with a mouldboard and disked two weeks before planting potato. The experimental area received a uniform application of 2.5 $t \cdot ha^{-1}$ of NPK (15:15:15) during planting. Nitrogen (Ammonium nitrate, 33.5%) was applied in bands 40 days after planting at a rate of 300 $kg \cdot ha^{-1}$. All plots were sprinkler-irrigated every six days to bring the soil back to field capacity (measured by using tensiometers at 0.33 bars suction). The standard certified potato cultivar "Spunta" was planted in both years at 3.0 $t \cdot ha^{-1}$. Potato rows were 0.70 m apart and within-the-row spacing was around 0.25 m. Plots were hilled 6 weeks after planting (standard practice by Lebanese farmers). To eliminate all weeds other than *P. aegyptiaca*, the entire experimental areas received a standard application of metribuzin (Sencor^R, 70%, PE) at 0.75 $kg \cdot ai \cdot ha^{-1}$ two weeks after potato sowing with a boom sprayer at a rate of 400 $L \cdot ha^{-1}$. Glyphosate was applied POE on potato plants at 20, 40, and 60 DAPE ((20, 20 + 40 and 20 + 40 + 60). Glyphosate was sprayed by a hand held CO_2-pressurized backpack sprayer that delivers 310 $L \cdot ha^{-1}$ at 138 Kpa through a Teejet 8002 flat fan spray tips. Irrigation water was withheld for two days after each spraying.

2.3. Field Experiment in Spring 2009

Ammonia gas and sub-lethal doses of glyphosate were the only treatments tested against *P. aegyptiaca*. Ammonia gas was injected in the soil (at field capacity) at 30 $kg \cdot ha^{-1}$ one week prior to planting potato to a depth of 30 - 35 cm. A tractor pulling a cylinder full with pressurized ammonia and connected to the injection system which released the ammonia gas at a pressure of 2 - 3 bars in the soil was used. The field was then lightly irrigated to avoid evaporation of the gas. During the initial potato growth and in other plots, sequential application of foliar sub-lethal doses of glyphosate at a rate of 125 $g \cdot ai \cdot ha^{-1}$ was applied on potato plants at 20, 40, and 60 DAPE (20, 20 + 40 and 20 + 40 + 60). Experimental plots (20 m long and 10 m wide) were arranged in a randomized complete block design with three replications. Each replicate or plot consists of 12 rows.

2.4. Field Experiment in 2010

Sub-lethal doses of glyphosate at 60, 80, and 100 $g \cdot ai \cdot ha^{-1}$ were the only treatments tested against *P. aegyptiaca*. These rates were selected according to the results obtained from previous field and greenhouse experiments. All rates were tested for single and sequential application at 20, 40, and 60 DAPE (20, 20 + 40 and 20 + 40 + 60). A factorial experiment was used. Experimental plots (6 m long and 2.1 m wide) were arranged in a randomized complete block design with four replications.

2.5. Experimental Measurements and Statistical Analyses

Phelipanche aegyptiaca infection in the greenhouse experiment was assessed by counting emerged shoots and underground attachments and measuring total dry weight per box. At the end of the experiment, emerged *P. aegyptiaca* shoots (above ground) were counted and pulled out. Underground shoots were counted by washing the potato roots with water and separating remaining *P. aegyptiaca* attachments from potato roots. *Phelipanche aegyptiaca* infection in the field experiments was assessed by only counting emerged shoots (above ground) and measuring their total dry weight per m^2 of the middle row in the field experiments. Potato data included number of plants per box/row (5 m), phytotoxicity visual rating according to the European Weed Research Council scoring system, height/two plants/box or ten plants/plot, total and marketable yield. Potato yield was determined by harvesting potato/box or part of the middle row in each plot (5 m). Yield quality was determined by separating harvested tubers into two classes: marketable (>6 cm diameter) and non-marketable tubers (<5 cm in diameter). Statistical analyses were performed using SAS 9.2. (SAS Institute Inc., Cary, North Carolina USA). Data were analyzed statistically using ANOVA and Protected Least Significant Difference (LSD) Test at $p = 0.05$ level of probability was used to determine significant differences between treatments means.

3. Results and Discussion

3.1. Greenhouse Experiment 2009/10

Methyl bromide and single or sequential application of sub-lethal doses of glyphosate at 125, and 150 g ai/ha applied 20, 40, and 60 DAPE significantly reduced *P. aegyptiaca* total shoot number (above and underground) and total shoot dry weight 100 DAPE compared to the control (**Table 1**) All these treatments reduced *P. aegyptiaca* infection by 100%. Sequential application of glyphosate at all tested rates significantly reduced shoot and dry weight of *P. aegyptiaca* 100 DAPE compared to the control. Sulfuric acid and phosphoric acid had no significant effect on *P. aegyptiaca* shoot number or total dry weight in comparison to the controls (**Table 1**).

Using sub-lethal doses of systemic herbicides as single or sequential applications have been found very effective against *Phelipanche* spp. in various crops. Split application with low rates of sulfunylurea herbicides inhibited *P. aegyptiaca* growth in potato [22] and tomato [12]. Sequential application of sub-lethal doses of glyphosate inhibited *Phelipanche* spp. growth in many crops [8] [14] [23].

The significance of using sequential application of systemic herbicides such as glyphosate on potato early in the season is to inhibit *P. aegyptiaca* growth prior to shoot emergence. *Phelipanche aegyptiaca* seeds are continuously induced to germinate by potato roots and develop the attachment organ, the germ tube or radicle. Sequential application of systemic herbicides may prevent the attachment of the organ or its differentiation and allow for early season control of the parasite.

Injection with methyl bromide was suggested as a control strategy for *Phelipanche* spp. by various scientists, showing effectiveness of methyl bromide in controlling and eradicating *P. ramosa* [24] [25] [26]. Although methyl bromide is very effective in eradicating and killing *Phelipanche* spp., it is not economical for broad application, and poses significant environmental impacts [27].

The observed low efficiency of phosphoric and sulfuric acid in mitigating against *P. aegyptiaca* infection could be because of low concentrations applied to the boxes or because of leaching from the boxes. However, these results contradict previous findings that report that soil pH affects seed germination and development [20]. Jain and Foy [28] found that preconditioning of *P. aegyptiaca* seeds with phosphoric acid (21 ml of 1 Molar stock solution) decreased the parasite seed germination in tomato.

Table 1. Effect of glyphosate, sulfuric acid, phosphoric acid, and methyl bromide on *Phelipanche aegyptiaca* above and underground shoot number (SN) and total shoot dry weight (TSDWT) 100 DAPE. Data represent average of four replicates (boxes). Means followed by the same letter, within each column, do not significantly differ at the 5% level according to the LSD test. DAPE: Days after potato emergence.

Treatment	Rate	Application time	SN				TSDWT	
		(DAPE)	above ground		underground		(g)	
Glyphosate ($g \cdot ai \cdot ha^{-1}$)	125	20	0	b	0	b	0	b
	135	20	5	ab	4.3	ab	2	ab
	150	20	0	b	0	b	0	b
	125	20/40	0	b	0	b	0	b
	135	20/40	0	b	0	b	0	b
	150	20/40	0	b	0	b	0	b
	125	20/40/60	0	b	0	b	0	b
	135	20/40/60	0	b	0	b	0	b
	150	20/40/60	0	b	0	b	0	b
Sulfuric acid	2L/box		14.8	a	14.8	a	2.6	ab
Phosphoric acid	2L/box		10.0	a	12.3	a	2.8	ab
Methyl Bromide	900 g/10 m^2		0.0	b	0.0	b	0.0	b
Control without *Phelipanche*	0		0.0	b	0.0	b	0.0	b
Control with *Phelipanche*	0		17.8	a	12.3	a	4.6	a

Single or sequential application of glyphosate at all tested rates were toxic to the potato plants and significantly reduced potato shoot height at 30 and 60 DAPE, compared to the control (**Table 2**). Sulfuric acid, phosphoric acid, and methyl bromide had no significant effect on potato plants compared to the control. Visual potato injury appeared 10 days after the first glyphosate application and included leaf yellowing, necrosis, plant stunting, and compact potato shoots. Phytotoxicity was clearly reflected in the tuber quality of potato yield grown for fresh market with a high incidence of deformed (cracked) and small tubers (**Figures 1(A)-(D)**).

It is well known that glyphosate translocates sympoplastically and apoplastically in plants [29] [30]. During initial tuber development, tubers accumulate photosynthetic assimilates produced by the leaves and other exogenous compounds such as glyphosate. Thus, the glyphosate effect may appear as yellowing or necrosis in young leaves and malformed tubers.

3.2. Field Experiment 2009

Data on the total number of *P. aegyptiaca* shoots indicate that sequential applications of glyphosate at 125 $g \cdot ai \cdot ha^{-1}$ was very effective in reducing its infection as compared to ammonia gas and/or the control (**Figure 2**). Unlike ammonia gas, glyphosate reduced *P. aegyptiaca* shoot number by 97% and 99%, after 60 and 80 DAPE respectively, compared to the control. In addition, the observed shoots were stunted and almost dead. It was shown that frequent spray of glyphosate completely controlled *Phelipanche* spp. in parsley [23] faba bean, tobacco and tomato [14].

The reasons for the low efficiency of ammonia gas in reducing *P. aegyptiaca* infection are not easily discernible. They could be because of escape of volatile ammonia from the soil upon injection or because of improper injection and distribution of the gas in the soil.

Results in **Table 3** show that glyphosate had significantly increased total potato yield compared to the ammonia treatment and the control. Moreover, the total number of potato tubers was also significantly higher by 63% in glyphosate treated plots than both the ammonia and the control plots. However, most potato tubers were small in size (non-marketable).

Table 2. Effect of glyphosate, sulfuric acid, phosphoric acid and methyl bromide on potato shoot height (PHT) and vigor (PVR). Data represent average of four replicates. Means followed by the same letter, within each column, do not significantly differ at the 5% level according to the LSD test. DAPE: Days after potato emergence; PVR: Crop phytotoxicity visual rating with 0 indicating complete death and 10 no injury.

Treatment rate	Rate	Application time	PHT (cm)				PVR (%)	
		(DAPE)	30 DAPE		60 DAPE		60 DAPE	
Glyphosate ($g \cdot ai \cdot ha^{-1}$)	125	20	71.3	b	70.8	b	6.5	b
	135	20	66.3	b	78.0	b	6.8	b
	150	20	65.0	b	70.8	b	6.5	b
	125	20/40	71.5	b	76.0	b	7.0	b
	135	20/40	76.8	b	82.8	b	7.5	b
	150	20/40	76.0	b	75.8	b	6.5	b
	125	20/40/60	77.5	b	80.3	b	7.5	b
	135	20/40/60	76.5	b	83.3	b	7.8	b
	150	20/40/60	64.3	b	69.8	c	6.5	b
Sulfuric acid	2L/box		92.3	a	104.8	a	9.0	a
Phosphoric acid	2L/box		86.8	a	99.3	a	8.8	a
Methyl Bromide	900 g/10 m^2		83.0	a	92.8	a	9.0	a
Control without *Phelipanche*	0		90.8	a	102.3	a	8.8	a
Control with *Phelipanche*	0		87.0	a	90.5	ab	9.0	a

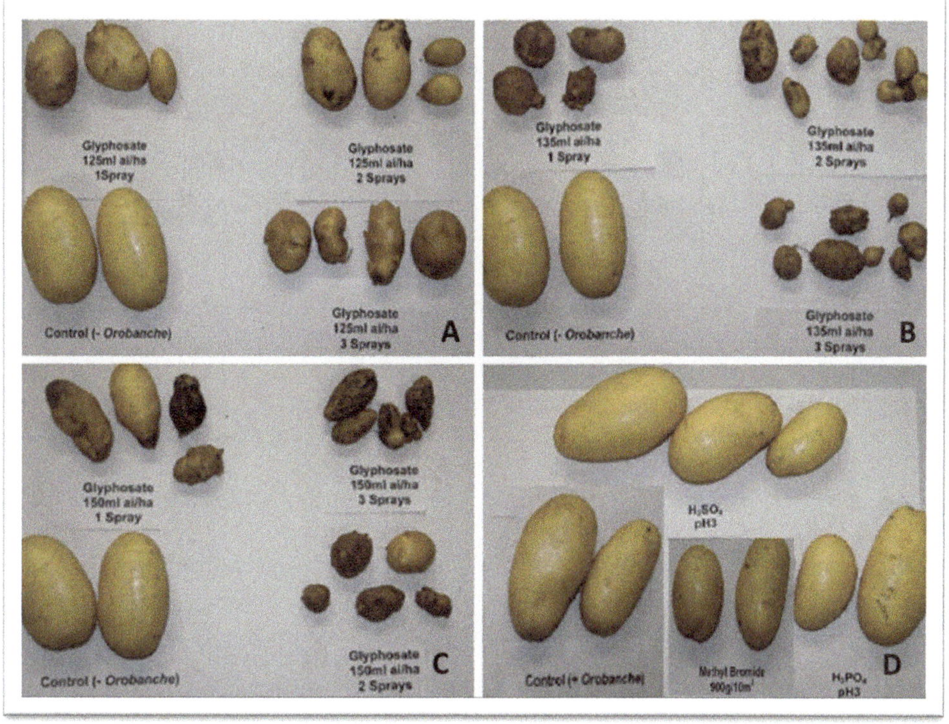

Figure 1. (A)-(C): Deformed potato tubers with cracks or fissures caused by a single or sequential application of glyphosate compared to the control or other chemicals (D).

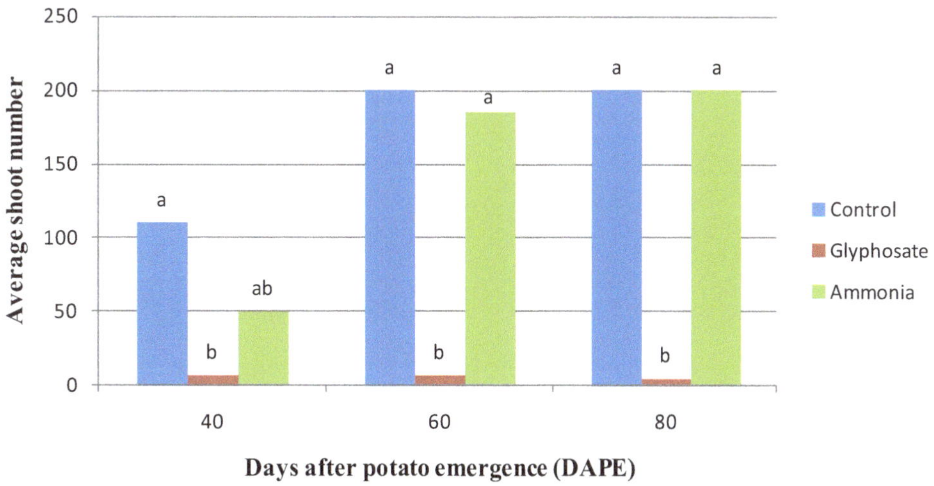

Figure 2. Effects of glyphosate and ammonia on *Phelipanche aegyptiaca* above ground total shoot number (SN). Data represent average of four replicates. Means followed by the same letter, within each column, do not significantly differ at the 5% level according to LSD test. DAPE: Days after potato emergence.

3.3. Field Experiments in 2010

Except for single application of glyphosate at 60 and 80 g·ai·ha^{-1} applied at 20 DAPE, all glyphosate treatments significantly reduced *P. aegyptiaca* infestation as compared to the control (**Table 4**). Single application of glyphosate at 100 g·ai·ha^{-1} at 20 DAPE resulted in significant reduction of *P. aegyptiaca* by 65% as compared to the control after 100 DAPE. Sequential application of glyphosate at 60, 80, or 100 g·ai·ha^{-1} applied at 20, 40, and 60 DAPE were the most effective treatments in reducing *P. aegyptiaca* shoot number after 100 DAPE and shoot dry weight. These treatments reduced *P. aegyptiaca* infestation by 100%, compared to the control (**Table 4**). These results demonstrate the previous observations that split application of sub lethal doses of glyphosate is recommended for the eradication of *Phelipanche* spp. Castejon-Munoz *et al.* [3] showed that glyphosate at 20 - 40 g·ha^{-1} at 12 and 14 days interval eradicated more than 80% of *Orobanche cemua* in sunflower.

Similar to the results of the green house experiment, all applications of glyphosate (60, 80, and 100 g·ai·ha^{-1}) were injurious to potato plants early in the growing season (after 60 DAPE) when compared to control (**Table 5**). Injuries appeared 10 days after first spraying and comprised of leaf yellowing, plant stunting, and compact plants. However, injury symptoms disappeared by 75 DAPE. Single or sequential application of glyphosate at all tested dosages had no significant effect on potato shoot height 60 DAPE. However, sequential application of glyphosate at 100 g·ai·ha^{-1} was toxic to potato plants and significantly reduced potato shoot height 75 DAPE.

Although single and sequential applications of glyphosate at 60 and 80 g·ai·ha^{-1} were injurious to potato plants early in the growing season, both rates had no negative effects on total potato quantity and quality (**Table 6**). While, glyphosate at 100 g·ai·ha^{-1} (Single or sequential) significantly reduced total potato yield and produced non-marketable tubers. Phytotoxicity was mostly reflected in the tuber quality of potato yield grown for fresh market with a high incidence of deformed and small tubers. These results are similar to the observations by Gilreath *et al.* [31] who observed that marketable yield of pepper declined with glyphosate dosages at 100 g·ai·ha^{-1}. Previous studies by Nadal *et al.* [32] concluded that sequential application of glyphosate at 67 g·ai·ha^{-1} completely controlled *Orobanche crenata* infestation and increased seed production of narbon bean (*Vicianar bonensis* L), compared to the control. Thus, selectivity remains the main obstacle in various crops and it could be mediated by the rate and time of application.

Table 3. Yield response of potato to single or sequential application of sub-lethal doses of glyphosate at 125 g·ai·ha^{-1} and injected ammonia gas at 30 kg·ha^{-1}. Data represent average of four replicates. Means followed by the same letter, within each column, do not significantly differ at the 5% level according to the LSD test. PTY: Potato tuber yield; TTN: Total tuber number; MTN: Marketable tuber number; NTN: Non-marketable tuber number.

Treatment	PTY		TTN		MTN		NTN	
	(Tons per ha)		(No. per ha)		(No. per ha)		(No. per ha)	
Glyphosate	16.2	a	153809.5	a	54761.90	a	99047.60	a
Ammonia	9.5	b	86666.65	b	39047.61	a	47619.04	b
Control	10.0	b	94761.89	b	38571.42	a	56190.47	b

Table 4. *Phelipanche aegyptiaca* above ground shoot number (SN) and total shoot dry weight (TSDWT) in response to sub-lethal doses of glyphosate after 100 days after potato emergence (DAPE). Data represent average of four replicates. Means followed by the same letter, within each column, do not significantly differ at the 5% level according to the LSD test.

Treatment	Rate	Application time	SN		TSDWT	
	(g·ai·ha^{-1})	(DAPE)	(per m^2)		(g per m^2)	
Glyphosate	60	20	31.3	a	20.5	a
	80	20	18.0	a	14.8	a
	100	20	9.0	b	7.8	b
	60	20/40	4.0	b	2.6	bc
	80	20/40	2.3	bc	1.4	bc
	100	20/40	1.0	c	0.7	bc
	60	20/40/60	0.0	c	0.0	c
	80	20/40/60	0.0	c	0.0	c
	100	20/40/60	0.0	c	0.0	c
Control			25.7	a	16.1	a

Table 5. Effect of glyphosate on potato shoot height (PHT) and vigor (PVR). Data represent average of four replicates. Means followed by the same letter, within each column, do not significantly differ at the 5% level according to the LSD test. DAPE: Days after potato emergence; PVR: Crop phytotoxicity visual rating with 0 indicating complete death and 10 no injury.

Treatment	Rate	Application time	PVR (%)				PHT (cm)			
	($g·ai·ha^{-1}$)	(DAPE)	60 DAPE		75 DAPE		60 DAPE		75 DAPE	
Glyphosate	60	20	7.6	bc	8.4	ab	58.3	a	78.4	ab
	80	20	7.1	bc	8.3	ab	59.4	a	78.2	ab
	100	20	6.6	c	7.5	bc	54.3	a	66.2	ab
	60	20/40	7.6	bc	7.6	bc	60.9	a	77.0	ab
	80	20/40	7.6	bc	7.3	bc	61.2	a	68.3	ab
	100	20/40	7.0	bc	6.1	d	56.7	a	64.0	b
	60	20/40/60	8.0	b	7.5	bc	64.5	a	80.7	ab
	80	20/40/60	7.5	bc	7.4	bc	62.3	a	66.5	ab
	100	20/40/60	7.1	bc	6.8	cd	56.1	a	64.0	b
Control			9.0	a	9.0	a	69.2	a	82.2	a

Table 6. Yield response of potato yield to glyphosate. Data represent average of four replicates. Means followed by the same letter, within each column, do not significantly differ at the 5% level according to the LSD test. DAPE: Days after potato emergence; TPY: Total potato yield; TPN: Total potato number; MPY: Marketable potato yield; NPY: Non-marketable potato yield.

Treatment	Rate	Application time	TPY		MPY		NPY	
	($g·ai·ha^{-1}$)	(DAPE)	(t/ha)		(t/ha)		(t/ha)	
Glyphosate	60	20	33.9	a	28.3	a	5.6	a
	80	20	29.9	a	24.0	a	5.9	a
	100	20	25.9	b	22.7	a	3.2	b
	60	20/40	36.0	a	30.9	a	5.1	a
	80	20/40	29.9	a	25.4	a	4.5	a
	100	20/40	24.9	b	19.9	b	5.0	a
	60	20/40/60	36.8	a	28.0	a	8.8	a
	80	20/40/60	30.2	a	23.5	a	6.7	a
	100	20/40/60	20.4	b	14.3	b	6.1	a
Control			26.9	a	20.6	a	6.3	a

4. Conclusion

Our results suggest that sequential foliar application (20 + 40 and 20 + 40 + 60 DAPE) of glyphosate at 60 - 80 $g·ai·ha^{-1}$ could be used selectively to reduce *P. aegyptiaca* infection in potato. The timing of glyphosate application is crucial for successful control of *P. aegyptiaca* and could vary among potato varieties and growing seasons. Additional research is required to determine the optimum timing of glyphosate application and duration of control based on estimated *P. aegyptiaca* phenological underground growth stages in the field.

References

[1] Abou-Jawdah, Y., Sobh, H. and Saad, A.T. (2001) Incidence of Potato Virus Diseases and Their Significance for a Seed Certification Program in Lebanon. *Phytopathologia Mediterranea*, **40**, 113-118.

[2] Joel, D.M., Hershenhorn, J., Eizenberg, H., Aly, R., Ejeta, G., Rich, P., Ransom, J., Sauerborn, J. and Rubiales, D.

(2007) Biology and Management of Weedy Root Parasites. *Horticultural Reviews*, **33**, 207-349. http://dx.doi.org/10.1002/9780470168011.ch4

[3] Parker, C. and Riches, C.R. (1993) Parasitic Weeds of the World: Biology and Control. Cab Intl, Wallingford, UK, 332.

[4] Sabra, A. and Haidar, M.A. (2012) Invasive Weed Mapping of Lebanon. *Journal of Agricultural Science and Technology B*, **2**, 1010-1015.

[5] Haidar, M. and Bibi, W. (1995) Common Weeds of Lebanon. Arab and Near East Plant Protection Newsletter. Abstract, 34.

[6] Eizenberg, H., Colquhoun, J.B. and Mallory-Smith, C.A. (2006) Imazamox Application Timing for Small Broomrape (*Orobanche minor*) Control in Red Clover. *Weed Science*, **54**, 923-927. http://dx.doi.org/10.1614/WS-05-151R.1

[7] Eizenberg, H., Hershenhorn, J., Ephrath, J. and Kanampiu, F. (2013) The Complexity of Chemical Control of Parasitic Weeds. In: Joel, D.M., Gressel, J. and Musselman, L.J., Eds., *Parasitic Orobanchaceae Parasitic Mechanisms and Control*, Springer, Berlin, 415-423. http://dx.doi.org/10.1007/978-3-642-38146-1_23

[8] Elzein, A. and Kroschel, J. (2003) Progress on Management of Parasitic Weeds. In: Labrada, R, Ed., *Weed Management for Developing Countries*, FAO Plant Production and Protection Paper, 109-143.

[9] Haidar, M. and Sidahmad, M. (2000) Soil Solarization and Chicken Manure for the Control of *Orobanche crenata* and Other Weeds in Lebanon. *Crop Protection*, **19**, 169-173. http://dx.doi.org/10.1016/S0261-2194(99)00083-6

[10] Haidar, M. and Sidahmed, M.M. (2006) Elemental Sulfur and Chicken Manure for the Control of Branched Broomrape (*Orobanche ramosa*). *Crop Protection*, **25**, 47-51. http://dx.doi.org/10.1016/j.cropro.2005.03.022

[11] Haidar, M., Bibi, W. and Sidahmad, M.M. (2003) Response of Branched Broomrape (*Orobanche ramosa*) Growth and Development to Various Soil Amendments in Potato. *Crop Protection*, **22**, 291-294. http://dx.doi.org/10.1016/S0261-2194(02)00150-3

[12] Hershenhorn, J., Plakhine, D., Goldwasser, Y., Westwood, J.H., Foy, C.L. and Kleifeld, Y. (1998) Effect of Sulfonylurea Herbicides on Early Development of Egyptian Broomrape (*Orobanche aegyptiaca*) in Tomato (*Lycopersicon esculentum*). *Weed Technology*, **12**, 108-114.

[13] Joel, D.M., Gressel, J. and Musselman, L.J. (2013) Parasitic Orobanchaceae: Parasitic Mechanisms and Control. Springer, Berlin, 513 p. http://dx.doi.org/10.1007/978-3-642-38146-1

[14] Kasasian, L. (1973) Control of *Orobanche*. *Proceedings of the National Academy of Sciences of the United States of America*, **19**, 368-371. http://dx.doi.org/10.1080/09670877309412782

[15] Castejon-Munoz, M., Romero-Munoz, F. and Garcia Torres, L. (1990) Control of Broomrape (*Orobanche cernua* Loelf) in Sunflower (*Helianthus annus* L.) with Glyphosate. *Crop Protection*, **9**, 332-336. http://dx.doi.org/10.1016/0261-2194(90)90002-O

[16] Kotoula-Syka, E. and Eleftherohorinos, I.G. (1990) *Orobanche ramosa* L. (Broomrape) Control in Tomatoes (*Lycopersicon esculentum* Mill.) with Chlorosulfuron, Glyphosate and Imazaquin. *Weed Research*, **31**, 19-27. http://dx.doi.org/10.1111/j.1365-3180.1991.tb01738.x

[17] Nandula, V., Foy, C. and Orcutt, D. (1999) Glyphosate for *Orobanche aegyptiaca* Control in *Vicia sativa* and *Brassica napus*. *Weed Science*, **47**, 486-491.

[18] Abu-Irmaileh, B. (1994) Nitrogen Reduces Branched Broomrape (*Orobanche ramosa*) Seed Germination. *Weed Science*, **42**, 57-60.

[19] Westwood, J.H. and Foy, C.L. (1999) Influence of Nitrogen on Germination and Early Development of Broomrape (*Orobanche* spp.). *Weed Science*, **47**, 2-7.

[20] Simpson, K. (1986) Fertilizers and Manures. Longman Group Limited, New York, 254.

[21] Gressel, J. and Joel, D.M. (2013) Weedy Orobanchaceae: The Problem. In: Joel, D.M., Gressel, J. and Musselman, L.J., Eds., *Parasitic Orobanchaceae*: *Parasitic Mechanisms and Control*, Springer, Berlin, 309-312. http://dx.doi.org/10.1007/978-3-642-38146-1_17

[22] Goldwasser, Y., Eizenberg, H., Golan, S., Hershenhorn, J. and Kleifeld, Y. (2001) *Orobanche aegyptiaca* Control in Potato. *Crop Protection*, **20**, 403-410. http://dx.doi.org/10.1016/S0261-2194(00)00162-9

[23] Goldwasser, Y., Eizenberg, H., Golan, S. and Kleifeld, Y. (2003) Control of *Orobanche crenata* and *Orobanche aegyptiaca* in Parsley. *Crop Protection*, **22**, 295-305. http://dx.doi.org/10.1016/S0261-2194(02)00152-7

[24] Nemli, Y., Emiroglu, U. and Kucukozden, R. (1991) Chemical Control of Broomrape (*Orobanche ramosa* L.) in Tobacco. In: Weymann, K. and Musselman, L.J., Eds., *Proceedings of the International Workshop on Orobanche Research*, Eberhard-Karls-Universitat, Tubingen, 191-199.

[25] Vasilakakis, C.B., Vezirtjoglou, R.M. and Hadzistavros, C.S. (1988) The Effect of Polydynamic Agrochemicals on the

Broomrape (*Orobanche ramose* L.) of Tobacco in the Field. *Bulletin d'Information*, **8**, 146-147.

[26] Zahran, M.K. (1970) Satisfactory Control of *Orobanche crenata* in Broad Beans by Soil Fumigation in UAR. *Proceedings of the British Weed Control Conference*, **10**, 680-684.

[27] Ruzo, L.O. (2006) Physical, Chemical and Environmental Properties of Selected Chemical Alternatives for the Pre-Plant Use of Methyl Bromide as a Soil Fumigant. *Pest Management Science*, **62**, 99-113. http://dx.doi.org/10.1002/ps.1135

[28] Jain, R. and Foy, C.L. (1992) Nutrient Effects on Parasitism and Germination of Egyptian Broomrape (*Orobanche aegyptiaca*). *Weed Technology*, **6**, 269-275.

[29] Dörr, I. and Kollmann, R. (1995) Symplasmic Sieve Element Continuity between *Orobanche* and Its Hosts. *Botanica Acta*, **108**, 47-55. http://dx.doi.org/10.1111/j.1438-8677.1995.tb00830.x

[30] Walker, E.R. and Oliver, L.R. (2008) Translocation and Absorption of Glyphosate in Flowering Sicklepod (*Senna obtusifolia*). *Weed Science*, **56**, 338-343. http://dx.doi.org/10.1614/WS-07-069.1

[31] Gilreath, J.P., Locascio, S.J. and Chase, C.A. (2000) Phytotoxic Effects of Glyphosate on Pepper (*Capsicum annuum*). *Weed Technology*, **14**, 488-494. http://dx.doi.org/10.1614/0890-037X(2000)014[0488:PEOGOP]2.0.CO;2

[32] Nadal, S., Moreno, M.T. and Roman, B. (2008) Control of *Orobanche crenata* in *Vicianar bonensis* by Glyphosate. *Crop Protection*, **27**, 873-876. http://dx.doi.org/10.1016/j.cropro.2007.10.009

17

Effect of Different Concentrations of Orange Juice for *in Vitro* Regeneration and Multiplication of Cocoyam (Taro)

Alfred O. Ubalua[1*], Ahamefula I. Ikpeama[1], Onyinyechi D. Okeagu[2]

[1]Plant Tissue Culture Unit, Biotechnology Research and Development Center, National Root Crops Research Institute (NRCRI) Umudike, Umuahia, Nigeria

[2]Yam Programme, National Root Crops Research Institute (NRCRI) Umudike, Umuahia, Nigeria

Email: [*]alfreduba@yahoo.com

Abstract

Taro (*Colocasia esculenta*) and tannia (*Xanthosoma sagittifolium*) are commonly referred to as cocoyam in Nigeria. They are cherished for their rich taste, nutritional and medicinal properties. Traditionally, cocoyams are vegetatively propagated from tuber fragments, a practice that encourages pathogen distribution. For rapid multiplication and production of quality planting materials, tissue culture technology offers promising alternative compared to the traditional production methods. In this study different concentrations of ripped sweet orange (*Citrus sinensis*) juice were screened for regeneration and multiplication of 2 months old *in vitro* cocoyam shoot explants. Among the concentrations, maximum numbers of roots (37 ± 5) were observed in Murashige and Skoog (MS) medium supplemented with 10% orange juice after 8 weeks in culture compared to 16 ± 4 (roots) observed in the control medium. On shoot multiplicity, 16 ± 3 shoots were induced in the control medium in contrast to 12 ± 0.8 shoots in the MS medium supplemented with 10% orange juice after 8 weeks of culture. Higher concentrations (25% and 35%) of the orange juice supplemented medium resulted in concomitant inhibition of all the growth parameters. The study successfully established that ripped orange juice could substitute the use of conventional growth hormones cytokinins (BAP) and auxins (NAA) in *in vitro* regeneration and rapid multiplication of cocoyam shoot explants.

Keywords

Taro, Orange Juice, Shoot Explants, *In Vitro* Regeneration, Rapid Multiplication

[*]Corresponding author.

1. Introduction

It is 113 years since the concept of isolation and growth of plant cells in artificial culture medium was proposed by Gotlieb Haberlandt. Eighty years later, precisely in 1983, the first transgenic plant was engineered. The two events (1902 and 1983), marked the beginning in our long scientific journey towards today's achievements in the plant and agricultural biotechnology revolution [1] [2]. The developments in plant tissue culture that have taken place in the last 45 years have revolutionized agricultural scene in a number of fronts including rapid production of mass quantity of uniform disease-free plants from a single explant, genetic engineering, callus induction, anther culture, protoplast culture, somatic embryogenesis, and genetic transformation of plants. Several researchers have reported on the production of improve taro cultivars through tissue culture [3]-[5]. *De novo* regeneration in taro has also been reported [6]-[9]. Production of improved taro varieties via genetic transformation offers an attractive alternative to conventional breeding. This technique was successfully carried out using both *Agrobacterium tumefaciens* and microprojectile bombardment of regenerable embryogenic suspension [9]-[12] and regenerable callus [13] [14]. Tissue culture techniques are now an attractive field of biotechnological research, and their roles are presently exploited in the areas of large scale clonal propagation, crop improvement, and conservation of plant genetic resources. Another modern extension of this technology is in phytosanitory germplasm transfer across international and inter-regional boundaries. It is therefore hopeful that these biotechnology alternatives may offer solutions to the contrasting reports on the dormancy and viability of taro seeds produced by sexual hybridization [15] [16].

Plant physiological processes are regulated by chemical signals known as plant growth regulators (PGRs). They are normally active at very low concentrations in plants. Indeed for the specific regulation of many plant processes and the differentiation of cells into specific plant parts, a variety of ratios and concentrations of these plant hormones are required rather than a single hormone acting alone [17]. They may act as both stimulators and inhibitors of growth, and could cause different plant parts (shoots, buds and roots) to respond differently, although the proportion of auxins (NAA) to cytokinins (BAP) determines the type and extent of organogenesis in plant cell cultures [18] [19]. Currently, great interest has been on developing technologies based on the use of natural fruit juices as growth hormones, carbon and vitamin sources for *in vitro* regeneration and propagation of plantlets in tissue culture media. Orange (*Citrus sinensis*) juice is tasty and nutritious. It is composed of vitamin C (82 mg), thiamin (0.28 mg), folate (45 mcg DFE), vitamin B6 (0.13 mg), vitamin A (194 IU), niacin (0.70 mg), potassium (473 mg), magnesium (27 mg), calcium (25/350 - 500 mg), iron (0.42 mg) and sodium (2 mg) [20]. The use of natural plant products such as fruit juices could enhance the usefulness of plants as renewable resources of valuable chemicals and can be optimized to play an increasingly significant role in commercial development of new products for regulating plant growth [21].

Production challenges confronting cocoyam cultivation in the last three decades in Nigeria ranges from scarcity of quality planting materials, low multiplication ratio to incidences of pests and diseases [22]. Traditionally, cocoyams are vegetatively propagated from tuber fragments, a practice that encourages pathogen distribution. Furthermore, cocoyam rarely flowers and when it flowers, the flowers are protogynous which makes the use of classical breeding difficult. Successful inductions of flowers, after treatment with gibberellic acid have been reported [23]-[26]. Presently tissue culture techniques offer exciting possibilities for sustainable production and rapid generation of millions of disease-free healthy and vigorous plantlets in one year irrespective of seasons and in a relatively small space. Fruit like sweet orange (*Citrus sinensis*) is abundant in Nigeria and a valuable percentage are perennially lost due to inadequate storage and processing. A ubiquitous and easy growing tree, but the fruits are prone to postharvest rot in a short period of about 2 weeks. In this study therefore, the potential of substituting cheap and locally available sweet orange juice excluding the conventional growth hormones (NAA and BAP) on *in vitro* regeneration and multiplication of cocoyam shoot explants is explored.

2. Materials and Methods

2.1. Extraction of the Orange Juice

Ripped oranges (*Citrus sinensis*) were washed in running tap water, followed by immersion in 70% ethanol for 20 min. Sterilization was carried out in 3.5% sodium hypochlorite solution in combination with 2 drops of Tween 20 for 15 min. This was followed by rinsing three times in sterile distilled water (10 min per each rinse) and allowed to dry at room temperature. The oranges were cut with sterile scalpel and hand-squeezed. The pulps were sieved through sterile cheesecloth and centrifuged for 20 minutes at 2000 rpm.

2.2. Culture Medium and Establishment of Cultures

The control medium was composed of Murashige and Skoog's basal medium supplemented with vitamins, myo-inositol, L-cysteine, BAP, NAA, gelrite and sucrose. Other media were composed of all the aforementioned components with different concentrations (10%, 15%, 25%, and 35%) of orange juice excluding the two growth hormones (BAP and NAA) (**Table 1**). All the media were supplemented with 3% sucrose, adjusted to pH 5.8 with 1N H_2SO_4 and/or 1N NaOH before adding 2 g/l of gelrite and sterilizing by autoclaving at 121°C for 20 minutes. Four weeks old cocoyam shoot explants (*Colocasia esculenta*) were aseptically trimmed with sterile forceps and scalpel in a laminar air flow hood. The base of the explants were neatly sliced-off and inoculated singly into sterile test tubes containing 10 ml of the sterile MS media. The test tubes were sealed properly with parafilm and clearly labeled before growing the cultures at a temperature of 28°C ± 2°C for 16 h photoperiod and 8 h darkness at light intensity of 2000 - 3000 lux for 8 weeks. All treatments consisted of three replicates and each replicate contained eight shoot explants. Shoot height, number of roots, number of shoots and leaves were recorded after every 2 weeks for 8 weeks in culture. Data were analyzed using DMRT test at 5% level of significance.

3. Results and Discussion

The results of *in vitro* regeneration and multiplication of young cocoyam shoot explants are as captured in **Tables 2-5**. Various concentrations of the orange juice in the culture media induced varied growth responses on

Table 1. Media compositions for *in vitro* regeneration and multiplication of cocoyam plantlets.

Media description	Components
Murashige and Skooge (MS) (control)	MS media, vitamin mixture, myo-inositol, L-cysteine, BAP, NAA, gelrite + sucrose
MS + 10% orange juice	MS media, vitamin mixture, myo-inositol, L-cysteine, 10% orange juice, gelrite + sucrose.
MS + 15% orange juice	MS media, vitamin mixture, myo-inositol, L-cysteine, 15% orange juice, gelrite + sucrose.
MS + 25% orange juice	MS media, vitamin mixture, myo-inositol, L-cysteine, 25% orange juice, gelrite + sucrose.
MS + 35% orange juice	MS media, vitamin mixture, myo-inositol, L-cysteine, 35% orange juice, gelrite + sucrose.

Table 2. Mean number of roots induced from the *in vitro* cocoyam shoot explants raised in the media.

Media description	2 weeks	4 weeks	6 weeks	8 weeks
MS (control)	0^b	0^b	8 ± 2^c	16 ± 4^c
MS + 10% orange juice	8 ± 2^a	17 ± 4^a	28 ± 6^a	37 ± 5^a
MS + 15% orange juice	0^b	0^b	12 ± 3^b	18 ± 4^b
MS + 25% orange juice	0^b	0^b	8 ± 1.6^c	14 ± 3^d
MS + 35% orange juice	0^b	0^b	6 ± 0.8^d	10 ± 0.9^e

Key: Values represents mean ± standard errors for three replications for each treatment. Mean in a column with the same superscript letter do not differ significantly according to DMRT at 5% level of probability.

Table 3. Mean number of leaves produced by the *in vitro* cocoyam shoot explants raised in the media.

Media description	2 weeks	4 weeks	6 weeks	8 weeks
MS (control)	4 ± 0.3^b	9 ± 2^b	16 ± 4^b	24 ± 5^b
MS + 10% orange juice	3 ± 0.1^c	7 ± 1.4^c	10 ± 2^d	12 ± 3^e
MS + 15% orange juice	8 ± 2^a	19 ± 4^a	26 ± 7^a	32 ± 6^a
MS + 25% orange juice	2 ± 0.2^d	6 ± 0.4^d	11 ± 2.5^c	19 ± 3.8^c
MS + 35% orange juice	2 ± 0.1^d	6 ± 0.7^d	9 ± 2^e	18 ± 4^d

Key: Values represents mean ± standard errors for three replications for each treatment. Mean in a column with the same superscript letter do not differ significantly according to DMRT at 5% level of probability.

Table 4. Mean shoot height (cm) produced by the *in vitro* cocoyam shoot explants raised in the media.

Media description	2 weeks	4 weeks	6 weeks	8 weeks
MS (control)	1.4 ± 0.1^{a}	2.15 ± 0.3^{a}	3.6 ± 0.04^{a}	5.15 ± 0.2^{a}
MS + 10% orange juice	1.2 ± 0.3^{a}	2.13 ± 0.1^{a}	3.8 ± 0.2^{a}	5.23 ± 0.03^{a}
MS + 15% orange juice	1.2 ± 0.04^{a}	2.13 ± 0.01^{a}	3.1 ± 0.4^{b}	5.09 ± 0.1^{a}
MS + 25% orange juice	0.87 ± 0.02^{a}	1.62 ± 0.03^{a}	3.2 ± 0.05^{b}	4.77 ± 0.06^{a}
MS + 35% orange juice	0.64 ± 0.03^{a}	1.52 ± 0.02^{a}	2.8 ± 0.1^{b}	4.33 ± 0.6^{ab}

Key: Values represents mean ± standard errors for three replications for each treatment. Mean in a column with the same superscript letter do not differ significantly according to DMRT at 5% level of probability.

Table 5. Mean number of multiple shoots induced from the *in vitro* cocoyam explants raised in the media.

Media description	2 weeks	4 weeks	6 weeks	8 weeks
MS (control)	2 ± 0.1^{b}	5 ± 0.2^{c}	8 ± 0.4^{b}	16 ± 3^{a}
MS + 10% orange juice	3 ± 0.2^{a}	6 ± 0.1^{b}	9 ± 0.3^{a}	12 ± 0.8^{b}
MS + 15% orange juice	3 ± 0.03^{a}	7 ± 0.4^{a}	9 ± 0.6^{a}	11 ± 2^{c}
MS + 25% orange juice	1 ± 0.02^{c}	3 ± 0.01^{e}	8 ± 0.2^{b}	11 ± 3.2^{c}
MS + 35% orange juice	1 ± 0.04^{c}	4 ± 0.3^{d}	8 ± 0.5^{b}	11 ± 2.4^{c}

Key: Values represents mean ± standard errors for three replications for each treatment. Mean in a column with the same superscript letter(s) do not differ significantly according to DMRT at 5% level of probability.

the *in vitro* culture of the cocoyam shoot explants. Regeneration and proliferation efficiency of the explants were higher in the MS media supplemented with 10% and 15% orange juice compared to the other treatments (**Tables 2-5**). All the culture media except those supplemented with 25% and 35% orange juice concentrations induced optimal responses on regeneration and propagation of the explants. In 10% and 15% orange juice supplemented media, there were significant improvement in roots and leaf growth compared to the control medium (**Table 2** & **Table 3**). Development and proliferation of new roots were greatest in 10% orange juice supplemented MS medium (**Table 2**). Low concentrations of the orange juice may have favoured the initiation of numerous roots and leaf production in cocoyam compared to 25% and 35% orange juice concentrations (**Table 2** & **Table 3** respectively). But at 8 weeks of culture the superiority of the 10% orange juice supplemented medium was further demonstrated with marked profused root induction (37 ± 5) against the other treatments (**Table 2**). Thus, the performance of the medium containing 10% orange juice concentration was significant over the other treatments including the control because at this concentration we observed the earliest root induction of (8 ± 2) in 2 weeks and a maximum proliferation of 37 ± 5 at 8 weeks (**Table 2**). The rooting efficiency and the number of roots regenerated per shoot were significantly ($P < 0.05$) higher in 10% orange juice supplemented medium followed by a decrease when the orange juice concentration was increased to 15%, 25%, and 35% (**Table 2**). The superior performance of the 10% orange juice supplemented medium could be that the plantlets will not need to expend energy to break down sucrose into utilizable monosaccharide [27]. Optimal growth and morphogenesis of tissues may vary for different plants according to their nutritional requirements [28]. Ikram-ul-Hag *et al.*, [29], achieved maximum *in vitro* growth of cucumber (*Cucumus sativus* L.) in MS medium supplemented with 158 ml orange juice excluding growth hormones compared to the control medium and other media formulation containing apple, red grapes and strawberry juices.

The superiority of the orange juice supplemented medium over the control medium was further demonstrated on leaf induction and growth after 2, 4, 6, & 8 weeks in culture (**Table 3**). At 15% orange juice concentration in the MS medium, significant improvement in leaf production was observed after 2, 4, 6, & 8 weeks in culture compared to the other treatments (**Table 3**). Leaf induction, shoot height development and shoot multiplication were observed after 2 weeks of culture (**Tables 3-5**). Although enhanced leaf induction were observed in the 2nd week in the medium containing 15% orange juice, further increase in the concentration of the orange juice in the medium resulted in the significant decrease in the number of leaves produced per explants (**Table 3**). Higher concentrations might be inhibitory and could reverse the beneficial effects of the juice on the regeneration and growth of the plantlets. Thus, higher concentrations of the orange juice at 15%, 25% and 35% in the media inhibited roots regeneration and proliferation (**Table 2**). Similarly, leaf induction and growth were also inhibited at 25% and 35% orange juice concentration (**Table 3**).

The effects of the orange juice supplemented media were observed to be insignificant on shoot height (**Table 4**) and shoot multiplicity compared to the enhanced multiple shoot induction in the control medium after 8 weeks of culture (**Table 5**). Manawadu *et al.*, [30] observed that the exact conditions that are necessary for regeneration of intact plantlets from cultured explants may vary for each plant species. In their *in vitro* study on shoot regeneration of *Raphanus sativus*, they achieved highest shoot height on MS medium containing 10% orange juice compared to 20% coconut water and 20% coconut milk. In this study we also observed highest shoot height after 6 and 8 weeks in culture at 10% orange juice concentration compared to the other treatments (**Table 4**). Although the control medium induced the highest number of multiple shoots (16 ± 3) after 8 weeks in culture compared to the other treatments (**Table 5**), the media containing 10% and 15% orange juice produced the highest number roots and leaves (**Table 2** & **Table 3** respectively). Besides delayed root induction in the control medium and in the media supplemented with 15%, 25% & 35% orange juice concentration (**Table 2**) other growth parameters like leaves, shoot height and number of shoots responded at 2 weeks of culture (**Tables 3-5**). Thus, 8 ± 2 (leaves), 1.4 ± 0.1 cm shoot height and 3 ± 0.03 number of shoots were observed in 15% orange juice supplemented medium, control medium, and 15% orange juice supplemented medium (**Tables 3-5**) respectively. However at 2 and 4 weeks of culture, the performances of the cocoyam shoot explants were significantly different at various treatments considered (**Table 5**). Thus, at 2, 4, 6, & 8 weeks of culture, 3 ± 0.03, 7 ± 0.4, 9 ± 0.3 and 16 ± 3 number of shoots were observed suggesting that the control medium supported maximum shoot multiplicity of the cocoyam explants at 8 weeks of culture.

Orange juice, a complex organic extract used in this study contained carbohydrates, protein, several vitamins, lower levels of some amino acids and organic acids [31]. Any of the aforementioned components or other(s) yet unknown substances, alone or in combination might be a factor(s) enhancing plantlet regeneration compared to the control [31]. The present study revealed that 10% - 15% orange juice concentrations in the MS media were more suitable for *in vitro* regeneration and multiplication of cocoyam plantlets. At these concentrations, the level of organic acids, vitamins or other unknown substance(s) which may be responsible for promoting the growth of *in vitro* cocoyam plantlets may be optimum. Higher concentrations might be inhibitory and could reverse the beneficial effects of the juice on the regeneration and growth of the plantlets. Amazingly, multiple shoot formation was observed on all the growing plantlets on all the media type (**Table 5**) implying that such a trait may be inherent in the cocoyam crop. Therefore, shoot multiplicity of the cocoyam shoot explants in culture is another advantage that recommends tissue culture technique for rapid multiplication of disease-free healthy cocoyam plantlets. Hence, tissue culture application could be a valuable tool in mass production of disease-free healthy cocoyam planting materials compared to the traditional methods. In addition, supplementation of orange juice extract in culture medium is not cumbersome, but advantageous and a convenient approach to reduce cost of culture media. We also observed that the promotive effect of the orange juice extract concentration varied with the growth of the different organs of the plantlets as typified with that of the roots and the leaves at 10% and 15% respectively. The 10% and 15% orange juice supplemented media performed convincingly better than the control especially on roots and leaves regeneration and development (**Table 2** and **Table 3**). Additionally, the incorporation of the orange juice did not trigger any form of morphogenic changes in the regenerated juvenile plantlets (**Figures 1(a)-(e)**) rather the regenerants on the orange juice supplemented media were more robust, greenish and even healthier (**Figure 1(c)** & **Figure 1(e)**). The study has successfully developed a simple, cheap and reliable protocol for *in vitro* regeneration and multiplication of cocoyam plantlets. Further experiment will be exclusively on orange juice extract and agar as culture medium for *in vitro* regeneration and multiplication.

4. Conclusion

At 10% and 15% orange juice concentration, the supplemented media produced the best result, thus recommending its application for *in vitro* regeneration and multiplication of cocoyam plantlets. We observed that higher concentration of the orange juice (25% & 35%) in the MS medium resulted in the concomitant decrease in the growth responses of roots, leaves, shoot height and shoot multiplicity. Based on our results, we therefore concluded that orange juice at 10% and 15% concentrations in the MS media is a potential alternative organic

(a) (b) (c) (d) (e)

Figure 1. *In vitro* regeneration and multiplication of cocoyam plantlets, (a) growing cocoyam shoot explants; (b) shoot multiplication in the control medium (MS); (c) shoot multiplication in MS medium augmented with 10% orange juice; (d) root proliferation in the control medium (MS); (e) root proliferation in the MS medium augmented with 10% orange juice.

additive to synthetic growth hormones in the culture medium for *in vitro* regeneration and multiplication of cocoyam. Impressively, one cocoyam shoot explant could develop multiples of 9 new shoots in culture within 4 weeks. This cheap and straight forward approach may be economically feasible for large scale production of disease-free cocoyam planting materials through tissue culture technology.

References

[1] Altman, A. (2003) From Plant Tissue Culture to Biotechnology: Scientific Revolutions, Abiotic Stress Tolerance and Forestry. *In Vitro Cellular Developmental Biology—Plant*, **39**, 75-84. http://dx.doi.org/10.1079/IVP2002379

[2] Ubalua, A.O. (2009) Transgenic Plants: Successes and Controversies. *Biotechnology and Molecular Biology Reviews*, **4**, 118-127.

[3] Jackson, G.V.H., Ball, E.A. and Arditi, J. (1977) Tissue Culture of Taro (*Colocasia esculenta* (L.) Schott). *Journal of Horticultural Science*, **52**, 373-382.

[4] Tuia, V.S. (1997) *In Vitro* Multiplication of Taro (*Colocasia esculenta* var. *esculenta* L. Schott), M. Agri.: University of the South Pacific.

[5] Deo, P.C., Tyagi, A.P., Taylor, M., Becker, D.K. and Harding, R.M. (2009) Improving taro (*Colocasia esculenta* var. *esculenta*) Production Using Biotechnological Approaches. *South Pacific Journal of Natural Science*, **27**, 6-13. http://dx.doi.org/10.1071/SP09002

[6] Yam, T.W., Ichihashi, S. and Arditi, J. (1991) Callus Growth and Plantlet Regeneration in Taro, *Colocasia esculenta* var. *esculenta* (L.) Schott (*Araceae*). *Annals of Botany*, **67**, 317-323.

[7] Thinh, N.T. (1997) Cryopreservation of Germplasm of Vegetatively Propagated Tropical Monocots by Vitrification. Doctoral Dissertation, Kobe University.

[8] Verma, V.M., Cho, J.J., Aikne, J. and David, J. (2004) High Frequency Plant Production of Taro (*Colocasia esculenta* (L.) Schott) by Tissue Culture. *Proceedings for the 4th International Crop Science Congress*, Brisbane.

[9] Deo, P.C. (2008) Somatic Embryogenesis and Transformation in Taro (*Colocasia esculenta* var. *esculenta*). Ph.D. Thesis, The University of the South Pacific, Suva.

[10] Aguado-Santacruz, G.A., Rascón-Cruz, Q., Cabrera-Ponce, J.L., Martínez-Hernández, A., Olalde-Portugal, V. and Herrera-Estrella, L. (2002) Transgenic Plants of Blue Grama Grass, *Bouteloua gracilis* (H.B.K.) Lag.ex. Steud., from Microprojectile Bombardment of Highly Chlorophyllous Embryogenic Cells. *Theoretical and Applied Genetics*, **104**, 763-771. http://dx.doi.org/10.1007/s00122-001-0803-2

[11] Sahrawat, A.K., Becker, D., Lütticke, S. and Lörz, H. (2003) Genetic Improvement of Wheat via Alien Gene Transfer, an Assessment. *Plant Science*, **165**, 1147-1168. http://dx.doi.org/10.1016/S0168-9452(03)00323-6

[12] ul-Haq, I. (2005) Callus Proliferation and Somatic Embryogenesis in Cotton (*Gossypium hirsutum* L.). *African Journal

of Biotechnology, **4**, 206-209.

[13] He, X., Miyasaka, S., Fitch, M.M., Zhu, Y.J. and Moore, P.H. (2004) Transformation of Taro (*Colocasia esculenta*) with a Rice Chitinase Gene (Abstract). *In Vitro Cellular and Developmental Biology-Plant*, **40**, 2104.

[14] He, X., Miyasaka, S., Fitch, M.M., Zhu, Y.J., Moore, P.H. and Zhu, Y.J. (2008) *Agrobacterium tumefaciens*-Mediated Transformation of Taro (*Colocasia esculenta* (L.) Schott) with a Rice Chitinase Gene for Improved Tolerance to a Fungal Pathogen *Sclerotium rolfsii*. *Plant Cell Reports*, **27**, 903-909. http://dx.doi.org/10.1007/s00299-008-0519-8

[15] Straus, M.S., Michaud, J.D. and Arditi, J. (1979) Seed Storage and Germination and Seedling Proliferation in Taro (*Colocasia esculenta* (L.) Schott). *Annals of Botany*, **43**, 603-612.

[16] Wang, J.K. (1983) Taro: A Review of *Colocasia esculenta* and Its Potential. University of Hawaii, Honolulu.

[17] Harms, C.L. and Oplinger, E.S. (1988) Plant Growth Regulators: Their Use in Crop Production. North Central Region Extension Publication 303, Wisconsin.

[18] Skoog, F. and Miller, R.A. (1957) Chemical Regulations of Growth and Organ Formation in Plant Tissue Culture *in Vitro*. *Symposia of the Society for Experimental Biology*, **11**, 118-130.

[19] Nair, R.R., Kavitha, M., Thilaga, S. and Ganesh, D. (2012) Conservation and *in Vitro* Multiplication of Highly Endangered Indian Traditional Medicinal Plant (*Morinda reticulata* Gamble) through Nodal Explants. *Plant Knowledge Journal*, **1**, 46-51.

[20] USDA (2008) USDA National Nutrient Database for Standard Reference, Release 21. NDB 09209.

[21] Ali, F.M., Ahsan, M., Saeed, N.A., Ahmed, M., Ali, Q., Kanwal, N., Tehseen, M.M., Ijaz, U., Bibi, I. and Niazi, N.K. (2014) Establishment and Optimization of Callus-to-Plant Regeneration System Using Mature Andimmature Embryos of Maize (*Zea mays*). *International Journal of Agricultural Biology*, **16**, 11-117.

[22] Ubalua, A.O. and Chukwu, L.I. (2008) Potentials and Constraints of Cocoyam Production in Nigeria. *Proceedings of the* 42*nd Annual Conference*, *Agricultural Society of Nigeria* (*ASN*), Ebonyi State University, Abakaliki, 19-23 October 2008.

[23] Wilson, J.E. (1979) Cocoyam Breeding by Flower Induction, Pollination, and Germination. Manual Ser No. 4, IITA, Ibadan.

[24] Alamu, S., McDavid, C.R. and Duncan, E.J. (1982) Production of Viable Gibberellic Acid-Treated Tannia (*Xanthosoma sagittifolium*) Plants. *Tropical Agriculture*, **59**, 323-334.

[25] Agueguia, A. and Nzietchueng, S. (1984) Production of Hybrid *Xanthosoma sagitifolium* and Test for Resistance to *Phythium myriotylium*. In: Terry, E.R., Doku, E.V., Arene, O.B. and Mahungu, N.M., Eds., *Tropical Root Crops*: *Production and Uses in Africa*, *Proceedings of the* 2nd *Triennial Symposium of the International Society for Tropical Root Crops*, Africa Branch, Douala, IDRC, Ottawa, Vol. 221, 169-171.

[26] Tambong, J.T. and Meboka, M. (1994) Cocoyam (*Xanthosoma sagittifolium* (L.)) Hybridization Studies: Pollen Viability and Seed Germination. *Acta Horticulturae*, **380**, 448-459. http://dx.doi.org/10.17660/ActaHortic.1994.380.69

[27] Alkhateeb, A. (2008) Comparison Effects of Sucrose and Date Palm Syrup on Somatic Embryogenesis of Date Palm (*Phoenix dactylifera* L.). *American Journal of Biotechnology and Biochemistry*, **4**, 19-23. http://dx.doi.org/10.3844/ajbbsp.2008.19.23

[28] Saad, A.I.M. and Elshahed, A.M. (2012) Plant Tissue Culture Media. In: Leva, A. and Rinaldi, L.M.R., Eds., *Recent Advances in Plant in Vitro Culture*, Chap. 2, InTech, Winchester, 29-40.

[29] Ikram-ul-Hag, A.Z., Taseer, M.G., Mukesh, D.M.U. and Ali, S.M. (2011) Effects of Different Fruit Juices Used as Carbon Source on Cucumber Seedling under *in Vitro* Cultures. *African Journal biotechnology*, **10**, 7404-7408.

[30] Manawadu, I., Dahanayake, N., Senanayake, S.G.N. (2014) Effects of Different Natural Organic Additives on *in Vitro* Shoot Regeneration of *Raphanus sativus* L. var. Beeralu. *Journal of Agricultural Science and Technology*, **A4**, 219-223.

[31] Rahman, A., Islam, M.O., Prodhan, A. and Ichihashi, S. (2004) Effects of Complex Organic Extracts on Plantlet Regeneration from PLBs and Plantlet Growth in the *Doritaenopsis* Orchid. *Japan Agricultural Research Quarterly*, **38**, 55-59. http://dx.doi.org/10.6090/jarq.38.55

18

In Vitro Regeneration of Endangered Medicinal *Hypoxis* Species

Busie E. B. Nsibande[1,2], Xueyuan Li[1], Annelie Ahlman[1], Li-Hua Zhu[1*]

[1]Department of Plant Breeding, Swedish Agricultural University, Alnarp, Sweden
[2]Mushroom Development Unit, National Plant Genetic Resources Centre, Malkerns, Swaziland
Email: [*]Li-Hua.Zhu@slu.se

Abstract

The genus *Hypoxis* contains some of southern Africa's most important wild medicinal species, a situation that has placed most of them amongst endangered species facing extinction. In this study, four *Hypoxis* species collected from Swaziland: *H. argentea*, *H. filiformis*, *H. acuminata* and *H. hemerocallidea*, were included to assess their potential for *in vitro* propagation in order to efficiently conserve these species in the future. Among all types of explants tested only corm explants and seeds gave rise to shoots under the *in vitro* conditions. For the corm establishment, *H. filiformis* was the most *in vitro* responsive species, which had 100% shoot regeneration with high shoot number when cultured on the MS medium supplemented with 3 mg·l^{-1} kinetin with a piece of shoot attached, while the highest mean shoot number (17) was obtained on the same medium, but with no shoot attachment to the corm explants. *H. argentea* was the next most responsive species *in vitro* with up to 70% establishment, while the other two species had only up to 20% of *in vitro* establishment. For the seed explants, up to 29% seed germination was obtained for *H. argentea* when the seed coats were crushed before the *in vitro* culture. The regenerated shoots were rooted *in vitro* and acclimatized in the greenhouse successfully

Keywords

Corm, Seed, Micropropagation, Plant Growth Regulators

1. Introduction

Species of the genus *Hypoxis*, belonging to the family Hypoxidaceae [1]-[4], are monocotyledons that are commonly distributed in the Southern Hemisphere [5] including sub-Saharan Africa, America, south-east Asia and Australia [6]. The genus *Hypoxis* consists of about 90 species, of which 30 are found in southern Africa [3]. The

[*]Corresponding author.

fleshy corm acts as a survival mechanism that enables the plant to withstand cold conditions, drought, and fire [6] and contains medicinal compounds. The corms of some species are also used as a source of food by some southern African tribes in times of famine [3].

The genus has generated pharmaceutical interest based on its use as a traditional medicine by indigenous people of eastern and southern Africa [7]. Eleven species are reported to be used in southern Africa for medicinal purposes, whilst *H. hemerocallidea* and *H. colchicifolia* are reportedly the most popular ones [3] [4], a situation that places these species under threat [3]. Extracts from the corms of *Hypoxis* are used as ingredients in a wide range of products such as anti-oxidants, anti-inflammatories, anti-diabetics and anti-convulsants [8] [9]. *H. hemerocallidea*, or African potato, is counted amongst the special indigenous medicinal species of commercial importance in southern Africa [10]. It derives its value from the hypoxoside content, a sterol that is converted by the human body to rooperol, a biologically active compound that balances the human immune system, and thus offering great potential as a source of new drugs with immune-modulatory properties [3]. African potato is used commonly as an immune booster to treat many ailments [3] [11] [12], and for its nutritional value [12].

Besides hypoxoside, acuminoside, a geraniol glycoside was also isolated from *H. acuminate* [13], while ellagic acid, a natural phenol antioxidant compound with a rigid structure and poor solubility, was found in the only monocotyledonous species *H. filiformis* [14].

Although *Hypoxis* species are naturally regenerated by seeds, they are difficult to propagate by seeds because the seeds seem to have a very strong seed coat or deep dormancy [15]. Vegetative propagation of *Hypoxis* species is still very uncommon in practice [4]. However, micropropagation may provide an alternative method for propagating species of this genus *en masse* to help prevent further depletion due to over-harvesting and preserve the valuable wild species. Very few studies on *in vitro* regeneration and establishment of *Hypoxis* have been reported, with varying levels of success: *H. hemerocallidea* (= *H. rooperii*) [2] [6] [16] [17] (Page and Van Staden, 1984, 1986; Appleton and Van Staden, 1995; Ndong *et al.*, 2006), *H. obtuse* [1] [6], *H. acuminata*and *H. rigidula* [6], *H. angustifolia* [15], and *H. colchicifolia* [4]. The regeneration frequency in these studies varied a lot depending on species and culture medium. Some reports did not provide detailed information about the shoot regeneration frequency thus making it difficult to understand how well the protocols worked. Moreover, there is very limited report on successful *in vitro* seed germination. Reference [4] reported that only one seed germinated *in vitro*, which died of browning shortly after. It is thus important to optimize the current protocols and to develop new protocols for new species. As *Hypoxis* is a large genus, developing efficient vegetative production systems or approaches involving more species will help not only alleviate the pressure on the natural populations, but also contribute to the conservation of many valuable wild *Hypoxis* species.

The aim of this study was to assess the potential for mass propagating four natural *Hypoxis* species which are widely growing in Swaziland through the *in vitro* regeneration approach. We have been able to successfully establish all the four species *in vitro* and three *ex vitro*; to the authors' knowledge no *in vitro* regeneration has previously been reported for the species *H. filiformis*.

2. Materials and Methods

2.1. Plant Materials

Four species of *Hypoxis* naturally growing in Swaziland were used in this study: *Hypoxis acuminata* Baker, *Hypoxis argentea* Harv. ex Baker, *Hypoxis hemerocallidea* Firsch and Ave'-Lall, and *Hypoxis filiformis* Baker. Specimens of the species were identified at the National Herbarium, Swaziland. Corms of the four species were planted in 2:1 planting soil to sand mixture in the greenhouse at the Swedish University of Agricultural Sciences (SLU), Alnarp, Sweden.

2.2. Preparation and Sterilization of Explants

Seeds, leaves, shoot apexes, basal parts of inflorescence peduncles, and flower buds were surface sterilized in 6% calcium hypochlorite ($Ca(ClO)_2$) solution with Tween 20 on a shaker for 15 min, followed by 3 rinses with sterile distilled water before being cultured on the culture medium. Roots were first washed at length in running tap water, followed by washing in 70% alcohol for 5 min and then with 6% $Ca(ClO)_2$ with Tween 20 for 15 min. For corms, a modification of the protocol described by [2] was used. The corms were first washed at length in running tap water with a detergent, followed by washing in 70% alcohol for 5 min, and then washing in 6% $Ca(ClO)_2$ with Tween 20 and shaking for 30 min. The outer epidermal tissues were then removed with a knife and the corms were cut into large pieces, approximately 3.0 cm^2 thick, and washed in 6% $Ca(ClO)_2$ with Tween 20 for 15 min, followed by 3 rinses with sterile distilled water. They were then cut into smaller pieces of approximately 0.5 - 1.0 cm^2 thick before being placed on the culture medium.

2.3. Culture Media and Treatment

MS [18] medium containing 30.0 $g \cdot l^{-1}$ sucrose, 8.0 $g \cdot l^{-1}$ Bacto agar, and 1.0% activated charcoal was used as the basal medium. Plant growth regulators (PGRs), mainly auxins indole-3-acetic acid (IAA), and naphthaleneacetic acid (NAA), and cytokininsbenzyladenine (BA) and kinetin (Kin), either alone or in combination, were added in the basal medium. Medium was changed regularly and the frequency depended on the requirement such as browning, contamination or infection of cultures. The pH was adjusted to 5.8 before adding agar and activated charcoal. Explants were either maintained in the same medium or transferred to other media (details are provided in the result section).

2.4. Culture of Seeds

Seeds were obtained from *H. argentea*, *H. filiformis*, and *H. hemerocallidea.* The seeds, harvested from the plants grown in the greenhouse were surface sterilized as described in 2.2 and subsequently cultured on the half-strength MS medium free of PGRs. Both intact seeds and seeds crushed with forceps were cultured. Once germinated the young seedlings were transferred to the full-strength MS without PGRs.

2.5. Leaf Culture

Young leaves of *H. acuminata*, *H. filiformis* and *H. hemerocallidea* were harvested, washed and cut into pieces before being surface sterilized as described in 2.2. Following sterilization, leaves were further cut into smaller pieces of approximately 0.5 cm^2 before being placed on the basal MS medium supplemented with 0.5 $mg \cdot l^{-1}$. Untreated leaves were used as the control.

2.6. Culture of Flower Buds

Unopened flower buds of *H. argentea*, *H. filiformis* and *H. hemerocallidea* were first washed and sterilized as described in 2.2 above. The buds were cut into half below the perianth segment and the top part was removed and the lower part cultured in double basal-strength MS containing 30.0 $g \cdot l^{-1}$ sucrose, 0.1 $g \cdot l^{-1}$ myo-inositol, 1.0 $g \cdot l^{-1}$ casein hydrolysate, a modification of the protocol used by [4]. Flower buds that were cultured on normal full-strength MS were used as the control.

2.7. Root Culture

Roots for *H. acuminata*, *H. filiformis*, and *H. hemerocallidea* were cleaned and sterilized using the protocol described in 2.2. The roots were then cut into small pieces of about 0.5 - 1.0 cm in length and cultured on the basal medium with supplemented with either no PGR, or 0.5 $mg \cdot l^{-1}$ BA, or 2.0 $mg \cdot l^{-1}$ BA or 1.0 $mg \cdot l^{-1}$ Kin, or 3.0 $mg \cdot l^{-1}$ Kin.

2.8. Culture of Corms

Corms were cleaned and prepared as described in 2.2 above before culturing. To save on plant material, a non-destructive strategy was devised where only part of the corm was used whilst the rest remained attached to the plant in the greenhouse and the plant survived and continued growing. Corm explants were either cultured with or without a piece of shoot attached on either the basal MS medium alone or supplemented with 1.0 $mg \cdot l^{-1}$ NAA, or 3.0 $mg \cdot l^{-1}$ BA, or 3.0 $mg \cdot l^{-1}$ kinetin, or 5.0 $mg \cdot l^{-1}$ BA, respectively. The explants were cultured on medium containing the same type and concentration of PGRs for the duration of the culture period or transferred to a medium of a different type and concentration.

2.9. *In Vitro* Rooting and Acclimatization

Well-developed shoots were either transferred to the basal MS supplemented with 2 $mg \cdot l^{-1}$ BA and 5 $mg \cdot l^{-1}$ IAA, or to PGR free basal medium. Rooted shoots and seedlings were acclimatized *in vitro* by opening the culture vehicle and leaving it open overnight in the growth chamber. The roots were washed with distilled water to remove all traces of agar before soaking in 0.1% mercuric chloride ($HgCl_2$) solution for 15 min and immersing in distilled water, leaving the leaves above the water level, for 5 days in the growth chamber.

2.10. Culture Conditions

All *in vitro* cultures were maintained in a growth chamber with the temperature of 25/18°C (day/night) and 16 h photoperiod at the light intensity of 40 $\mu mol \cdot m^{-2} \cdot s^{-1}$ provided by cool white fluorescent tubes.

2.11. Acclimatization in Greenhouse

Shoots were planted in compost soil mixed with sand (2:1) in the greenhouse and covered with transparent plastic cups for 7 to 10 days; thereafter the cups were partially lifted to allow the plantlets to adjust to the new conditions before being completely removed. Plantlets were watered every two days without wetting the leaves until they became hardened and well established.

2.12. Statistical Analysis

An analysis of variance and the Chi-square Two-way test were conducted using Minitab 16, as well as the Fisher's test for multiple comparison analysis. The number of explants ranged from 10 to 100, depending on material availability and all the experiments were repeated twice.

3. Results

3.1. Culture of Seeds

Seeds only germinated when the seed coat was crushed with forceps before being cultured (**Figure 1**), while intact seeds failed to germinate. Seed infection differed significantly among species with seeds of *H. filiformis* having the highest infection. Among the three species tested, the highest germination of 29.0% was obtained with *H. argentea*; whilst germination of this species commenced about two weeks after culture (**Table 1**), it took the other two species about one month to germinate. Some seeds of *H. argentea* and *H. filiformis* with the crushed seed coats formed callus that gave rise to multiple shoots. Since these shoots came from a single seed they were treated as one germinated seed. Up to 8 shoots were recorded arising from the callus of one seed in *H. filiformis*.

Figure 1. Germinating seeds of *H. argentea*with the seed coat crushed grown initially on half-strength MS (left) and 5-week-old seedlings on full-strength MS (middle and right).

Table 1. Corm establishment of Table 1: Seed germination of *Hypoxis* species with crushed seed coats cultured on half-strength MS.

Species	Total seed no.	Infection (%)	Germination (%)
H. argentea	100	4 b	29.0 a
H. filiformis	100	18 a	12.0 b
H. hemerocallide	15	0 b	13.3 b

Figures sharing different letters within the same column indicate a significant difference at p = 0.05.

3.2. Culture of Corms

3.2.1. *H. acuminata*

Nine out of the ten treatments responded to *in vitro* by producing callus, whilst six of them resulted in shoot regeneration, only one treatment produced shoots directly from the explants. Shoot regeneration varied from 0% to 20.0% among the different treatments, but there was no statistic difference among the treatments. It appears that this species only responded to Kin and BA for shoot regeneration, but not to NAA (**Table 2**). Infection ranged from 0% to 80%, while callus formation was 0% to 50%. Both infection and callus formation frequencies are not clearly related to the type and concentration of PGRs used (**Table 2**). Direct regeneration of explants cultured on the MS medium supplemented with 3 mg·l^{-1} BA was observed, whereas most of the treatments produced shoots via callus. In some treatments explants produced callus only, with no shoots or roots. Explants cultured on medium with a higher concentration of Kin (5 mg·l^{-1}) produced roots relatively earlier, and in most instances before shoots were formed (**Figure 2**). No significant difference was found in shoot regeneration from the corm explants cultured with shoot attached compared to those cultured without shoot on the same medium regardless of the presence or absence of PGR (compare treatments 1 and 2; 7 and 8; 9 and 10). Moreover, there was no significant difference either ($p > 0.05$) among the treatments regarding the number of shoots produced per regenerated explant (**Table 2**).

3.2.2. *H. argentea*

Infection ranged from 5.6% to 70%, while the callus formation frequency ranged from 12.5% to 80% for all

Table 2. Corm establishment of *H. acuminata* on different media after 24 weeks *in vitro* culture.

Code	Treatment	Infection (%)	Explants with callus (%)	Explants with shoots (%)	Shoot no/explant (means ± SE)
1	MS	24.0 b	50.0 a	0 a	-
2	MS + shoot	18.8 b	12.5 b	6.3 a	1.0 ± 0 a
3	MS → 3 Kin	80.0 a	10.0 b	20.0 a	2.0 ± 0 a
4	MS → 5 Kin	80.0 a	20.0 ab	20.0 a	1.5 ± 0.5 a
5	NAA	33.3 b	16.7 ab	0 a	-
6	NAA → 3 Kin	8.3 b	25.0 ab	0 a	-
7	BA	8.3 b	50.0 a	16.7 a	2.0 ± 1 a
8	BA + shoot	80.0 a	0 b	10.0 a	7.0 ± 0 a
9	Kin	8.3 b	50.0 a	8.3 a	1.0 ± 0 a
10	Kin + shoot	16.7 b	25.0 ab	16.7 a	4.0 ± 3 a

Note: 1 = PGR free MS; 2 = PGR free MS with shoot attached; 3 = PGR free MS to 3 mg·l^{-1} kinetin (Kin); 4 = PGR free MS to 5 mg·l^{-1} Kin; 5 = 1 mg·l^{-1} NAA; 6 = 1 mg·l^{-1} NAA for 3 weeks before being transferred to 3 mg·l^{-1} kinetin; 7 = 3 mg·l^{-1} BAP; 8 = 3 mg·l^{-1} BAP with shoot attached; 9 = 3 mg·l^{-1} Kin; 10 = 3 mg·l^{-1} Kin with shoot attached. Figures sharing the same letter indicate that the treatments were not significant at $p = 0.05$.

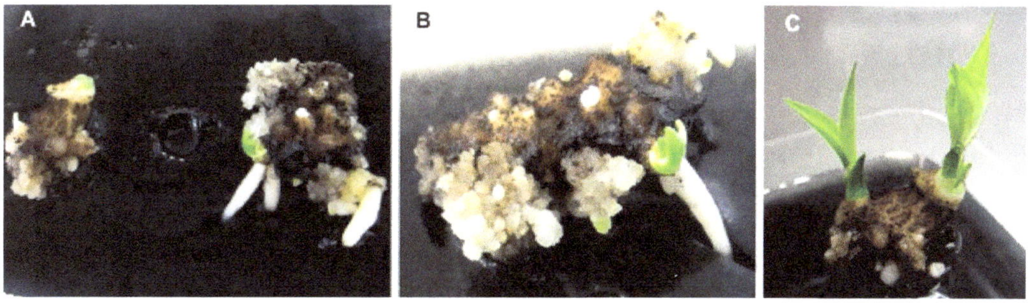

Figure 2. Corm explants of *H. acuminata* cultured on medium with 5 mg·l^{-1} kinetin at different stages of development. A: Calli and roots; B: Calli, roots and emerging shoots; C: 9-week-old shoots through indirect regeneration via callus.

treatments. The highest shoot regeneration (70%), which was significantly higher than most of the other treatments, was obtained when the explants were cultured with a piece of shoot attached on the MS medium containing 3 mg·l^{-1} BA. This treatment also resulted in higher shoot number per regenerated explant compared to the majority of the treatments. The highest shoot frequency (7 shoots/explant) though was obtained from the explants cultured on the MS medium supplemented with 1 mg·l^{-1} NAA (**Table 3**). There was no significant difference in the shoot regeneration frequency between explants cultured with or without shoot attachment on the MS medium free of PGRs (treatments 1 and 2), however, a significant regeneration was observed between the two types of explants in the presence of BA on the medium (treatments 3 and 4), but not in the presence of NAA (treatments 5 and 6) (**Table 3**). Generally regenerated shoots were formed from callus (**Figure 3**).

3.2.3. *H. filiformis*

Ten out of eleven treatments, including PGR free MS medium, produced shoots from corm explants. There were significant differences ($p < 0.05$) in shoot regeneration among the treatments, and in the mean number of shoots produced per regenerated explant (**Table 4**). One hundred percent regeneration was obtained from corm explants cultured with a piece of shoot attached on the MS medium supplemented with 3 mg·l^{-1} Kin (treatment 4), whereas the highest mean number (17) of shoots per regenerated explant was obtained on the same medium, but

Figure 3. Shoots of *H. argentea* regenerated from corm explants cultured on MS medium supplemented with different PGRs. A: Callus and indirect shoots formed on medium with 3 mg·l^{-1} BA; B and C: Shoots formed on medium with 1 mg·l^{-1} NAA for 5 weeks, followed 3 mg·l^{-1} BA (B) or 3 mg·l^{-1} kinetin (C).

Table 3. Corm establishment of *H. argentea* on different media after 24 weeks *in vitro* culture.

Code	Treatment	Infection (%)	Explants with callus (%)	Explants with shoots (%)	Shoot no/explant (means ± SE)
1	MS	5.6 bcde	27.8 bc	22.2 bc	1.8 ± 0.5 b
2	MS + shoot	40.0 abcd	30.0 bc	40.0 ab	1.3 ± 0.3 b
3	BA	10.0 de	15.0 c	0 c	-
4	BA + shoot	30.0 bcde	70.0 a	70.0 a	4.1 ± 0.9 ab
5	NAA	41.7 abc	50.0 ab	41.7 ab	7.2 ± 3.0 a
6	NAA + shoot	70.0 a	30.0 bc	0 c	-
7	NAA → BA	60.0 ab	30.0 bc	20.0 bc	4.5 ± 2.5 ab
8	NAA → Kin	25.0 cde	12.5 c	18.8 bc	3.0 ± 1.0 ab
9	Kin	33.3 bc	16.7 bc	22.2 bc	2.5 ± 1.2 b

Note: 1 = PGR free MS; 2 = PGR free MS with shoot attached; 3 = 3 mg·l^{-1} BA; 4 =3 mg·l^{-1} BA with shoot; 5 = 1 mg·l^{-1} NAA; 6 = 1 mg·l^{-1} NAA with shoot attached; 7 = 1 mg·l^{-1} NAA for 3 weeks before being transferred to 3 mg·l^{-1} BA; 8 = 1 mg·l^{-1} NAA for 3 weeks before being transferred to 3 mg·l^{-1} kinetin (Kin); 9 = 3 mg·l^{-1} Kin. Figures sharing different letters within the same column indicate a significant difference at p = 0.05.

Table 4. Corm establishment of *H. filiformis*corm explants on different media after 24 weeks *in vitro* culture.

Code	Treatment	Infection (%)	Explants with callus (%)	Explants with shoots (%)	Shoot no/explant (means ± SE)
1	MS	16.7 c	0 e	22.2 de	1.5 ± 0.5 d
2	MS + shoot	50.0 ab	50.0 bc	50.0 bcd	8.0 ± 2.9 abc
3	Kin	60.0 a	11.1 de	11.1 de	17.0 ± 14.0 a
4	Kin + shoot	0 c	100.0 a	100.0 a	4.0 ± 0.7 bc
5	BA	21.4b c	28.6 cde	0 e	-
6	BA + shoot	0 c	80.0 ab	80.0 abc	4.3 ± 0.5 bc
7	NAA	29.4 abc	35.3 cd	58.8 bc	4.6 ± 1.3 bcd
8	NAA → BA	21.1 bc	73.7 ab	68.4 b	6.4 ± 1.3 bc
9	NAA + BA → BA	30.0 abc	70.0 ab	70.0 ab	10.0 ± 3.7 ab
10	NAA → 5 BA	21.4 bc	30.0 cde	30.0 cde	14.7 ± 2.0 a
11	NAA → Kin	16.7 c	60.0 bc	10.0 e	10.0 ± 0 ab

Note: 1 = PGR free MS; 2 = PGR free MS with shoot attached; 3 = 3 mg·l^{-1} kinetin (Kin); 4 = 3 mg·l^{-1} Kin with shoot attached; 5 = 3 mg·l^{-1} BA; 6 = 3 mg·l^{-1} BA with shoot attached; 7 = 1 mg·l^{-1} NAA; 8 = 1 mg·l^{-1} NAA for 3 weeks before being transferred to 3 mg·l^{-1} BA; 9 = 1 mg·l^{-1} NAA + 3 mg·l^{-1} BA for 3 weeks before being transferred to 3 mg·l^{-1} BA; 10 = 1 mg·l^{-1} NAA to 5 mg·l^{-1} BA; 11 = 1 mg·l^{-1} NAA to 3 mg·l^{-1} Kin. Figures sharing different letters within the same column indicate a significant difference at p = 0.05.

without shoot attachment to the corm (treatment 3). This medium had only 11% explants giving rise to shoots (**Figure 4**). On both media, regenerated shoots were produced from callus which had equally the same callus and shoot formation frequencies of 100% and 11.1%, respectively. One explant had up to 32 shoots formed when initially cultured on the MS supplemented with 1 mg·l^{-1} NAA and 3mg·l^{-1} BA for 3 weeks, followed by culture on MS supplemented with 3 mg·l^{-1} BA (treatment 9). Notable differences in shoot regeneration and shoot frequency were also observed between corm explants cultured with or without shoot attachment in the absence of PGRs; explants with shoot attachment performed relatively better than those without and the shoot frequency was significantly higher (treatments 1 and 2). Additionally, the regeneration frequency was significantly higher for the explants with shoot attached than those without shoot regardless of PGR type (treatments 3 and 4; 5 and 6). Infection of explants during culture ranged between 0% and 60% with the largest number of shoots per regenerated explant arising from the treatment having the highest infection rate.

3.2.4. *H. hemerocallidea*

Corm explants of *H. hemerocallidea* took a very long time to respond to *in vitro* culture compared to the other three species, and the response was relatively poor with the majority of the treatments with no callus formation. Shoots were only produced when the explants were first cultured on the MS medium supplemented with 1 mg·l^{-1} NAA for 9 weeks before culturing on the MS medium supplemented with either 3 mg·l^{-1} BA, or 3 mg·l Kin, or 5 mg·l Kin. The shoot number per regenerated explant was also low (**Table 5**). Direct shoots started developing 10 weeks after transfer to the MS medium supplemented with cytokinins and growth of the shoots was extremely slow. Explants of this species had the least infection compared to the other species with 4 treatments having 0% infection. Even though the least regeneration was achieved, most explants remained fresh and intact during the culture period.

3.3. Culture of Other Explants

All other types of explants including leaves, flower buds, roots, and inflorescence peduncles failed to regenerate in any of the media tested in all four species. The explants usually turned brown and had to be sub-cultured very frequently. Although the browning could be notably reduced when 1.0% activated charcoal was added in the media, still no any regeneration occurred during the culture period.

3.4. Rooting of Regenerated Shoots *in Vitro* and Acclimatization *in Vitro* and *ex Vitro*

Most shoots produced roots when transferred to either the PGR free MS medium or to the MS medium supple-

Figure 4. Shoots of *H. filiformis* regenerated from corm explants cultured on the MS medium supplemented with different PGRs. A: Callus and multiple shoots produced on medium supplemented with 1 mg·l^{-1} NAA; B: Shoots produced on medium with 3 mg·l^{-1} BA following an initial culture with 1 mg·l^{-1} NAA; C: Callus and shoots formed on medium with 5 mg·l^{-1} BA following an initial culture with 1 mg·l^{-1} NAA; D: Corm explant with callus and multiple shoots on medium with 3 mg·l^{-1} kinetin; E: 17-week-old shoots with long roots on medium with 3 mg·l^{-1} BA following an initial culture with 1 mg·l^{-1} NAA; F: Shoots rooted on medium with 2 mg·l^{-1} BA and 5 mg·l^{-1} IAA.

Table 5. Response of *H. hemerocallidea* corm explants to different media after 24 weeks *in vitro* culture.

Code	Treatment	Infection (%)	Explants with callus (%)	Explants with shoots (%)	Shoot no/explant (means ± SE)
1	MS	0 b	0 b	0 b	-
2	MS + shoot	10.0 b	0 b	0 b	-
3	NAA	0 b	0 b	0 b	-
4	NAA → kin	70.0 a	0 b	10.0 ab	1 ± 0 a
5	NAA → 5 kin	50.0 a	0 b	20.0 a	1 ± 0 a
6	NAA → BA	0 b	7.1 ab	7.1 ab	2 ± 0 a
7	BA	8.3 b	0 b	0 b	-
8	Kin	8.3 b	0 b	0 b	-
9	Kin + shoot	0 b	20.0 a	0 b	-

Note: 1 = PGR free MS; 2 = PGR free MS with shoot attached; 3 = 1 mg·l^{-1} NAA; 4 = 1 mg·l^{-1} NAA for 3 weeks before being transferred to 3 mg·l^{-1} kinetin (Kin); 5 = 1 mg·l^{-1} NAA for 3 weeks before being transferred to 5 mg·l^{-1} Kin; 6 = 1 mg·l^{-1} NAA for 3 weeks before being transferred to 3 mg·l^{-1} BA; 7 = 3 mg·l^{-1} BA; 8 = 3 mg·l^{-1} Kin; 9 = 3 mg·l^{-2} Kin with shoot attached. Figures sharing different letters within the same column indicate a significant difference at p = 0.05.

mented with 2 mg·l^{-1} BA and 5 mg·l^{-1} IAA (**Figure 4E** and **Figure 4F**). Shoots of *H. filiformis* responded fairly well in both types of media, whereas those of *H. acuminata* and *H. argentea* turned yellowish after a few days on the PGR free medium. When kept on the same media for a prolonged period the shoots formed cormlets, which could be cut off for generating more shoots. Seedlings also produced cormlets when maintained on the MS medium without PGRs. Acclimatized shoots and seedlings were well established within 3 weeks of transplanting (**Figure 5**). Shoots of *H. filiformis* seemed to adapt better *ex vivo* compared to the other two species with 100% of establishment, while 75% for *H. argentea* and 16.7% for *H. acuminata*.

Figure 5. Regenerated plantlets of *H.argentea*and *H. filiformis* 3 weeks after planting (left) and fully grown *Hypoxis* plants (right) in greenhouse from *in vitro* culture.

4. Discussion

Endogenous pathogens are often a major problem causing failure in *in vitro* establishment of *Hypoxis* species, especially when using corms as explants [4] [16] [19]. Our study shows that endogenous pathogens and browning of explants caused by phenolic exudation are the two major sources for the majority of loss of explants during *in vitro* establishment and regeneration. Once placed on a solidified medium, the explants turned brown or even black rapidly and the medium turned brownish. To prevent browning activated charcoal was added in the medium, and in this study, adding 1.0% activated charcoal had proved to be effective in reducing the build-up of phenolic exudation in the culture media. In this study, endogenous pathogens of explants contributed to the most loss of the explants during the whole experiment period, especially for corms which were more difficult to decontaminate. Washing infected explants and shoots with a solution of 0.1% $HgCl_2$ for 5 min helped in clearing almost all noticeable infections, but this treatment can negatively affect growth of the explants.

Germination of *Hypoxis* is a common problem in practice, apparently due to the hard seed coat or deep dormancy. However, seeds can be a good choice of explants for *in vitro* establishment as they are relatively easier to be decontaminated compared to corms. Reference [4] reported that seeds of *H. colchicifolia* failed to establish *in vitro* with only one seed germinating, which died of browning soon after. In this study, most of the dormancy breaking efforts applied failed to induce seed germination in a preliminary test (unpublished). The only successful attempt for seed germination was crushing the seed coat using forceps, indicating that the hard seed coat is one of the major limiting factors for *in vitro* germination of *Hypoxis* species. With this method up to 29.0% germination was obtained for *H. argentea*. Apart from the hard seed coat, there might be some other endogenous inhibitory factors that affect seed viability as even with the crushed seed coat the germination percentage was still lower than 50%. Previous studies had shown that the viability of *Hypoxis* seeds varied from season to season [20] [21]. In this study, the seeds used were collected from the plants grown in the greenhouse where the conditions are surely different from their natural growing conditions in southern Africa, and the seeds obtained under the greenhouse conditions might not match the best natural conditions for seed germination.

In vitro establishment and regeneration are affected by types of explants and are genotype/species-dependent. In this study, among all the explant types tested, *in vitro* establishment and regeneration were obtained only from corms and seeds. Even for the corms the *in vitro* response varied considerably among the four *Hypoxis* species, for example, *H. filiformis* was the most responsive as direct shoots were produced from corm explants only 3 - 5 weeks after culture initiation. In contrast, *H. hemerocallidea* showed a much less and slow *in vitro* response under the same conditions as shoots were obtained 19 weeks after culture initiation, and even when shoots were produced they exhibited an extremely slow growth rate. Reference [1] conducted some regeneration tests on three *Hypoxis* species, *H. obtusa*, *H. nyasica* and *H. anguistifolia*, but only managed to successfully regenerate shoots from corms of *H. obtusa*, suggesting a strong species or genotype dependency for *in vitro* regeneration of *Hypoxis* species in general. Although *in vitro* regeneration has been previously reported from flower buds, in this study flower buds failed to establish *in vitro*. According to our results, among all types of explants tested, corms appear to be the most suitable explants for *in vitro* regeneration, followed by seeds.

The presence and concentrations of PGRs in the media were at most influential during culture initiation and establishment as well as for shoot multiplication. Corm explants of some species could produce shoots on the MS without PGRs, for instance, regeneration of corm explants was obtained in *H. filiformis* and in *H. argentea.*

However, the higher multiplication rate in corm explants could only be achieved with the addition of PGRs; in some instances, like in the case of *H. acuminata*, first culturing corm explants on PGR free basal MS medium before transferring to PGR containing medium seemed to positively influence the regeneration process. Almost all corm explants responded fairly well to *in vitro* regeneration when cultured with a piece of shoot attached in the presence of PGRs; 100% and 80% indirect regeneration was achieved in *H. filiformis* when explants were cultured on MS supplemented with 3 $mg \cdot l^{-1}$ kinetin and 3 $mg \cdot l^{-1}$ BAP, respectively. The highest regeneration of 70% in *H. argentea* was also obtained when 3 $mg \cdot l^{-1}$ BAP was added into the basic MS medium; however, this was not the case with the other two species tested in this study. Even without the presence of PGRs, corm explants cultured with shoot performed better compared to those cultured without in *H. acuminata*, *H. argentea*, and *H. filiformis*. These results clearly indicate that plant growth regulators play a crucial role in the *in vitro* regeneration of *Hypoxis* species, and such effects are species or genotype-dependent. In addition, leaving a piece of shoot attached to the corm relatively enhanced regeneration.

In vitro regeneration using corm explants could occur by direct shoot formation or indirect multiple shoot formation *via* shoot organogenesis from calli. In view of mass propagation, indirect shoot formation is desirable as a high multiplication frequency could be achieved on condition that somaclonal variation is not present. As high concentrations of cytokinins combined with auxins usually help to induce callus production, a two-step procedure had been used in previous studies, that is, induction of organogeniccalli on callus induction medium (CIM) followed by shoot production on the shoot induction medium (SIM). Using this procedure, [6] Appleton and van Staden (1995) reported that high growth response was achieved when *H. hemerocallidea* corm explants were cultured on the initiation medium comprising MS with 1 $mg \cdot l^{-1}$ NAA and 1 $mg \cdot l^{-1}$ BA, but they did not provide any information about the shoot regeneration frequency. Reference [2] were able to achieve 100% regeneration efficiency in corm explants of *H. hemerocallidea* with 5 - 8 direct shoots on the MS supplemented with 3 $mg \cdot l^{-1}$ Kin, while more efficient indirect regeneration was achieved by the two step cultures, *i.e.* on CIM with 3 $mg \cdot l^{-1}$ Kin and 1 $mg \cdot l^{-1}$ NAA, and then on SIM with 1.5 $mg \cdot l^{-1}$ BA and 0.5 $mg \cdot l^{-1}$ NAA. In this study no shoots were produced when *H. hemerocallidea* corm explants were directly cultured with either 3 $mg \cdot l^{-1}$ BA or Kin, whilst direct regeneration of 20% with low shoot number was achieved in corm explants when initially cultured on the basal medium supplemented with 1 $mg \cdot l^{-1}$ NAA followed by culture on 5 $mg \cdot l^{-1}$ Kin. This discrepancy might be due to the different physiological status of initial materials used in both experiments. Alternatively it could also be due to the genotype difference, which often affects *in vitro* regeneration greatly. Results from *H. filiformis* show that the highest regeneration (100.0%) was obtained from the basal medium supplemented with 3 $mg \cdot l^{-1}$ Kin, however, the highest shoot multiplication ratio (17 shoots/explant) was also achieved in the absence of NAA, when explants were directly cultured on the MS medium containing 3 $mg \cdot l^{-1}$ Kin (**Table 4**) suggesting that for some *Hypoxis* species addition of NAA may not be necessary for initiating regeneration. This indicates that the endogenous auxin level might be sufficient for initiating the regeneration process.

5. Conclusion

In conclusion, we have shown a successful establishment of *in vitro* materials for all 4 *Hypoxis* species, in which *H. filiformis* has not been reported earlier for *in vitro* regeneration. Our results have shown that *in vitro* response of *Hypoxis* is species or genotype dependent with *H. filiformis* being the easiest one to establish, and *H. hemerocallidea* the most difficult. Among all types of explants tested, corms showed a great potential for *in vitro* mass propagation as they showed high shoot regeneration frequency and produced multiple shoots for most of the species tested. Seeds could be established *in vitro* after crushing the seed coat and could be used as explants for micropropagation. Addition of auxins or cytokinins is necessary for shoot regeneration and multiplication in most cases. Attachment of shoot to explants enhanced *in vitro* establishment in some instances, suggesting that this effect is genotype or species dependent. The rooted and acclimatized shoots from *in vitro* culture were successfully established in the greenhouse. The successful *in vitro* and *ex vitro* establishments of these *Hypoxis* species indicate clearly the great potential of this vegetative propagation method to conserve them and to prevent their imminent depletion in the future.

Acknowledgements

We wish to thank Mr. Zacharia Dlamini, National Herbarium Curator, Ministry of Agriculture, Swaziland for his assistance with identification of the *Hypoxis* species; the staff of the National Plant Genetic Resources Centre and Mushroom Development Unit, Swaziland with assistance in collecting plant material; and Dr. Moneim Fatih, Project Coordinator in SLU for suggesting the research problem and his support to this study. This study is financed by Sida (Swedish International Development Cooperation Agency) as part of the Nordic support to the SADC Plant Genetic Resources Project.

References

[1] Vinesi, P., Serafini, M., Nicoletti, M., Spanò, L. and Betto, P. (1990) Plant-Regeneration and Hypoxoside Content in *Hypoxis*-Obtusa. *Journal of Natural Products*, **53**, 196-199. http://dx.doi.org/10.1021/np50067a031

[2] Ndong, Y.A., *et al.* (2006) Efficient *in Vitro* Regeneration of Fertile Plants from Corm Explants of *Hypoxis* Hemerocallidea Landrace Gaza—The "African Potato". *Plant Cell Reports*, **25**, 265-273. http://dx.doi.org/10.1007/s00299-005-0060-y

[3] Singh, Y. (2007) *Hypoxis* (Hypoxidaceae) in Southern Africa: Taxonomic Notes. *South African Journal of Botany*, **73**, 360-365. http://dx.doi.org/10.1016/j.sajb.2007.02.001

[4] Appleton, M.R., Ascough, G.D. and Staden, J.V. (2012) *In Vitro* Regeneration of *Hypoxis colchicifolia* Plantlets. *South African Journal of Botany*, 10.

[5] Kocyan, A., *et al.* (2011) Molecular Phylogenetics of Hypoxidaceae—Evidence from Plastid DNA Data and Inferences on Morphology and Biogeography. *Molecular Phylogenetics and Evolution*, **60**, 122-136. http://dx.doi.org/10.1016/j.ympev.2011.02.021

[6] Appleton, M.R. and van Staden, J. (1995) Micropropagation of Some South African *Hypoxis* Species with Medicinal and Horticultural Potential. *Ornamental Plant Improvement*, 75-77.

[7] Bredenkamp, M.W., Drewes, S.E. and Wentler, G.L. (1989) Medicinal-Plants of Southern-Africa. 3. A Geraniol Glycoside from *Hypoxis*-Acuminata. *Phytochemistry*, **28**, 263-265. http://dx.doi.org/10.1016/0031-9422(89)85053-8

[8] Drewes, S.E., *et al.* (2008) *Hypoxis* Hemerocallidea—Not Merely a Cure for Benign Prostate Hyperplasia. *Journal of Ethnopharmacology*, **119**, 593-598. http://dx.doi.org/10.1016/j.jep.2008.05.027

[9] Van Wyk, B.E. (2011) The Potential of South African Plants in the Development of New Medicinal Products. *South African Journal of Botany*, **77**, 812-829. http://dx.doi.org/10.1016/j.sajb.2011.08.011

[10] van Wyk, B.E. (2008) A Broad Review of Commercially Important Southern African Medicinal Plants. *Journal of Ethnopharmacology*, **119**, 342-355. http://dx.doi.org/10.1016/j.jep.2008.05.029

[11] Nair, V.D.P., Dairam, A., Agbonon, A., *et al.* (2007) Investigation of the Antioxidant Activity of African Potato (*Hypoxis hemerocallidea*). *Journal of Agricultural and Food Chemistry*, **55**, 1707-1711. http://dx.doi.org/10.1021/jf0619838

[12] Boukes, G.J., van de Venter, M. and Oosthuizen, V. (2008) Quantitative and Qualitative Analysis of Sterols/Sterolins and Hypoxoside Contents of Three *Hypoxis* (African Potato) Spp. *African Journal of Biotechnology*, **7**, 1624-1629.

[13] Bredenkamp, M.W., Drewes, S.E. and Wentler, G.L. (1989) A Geraniol Glycoside from *Hypoxis acuminata*. *Phytochemistry*, **28**, 263-265. http://dx.doi.org/10.1016/0031-9422(89)85053-8

[14] Yang, S.-W., Zhou, B.-N., Wisse, J.H., *et al.* (1998) Three New Ellagic Acid Derivatives from the Bark of *Eschweilera coriacea* from the Suriname Rainforest1. *Journal of Natural Products*, **61**, 901-906. http://dx.doi.org/10.1021/np980046u

[15] Appleton, M.R. and van Staden, J. (1995) *In Vitro* Propagation of a Miniature *Hypoxis*, *H. angustifolia*. *Ornamental Plant Improvement*, **420**, 95-97.

[16] Page, Y.M. and Vanstaden, J. (1984) *In Vitro* Propagation of *Hypoxis rooperi*. *Plant Cell, Tissue and Organ Culture*, **3**, 359-362. http://dx.doi.org/10.1007/BF00043088

[17] Page, Y.M. and Vanstaden, J. (1986) *In Vitro* Propagation of *Hypoxis Rooperi* from Flower Buds. *South African Journal of Botany*, **52**, 261-264.

[18] Murashige, T. and Skoog, F. (1962) A Revised Medium for Rapid Growth and Bio Assays with Tobacco Tissue Cultures. *Physiologia Plantarum*, **15**, 473-497. http://dx.doi.org/10.1111/j.1399-3054.1962.tb08052.x

[19] Suri, S.S., Jain, S. and Ramawat, K.G. (1999) Plantlet Regeneration and Bulbil Formation *in Vitro* from Leaf and Stem Explants of *Curculigo orchioides*, an Endangered Medicinal Plant. *Scientia Horticulturae*, **79**, 127-134. http://dx.doi.org/10.1016/S0304-4238(98)00118-6

[20] Hammerton, R.D. and Vanstaden, J. (1988) Seed-Germination of *Hypoxis hemerocallidea*. *South African Journal of Botany*, **54**, 277-280.

[21] Gillmer, M. and Symmonds, R. (1999) Seed Collection and Germination: *Hypoxis hemerocallidea* (Hypoxidaceae). *Plantlife*, **21**, 36-37.

19

Chemical Characterization and *in Vitro* Antitumor Activity of the Essential Oils from the Leaves and Flowers of *Callistemon viminalis*

Christiane Maria de Oliveira[1], Maria das Graças Cardoso[1*], Marisa Ionta[2], Marisi Gomes Soares[3], Juliana de Andrade Santiago[1], Guilherme Álvaro Ferreira da Silva[2], Maria Luisa Teixeira[1], Danúbia Aparecida de Carvalho Selvati Rezende[1], Rafaela Vieira de Souza[1], Luana Isac Soares[1], David Lee Nelson[4], Marcos Schleiden Sousa Carvalho[5]

[1]Department of Chemistry, Federal University of Lavras, Lavras, Brazil
[2]Institute of Biomedical Sciences, Federal University of Alfenas, Alfenas, Brazil
[3]Department of Chemistry, Federal University of Alfenas, Alfenas, Brazil
[4]Federal University of Vales de Jequitinhonha e Mucuri, Diamantina, Brazil
[5]Department of Agriculture, Federal University of Lavras, Lavras, Brazil
Email: [*]mcardoso@dqi.ufla.br

Abstract

The objective of this study was to characterize and verify the *in vitro* antitumor activity of essential oils (EOs) extracted from the leaves and flowers of *Callistemon viminalis*. The EOs were extracted by hydrodistillation using a modified Clevenger apparatus. The identification and quantification of constituents were performed on a gas chromatograph coupled to a mass spectrometer and a gas chromatograph with a flame ionization detector. The antitumoral activity was evaluated by a colorimetric assay (MTS) using different cell lines derived from human tumors (breast, lung, glioblastoma, and melanoma). The major constituents of the EOs of leaves and flowers were similar, only quantitative differences being observed. The compounds 1,8-cineole, α-pinene and α-terpineol were found in concentrations of 50.4%, 25.8% and 8.7% in the EOs obtained from the leaves and 48.8%, 24.5% and 3.9% in the EOs obtained from the flowers, respectively. The cytotoxic activity of the EOs was observed only in melanoma cultures (HT144). Cultures treated for 48 h with EOs from leaves and flowers (200 $\mu g \cdot mL^{-1}$) reduced the viability by 40% and 25%, respectively. Thus, the antiproliferative activity of the EO from leaves was more pronounced than the EO from flowers in cells derived from melanoma.

*Corresponding author.

Keywords

Cancer, Volatile Oil, Escova-de-Garrafa (Bottle Brush)

1. Introduction

Since ancient times, plants have provided a rich source of compounds with applications in the fields of medicine, pharmacy and biology. Molecules of plant origin, including alkaloids, saponins, triterpenes, glycosides and polyphenols, have shown promising antitumor activity in experimental models *in vitro* and *in vivo* [1]. Vinca alkaloids (vincristine and vinblastine), epipodophyllotoxin derivatives, camptothecin, etoposide and paclitaxel represent classic examples of compounds with therapeutic application in cancer [1]. However, the high toxicity of these compounds and the resistance commonly observed in some types of tumors have motivated the search for new therapeutic approaches.

Cancer is among the leading causes of mortality and morbidity in the world, and its incidence has increased because of multiple factors, including the increase in life expectancy of the population and changes in lifestyle. It is estimated that about 14 million new cases and 8.2 million deaths occurred in the world in 2012 [2]. According to the National Cancer Institute (INCA), it is estimated that about 580,000 new cases of cancer will be diagnosed in Brazil in 2015 [3], representing a major public health problem. Thus, the identification of new compounds with antiproliferative activity plays an important role in the prevention and treatment of cancer.

In recent years, essential oils (EOs) have been widely investigated for their various activities because they contain a wide variety of chemical components [4]-[6]. Studies have shown that many of the components in EOs may be useful in cancer therapy. In this context, this study sought to describe the chemical composition of the essential oils from the leaves and flowers of *Callistemon viminalis* and their antitumor activities against various cell lines of human cancer.

2. Material and Methods

2.1. Collection, Identification and Extraction of the Essential Oils

C. viminalis was collected in March 2014, in the morning, on the campus of the Federal University of Lavras (UFLA) in Lavras, MG, Brazil (21°13'S latitude and 44°58'W longitude; average altitude, 915 m) on a day with mild weather and no rainfall. The parts of the plant used separately for the extraction of essential oils were the leaves and the flowers. They were cleaned, weighed, and minced, and the essential oils were immediately extracted. A voucher specimen was deposited in the ESAL Herbarium in UFLA under the registration number 26,624.

The essential oils from leaves and flowers were extracted in the Organic Chemistry Laboratory–Essential Oils of the Department of Chemistry, UFLA. The plant materials (300 g each) were subjected to hydrodistillation for 2 hours using a modified Clevenger apparatus [7]. After extraction, the oils were separated from the hidrolacts by centrifugation using a benchtop centrifuge with a horizontal crosspiece (Fanem Model 206 Baby®I BL) at 965.36 g for 5 min. Subsequently, the essential oils were stored in amber containers under refrigeration [8].

To determine the moisture content for subsequent calculation of the yield of the oil present in the fresh plant, 5 g of the chopped plant material was extracted with 80 mL of cyclohexane, employing a Dean-Stark extractor. The moisture content was determined in triplicate according to the method described by Pimentel *et al.* [9].

2.2. Identification and Quantification of Essential Oil Components

Quantitative analysis and the identification of the constituents of the essential oils were accomplished at the Laboratory of Phytochemistry and Medicinal Chemistry (LFQM), Federal University of Alfenas (UNIFAL).

2.2.1. Analysis by GC-FID and GC-MS

The GC analyses were performed on a Hewlett-Packard 5890–Series II chromatograph (carrier gas, He) equipped with a FID detector, HP-5 capillary column (5% diphenyl in 95% of dimethylpolysiloxane–length, 30 m, internal diameter, 0.32 mm, film thickness, 0.25 mm), an HP 7673 automatic injector (split 1:20) and an HP

3396A electronic integrator. The oven temperature was programmed from 60°C - 280°C with a heating rate of 3°C min^{-1}, and the temperature was maintained for 10 min at 280°C. The temperatures of the injector and detector were maintained at 220°C and 280°C, respectively. The concentrations of the components were calculated on the basis of the peak areas in the chromatograms in order of elution from the HP-5 column. Analysis by GC-MS was performed on a Hewlett-Packard HP-6890 chromatograph (HP-5 column–5% diphenyl in 95% of dimethylpolysiloxane–length, 30 m, internal diameter, 0.32 mm, film thickness, 0.25 mm, carrier gas, He) coupled to a Hewlett-Packard HP-5973 mass spectrometer (operating by electron impact, 70 ev) under the same temperature conditions described above.

The identification of the components was based on the comparison of their mass spectra with those of the spectrometer database (Wiley and NIST Libraries) and confirmed by determination of their Kovats indices, which were calculated relative to the retention times of a series of n alkanes [10].

2.2.2. Analysis by 1H- and 13C-NMR

1H- and 13C-NMR spectra were recorded at 500 and 125 Mhz, respectively, on a Bruker model DRX-500 spectrometer with the sample dissolved in CDCl3 containing 1% TMS (Aldrich).

2.3. *In Vitro* Evaluation of Anti-Tumor Activity

2.3.1. Cell Lines and Culture Conditions

The tests were conducted on cell lines derived from human tumors (**Table 1**). Cell cultures were maintained in DMEM (Dulbecco's Modified Eagle's Medium; Sigma, CA, USA) supplemented with 10% fetal bovine serum (Vitrocell, Campinas, Brazil). The cultures were maintained under controlled atmosphere (95% air and 5% CO_2) in an incubator at 37°C.

2.3.2. Treatment Schedule

Cells were seeded in 96 well plates at a density of 5 × 103 (U251MG and A549), 1 × 104 (HT144) or 2 × 104 (HepG2 and MCF7) cells/well. After adherence (24 h), the cells were treated for 48 h with EOs extracted from the leaves and flowers of *C. viminalis*. The oils were dissolved in DMSO to obtain the stock-solution (100 $mg{\cdot}mL^{-1}$).

Initially, essential oils were tested on the different cell lines derived from human tumors at a concentration of 50 $\mu g{\cdot}mL^{-1}$ for observation of the activity profile. Subsequently, dose-response curves were constructed using the HT144 cell line and different concentrations of essential oil extracted from the leaves (5 to 200 $\mu g{\cdot}mL^{-1}$). The final DMSO concentration did not exceed 0.5% (v/v) and did not alter the cell viability that was determined by MTS.

2.3.3. Cell Viability

Cell viability was assessed by the colorimetric assay (MTS) using the Cell Titer 96TM kit from Promega®. This assay is based on the conversion of 3-(4,5-dimethylthiazol-2-yl)-5-(3-carboxymethylphenyl)-2-(4-sulfophenyl-2H-tetrazolium) bromide (MTS-yellow) to formazan (brown) by dehydrogenase enzymes present in the mitochondria of viable cells.

The production of formazan, a compound soluble in the culture medium that absorbs light at 490 nm, is directly

Table 1. Cell lines derived from human tumors used in the *in vitro* assay of antiproliferative activity.

Code	Name
A549	lung adenocarcinoma
MCF7	breast adenocarcinoma
HepG2	liver carcinoma
U251MG	glioblastoma
HT144	melanoma

proportional to the number of living cells [11]. Thus, this reaction occurs only with viable and metabolically active cells and provides a versatile and quantitative assay [12]. Viability was determined by comparing the absorbance values of the treated samples with the absorbance values obtained in control samples. The experiments were performed in triplicate and data were presented as the mean ± standard deviation.

$$\text{Cell viability \%} = \left(\text{Abs}_{\text{sample}} \times 100\right) / \text{Abs}_{\text{control}}$$

2.4. Statistical Analysis

The ANOVA test and Tukey's post-test were applied. Differences were considered significant when the p value was less than 0.05 ($p < 0.05$). The analysis was performed using the GraphPad Prism software (GraphPad Software, Inc., San Diego, CA, USA).

3. Results and Discussion

The same constituents were identified in the essential oils extracted from *C. viminalis* leaves and flowers, but in different percentages. Nineteen chemical compounds were identified in the essential oils from both the leaves and flowers. 1,8-cineole, α-pinene and α-terpineol were the major components at concentrations of 50.4%, 25.8% and 8.7% for the leaves and 48.8%, 24.5% and 3.9% for the flowers, respectively. The monoterpene 1,8-cineole was identified as the major component in the essential oils from both the leaves (50.4%) and the flowers (48.8%) by GC-MS analysis of the crude oil in association with analysis by 1H and 13C-NMR, followed by comparison with literature data [13] [14].

Oyedeji *et al.* (2009) [15] identified 20 constituents in the essential oil from *C. viminalis* leaves collected in South Africa and reported 1,8-cineole (83.2%), α-pinene (6.4%) and α-terpineol (4.9%) as the main components. Silva *et al.* (2010) [16] evaluated the chemical composition of the essential oil from *C. viminalis* collected in Brazil and identified 1,8-cineole (65%), α-pinene (12%) and α-terpineol (13%) as the principal components.

Pires *et al.* (2013) [17] evaluated the chemical composition of the essential oils from *C. viminalis* leaves and flowers collected in Brazil, and observed that 1,8-cineole was the major component in the essential oil from fresh flowers and leaves (70.43% and 42.39%, respectively), followed by tricicleno (12.97% and 28.08%, respectively) and α-terpineol (6.56% and 6.67%). However, tricicleno was not found in the essential oils from the leaves and flowers in the current study.

The production of secondary metabolites of a species can be influenced by several factors, including the physiological demand of the plant (growth, defense and reproduction), as well as genetic and environmental factors (soil composition, rainfall and climate). Thus, minor quantitative and qualitative variations in the chemical composition of the EOs of *C. viminalis* may occur [18]-[20]. Nevertheless, most studies indicate that 1,8-cineole, α-pinene and α-terpineol are the principal components, so the results of this study corroborate those described in the literature.

The essential oils extracted from the leaves and flowers of *C. viminalis* were tested for *in vitro* antitumor activity. Initially, the different cell lines derived from human tumors were treated with EOs at a concentration of 50 $\mu g \cdot mL^{-1}$. The results showed that EO extracted from *C. viminalis* flowers presented no antiproliferative activity against the cell lines tested, there being no significant reduction in cell viability in treated cultures compared to the control (**Figure 1**).

However, a 20% reduction in viability was observed in cultures of melanoma (HT144) treated for 48 h with EO extracted from the leaves of *C. viminalis* (**Figure 2**).

On the basis of these results, the HT144 cell line was selected for further testing. The results showed that EOs exert a dose-dependent effect on the HT144 cells because the viability in treated cultures progressively decreased as the concentration of EOs increased (**Figure 3** and **Figure 4**). The antiproliferative effect of the EO from the leaves was observed to be more active than that from the flowers. The viabilities of cultures treated for 48 hours with EOs from the leaves and flowers of *C. viminalis* at the concentration of 200 $\mu g \cdot mL^{-1}$ decreased by about 40% and 25%, respectively.

The lipophilic nature of the EOs allows them to easily cross cell membranes and reach the interior of the cells. The cytotoxic effects of several compounds may be related to damage or changes in cell membrane permeability. Thus, several studies have reported the potency of the constituents of EOs in the prevention and treatment of cancer [21]. Sobral *et al.* (2014) [22] showed, through studies performed *in vitro* and *in vivo*, that the antitumor

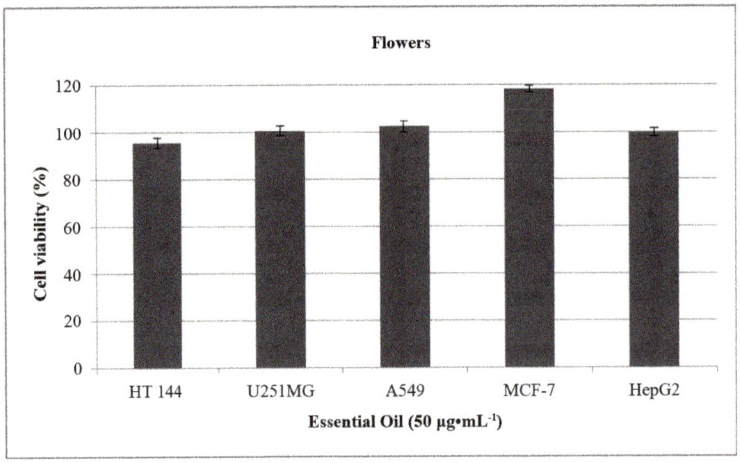

Figure 1. Cell viability (%) determined by the MTS assay in cell cultures treated with the essential oil from *C. viminalis* flowers at a concentration of 50 $\mu g{\cdot}mL^{-1}$.

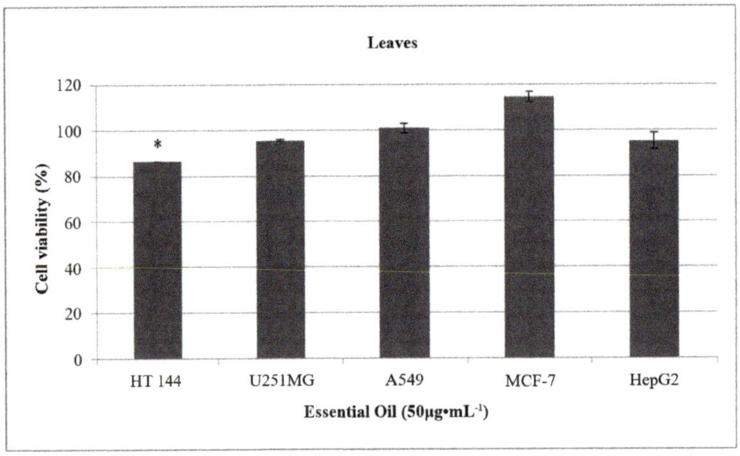

Figure 2. Cell viability (%) determined by the MTS assay in cell cultures treated with the essential oil from *C. viminalis* leaves at a concentration of 50 $\mu g{\cdot}mL^{-1}$. $^{*}(p < 0.05)$.

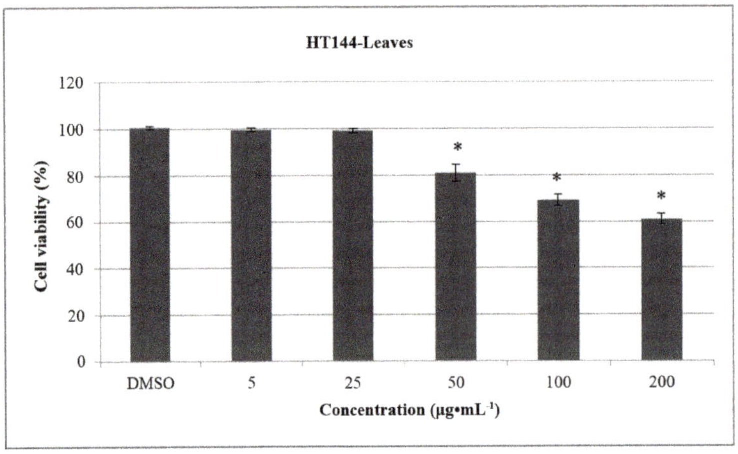

Figure 3. Cell viability (%) determined by the MTS assay in HT144 cell cultures treated for 48 h with different concentrations of the essential oil from *C. viminalis* leaves. $^{*}(p < 0.05)$.

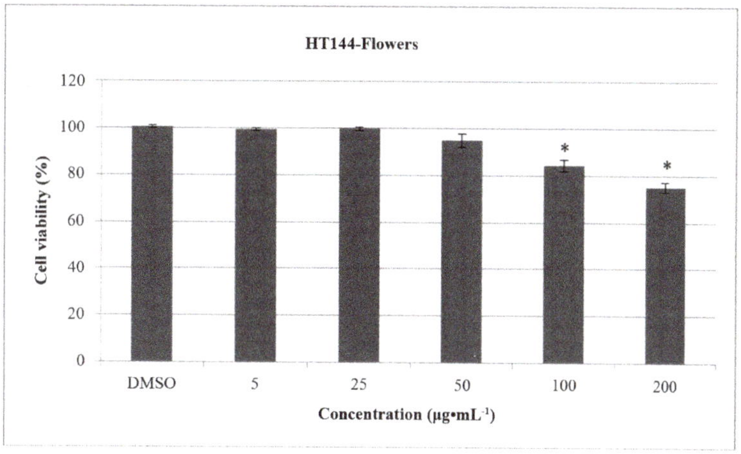

Figure 4. Cell viability (%) determined by the MTS assay in HT144 cell cultures treated for 48 h with different concentrations of the essential oil from *C. viminalis* flowers. *($p < 0.05$).

activity of EOs of many species has been linked to the presence of monoterpenes, the class of compounds also found in *C. viminalis* oils.

The cytotoxicity of the essential oil extracted from *Rosmarinus officinalis* L. and its three main components [1,8-cineole (27.23%), α-pinene (19.43%) and β-pinene (6.71%)] was investigated in cell lines derived from human ovarian (SK-OV-3 and HO-8910) and liver (Bel-7402) cancers. The EO from *R. officinalis* exhibited a pronounced cytotoxicity to the three cell lines tested. The IC_{50} values were 0.025; 0.076 and 0.13‰ (v/v), respectively. The monoterpene 1,8-cineole, which is also the major compound found in essential oils from *C. viminalis* leaves and flowers, presented moderate cytotoxicity to those lines [23]. Moderate cytotoxicity of 1,8-cineole and α-terpineol, major constituents also found in this study, was observed in the HepG2 (liver carcinoma) and K-562 (leukemia) lines [24].

Lampronti, Saab and Gambari (2006) [25] showed that the EO from *Satureja montana*, whose main component is α-terpineol, exhibited considerable cytotoxicity to K562 cells (human erythroleukemic). Hassan *et al.* (2010) [26] suggested that α-terpineol inhibits tumor cell growth by a mechanism that involves inhibition of the NF-kB (nuclear factor kappa B). Jin *et al.* (2010) [27] showed that α-pinene, also found in this study, triggered oxidative stress and related signaling pathways in the HepG2 and A549 tumor cells.

In the present study, it was shown that EOs from the leaves and flowers of *C. viminalis* exerted moderate cytotoxicity in the HT144 cell line. Considering that the melanoma is high lethal [3], the results obtained in the present study are very promising and further studies will be performed to verify whether the cytotoxic activity previously observed is related to pro-apoptotic potential of the essential oils. Reduction of cell viability in HT144 cultures may be related to the presence of 1,8 cineole, α-pinene and α-terpineol, but considering that essential oils have a complex chemical composition, it is possible that the reduction in viability was the result of synergistic action of the constituents found in the EOs [28]. The monoterpenes are considered to be the active compounds, and, therefore, additional studies should be conducted to further investigate the relationship of monoterpenes with antiproliferative or cytotoxic effects of EOs on melanoma cells.

4. Conclusions

The principal constituints found in the essential oils from *Callistemon viminalis* were 1,8 cineole, α-pinene and α-terpineol in the concentrations of 50.4%, 25.8% and 8.7% for the leaves; and 48.8%, 24.5% and 3.9% for flowers, respectively.

The cytotoxic activity of the essential oils was observed only in melanoma cultures (HT144); cultures treated for 48 h with EOs leaves and flowers (200 $\mu g \cdot mL^{-1}$) reduced the viability by 40% and 25%, respectively.

Acknowledgements

The authors thank the Conselho Nacional de Desenvolvimento Científico e Tecnológico (CNPq), the Coordenação de Aperfeiçoamento de Pessoal de Nível Superior (CAPES) and the Fundação de Amparo à Pesquisa do Estado de Minas Gerais (FAPEMIG) for financial support, a scholarship and a PVNS fellowship (D.L. Nelson).

References

[1] Cragg, G.M. and Newman, D.J. (2005) Plants as a Source of Anti-Cancer Agents. *Journal of Ethnopharmacology*, **100**, 72-79. http://dx.doi.org/10.1016/j.jep.2005.05.011

[2] Torre, L.A., Bray, F., Siegel, R.L., Ferlay, J., Lortet-Tieulent, J. and Jemal, A. (2015) Global Cancer Statistics, 2012. *CA: A Cancer Journal for Clinicians*, **65**, 87-108. http://dx.doi.org/10.3322/caac.21262

[3] Instituto Nacional de Câncer (INCA) José Alencar Gomes da Silva (Brasil) Disponível em: http://www.inca.gov.bracesso

[4] Gomes, M.S., Cardoso, M.G., Soares, M.J., Batista, L.R., Machado, S.M.F., Andrade, M.A., Azeredo, C.M.O., Resende, J.M.V. and Rodrigues, L.M.A. (2014) Use of Essential Oils of the Genus Citrus as Biocidal Agents. *American Journal of Plant Sciences*, **5**, 299-305. http://dx.doi.org/10.4236/ajps.2014.53041

[5] Silva, L.F., Cardoso, M.G., Batista, L.R., Gomes, M.S., Rodrigues, L.M.A., Carvalho, D.A.S.R., Teixeira, M.L., Carvalho, M.S.S., Santiago, J.A. and Nelson, D.L. (2015) Chemical Characterization, Antibacterial and Antioxidant Activities of Essential Oils of *Mentha viridis* L. and *Mentha pulegium* L. (L). *American Journal of Plant Sciences*, **6**, 666-675. http://dx.doi.org/10.4236/ajps.2015.65072

[6] Teixeira, M.L., Cardoso, M.G., Figueiredo, A.C.S., Moraes, J.C., Assis, F.A., Andrade, J., Nelson, D.L., Souza, M.G., Souza, J.A. and Albuquerque, L.R.M. (2014) Essential Oils from *Lippia origanoides* Kunth. and *Mentha spicata* L.: Chemical Composition, Insecticidal and Antioxidant Activities. *American Journal of Plant Sciences*, **5**, 1181-1190. http://dx.doi.org/10.4236/ajps.2014.59131

[7] BRASIL. Agência Nacional De Vigilância Sanitária. Farmacopeia brasileira. 5th Edition, Brasília, 198-199, 2010.

[8] Oliveira, C.M., Cardoso, M.G., Carvalho, M.L.M., Figueiredo, A.C.S., Miranda, C.A.S.F., Albuquerque, L.R.M., Nelson, D.L., Gomes, M.S., Silva, L.F., Santiago, J.A., Teixeira, M.L. and Brandão, R.M. (2014) Chemical Composition and Allelopathic Activity of the Essential Oil from *Callistemon viminalis* (Myrtaceae) Blossoms on Lettuce (*Lactuca sativa* L.) Seedlings. *American Journal of Plant Sciences*, **5**, 3551-3557. http://dx.doi.org/10.4236/ajps.2014.524371

[9] Pimentel, F.A., Cardoso, M.G., Salgado, A.P.S.P., Aguiar, P.M., Silva, V.F., Morais, A.R. and Nelson, D.L. (2006) A Convenient Method for the Determination of Moisture in Aromatic Plants. *Química Nova*, **29**, 373-375. http://dx.doi.org/10.1590/S0100-40422006000200031

[10] Adams, R.P. (2007) Identification of Essential Oils Components by Gas Chromatografhy/Mass Spectroscop. 4th Edition, Allured, Carol Streem, 804 p.

[11] Barltrop, J.A., Owen, T.C., Cory, A.H. and Cory, J.G. (1991) 5-(3-Carboxymethoxyphenyl)-2-(4,5-dime-thylthiazolyl)-3-(4-sulfophenyl)tetrazolium, Inner Salt (MTS) and Related Analogs of 3-(4,5-dimethylthia-zolyl)-2,5-diphenyltetrazolium Bromide (MTT) Reducing to Purple Water-Soluble Formazans as Cell-Viability Indicators. *Bioorganic & Medicinal Chemistry Letters*, **1**, 611-614.

[12] Mosmann, T. (1983) Rapid Colorimetric Assay for Cellular Growth and Survival: Application to Proliferation and Cytotoxicity Assays. *Journal of Immunological Methods*, **65**, 55-63. http://dx.doi.org/10.1016/0022-1759(83)90303-4

[13] Brochini, C.B., Nuñez, C.V., Moreira, I.C., Roque, N.F., Chaves, M.H. and Martins, D. (1999) Identificação de componentes de óleos voláteis: Análise espectroscópica de misturas de sesquiterpenos. *Química Nova*, **22**, 37-40. http://dx.doi.org/10.1590/S0100-40421999000100008

[14] Adio, A.M., Paul, C., Kloth, P. and König, W.A. (2004) Sesquiterpenes of the Liverwort Scapania Undulate. *Phytochemistry*, **65**, 199-206. http://dx.doi.org/10.1016/j.phytochem.2003.10.018

[15] Oyedeji, O.O., Lawal, O.A., Shode, F.O. and Oyedeji, A.O. (2009) Chemical Composition and Antibacterial Activity of the Essential Oils of *Callistemon citrinus* and *Callistemon viminalis* from South Africa. *Molecules*, **14**, 1990-1998. http://dx.doi.org/10.3390/molecules14061990

[16] Silva, C.J., Barbosa, L.C.A., Pinheiro, A.L. and Andrade, N.J. (2010) Chemical Composition and Antibacterial Activities from the Essential Oils of Myrtaceae Species Planted in Brazil. *Química Nova*, **33**, 104-108. http://dx.doi.org/10.1590/S0100-40422010000100019

[17] Pires, C.H., Paula, J.A.M., Tresvenzol, L.M.F., Ferri, P.H., Paula, J.R., Fiuza, T.S. and Bara, M.T.F. (2013) Composição química e atividade antimicrobiana dos óleos essenciais das folhas e flores de Callistemon viminalis (sol. ex Gaertn.) G. Don ex. Loudon (Myrtaceae). *Revista de Ciências Farmacêuticas Básica e Aplicada*, **34**, 597-601.

[18] Deschamps, C., Zanatta, J.L., Roawalka, L., Oliveira, M.C., Bizzo, H.R. and Alquini, Y. (2012) Densidade de tricomas glandulares e produção de óleo essencial em *Mentha arvensis* L., *Mentha x piperita* L. e *Mentha cf. aquatica* L. *Ciência e Natura*, **28**, 23-34.

[19] Masetto, M.A.M., Deschamps, C., Mógor, A.F. and Bizzo, H.R. (2011) Teor e composição do óleo essencial de inflorescências e folhas de *Lavandula dentata* L. em diferentes estádios de desenvolvimento floral e épocas de colheita. *Revista Brasileira de Plantas Medicinais*, **13**, 413-421. http://dx.doi.org/10.1590/S1516-05722011000400007

[20] Gobbo-Neto, L. and Lopes, N.P. (2007) Plantas Medicinais: Fatores de influência no conteúdo de metabólitos secundários. *Química Nova*, **30**, 374-381. http://dx.doi.org/10.1590/S0100-40422007000200026

[21] Gautam, N., Mantha, A.K. and Mittal, S. (2014) Essential Oils and Their Constituents as Anticancer Agents: A Mechanistic View. *BioMed Research International*, **2014**, Article ID: 154106. http://dx.doi.org/10.1155/2014/154106

[22] Sobral, L.M., Sousa, L.O., Coletta, R.D., Cabral, H., Greene, L.J., Tajara, E.H., Gutkind, J.S., Curti, C. and Leopoldino, A.M. (2014) Stable SET Knockdown in Head and Neck Squamous Cell Carcinoma Promotes Cell Invasion and the Mesenchymal-Like Phenotype *in Vitro*, as Well as Necrosis, Cisplatin Sensitivity and Lymph Node Metastasis in Xenograft Tumor Models. *Molecular Cancer*, **13**, 1-13. http://dx.doi.org/10.1186/1476-4598-13-32

[23] Wang, W., Li, N., Luo, M., Zu, Y. and Efferth, T. (2012) Antibacterial Activity and Anticancer Activity of *Rosmarinus officinalis* L. Essential Oil Compared to That of Its Main Components. *Molecules*, **17**, 2704-2713. http://dx.doi.org/10.3390/molecules17032704

[24] Hayes, A.J., Leach, D.N., Markham, J.L. and Markovic, B. (1997) *In Vitro* Cytotoxicity of Australian Tea Tree Oil Using Human Cell Lines. *Journal of Essential Oil Research*, **9**, 575-582. http://dx.doi.org/10.1080/10412905.1997.9700780

[25] Lampronti, I., Saab, A.M. and Gambari, R. (2006) Antiproliferative Activity of Essential Oils Derived from Plants Belonging to the Magnoliophyta Division. *International Journal of Oncology*, **29**, 989-995. http://dx.doi.org/10.3892/ijo.29.4.989

[26] Hassan, S.B., Gali-Muhtasib, H., Goeransson, H. and Larsson, R. (2010) Alpha Terpineol: A Potential Anticancer Agent Which Acts through Suppressing NF-κB Signaling. *Anticancer Research*, **30**, 1911-1919.

[27] Jin, K., Bak, M., Jun, M., Lim, H., Jo, W. and Jeong, W.S. (2010) α-Pinene Triggers Oxidative Stress and Related Signaling Pathways in A549 and HepG2 Cells. *Food Science and Biotechnology*, **19**, 1325-1332. http://dx.doi.org/10.1007/s10068-010-0189-5

[28] Fontes, J.E.N., Ferraz, R.P.C., Britto, A.C.S., Carvalho, A.A., Moraes, M.O., Pessoa, C., Costa, E.V. and Bezerra, D.P. (2013) Antitumor Effect of the Essential Oil from Leaves of *Guatteria pogonopus* (Annonaceae). *Chemistry & Biodiversity*, **10**, 722-729. http://dx.doi.org/10.1002/cbdv.201200304

20

Canopy Light Interception of a Conventional and an Erect Leafed Sorghum

Dennis C. Gitz III*, Zhanguo Xin, Jeffrey T. Baker, Robert J. Lascano, John J. Burke

Cropping Systems Research Laboratory, USDA-ARS#, Lubbock, TX, USA

Email: *dennis.gitz@ars.usda.gov

Abstract

Two sorghum lines, an erect leafed mutant sorghum and the wild type from which the mutant was generated, were field grown in rectilinear arrays at low (10 plants m^{-2}) and high (23 plants m^{-2}) population densities. Canopy light interception, biomass accretion and yield were measured. Photosynthetically active radiation under the canopy at ground level and midway through the canopy were higher in the erect leafed line, as compared to the normal leafed line. Planting density had less effect on mean grain yield and biomass in the erect leaf line than in the wild type. Though not conclusive, when taken together, the results suggested that optimal planting densities are higher for the erect leaf line and that the erect leaf trait could be useful for incorporation into breeding programs.

Keywords

Sorghum, Erectophile, Mutant, Erect Leaf

1. Introduction

The eu-sorghums (*Sorghum bicolor* L.) were domesticated between 2000-5000 years ago. While there is some evidence that sorghum could have existed in neolithic China [1], it is now generally accepted that the plant's

*Corresponding author.

#The US Department of Agriculture (USDA) prohibits discrimination in all its programs and activities on the basis of race, color, national origin, age disability, and where applicable, sex, marital status, familial status, parental status, religion, sexual orientation, genetic information, political beliefs, reprisal, or because all or part of an individual's income is derived from any public assistance program. (Not all prohibited bases apply to all programs.) Persons with disabilities who required alternative means for communication of program information (Braille, large print, audiotape, etc.) should contact USDA's TARGET Center at (202) 720-2600 (voice and TDD). To file a complaint of discrimination, write to USDA, Office of Civil Rights, 1400 Independence Avenue, S.W., Washington, D.C. 20250-9410, or call (800) 795-3272 (voice) or (202) 720-6382 (TDD). USDA is an equal opportunity provider and employer.

origin is Africa where the wild representatives are exclusively found. There is no discrete center of sorghum distribution; rather, landraces are found across the African continent [2]. Extant sorghum races can freely interbreed and are found in semi-arid to wet and from tropical to temperate climes. The lack of a center of domestication, the expansive spatial and climatic ranges, and the ability to freely interbreed between both cultivars and landraces have all contributed to the considerable genetic diversity of this plant. There are tens of thousands of sorghum landraces, mutant genotypes, classical cultivars, and modern commercial hybrids; and this does not consider the closely related species that could be exploited for inclusion in classical breeding or in germplasm maintenance operations. The United States sorghum germplasm collection alone has more than 40,000 entries, and many other collections exist [3].

Sorghum has been used as a stress tolerant crop plant since being brought to the new world [4]. Landraces from hot dry regions of Africa were brought to the Americas during the mid 1800's where the resiliency of the crop was exploited. Because of sorghum's stress tolerance, including drought stress tolerance, North American bankers would require that sorghum be planted alongside higher value but drought sensitive grain crops such as corn (*Zea mays* L.) to minimize credit default risk in the early 1900's. Attempts to further enhance drought tolerance by exploiting sorghum's diversity in selective breeding programs were proposed at the turn of the last century [5] [6]. However, interest in such programs waned when internal combustion engine powered irrigation insured reduced variability in production of corn, and of other more marketable and hence valuable crops [7]. Continually increasing demand for such crops has resulted in expansion of irrigated crop production systems. As a result demand for finite irrigation water continues to increase. Because of increased demand and because irrigation water is a finite resource, irrigation water availability is becoming increasingly limited worldwide. Irrigation water limitation has led to renewed interest in selection of drought tolerant, or more water use efficient sorghum (e.g., [8]-[11]).

Despite the diversity in sorghum germplasm, breeding programs have met with limited success exploiting existing traits for enhanced crop water use efficiency and drought tolerance. For such reasons, mutant libraries have been created to compliment the considerable diversity of landraces and cultivars in modern germplasm collections. These mutant libraries (putatively) offer an array of mendelian heritable traits that could be exploited in breeding programs attempting to exploit characteristics conferring increased productivity or at least maintaining agricultural productivity in the face of decreasing supplies of water suitable for crop irrigation.

Earlier, a mutant sorghum population was created and described which displayed several traits that could prove useful for incorporation into grain, forage, and biomass production systems [12]. Among the traits expressed in this population were brown midrib, multiple tillers, late flowering, increased and reduced epicuticular wax production, and erect leafed sorghum plants. Erect leafed sorghum is especially attractive since exposing both leaf surfaces to sunlight leads to more efficient radiation use and the smaller footprint allows for higher population densities, both of which could result in higher yields on a land area basis and increased crop water use efficiencies. Herein, we report the effect of stand density upon biomass accretion, yield, and light interception of a recently described erect leafed mutant sorghum as compared to the wild type from which it was derived [13].

2. Materials and Methods

2.1. Overview

Two sorghum lines from the USDA sorghum mutant library [13] were planted and allowed to develop. Light within the canopies was quantified diurnally. A preliminary experiment was done in a controlled environment polyhouse to determine the upper planting density to use in subsequent field work in which yield parameters were additionally quantified.

2.2. Plant Material and Culture

Sorghum (*Sorghum bicolor* L.) seeds were planted in a climate controlled polyhouse or in the field at the USDA nursery in Lubbock, TX (33˚35'38.00"N, 101˚54'10.50"W). Two sorghum lines were used in the experiments, BTx623 [14] and an erect leafed mutant derived and isolated from the BTx623 parent line, ERL-1a [13].

In the polyhouse (30˚/18˚, Day/Night; uncontrolled humidity; no supplemental lighting), black polyethylene containers (20 cm diameter, 4L capacity) were filled with a soil-less peat based commercial horticultural potting mix (Sunshine Traditional Loose Fill mix; Sungro Horticultural Products[1], Bellvue, WA; www.sungro.com), the rooting medium was saturated to field capacity by irrigating to runoff several times over about an hour, and several [3]-[5] sorghum seeds were then pushed about 2.5 cm into the growth medium into the center of each container. After about seven days the first true leaf was clearly visible and the group of plants was thinned to a single individual. The plants were fertigated daily with an automated horticultural drip delivery system to well past the drip point with a complete nutrient solution. The nutrient solution consisted of 11 kg of Peter's ExCel Multipurpose fertilizer (Scotts Company LLC, Marysville OH) brought to 189 liters and subsequently injected into the irrigation water at a ratio of 1:150 (v:v, fertilizer solution:municipal water). The plants were allowed to develop until they were between the milk and soft dough developmental stages [15] when light interception was characterized.

In the field, seeds were planted into 5 × 10 m plots with a garden planter (Model 1001-B, Earthway®, Bristol IN) using a radish seed plate (Pt. No. 1002-5) that had been modified to accept sorghum seeds by carefully enlarging the with a small spherical high speed rotary file. Seeds were planted on DOY 166 in the 2011 growing season in 19 cm (High Density) or 38 cm (Low Density) east-west rows. In an attempt to insure even emergence and minimize developmental variation along and across treatments, the plots were irrigated through subsurface drip irrigation (SDI) lines running north-south, perpendicular to the planting rows, spaced at 1 m and 20 - 25 cm below the soil surface. Immediately after planting 2.5 cm of water was applied to the field through SDI was used to "wet up" the soil. After emergence the stands were thinned so remaining plants were equidistantly spaced in either a 34 cm × 34 cm or 17 × 17 cm rectilinear arrays. There were sixteen plots with four plots per genotype × treatment (n = 4). Actual measured populations after thinning operations and establishment were determined at harvest to be 10.0 ± 0.4 plants m^{-2} and 22.9 ± 1.7 plants m^{-2} (mean ± S.E.) A small stand of sweet corn at the far end of the field planted over the same SDI drip tubing running under the sorghum plots. Because the sorghum and corn plots were simultaneously irrigated and because corn is considerably more responsive to drought stress, corn was used as bio-indicator of soil moisture through the season. Irrigations were applied to prevent the corn from exhibiting symptoms of water stress such as erect rolled leaves and a lighter green color. The sorghum exhibited no obvious symptoms of drought stress during the growing season. The soil at the Lubbock USDA location is an Amarillo fine sandy loam (fine-loamy, mixed, superactive, thermic Aridic Paleustalfs).

2.3. Radiation Measurement

In the polyhouse photosynthetically active radiation (PAR) was quantified within canopies simulated by arranging container grown plants along a bench. This was done to estimate how population density might affect light within the canopy micro-environment in field grown plants, and to help select population densities for field experimentation. The container grown conventional or erect leafed mutant sorghum lines were arranged along two benches centered within and along the north-south axis of the polyhouse. Forty eight plant containers were arranged in a 6 × 8 rectangular array with the long axis of the array oriented north-south. The benches were placed along the center-line of the polyhouse one to the north of the other with a four meter space between the two canopies to minimze shading. Linear quantum sensors were horizontally and equidistantly suspended from a structural central beam within each canopy at 30, 60, 90, and 120 cm above the surface of the rooting medium to quantify PAR. A separate linear PAR sensor was mounted above the canopy on the horizontal lower chord of the polyhouse roof truss. The ambient light sensor was oriented, east-west, perpendicular to the canopy PAR sensors.

In the field, PAR within the canopy was measured at ground level, and midway up the canopy at the onset of boot on a clear day, DOY 221, with a device assembled in the lab (**Figure 1**). The device used two 1m linear PAR sensors (Model LI-191 Line Quantum Sensor, LiCor® Environmental Inc., Lincoln NE) held perpendicular to a 2 m extruded aluminum staff. One of the linear sensors was mounted permanently at ground level and the support for the other sensor's height above the ground was adjustable. The height of the second linear PAR sensor was adjustable and was set so that the sensor height was at about half the canopy height. One point PAR sensor (LiCor® Model LI-190 Quantum Sensor) above the canopy to measure radiation above the canopy. Data were recorded using a datalogger (LI-1400 Data Logger). The PAR readings were taken along the north side of where the instrument was inserted into the canopy towards the south. After inserting the probes into the canopy,

[1]Mention of trade names or commercial products in this publication is solely for the purpose of providing specific information and does not imply recommendation or endorsement by the US Department of Agriculture.

Figure 1. Apparatus for measurement of light within field grown sorghum canopies.

the staff was held vertical as indicated by spirit levels mounted on the sides of the staff and the outputs from all three PAR sensors simultaneously recorded as $\mu mol \cdot m^{-2} \cdot s^{-1}$ of PAR, along with the time of the measurements. Care was taken to avoid inserting the probes directly along the solar azimuth to avoid shading of the ground level sensor by the mid-canopy sensor, especially around solar noon. This was accomplished by offsetting the probe away from the solar azimuth by about 15˚; inserting the probe towards the SSW or SSE just before or after solar noon, respectively. Averages from three locations in each plot were taken. Measurements were started each hour from sunrise to sunset. Each pass through the field was completed in one half hour; about 30 minutes. For presentation, the % transmitted radiation that is, the percentage of ambient radiation incident upon the linear PAR sensors plus or minus standard deviation about the mean was calculated and plotted as a function of the time of day; where the time of day was defined as the average of individual sampling times. Transmitted radiation was plotted for presentation rather than intercepted radiation because interpretation of the data was found to be more intuitive.

2.4. Canopy Development and Yield

The Leaf Area Index (LAI) of the plants, the area of the leaves within the crop canopy above the soil surface area (leaf area/soil surface area), was measured *in situ* with a fisheye lens canopy analyzer (LAI-2200, LiCor® Inc.) shortly after dawn on DOY 222. For each plot, the average of four measurements was calculated and recorded. Leaf area of five individual plants within each plot was measured with a LI-3100 benchtop leaf area meter. Above ground plant biomass was determined by drying plants at 60˚C for several days and noting the mass.

3. Results

The PAR within the synthetic canopy, constructed by arranging container grown sorghum plants in a 20 cm × 20 cm array in the polyhouse on a clear day, is shown in **Figure 2**. Note that relative transmittance (%) of ambient light and absolute light levels ($\mu mol \cdot m^{-2} \cdot s^{-1}$) within the greenhouse were plotted in the figure(s) rather than intercepted radiation. Consequentially, the uppermost curves resulted from the uppermost linear PAR sensors. At times, the calculated transmittance of the uppermost line sensors within the canopies was higher, up to 105%, that of the line sensor used for ambient values. This resulted from occasional shading of the ambient PAR sensor by the polyhouse architecture. Rather than assume 100% transmittance, these data were deleted from the presentation. A mid-day decrease in measured radiation within the canopy was observed. This radiation decrease

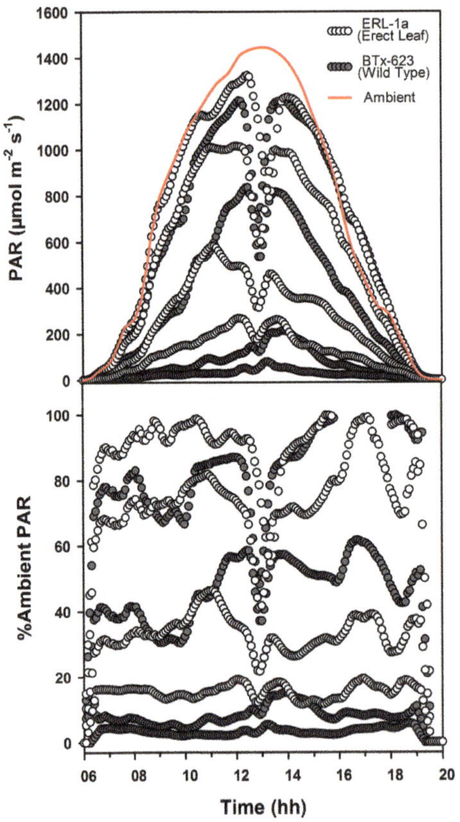

Figure 2. Photosynthetically active radiation (Upper Pane) and calculated Transmittance (Lower Pane) of light within an artificially constructed sorghum canopy in a greenhouse. Open symbols are measurements from within erect leaved sorghum. Closed symbols are from within conventional sorghum. Upper curves are from higher in canopy and lower curves are from deeper within the canopy. Red line is PAR above canopies.

occurred at solar noon and was observed in all sensors in both canopies and resulted from the direct component of solar radiation being occluded by a central North-South roof support beam from which the linear PAR sensors were suspended. Because the ambient line sensor was mounted elsewhere and above beams that could cast such shadows, the calculated transmission at solar noon is a measure of the indirect component of radiation moving through the canopy. While the translucent polyethylene roof of the polyhouse seemed to diffuse radiation similar to a high clouds in a field setting, it is unknown how the angular distribution of PAR within the polyhouse compares to such conditions. Angular distribution of PAR was not quantified in the field or in the polyhouse. Even so, PAR clearly decreased as one moved deeper into both the wild type and erect leafed canopies throughout the day and measured PAR was consistently greater within the erect leafed ERL-1a sorghum canopy as compared to that within the Btx-623 canopy.

A view of the field grown wild type (Btx-623) and mutant (ERL-1a) sorghum lines is shown in **Figure 3**. Visual inspection reveals a considerably more open canopy in the erect leafed line as compared to the conventional sorghum. The leaves appeared to adopt an erect orientation as might be expected from corn leaves when rolling in response to water deficit. However, the leaves of the erect sorghum were not rolled but appeared corrugated.

Measurements of ambient radiation and of light within field plots are shown in **Figure 4**. The sky was clear throughout the day on which the measurements were made. Ambient PAR exhibited the classical sinusoidal temporal distribution centered about solar noon expected from whole sky solar radiation with minimal indirect component (**Figure 4(A)**). As expected, PAR within densely planted sorghum (High Density) canopies was less

Figure 3. (A) conventional and (B) erect leafed sorghum in field.

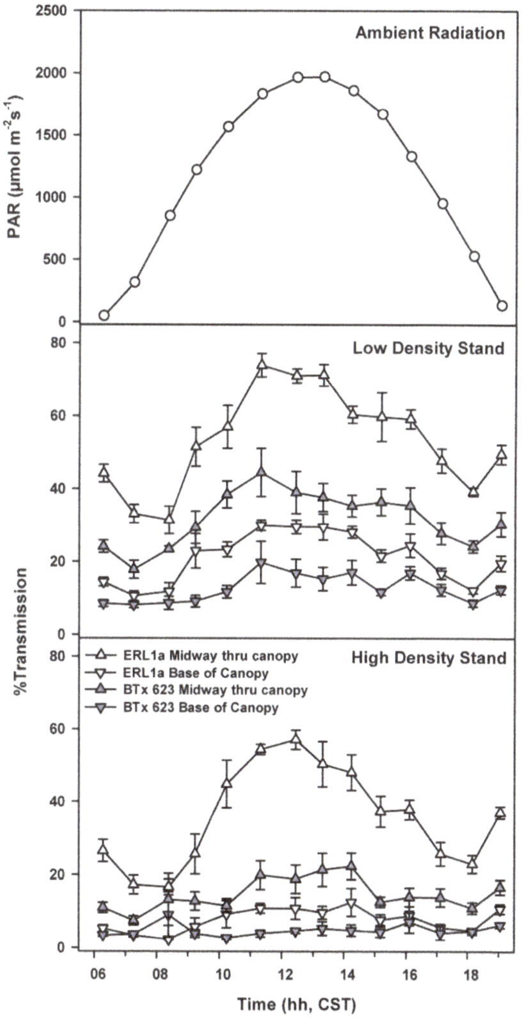

Figure 4. Ambient radiation (top pane) and percentage of ambient radiation with canopies of erect leafed and conventional sorghum planted at "Low Density" (10 plants m^{-2}, middle pane) and "High Density" (23 plants m^{-2}, bottom pane).

than that of less densely planted (Low Density) canopies and decreased as one moved deeper within the canopy (**Figure 4(B)** & **Figure 4(C)**). Erect leaved sorghum allowed a greater amount of PAR to pass through the canopy at each planting density and at each level within the canopy especially near solar noon when the solar angle was not as oblique and canopy transmission was less affected by plant to plant shading.

In general, every measured yield parameter measured was lower as compared to the conventional sorghum, except for seed size which was nearly identical between lines (not shown). Yield differences observed resulted from differences in the numbers of seeds rather than individual seed size. The leaf area of individual plants measured by detaching leaves and individually feed them through a leaf area meter decreased by about 50% with a doubling of plant population density (**Figure 5(A)**). Biomass (**Figure 5(C)**) and yield (**Figure 5(D)**) exhibited similar responses to density. However, the partitioning of accreted dry matter towards grain production was enhanced at the higher planting densities. That is, grain yield decreased less than yield. Yield decreases resulted from a decrease in seed numbers and not from a decrease in seed size (not shown). Leaf area index measured by optical methods was indistinguishable within planting density treatments, but exhibited higher variation with the erect leaf variety. This was attributed to the "clumpy" nature of the ERL-1a canopy.

4. Discussion

The idea of a high yielding wheat ideotype having several morphological characteristics including erect leaves was discussed about a half century ago [16]. Since then, results from work comparing erect leaf phenotypes to more horizontal leaf orientations have yielded mixed results. Studies with barley [17] provided no clear evidence that erect leaf angle enhanced grain yield while others reported yield increases of 25% in rice [18] and a 40% increase has been reported for corn [19]. Indirect solar radiation is the component of whole sky radiation that erect leaves are thought to best take advantage. Differences in the proportion of direct to indirect radiation might explain the differences in results between researchers attempting to evaluate the effect of the erect leaf phenotype on crop yield. The present work was conducted in a semi-arid region with a large number of very clear days with no perceptible haze. The contribution of direct radiation to whole sky solar irradiance is much

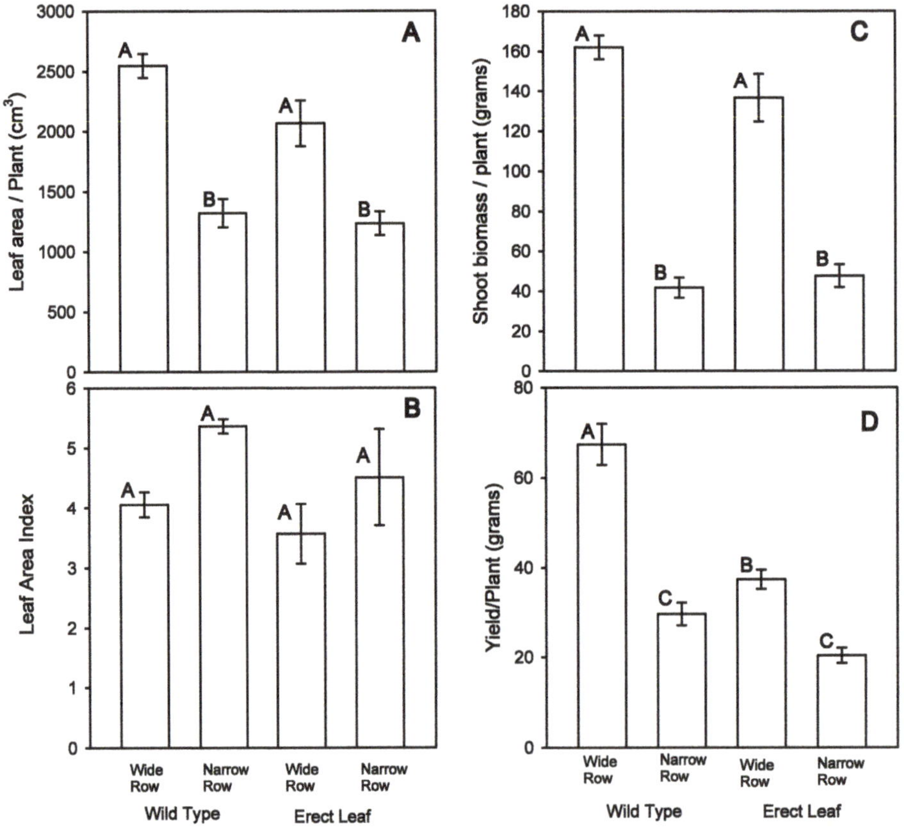

Figure 5. Canopy leaf area indices (left) and yield response (right) of erect leafed and conventional sorghums. (A) Leaf area of individual plants; (B) leaf area indices of canopies measured by optical methods; (C) whole above ground plant biomass; (D) seed yield of individual plants. Bars are standard error about the mean. Results of means separation by ANOVA at 0.95% error with Bonferroni correction are shown to left of bars.

greater than in other regions. Even though direct experimental comparisons between erectophiles and more conventional or "lax leaved" crops have yielded mixed results, many modern crop cultivars have erect leaves. The mechanistic underpinnings for potential yield increases in erectophiles remain an active research topic. It is unknown whether reported yield increases, or the relative insensitivity of yield to crop density, have resulted from more complete interception of radiation [20], altered distribution of radiation within the canopy leading to altered distribution of photosynthesis within the canopy [21], or reduction in leaf sensecence from increased light penetration. Moreover it may not be as simple as selecting for erect leaves throughout the canopy. Selecting for erect upper leaves and more prostrate interior leaves has been attempted in maize breeding programs [22]. Nevertheless, it is generally accepted that increasing leaf angle from horizontal to verticle results in higher canopy photosynthesis. It has been estimated that increasing leaf angle from 30˚ to 60˚ can result in canopy photosynthesis increase of up to 15% in corn [23].

In the present study it is difficult to draw inferences with regard to the effect of the erect leaf trait on optimal planting density since the trait is associated with a reduction seed yield even at lower population densities. Based on seed yield data alone it would appear that the trait is not useful for incorporation into breeding programs. However, while seed yield of the erect leaf mutant was intrinsically less than that of the conventional line, erect leaf seed yield was also less responsive to population density. That is, seed yield decreased less with the erect leafed sorghum as compared to the conventional leafed sorghum. This suggests that the optimal plant population density of the erect leafed plant is higher than that of the conventional line. While this makes sense from a simple radiation interception perspective, another result of this was that carbon partitioning as grain/shoot biomass approached nearly 50% with the erect leafed sorghum at higher planting densities (**Figure 5(C)** & **Figure 5(D)**). It is generally accepted that source-sink relationships are involved with the biochemical regulation of photosynthesis. For such reasons increased sink size might also be associated with enhanced water use efficiencies at higher population densities.

Acknowledgements

We thank Dr. Lan Liu-Gitz for helpful discussions and suggestions through the course of this work and also Messrs. Cody Byars and Peyton Schuman for their technical assistance during the experimental phase of the work. This project was funded in part by the USDA-ARS Ogallala Aquifer Program.

References

[1] De Candolle, A. (1886) Plants Cultivated for Their Seeds. In: *Origin of Cultivated Plants*, 2nd Edition, International Scientific Series Vol. XLIX, Kegan Paul Trench & Co., London, 376-384. https://archive.org/download/originofcultivat00cand/originofcultivat00cand.pdf

[2] Harlan, J.R. (1971) Agricultural Origins: Centers and Non-Centers. *Science*, **174**, 468-474. http://dx.doi.org/10.1126/science.174.4008.468

[3] Dahlberg, J., Berenji, J., Sikora, V. and Latkovic, V. (2011) Assessing Sorghum [*Sorghum bicolor* (L.) Moench] Germplasm for New Traits: Food, Fuels & Unique Uses. *Maydica*, **56**, 85-92. http://www.maydica.org/articles/56_165.pdf
Hu, T. and Desai, J.P. (2004) Soft-Tissue Material Properties under Large Deformation: Strain Rate Effect. *Proceedings of the 26th Annual International Conference of the IEEE EMBS*, San Francisco, 1-5 September 2004, 2758-2761.

[4] Borman, T.A. (1914) Sorghums: Sure Money Crops. The Kansas Farmer Company, Topeka. http://dx.doi.org/10.5962/bhl.title.43352

[5] Ball, C.R. (1913) The Grain Sorghums: Immigrant Crops That Have Made Good. In: *Yearbook of Agriculture*, US Department of Agriculture, 221-238. http://naldc.nal.usda.gov/naldc/download.xhtml?id=IND43747834&content=PDF

[6] Martin, J.H. (1936) Sorghum Improvement. In: *Yearbook of Agriculture*, US Department of Agriculture, 523-560.

[7] Musick, J.T., Pringle, F.B., Harman, W.L. and Stewart, B.A. (1990) Long-Term Irrigation Trends—Texas High Plains. *Applied Engineering in Agriculture*, **6**, 717-724. http://dx.doi.org/10.13031/2013.26454

[8] Rosenow, D.T., Quisenberry, J.E., Wendt, C.W. and Clark, L.E. (1983) Drought Tolerant Sorghum and Cotton Germplasm. *Agricultural Water Management*, **7**, 207-222. http://dx.doi.org/10.1016/0378-3774(83)90084-7.

[9] Blum, A., Mayer, J. and Golan, G. (1989) Agronomic and Physiological Assessments of Genotypic Variation for Drought Resistance in Sorghum. *Australian Journal of Agricultural Research*, **40**, 49-61. http://dx.doi.org/10.1071/ar9890049

[10] Fussell, L.K., Bidinger, F.R. and Bieler, P. (1991) Crop Physiology and Breeding for Drought Tolerance: Research and Development. *Field Crops Research*, **27**, 183-199. http://dx.doi.org/10.1016/0378-4290(91)90061-y

[11] Mutava, R.N., Prasad, P.V.V., Tuinstra, M.R., Kofoid, K.D. and Yua, J. (2011) Characterization of Sorghum Genotypes for Traits Related to Drought Tolerance. *Field Crops Research*, **123**, 10-18. http://dx.doi.org/10.1016/j.fcr.2011.04.006

[12] Xin, Z., Wang, M.L., Burow, G. and Burke, J.J. (2009) An Induced Sorghum Mutant Population Suitable for Bioenergy Research. *BioEnergy Research*, **2**, 10-16. http://dx.doi.org/10.1007/s12155-008-9029-3

[13] Xin, Z., Gitz, D.C., Chen, J., Burow, G.B., Hayes, C. and Burke, J.J. (2015) Registration of Two Allelic Erect Leaf Mutants of Sorghum. *Journal of Plant Registrations*, **9**, 254-257. http://dx.doi.org/10.3198/jpr2014.09.0060crgs

[14] Frederiksen, R.A. and Miller, F. (1972) Proposal for Release and Increase ATx622, BTx622, ATx623, BTx623, ATx624, BTx624. TAES Form 96-72: Seed Release Committee of the Texas Agricultural Experiment Station College Station.

[15] Vanderlip, R.L. and Reeves, H.E. (1972) Growth Stages of Sorghum [*Sorghum bicolor*, (L.) Moench.]. *Agronomy Journal*, **64**, 13-16. http://dx.doi.org/10.2134/agronj1972.00021962006400010005x

[16] Donald, C.M.T. (1968) The Breeding of Crop Ideotypes. *Euphytica*, **17**, 385-403. http://dx.doi.org/10.1007/bf00056241

[17] Tungland, L., Chapko, L.B., Wiersma, J.V. and Rasmusson, D.C. (1987) Effect of Erect Leaf Angle on Grain Yield in Barley. *Crop Science*, **27**, 37-40. http://dx.doi.org/10.2135/cropsci1987.0011183X002700010009x

[18] Khush, G.S. (1995) Breaking the Yield Frontier of Rice. *GeoJournal*, **35**, 329-332. http://dx.doi.org/10.1007/BF00989140

[19] Pendleton, J.W., Smith, G.E., Winter, S.R. and Johnston, T.J. (1968) Field Investigations of the Relationships of Leaf Angle in Corn (*Zea mays* L.) to Grain Yield and Apparent Photosynthesis. *Agronomy Journal*, **60**, 422-424. http://dx.doi.org/10.2134/agronj1968.00021962006000040027x

[20] Feil, B. (1992) Breeding Progress in Small Grain Cereals—A Comparison of Old and Modern Cultivars. *Plant Breeding*, **108**, 1-11. http://dx.doi.org/10.1111/j.1439-0523.1992.tb00093.x

[21] Angus, J.F., Jones, R. and Wilson, J.H. (1972) A Comparison of Barley Cultivars with Different Leaf Inclinations. *Crop and Pasture Science*, **23**, 945-957. http://dx.doi.org/10.1071/ar9720945

[22] Ghannoum, O. (2013) Potential Contribution of Increased Photosynthetic Efficiency to Increased Yield Potential of Maize. Applying Photosynthesis Research to Improvement of Food Crops, ACIAR Proceedings #140, Proceedings of a Workshop Held at the Australian National University, Canberra, Australian Capital Territory, Australia, 2-4 September 2009, 70-76. http://aciar.gov.au/files/pr140_web.pdf

[23] Tollenaar, M. and Dwyer, L.M. (1999) Physiology of Maize. In: Smith, D.L. and Hamel, C., Eds., *Crop Yield, Physiology and Processes*, Springer Verlag, Berlin, 169-204. http://dx.doi.org/10.1007/978-3-642-58554-8_5

21

Isolation and Screening of Silicate Bacteria from Various Habitats for Biological Control of Phytopathogenic Fungi*

Zakira Naureen[1,2#], Muhammad Aqeel[2], Muhammad Nadeem Hassan[2], Syed Abdullah Gilani[1], Nahla Bouqellah[3], Fazal Mabood[1], Javid Hussain[1], Fauzia Y. Hafeez[2]

[1]Department of Biological Sciences and Chemistry, University of Nizwa, Birkat Al Mawz, Oman
[2]COMSATS University, Chak Shehzad, Islamabad, Pakistan
[3]Department of Biology, Taibah University Madina, Saudi Arabia
Email: #zakira@unizwa.edu.om

Abstract

Silicate solubilizing bacteria (SSB) can play an efficient role in soil by solubilizing insoluble forms of silicates. In addition to this some SSB can also solubilize potassium and phosphates, hence increasing soil fertility and enhancing plant defense mechanisms. A total of 111 bacterial strains were isolated from various habitats of Pakistan and screened for solubilization of silicate, phosphate and potassium on respective media. Out of these, 35 bacterial isolates were capable of solubilizing either silicate, phosphate or potassium. Amongst these 7 bacterial isolates were capable of solubilizing all three minerals tested. The highest silicate (zone diameter 54 mm) and phosphate solubilization (zone diameter 55 mm) was observed for bacterial isolate NR-2 while the highest potassium solubilization was observed for NE-4b (zone diameter 11 mm). Dual culture antagonistic assays were carried out by using these bacterial isolates against four plant pathogenic fungi *Magnaporthae grisae*, *Rhizoctonia solani*, *Altarnaria alternata* and *Macrophomina pheasolina*. Mean zone of inhibition of these bacterial isolates against the four pathogenic fungi ranged between 4 mm to 39 mm. The largest zone of inhibition against all four bacterial strains was recorded for bacterial isolate NR-2 followed by NE-4b. These strains will be further investigated for their plant growth promoting activities in the future.

Keywords

Silicate Solubilizing Bacteria, Biocontrol, Phytopathogenic Fungi

*Biocontrol of phytopathogenic fungi by silicate solubilizing bacteria.
#Corresponding author.

1. Introduction

Silicon (Si) enhances the growth, development and yield of many plants and is reported to decrease the incidence of many fungal diseases in different pathosystems by strengthening the cell walls and especially the outer membrane of epidermal cells in leaves thus increasing resistance to the penetration of pathogenic fungi [1] [2]. Silicon benefits the plants in several other ways by accelerating growth, conferring rigidity to leaves thus maximizing leaf surface area for photosynthesis and mitigating the effects of abiotic stresses like drought, salt and metal toxicity in several plants including wheat, rice, sugarcane, cucumber, tomato, citrus and barley [3]-[8]. Despite of its abundance in earth's crust it is mostly present in insoluble forms that cannot be readily absorbed by plant roots [9] [10]. It remains in insoluble form unless solubilized by weathering action of rocks or biological activity of plant roots and microorganisms. Silicate solubilizing bacteria (SSB) can play an efficient role here by solubilizing insoluble forms of silicates hence increasing soil fertility and enhancing plant defense mechanisms [9]. The solubilized Si in the form of orthosilicic acid (H_4SiO_4) is absorbed along with water [10]. Si is accumulated in the form of silica gel and is deposited in epidermal cells, scelerenchyma, vascular bundles and in florescence brackets in cereals [11]. The accumulated Si not only improves growth and yield of these plants but is also involved in induction of systemic resistance (ISR) against pests and diseases [12].

The present study was carried out to isolate and characterize silicate solubilizing bacteria from various habitats and evaluation of antagonistic activity against four major crop pathogens *Magnaporthae grisae*, *Rhizoctonia solani*, *Altarnaria alternata* and *Macrophomina pheasolina.*

2. Material and Methods

2.1. Collection of Soil Samples from Different Locations

Soil samples were collected from various habitats in Pakistan (**Table 1**). Soil samples were also collected from fields which had been under routine cultivation of rice, wheat and sugarcane (**Table 1**) as these crops are high silicate accumulators. Half Kg of each soil sample was collected at a depth of 10 cm from top soil. The samples were then air dried and ten gram of each air-dried soil sample was transferred to 100 mL of sterile distilled water in a 250 ml Erlenmeyer flask separately. The samples were than incubated on a rotary shaker (120 rpm) for 30 min at room temperature as described by [13]. Serial dilutions of each soil sample, ranging from 10^{-1} to 10^{-5} was prepared in sterile water. One mL aliquot of the appropriate dilution (10^{-5}) was spread on plates containing Luria Bertani (LB) agar. Plates were incubated at 30˚C ± 1˚C for 24 - 48 h and colonies were examined. Morphologically distinct colonies were counted, selected, purified and maintained in LB broth amended with glycerol [14].

2.2. Morphological Characterization

Selected bacterial isolates were purified and plated on LB agar plates. Morphological characteristics were studied for pure single colonies [14].

2.3. Silicate Solubilization

Bacterial isolates from different soil samples were then spread on agar plates containing silicate medium with 0.25% insoluble magnesium trisilicate [15]. Plates were incubated at a temperature of 30˚C ± 1˚C for 36 - 72 h. After the incubation period, the bacterial colonies exhibiting clear zones around them were selected for further investigations.

Percentage of silicate solubilizing bacteria was checked for each soil sample and calculated as follows.

% age silicate solubilizers = Number of silicate solubilizing bacteria/total bacteria × 100

2.4. Phosphate Solubilization

All of the isolated bacterial isolates were checked for phosphate solubilization on Pikoviskaya agar plates [16]-[18]. Clear zone diameter was recorded after 24 to 72 h.

% age Phosphate solubilizers = Number of silicate solubilizing bacteria/total bacteria × 100

Table 1. Morphological characterization, Si, P and K solubilization of 35 bacterial isolates from various habitats in Pakistan.

Serial No.	Strain	Origin/ Location	Colony Morphology	Gram Reaction	Cell Morphology	Silicate Solubilization Zone (mm)	P-Solubilization Zone (mm)	K-Solubilization (mm)
1	NE-1	Wheat root endosphere, Narowal	White, irregular, umbonate	+ve	Rod	0	0	4
2	NE-2	Wheat root en-dosphere, Narow-al	White, irregular convex	+ve	Long rods	12	25	3
3	NE-3	Wheat root endosphere, Narowal	Off white, circular, raised	+ve	Short rods	6	50	6
4	NE-4	Wheat root endosphere, Narowal	Fluorescent yellow, smooth circular colonies	−ve	Long rods	17	24	2
5	NE-4b	Wheat root endosphere, Narowal	Yellow, irregular, Lobate, raised	+ve	Rods	34	28	11
6	NE-5	Wheat root endosphere, Narowal	White, smooth round, entire	+ve	Rods present in chains	7	28	5
7	NR-2	Wheat rhizos-phere Narowal	White, irregular, large, flat	+ve	Rods medium sized	54	55	7
8	NR-3	Wheat rhizos-phere Narowal	Off white, irregular medium sized undulate	+ve	Rods	9	12	6
9	R-1	Soil sample Rawalpindi	White, circular, elevated	−ve	Cocci	5	0	4
10	R-2	Soil sample Rawalpindi	Milky, irregular, flat	+ve	Rods	0	0	4
11	R-6	Soil sample Rawalpindi	White, circular, serrate,	−ve	Rods	0	0	8
12	R-8	Soil sample Rawalpindi	White, circular, raised	−ve	Short rods	0	10	0
13	WE 1	Wheat root Khushab	Opaque, irregular, convex	−ve	Long rods	0	20	0
14	WE 2	Wheat root Khushab	Creamy, rhizoid, lobate	+ve	Short rods	0	15	0
15	WE 3	Wheat root Khushab	Circular, centrally dotted	+ve	Cocci	20	25	0
16	EM 5	Endosphere of mint Morkallan	White, circular, entire, raised	−ve	Rods	15	10	0
17	EM 6	Endosphere of mint Morkallan	Pink, circular, entire	+ve	Filamentous rods	10	7	0
18	EM 7	Endosphere of mint Morkallan	Off white circular, entire	−ve	Rods	25	18	0

Continued

19	RM 1	Rhizosphere of mint Morkallan	Dirty white, irregular, flat	−ve	Short rods	30	22	0
20	RM 2	Rhizosphere of mint Morkallan	Circular, gummy, elevated	+ve	Short rods	12	21	0
21	RM 3	Rhizosphere of mint Morkallan	White, irregular, elevated	+ve	Rods medium sized	14	10	0
22	RM 4	Rhizosphere of mint Morkallan	White, small circular, flat	−ve	Cocci	10	5	0
23	AYM-1	Soil sample Ayubia	White, round smooth	Gram − ve	Cocci, chains	22	10	0
24	AYM-2	Soil sample Ayubia	White, irregular, smooth, small sized colonies	Gram − ve	Cocci	12	18	0
25	AYM-3	Soil sample Ayubia	White, round elevated, medium sized	Gram − ve	Cocci, bunches, isolated in the form of pairs also.	10	24	0
26	SLB-1	Soil sample Nathia Gali	circular pointed small white gummy colonies	Gram + ve	Cocci, present in the form of chains	24	19	0
27	SLB-2	Soil sample Nathia Gali	Round small white gummy colonies	Gram − ve	Cocci, branched chains	30	22	0
28	SLB-3	Soil sample Nathia Gali	Round elevated small white gummy colonies	Gram + ve	Cocci, in the form of pair and also present in the form of bunches	8	5	0
29	SLB-4	Soil sample Nathia Gali	Small white gummy colonies	Gram − ve	Short rods	17	21	0
30	SLB-5	Soil sample Nathia Gali	Dot size small white gummy colonies	Gram − ve	Cocci, present in the form of pairs, chains, bunches.	13	20	0
31	SLB-6	Soil sample Nathia Gali	Round pointed small white gummy colonies	Gram − ve	Cocci, bunches, pair and isolated forms.	14	22	0
32	SLB-7	Soil sample Nathia Gali	Round pointed small off white gummy colonies	Gram + ve	Cocci, randomly present.	10	13	0
33	SLB-8	Soil sample Nathia Gali	Round pointed small white gummy colonies	Gram + ve	Cocci in the form of groups.	10	0	6
34	PTN 8	Potato field Nathya Gali	White, irregular elevated	Gram − ve	Cocci, chains	8	17	0
35	PTN 24	Potato field Nathya Gali	White, circular, entire, elevated	Gram − ve	Rods	6	0	0

2.5. Potash Solubilization

Bacterial isolates were checked for potassium solubilization on plates containing Alexandrov's medium containing sucrose (5 g), Na_2PO_4 (2 g), $MgSO_4 \cdot 7H_2O$ (0.5 g), $FeCl_3$ (0.005 g), $CaCO_3$ (0.1 g), glass beads (1 g) in 1 L of deionized distilled water at a pH of 7.4 [19]. Clear zone formation around the bacterial growth was observed after 48 - 72 h.

% age Potash solubilizers = Number of potash solubilizing bacteria/total bacteria × 100

2.6. Semi Quantitative Screening of the Bacterial Isolates

All of the selected bacterial isolates were subjected to semi quantitative mineral (Si, P and K) solubilization assay in *in-vitro* conditions. Each bacterial isolates was inoculated on three different bacterial agar media as above and after the incubation period (36 - 72 h) at a temperature of 30˚C ± 1˚C, the area of clearing zones was calculated and recorded in mm. Each experiment considering a single bacterial isolate was repeated at least three times.

2.7. Acid Production by Bacterial Isolates

Freshly prepared broth cultures of selected bacterial isolates were spread on LB agar plates amended with bromophenol blue and incubated for 24 - 72 h at a temperature of 30˚C ± 1˚C. Acid production was detected as blue media turned yellow around bacterial colonies.

2.8. Antagonistic Assays

Antagonistic activity of selected bacterial isolates was checked against selected pathogenic fungi including *Magnaporthae grisae*, *Rhizoctonia solani*, *Altarnaria alternata* and *Macrophomina pheasolina* using dual culture assays as described by Naureen *et al.* (2009) and Hassan *et al.* (2010) [14] [18]. Briefly, a 10 mm disc of an actively growing pure culture of pathogenic fungi grown on potato dextrose agar (PDA) was placed at the centre of a Petri dish containing appropriate test medium (rice flour agar RFA; PDA; Tomato juice agar etc.). A circular inoculum of selected bacterial isolates was made with a 6 cm diameter Petri dish dipped in a suspension of bacterial broth culture (5×10^9-cfu·mL^{-1}) and placed around the fungal disc. Plates were incubated for one week at 30˚C ± 1˚C and mean zone of inhibition was recorded after 7 days.

Each experiment considering a single bacterial isolate was repeated at least three times.

2.9. Statistical Analysis

Each experiment was run thrice and means were calculated.

3. Results

Soil samples were collected during April to June 2011 from different habitats in Pakistan including Narowal, Khushab, Kallar Kahar, Chakwal, Bochaal, Miyani, Rawalpindi, Murree, Jhika Gali, Nathia Gali, Ayubia, Barian and Morkallan (Mansehra). Samples were collected either from fields that had been long under cultivation of rice, wheat or sugar cane or from mountainous regions which are rich in minerals. A total of 111 bacterial isolates were isolated from these soil samples and maintained as pure cultures.

Each bacterial strain was checked for silicate, phosphate and potash solubilization on respective medium. Out of 111 bacterial isolates 35 bacterial isolates that showed solubilization of any of the three minerals were selected for further studies. Twenty-nine bacterial isolates were capable of solubilizing magnesium trisilicate, 28 were capable of solubilizing phosphates and 12 were observed to solubilize potash (**Table 1**). Amongst these 7 bacterial isolates were capable of solubilizing all three minerals tested. The highest silicate (zone diameter 54mm) and phosphate solubilization (zone diameter 55 mm) were observed for bacterial isolate NR-2 while highest potassium solubilization was observed for NE-4b (zone diameter 11 mm). The highest percentage of silicate solubilizing bacteria was observed in soil samples collected from Ayubia and Nathia Gali mountains where 100% bacterial isolates were silicate solubilizers while 64% of the total bacterial population was silicate solubilizers in soil samples obtained from Narowal fields which had been under routine cultivation of wheat and rice (**Figure 1**).

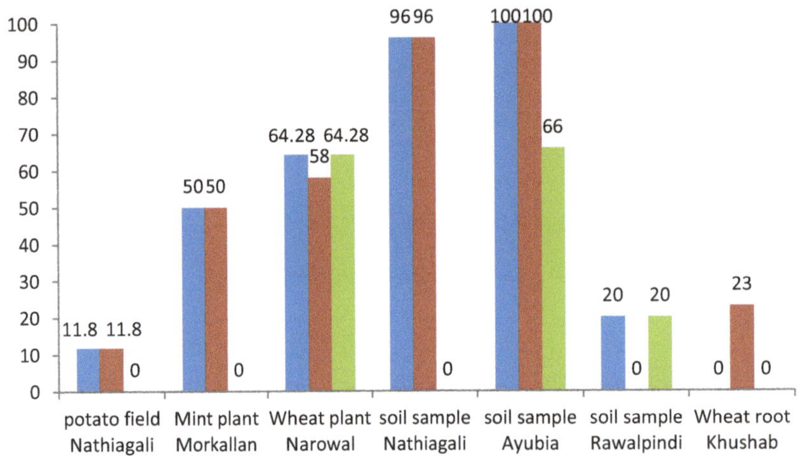

Figure 1. Percentage of silicate, phosphate and potash solubilizing bacteria in plant root and soil samples collected from different area of Pakistan.

A few silicate solubilizers were seen in samples from potato field (Nathia Gali) and wild Mint plant rhizosphere and endosphere which were taken from Morkallan village near Mansehra (**Figure 2**, **Figure 3**).

Similarly the highest percentage of phosphate solubilizing bacteria was seen in soil samples from Ayubia and Nathia Gali Mountains and almost all of the silicate solubilizers in these samples were phosphate solubilizers (**Figure 3**). The highest population of potash solubilizers was seen in soil samples taken from Ayubia Mountains followed by wheat rhizosphere samples taken from Narowal and a few potash solubilizers were seen in soil samples taken from Rawalpindi.

The selected 35 bacterial isolates were checked for antagonistic activity against plant pathogenic fungi such as *Magnaporthae grisae*, *Rhizoctonia solani*, *Altarnaria alternata* and *Macrophomina pheasolina* and zone of inhibition recorded (**Table 2**). Almost all of the selected isolates exhibited at least some antagonistic activities against the four pathogens. Mean zone of inhibition of these bacterial isolates against the four pathogenic fungi ranged between 4 mm to 39 mm. Highest antagonistic activity was observed in case of bacterial strain NR-2, isolated from wheat rhizosphere, against all fungal pathogen followed by NE-4b, isolated from wheat root from Narowal region.

4. Discussion

Bacterial isolates can solubilize insoluble minerals such as silicates, phosphates and potash into soluble form by production of organic acids such as 2 keto-gluconic acid, alkalis and polysaccharides [20]. Bacteria have been reported to accelerate the dissolution of silicates by producing excess proton, organic ligands, hydroxyl anion, extra cellular polysaccharides (EPS) and enzymes [21]-[24]. Despite abundance of microorganisms in soil only a few are capable of solubilizing silicates and our results indicated that out of a total of 111 bacterial isolates only 29 were capable of solubilizing silicates. Similarly, phosphate solubilization by bacterial isolates is also attributed to production of gluconic acid, 2 keto-gluconic acid and phosphatases. Our results indicated that almost all of the bacterial isolates capable of solubilizing phosphates and silicates produced acid as detected by yellow halo formation on media containing bromophenol blue and low pH in broth cultures.

Efficient silicate solubilizing bacteria can help release other essential nutrients in soil [25]. This can be directly due to solubilization of other minerals by silicate solubilizing bacteria or indirectly due to solubilized silicon. We also observed that most of the silicate solubilizers were able to solubilize phosphates with a few exceptions such as bacterial strain NE-1, R1, R2 and AYM-1 which solubilized silicates but not phosphates. It has been previous reported that the solubilized Si improves availability of phosphorus to plants by competing with P fixation sites in soil [25]. Thus Si acts as a substitute for P in plant system. Silicon is reported to increase the availability of Phosphorus indirectly by decreasing the availability of Fe and Mn in plants [26].

Potassium is also one of the major micronutrients required by the plant for optimum growth and yield. Potassium plays a regulatory role in plant being a constituent in 60 different enzymes in plants involved in drought tolerance and water use efficiency. Recent studies have shown that besides drought tolerance potassium plays an

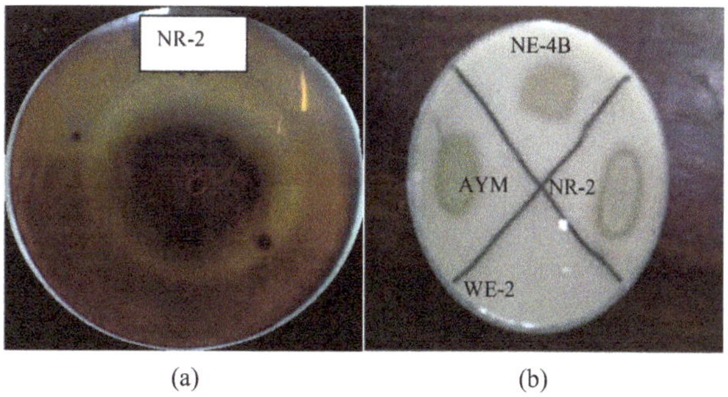

Figure 2. (a) Acid production by silicate solubilizing bacterial isolate NR-2 (b) Solubilization of silicate on media amended with 0.25 % Magnesium trisilcate.

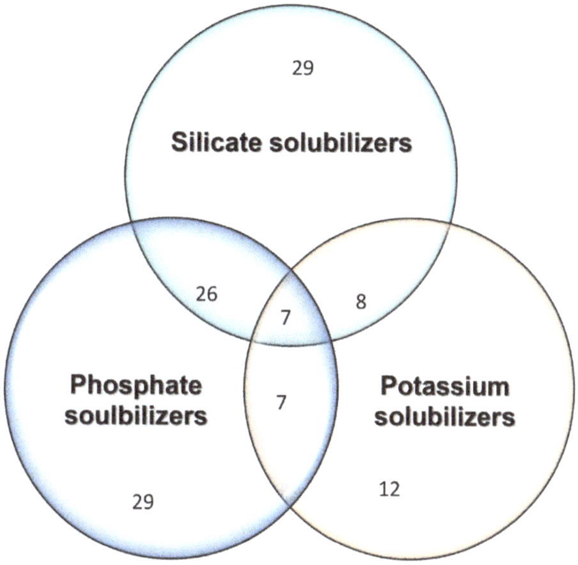

Figure 3. Venn diagram illustrating solubilization of Si, P and K by 35 selected bacterial isolates *Values represent actual numbers of bacterial isolates.*

important role in various metabolic processes and disease resistance in plants, almost all crops require potassium for optimum growth [27]. Deficiency of potassium leads to symptoms like chlorosis and leaf fall, unhealthy root system and then reduction in nutrient uptake and reduced cytokinin production in roots. This deficiency of potassium in soil can be successfully managed by application of microbial fertilizers based on bacterial isolates that can solubilize potassium in soil. Potassium is mostly present in soil in the form of silicates hence silicate solubilizing bacteria can play an efficient role here by liberating soluble potassium in soil. Our results indicated that some of silicate and phosphate solubilizing bacterial isolates were capable of solubilizing potassium as well while a few, such as R1 and R2, were found to solubilize silicates and potassium but not phosphate as detected by plate assays (**Table 1**, **Figure 3**).

In addition to increasing the availability of Si, P and K in soil, the silicate solubilizing bacteria provide a very efficient system of biological control of plant pathogenic fungi. These bacteria can not only directly combat phyto-pathogenic fungi but indirectly by release of Si in soil which in turn induces disease resistance in plants either by acting as a physical barrier or by acting as a modulator of host resistance to pathogen. Si is deposited beneath the cuticle to form a cuticle-Si double layer which mechanically impedes penetration of fungi and thus disrupts the infection process [26]-[30].

Almost all of the selected bacterial isolates showed at least some antagonistic activities against the four fungal pathogens used. Highest antagonistic activity was observed in case of bacterial isolate NR-2 as shown by maximum zone of inhibition observed against all of the fungal pathogens tested followed by bacterial isolate NE-4b and R-6 respectively. There are several ways by which these silicate solubilizing bacteria can antagonize fungal

Table 2. Inhibition of phytopathogenic fungi by selected Si, P and K solubilizing bacterial isolates.

Bacterial Isolates	Mean Zone of Inhibition after 7 Days (mm)			
	Magnaporthae grisae	*Rhizoctonia solani*	*Altarnaria alternata*	*Fusarium moniliformae*
NE-1	12	11	4	8
NE-2	18	18	14	17
NE-3	13	7	8	12
NE-4	20	18	16	17
NE-4b	34	29	23	21
NE-5	4	6	7	4
NR-2	39	33	25	31
NR-3	7	8	12	8
R-2	18	16	17	17
R-6	29	18	19	13
R-8	18	21	17	14
WE 1	17	11	9	10
WE 2	13	12	11	8
WE 3	14	14	8	15
EM 5	10	7	17	13
EM 6	16	5	9	10
EM 7	12	11	4	8
RM 1	18	18	14	10
RM 2	13	17	27	12
RM 3	20	18	16	17
RM 4	12	15	14	18
AYM-1	18	18	14	17
AYM-2	13	27	28	12
AYM-3	20	18	16	17
SLB-1	11	11	4	8
SLB-2	18	18	14	17
SLB-3	13	7	8	12
SLB-4	20	18	16	17
SLB-5	12	11	4	8
SLB-6	13	12	11	8
SLB-7	14	14	8	15
SLB-8	10	7	17	13
R1	13	12	11	8
PTN 8	14	14	8	15
PTN 24	10	7	17	13

pathogens. These include production of hydrolytic enzymes, siderophores, HCN and antibiotics [13] [14] [18]. The antagonistic potential of these bacteria is an important ability of practical utility in development of biopesticides against the fungal pathogens.

These bacterial isolates will be further investigated for combating plant pathogens in green house experiment for a possible development of bioformulations for commercial purpose.

Acknowledgements

We are grateful to the Pakistan Higher Education Commission for funding this project.

References

[1] Francois, F., Wilfried, R.B., James, G.M. and Richard, R.B. (2005) Silicon and Plant Disease Resistance against Pathogenic Fungi. *FEMS Microbiology Letters*, **249**, 1-6. http://dx.doi.org/10.1016/j.femsle.2005.06.034

[2] Carver, T.L.W., Zeyen, R.J. and Ahlstrand, G.G. (1987) The Relationship between Insoluble Silicon and Success or Failure of Attempted Primary Penetration by Powdery Mildew (*Erysiphe graminis*) Germlings on Barley. *Physiological and Molecular Plant Pathology*, **31**, 133-148. http://dx.doi.org/10.1016/0885-5765(87)90012-9

[3] Adatia, M.H. and Besford, R.T. (1986) The Effects of Silicon on Cucumber Plants Grown in Recirculating Nutrient Solution. *Annals of Botany*, **58**, 343-351.

[4] Raid, R.N., Annderson, D.L. and Ulloa, M.F. (1991) Influence of Cultivar and Soil Amendment with Calcium Silicate Slag on Foliar Disease Development and Yield of Sugarcane. *Florida Agricultural Experimental Station Journal*, Ser. N R-01689.

[5] Winslow, M.D. (1992) Silicon, Disease Resistance, and Yield of Rice Genotypes under Upland Cultural Conditions. *Crop Science*, **32**(5), 1208-1213. http://dx.doi.org/10.2135/cropsci1992.0011183X003200050030x

[6] Datnoff, L.E., Snyder, G.H. and Deran, C.W. (1992) Influence of Silicon Fertilizer Grade on Blast and Brown Spot Development in Rice Yields. *Plant Disease*, **76**, 1182-1184. http://dx.doi.org/10.1094/PD-76-1182

[7] Bélanger, R.R., Benhamou, N. and Menzies, J.G. (2003) Cytological Evidence of an Active Role of Silicon in Wheat Resistance to Powdery Mildew (*Blumeria graminis* f. sp. *tritici*). *Phytopathology*, **93**, 402-412. http://dx.doi.org/10.1094/PHYTO.2003.93.4.402

[8] Ma, J.F. and Yamaji, N. (2006) Silicon Uptake and Accumulation in Lower Plants. *Trends in Plant Science*, **11**, 392-397. http://dx.doi.org/10.1016/j.tplants.2006.06.007

[9] Vasanthi, N., Saleena, L.M. and Raj, S.A. (2012) Silicon in Day Today Life. *World Applied Sciences Journal*, **17**, 1425-1440.

[10] Rodrignes, F.A. and Datnoff, L.E. (2005) Silicon and Rice Disease Management. *Fitopatologia Brasileira*, **30**, 457-469. http://dx.doi.org/10.1590/S0100-41582005000500001

[11] Lanning, F.C., Ponnaiya, B.W.X. and Crumpton, C.F. (1958) The Chemical Nature of Silica in Plants. *Plant Physiology*, **33**, 339-343. http://dx.doi.org/10.1104/pp.33.5.339

[12] Vijayapriya, M. and Muthukkaruppan, S.M. (2010) Isolation and Screening of Silicate Solubilizing Bacteria and Its Biocontrol Nature against *Pyricularia oryzae*. *International Journal of Recent Scientific Research*, **4**, 87-91.

[13] Naureen, Z., Hafeez, F.Y., Hussain, J., Al Harrasi, A., Bouqellah, N. and Roberts, M.R. (2015) Suppression of Incidence of *Rhizoctonia solani* in Rice by Siderophore Producing Rhizobacterial Strains Based on Competition for Iron. *European Scientific Journal*, **11**, 186-207.

[14] Naureen, Z., Price, A.H., Wilson, M.J., Hafeez, F.Y. and Roberts, M.R. (2009) Suppression of Rice Blast Disease by Siderophore-Producing Bioantagonistic Bacterial Isolates Isolated from the Rhizosphere of Rice Grown in Pakistan. *Crop Protection*, **28**, 1052-1060. http://dx.doi.org/10.1016/j.cropro.2009.08.007

[15] Bunt, J.S. and Rovira, A.D. (1955) Microbiological Studies of Some Sub Antarctic Soils. *Journal of Soil Science*, **6**, 119-128. http://dx.doi.org/10.1111/j.1365-2389.1955.tb00836.x

[16] Yasmin, S., Baker, M.A.R., Malik, K.A. and Haffez, F.Y. (2004) Isolation, Characterization and Beneficial Effects of Rice-Associated Plant Growth-Promoting Bacteria from Zanzibar Soils. *Journal of Basic Microbiology*, **44**, 241-252. http://dx.doi.org/10.1002/jobm.200310344

[17] Naureen, Z., Hameed, S., Yasmin, S., Malik, K.A. and Hafeez, F.Y. (2005) Characterisation and Screening of Bacteria from Maize Grown in Indonesian and Pakistani Soils. *Journal of Basic Microbiology*, **45**, 447-459. http://dx.doi.org/10.1002/jobm.200510566

[18] Hassan, M.N., Afghan, S. and Hafeez, F.Y. (2010) Suppression of Red Rot Caused by *Colletotrichum falcatum* on Su-

garcane Plants Using Plant Growth-Promoting Rhizobacteria. *BioControl*, **55**, 531-542. http://dx.doi.org/10.1007/s10526-010-9268-z

[19] Lü, C. and Huang, B. (2010) Isolation and Characterization of Azotobacteria from Pine Rhizosphere. *African Journal of Microbiology Research*, **4**, 1299-1306.

[20] Joseph, M.H., Dhargave, T.S., Deshpande, C.P. and Srivastava, A.K. (2015) Microbial Solubilisation of Phosphate: Pseudomonas versus Trichoderma. *Annals of Plant and Soil Research*, **17**, 227-232.

[21] Berthelin, J. and Belgy, G. (1979) Microbial Degradation of Phyllosilicates during Simulated Podzolization. *Geoderma*, **21**, 297-310. http://dx.doi.org/10.1016/0016-7061(79)90004-1

[22] Malinovskaya, I.M., Kosenko, L.V., Votselko, S.K. and Podgorskii, V.S. (1990) Role of *Bacillus Mucilaginosus* Polysaccharide in Degradation of Silicate Minerals. *Mikrobiologiya*, **59**, 49-55.

[23] Hiebert, F.K. and Bennett, P.C. (1992) Microbial Control of Silicate Weathering in Organic-Rich Ground Water. *Science*, **258**, 278-281. http://dx.doi.org/10.1126/science.258.5080.278

[24] Barker, W.W., Welch, S.A., Chu, S. and Baneld, J.F. (1998) Experimental Observations of the Effects of Bacteria on Aluminosilicate Weathering. *American Mineralogist*, **83**, 1551-1563.

[25] Muralikannan, N. and Anthomiraj, S. (1998) Occurrence of Silicate Solublising Bacteria in Rice Ecosystem. *The Madras Agricultural Journal*, **85**, 47-50.

[26] Sahebi, M., Hanafi, M., Siti, A., Akmar, N., Rafii, M.Y., Azizi, P., Tengoua, F.F., Azwa, J.N.M. and Shabanimofrad, M. (2015) Importance of Silicon and Mechanisms of Biosilica Formation in Plants. *Biomed Research International*, **2015**, Article ID: 396010. http://dx.doi.org/10.1155/2015/396010

[27] Simonsson, M., Andersson, S., Andrist-Rangel, Y., Hillier, S., Mattsson, L. and Oborn, I. (2007) Potassium Release and Fixation as a Function of Fertilizer Application Rate and Soil Parent Material. *Geoderma*, **140**, 188-198. http://dx.doi.org/10.1016/j.geoderma.2007.04.002

[28] Chinnasami, K.N. and Chandrasekaran, S. (1978) Silica Status in Certain Soils of Tamil Nadu. *The Madras Agricultural Journal*, **65**, 743-746.

[29] Sugumaran, P. and Janarthanam, B. (2007) Solubilization of Potassium Containing Minerals by Bacteria and Their Effect on Plant Growth. *World Journal of Agricultural Sciences*, **3**, 350-355.

[30] Bowen, P., Menzies, J. and Ehret, D. (1992) Soluble Silicon Sprays Inhibit Powdery Mildew Development on Grape Leaves. *Journal of the American Society for Horticultural Science*, **117**, 906-912.

22

Quantitative Genetic Variation in Gum Arabic (*Acacia senegal* (L) Willd) Provenances

N. M. Fakuta[1*], I. F. Ojiekpon[1], I. B. Gashua[2], O. C. Ogunremi[3]

[1]Rubber Research Institute of Nigeria, Gum Arabic Sub-Station, Gashua, Nigeria
[2]Department of Science Laboratory, Technology Federal Polytechnic, Damaturu, Nigeria
[3]National Biotechnology Development Agency, Abuja, Nigeria
Email: *fakuta1st@gmail.com

Abstract

Twelve (12) *Acacia senegal* provenances were evaluated at the Rubber Research Institute of Nigeria, Gum Arabic Sub-Station Gashua, Yobe state. Gum yield and some morphological data were collect in three years to estimate mean effects, genetic variability, heritability and genetic advance in *Acacia senegal* provenances. The analysis of variance revealed highly significant difference for all traits. Genetic component analysis showed that both genotypic coefficient of variation (GCV) and phenotypic coefficient of variation (PVC) were high for all the traits except bark thickness. PCV was higher in proportion than GCV with respect to all the traits. Gum yield gave the highest relative difference (RD) over other traits. Heritability estimates were high for bark thickness (77.36) and stem diameter (61.75). Expected genetic advance was high in number of tree incision (1.29), stem diameter (1.54) and yield in kg/ha (22.31). This study indicated that there was sufficient genetic variability in *Acacia senegal* provenances that could be used as a base for improvement of yield and other component attributes of this species through direct selection and hybridization.

Keywords

Gum Arabic, Genotypic Coefficient of Variation, Phenotypic Coefficient of Variation, Genetic Advance, Provenances

1. Introduction

Acacia senegal is a leguminous multipurpose African tree species belonging to the family *Fabaceae* [1]. The

*Corresponding author.

plant is a deciduous tree up to 12 m tall with short thorns of less than 1 cm in length located below the nodes and occurred in sets of three with the middle one curving downward. The leaves are small 3 - 8 pairs of pinnae each with 7 - 25 pairs of leaflets clustered together and pale green in colour [1] [2]. Tapping commences when the trees reach a height of 1.2 - 1.5 m with a main stem diameter of about 5 cm from 5 - 7 years of age depending on management practice [3]. The trees are usually tapped from the main stem and branches using a special debarking tool called axe which is used in removing the bark, while taking care not to damage the cambium [4]. Gum yield of about 300 g/tree/annum is common, however, yield of about 900 g/tree/annum or 1 ton/ha is possible among older trees [5]. The tree is highly valued for gum Arabic production utilized as a multifunctional food additive, confectioneries, pharmaceuticals, food and beverages, brewing among others [5] [6]. The ultimate goal of *Acacia senegal* breeding is to improve the species earliness to tapping maturity, gum quantity and quality which largely depend upon the genetic variability, heritability and genetic advance of desirable traits. The magnitude and type of genetic variability help breeders to determine the selection criteria and breeding schemes to be deployed for improvement purposes. This investigation was therefore undertaken to estimate the variability, heritability and genetic advance in twelve (12) *Acacia senegal* provenances for improvement purposes.

2. Material and Methods

2.1. Germplasm Collection

Twelve *Acacia senegal* provenances were collected from two major ecological zones (Sudan and Sahel) with strong evidence of the species in Yobe state, Nigeria in 2006 (**Table 1**). Collection was specifically made at Yusufari, Jajimaji Bulkati and Usur (Sahelian ecology) and Damaturu, Ngaburwa, Fune and Gujba (Sudan ecology). The selection criteria were based on the provenance trial recommendations of Burley and Wood [7], where provenance nomenclature was standardized using locational details with a minimum interval of 100 meters apart to avoid inclusion of likely half-sibs in the collection. Healthiest pod in the best stage of maturity was selected from 25 trees in each provenance because of heterogeneous nature of the population.

2.2. Provenance Trial and Design

The provenance trial was established at the experimental field of the Rubber Research Institute of Nigeria, Gum Arabic Sub-Station, Gashua, located at latitude 12˚46'N, longitude 11˚00'E and altitude 360 m above sea level. Seedlings of the twelve (12) *Acacia senegal* provenances were nurtured in the nursery for four months before transplanting to the field in August, 2008 in a Randomized Complete Block Design (RBCD) with three replications. Each provenance consisted of a square plot with sixteen trees (4 × 4) row spaced at 3 m × 3 m [8].

Table 1. Description of twelve *Acacia senegal* provenances and their geographical locations, Yobe state.

Ecology	Location	Provenance code	Latitude (N)	Longitude (E)	Altitude (M)
Sahel	Yusufari 1	P1	13˚2'6"	11˚8'44"	370
	Yusufari 2	P2	13˚2'2"	11˚8'37"	371
	Yusufari 3	P3	13˚1'59"	11˚9'4"	375
	Jajimaji	P4	12˚53'45"	1˚52'38"	370
	Bulkati	P5	12˚53'5"	1˚53'15"	364
	Usur	P6	12˚53'1"	1˚59'32"	360
Sudan	Damaturu	P7	11˚41'27"	11˚56'39"	410
	Ngaburwa	P8	11˚45'34"	12˚7'56"	409
	Fune	P9	11˚41'43"	11˚41'46"	435
	Gujba1	P10	11˚23'8"	11˚58"8"	450
	Gujba2	P11	11˚24'27"	11˚59'27"	431
	Gujba3	P12	11˚24'39"	11˚59'38"	456

2.3. Provenance Evaluation

Each provenance was exploited for gum yield at four years after establishment using tapping knife to make an incision of 3 cm × 15 cm [9]. Number of incision was determined by counting the number of tapping cuts. Bark thickness (cm) and stem diameter (cm) were determined by using Venier caliper. Gum yield (kg/ha) was determined by harvesting the gum and shade drying to 15% moisture content before weighing on a digital weighing balance (Hecky High Precision). Analysis of morphological traits and gum yield were carried out for three years (2011 to 2013). Each year number of tree incisions, bark thickness and stem diameter were taken at the commencement of tapping while number of gum bearing incisions and gum yield were taken monthly for three month.

2.4. Data Analysis

Data collected were subjected to analysis of variance; F-test was used to test for significant differences among the measured traits [10]. Provenance means were compared using Duncan's Multiple Range Test [11]. The data were analyzed using the general linear model procedure of statistics analysis system software version 9.0 [12]. The mean squares were used to estimate phenotypic variance (σ_p^2) and genotypic variances (σ_g^2) [13]. The methods of Burton and Devane were used in calculating the genotypic and phenotypic coefficient of variations [14], as: GCV(%) = σ_g^2 × 100/grand mean and PCV (%) = σ_p^2 × 100/grand mean. (σ_p^2) = genetic variance (σ_g^2) + environment variance ($\sigma^2 e$) by Acquaah (2007). The recommendation of Nechif and co-workers were followed in the determination of Relative difference (RD %) between PCV (%) and GCV (%) = [100 (PCV − GCV)/PCV [15]. Broad sense heritability $\left(h_b^2\right) = \sigma_g^2 / \sigma_p^2$ and genetic advance $(\text{GA}) = \Delta\text{S} \times h_b^2 \times \sqrt{\sigma_p^2}$ were calculated by using the methods adopted by Acquaah and Falconer & Mackay respectively [16] [17].

3. Results and Discussion

3.1. Mean Effects of Gum Yield and Other Morphological Traits in *A. senegal* Provenances

Provenances 11 and 12 though statistically similar, had superior growth traits compared to the rest of the provenances most times except gum yield averaged over 3 years tapping season as presented in **Table 2**. The highest gum yield was achieved with P9 while the lowest was recorded by P6 and P3. Provenance three (P3) often resulted in significantly lower morphological and yield traits. The superior growth put up by provenances of Gujba origin is not surprising as variability in growth among provenance population of *Acacia senegal* has been very well documented [18]. Also, the higher gum yield recorded with P9 is in order because Fune vegetation type is reminiscent of sahelian agro-ecological zone. The specie has been reported to put strong growth and yield performance under hash environmental conditions [19].

The analysis of variance result is shown in **Table 3**. The provenances showed significant differences for all traits measured in the study, indicating the presence of an adequate variability among the provenances for selection and identification of superior genotypes. Year effects were also significant for all traits except bark thickness probably due to yearly differences in weather variables.

3.2. Genetic Components of Variance, Heritability and Genetic Advance

As shown earlier in **Table 2**, the twelve *A. senegal* provenances showed wide range of variability in most of the traits studied. Tree incisions, number of gum bearing incisions and yield were relatively low. This investigation revealed considerable amount of variations for traits like barks thickness and stem diameter (**Table 4**). Such wide variations indicate prospects for improvement in these traits. Provenance variations in *A. senegal* growth traits which include plant height, canopy spread and stem diameter were earlier reported [20].

The phenotypic variance (σ_p^2) was slightly higher than the genotypic variance (σ_g^2) for all traits except bark thickness that gave the same estimate. However, phenotypic coefficient of variation (PCV) was higher in proportion than genotypic coefficient of variation (GCV) for all the traits suggesting the presence of environmental influence in the expression of these traits. GCV and PCV values greater than 20% are regarded as high, whereas values less than 10% are considered to be low [21]. Also, values between 10 and 20% are considered to be moderate. High GCV and PCV values were recorded for number of tree incision, stem diameter and gum yield while GCV was moderate and PCV high for number of gum bearing incision. These offer a wide range of

Table 2. Mean effects of gum yield and other morphological traits in *A. senegal* provenances evaluated across 3 years in Gashua.

Provenance code	TI (no)	BT (cm)	SD (cm)	NGBI (no)	GY (kg/ha)
P1	2.61cd	0.38cd	5.27cd	1.63b-e	78.77abc
P2	2.22de	0.36cde	4.92de	1.42b-e	125.96ab
P3	1.61e	0.326de	4.15e	1.09e	55.67c
P4	2.78cd	0.33cde	5.08cd	1.82a-d	118.83ab
P5	2.94bcd	0.39bc	5.77bc	1.39cde	96.60abc
P6	2.20de	0.31e	4.74de	1.24de	55.58c
P7	2.46cde	0.476a	5.57cd	1.57b-e	109.19abc
P8	3.11abc	0.45ab	5.49cd	1.78bcd	80.74abc
P9	3.28abc	0.47a	5.83bc	1.92abc	135.78a
P10	2.78cd	0.48a	5.41cd	1.67b-e	69.65bc
P11	3.86a	0.51a	6.46ab	2.04ab	102.60abc
P12	3.75ab	0.51a	6.70a	2.43a	116.26ab
Mean	2.80	0.42	5.45	1.67	95.74
S.E±	0.524	0.035	0.507	0.384	36.435

Means followed by the same letter within a column of a treatment group are not significantly different statistically at 5% level of probability using DMRT. TI: tree incision, BT: bark thickness; SD: stem diameter; NGBI: number of gum bearing incision; GY: gum yield.

Table 3. Analysis of variance for gum yield and other morphological traits in *A. senegal* provenances evaluated across 3 years in Gashua.

Source of variation	df	Mean squares				
		TI (no)	BT (cm)	SD (cm)	NGBI (no)	GY (Kg/ha)
Year	2	50.05**	0.42	46.98**	22.59**	24965.12**
Rep (Year)	6	0.88	0.01*	0.87	0.60	1551.87
Provenance	11	3.79**	0.05**	4.50**	1.21**	6635.32*
Year × Provenance	22	0.47	0.0031	0.20	0.37	1857.30
Pooled error	66	0.82	0.0044	0.77	0.44	3982.44

*, **: Significant at $P < 0.05$, $P < 0.01$, respectively (F-test): TI: tree incision, BT: bark thickness; SD: stem diameter; NGBI: number of gum bearing incision; GY: gum yield.

Table 4. Descriptive statistics and genetic parameters of yield and other morphological traits of *A. senegal* evaluated across 3 years in Gashua.

Traits	Range	σ_g^2	σ_p^2	GCV (%)	PCV (%)	RD (%)	h_b^2 (%)	GA
NTI	1.61 - 3.86	0.99	1.81	35.30	64.72	45.46	54.54	1.29
BT(cm)	0.31 - 0.51	0.02	0.02	3.66	4.73	22.64	77.36	0.19
SD(cm)	4.15 - 6.70	1.24	2.01	22.82	36.96	38.25	61.75	1.54
NGBI	1.09 - 2.43	0.26	0.70	15.35	41.94	63.40	36.60	0.54
GY(kg/ha)	55.58 - 135.8	884.3	4866.7	926.3	5097.6	81.8	18.2	22.3

TI: tree incision, BT: bark thickness; SD: stem diameter; NGBI: number of gum bearing incision; GY: gum yield.

opportunities to carry out selection for the traits. High GCV and PCV recorded for number of tree incision, stem diameter and gum yield signifying that these traits are vital for gum Arabic improvement. RD estimates of GCV in relation to respective PCV were determined to judge the extent of genetic variation in these traits. These however, suggest that the observed variations in number of gum bearing incision (63.40%) and gum yield (81.17%) were mostly due to environmental factors.

The estimated heritability ranges from 18.17% for gum yield (kg/ha) to 77.36% for bark thickness. Heritability estimates were earlier classified as low (0 to 30%), moderate (30% to 60%) and high (above 60%). Considering this benchmark, heritability was high for bark thickness and stem diameter indicating a satisfactory possibility of genetic improvement for these traits through selection. High heritability values for bark thickness and stem diameter have been reported [22]. Low estimation of heritability was observed for gum yield(kg/ha) whereas the number of tree incision and number of gum bearing incision showed moderate heritability values which indicated that total variability was due to genetic and environment effects. This result is similar to previous study which observed that heritability estimates for traits of most forest tree species ranged from low to moderate and largely due to dominance, epistatic and environmental effects [23].

The genetic advances (GA) at 10% selection intensity for 5 traits are presented in **Table 2**. Estimate of GA for gum yield was 22.31, indicating that whenever 10% selection is made among high yielding genotypes as parents, mean yields of progenies could be improved by 22.31 kg/ha for the first cycle; meaning that the mean genotypic value of the new population for gum yield could be improved from 95.47 to 117.78 kg/ha. Bark thickness and number of gum bearing incision expressed low genetic advance. Thus high heritability coupled with moderate genetic advance observed for stem diameter and number of tree incision indicated that the traits were mainly controlled by additive genes and that direct selection for this trait could be effective whereas moderate or high heritability coupled with low genetic advance expressed by the remaining traits signified that the traits were largely under non additive gene control (dominance and/or epistasis) and environmental.

3.3. Conclusion

This study revealed that there was sufficient genetic variability in *A. senegal* provenances available in the germplasm that could be used for improvement in gum yield and other component attributes of the species through direct selection and hybridization.

Acknowledgements

The authors wish to express their sincere appreciation to the Executive Director of Rubber Research Institute of Nigeria for his support and the entire research staff of Gum-Arabic Sub-Station, Gashua for their cooperation during the course of the project.

References

[1] Dorthe, J. (2000) *Acacia senegal* (L) Willd Seed Leaflet. Danida Forest Seed Center, Denmark, No. 5, 1-2.

[2] Duke, J.A. (1981) Handbook of LEGUMES of World Economic Importance. Plenum Press, New York. http://dx.doi.org/10.1007/978-1-4684-8151-8

[3] Jamal, A. and Huntsinger, L.C. (1993) Deterioration of Sustainable Agro-Silvi-Pastoral System in the Sudan: The Gum Gardens of Kordofan. *Agroforestry Systems*, **23**, 23-38. http://dx.doi.org/10.1007/BF00704849

[4] Awouda, H.M. (1989) Outlook for Gum Arabic: Production and Supply. Report to the Gum Arabic Company, Khartoum, 18 p.

[5] Anderson, D.M. (1993) Some Factors Influencing the Demand of Gum Arabic (*Acacia senegal* (L.) Willdnd Other Water-Soluble Tree Exudates. *Forest Ecology and Management*, **58**, 1-18. http://dx.doi.org/10.1016/0378-1127(93)90127-9

[6] Leung, A.Y. (1980) Encyclopedia of Common Natural Ingredient Used in Food, Drug and Cosmetic. John Wiley and Sons, New York.

[7] Burley, J. and Wood, P.J. (1978) A Manual on Species and Provenance Research with Particular Reference to the Tropics. Tropical Forestry Papers No. 10, 34-61.

[8] FDA (2002) Baseline Survey of Gum Arabic Production in Nigeria. Federal Department of Agriculture in Collaboration with Forestry Research Institute of Nigeria and Rubber Research Institute of Nigeria, p. 126.

[9] Ojiekpon, I.F. (2005) Proper Exploitation of Gum Arabic for Quality Control. In: Adegbehin, J.O., Ed., *Training Manual on Gum Arabic Production, Proceedings of a Workshop*, Gum Arabic Sub-Station, Tajuwa Village, Gashua, 18-20 October 2005, 21-30.

[10] Snedecor, G.W. and Cochran, W.G. (1987) Statistical Method. 5th Edition, Iowa State University Press, 456.

[11] Duncan, D.B. (1955) Multiple Range and Multiple F-Tests. *Biometrics*, **11**, 1-42. http://dx.doi.org/10.2307/3001478

[12] SAS Institute (2000) SAS/STAT Guide for Personal Computer: Version 9.0. SAS Institute Inc., Cary.

[13] Johnson, H.W., Robinson, H.F. and Comstock, R.E. (1955) Estimates of Genetic and Environmental Variability in Soya Bean. *Agronomy Journal*, **47**, 318-324. http://dx.doi.org/10.2134/agronj1955.00021962004700070009x

[14] Burton, G.N. and Devane, E.M. (1953) Estimating Heritability in Fall Fescue (*Festuca arundiancea* L.) from Replicated Clonal Materials. *Agronomy Journal*, **45**, 478-481. http://dx.doi.org/10.2134/agronj1953.00021962004500100005x

[15] Nechif, O.R., Filimon, R.B. and Szilagyi, L. (2011) Genetic Variability, Heritability and Expected Genetic Advance as Indices for Yield and Yield Components Selection in Common Bean (*Phaseolus vulgaris* L.). Scientific Papers, UASVM Bucharest, Series A, 1222-5339.

[16] Acquaah, G. (2007) Principles of Plant Genetics and Breeding. Blackwell Publishers, Malden, 142-145.

[17] Falconer, D.S. and Mackay, T.F.C. (1996) Introduction to Quantitative Genetics. 4th Edition, Benjamin Cummings, England.

[18] Anders, R., Lars, G. and Lambart, G.O. (2003) Evaluation of Provenance Trial of *Acacia senegal* at Gonse Burkina Faso. Trial No. 12 in the Arid Zone Series, Danida Forest Seed Centre, Denmark and Food and Agriculture Organization, Rome., No. 5, 1- 41.

[19] Jean-Michel, H., Mama, N., Bertrand, M., Clement, F.N., Jean-Marie, T., Jean-Christophe, B., Philippe, T. and Régis, P. (2012) Gum Arabic Production in *Acacia senegal* Plantations in the Sudanian Zone of Cameroon: Effects of Climate, Soil, Tapping Date and Tree Provenance. *Bois et Forêts Des Tropiques*, No. 311, 21-33.

[20] Fakuta, N.M., Ojiekpon, I.F., Waizah, Y. and Gurin, A.S. (2010) Variability in Pod and Seed Traits on Juvenile Growth Performance of *Acacia senegal* (L) Willd Provenances of Sudan and Sahelian Origin of Yobe State, Nigeria. *Proceedings of the* 44*th Annual Conference of the Agricultural Society of Nigeria*, Ladoke Akintola University of Technology, Ogbomoso, 18-22 October 2010.

[21] Deshmukh, S.N., Basu, M.S. and Reddy, P.S. (1992) Genetic Variability, Character Associations and Path Coefficient Analysis of Quantitative Traits in Virginia Bunch Varieties of Ground Nut. *Indian Journal of Agricultural Sciences*, **56**, 515-518.

[22] Fakuta, N.M., Ojiekpon, I.F., Simon, S.Y. and Gani, M. (2013) Genetical Studies for Seedling Growth Traits in Relation to Early Tapping Maturity of Gum Arabic (*Acacia senegal* (L) Willd). *World Journal of Agricultural Sciences*, **1**, 248-251.

[23] Cornelius, J. (1994) Heritabilities and Additive Genetic Coefficients of Variation in Trees. *Canadian Journal of Forest Research*, **24**, 372-379. http://dx.doi.org/10.1139/x94-050

23

Natural Regeneration and Biological Invasion by *Pinus caribaea* Morelet in Two Vereda Sites: Woody Vegetation Response

Danúbia Magalhães Soares[1], André R. Terra Nascimento[2*], Lorena Cunha Silva[2], Gastão Viégas de Pinho Júnior[3]

[1]Graduate Degree Program in Ecology and Natural Resources Conservation, Universidade Federal de Uberlândia, Uberlândia, Minas Gerais
[2]Instituto de Biologia, Universidade Federal de Uberlândia, Uberlândia, Minas Gerais
[3]Elementary/High Schools in Uberlândia, Uberlândia, Minas Gerais
Email: *arnterra@gmail.com

Abstract

Veredas are regarded as humid complexes of ecological, water and scenery significance in Cerrado biome related to water table outcrop. Direct alterations in these environments can lead to species diversity reduction and also facilitate biological invasion processes. Thus our study aims to investigate the natural regeneration, the coverage of woody vegetation and the invasive species in two sites in secondary successional stage in the central part of Cerrado biome. Our study was carried out at two sites located at Minas Gerais state, Brazil. We established plots randomly throughout transects laid perpendicularly to the vereda's drain line. We used the line intercept method to analyze the vegetation coverage. We set five 10-m width transects in each of both areas, perpendicular to the vereda's drain line to evaluate the distribution of invasive alien species in the two areas. The richness did not vary between communities ($U = 67.50$; $p = 0.061$); on the other hand we found significant difference in density estimates ($U = 36.50$; $p = 0.002$). We found significant difference between categories for soil coverage in both sites. Grasses showed the highest coverage when compared to the other categories ($K = 45.06$; $p < 0.001$ for site 1 and $K = 67.47$; $p < 0.001$ for site 2). We observed a zonation tendency in site 1, with a higher native grass coverage at middle and backwards zones ($K = 7.734$, $p < 0.05$). We only found *Pinus caribaea* Morelet individuals in the invasive species survey. We recorded 46 *Pinus caribaea* individuals at site 1 and 116 *Pinus caribaea* individuals at site 2. From the latter group we found 13 individuals within the vereda and 103 in the surrounding vegetation. Both sites have a high natural woody plant regeneration potential although the presence of alien grasses in the veredas' edge and *Pinus caribaea* all over the humidity gradient highlights the vulnerability of these areas to biological invasion.

*Corresponding author.

Keywords

Woody Species, Ecological Patterns, Invasive Potential, Humid Areas

1. Introduction

Veredas are regarded as humid complexes of ecological, water and scenery significance in Cerrado biome (**Figure 1**). A vereda is in fact a multiform area showing niches and habitats heterogeneity important to terrestrial and water fauna, plus contributing to the watercourses perenniality [1] [2]. This physiognomy only occurs when there is a water table outcrop surrounded by *Mauritia flexuosa* L.f. (buriti), palm trees [3]. Nonetheless, since the tree-shrub layer is prevailing, this palm tree species occurrence is questionable when it comes to identify this sort of environment, so there might be areas bearing similar structure and plant composition in which this species does not occur [4].

Distinct flood levels and edaphic characteristics can generate distinct plant species gradients which vary as much as their coverage and soil conditions tolerance [1] [5]. Thus veredas can show three distinct floristic areas: edge, middle and backwards [1]. Although reproductive events of many species present in these environments rely on environmental disturbances such as fire and light condition variations [6], peculiar environmental conditions give veredas a low resilience, thus human activities can lead to severe disturbances in these environments. The soil compaction, water table lowering and biological invasion are challenges that must be overcome in order to maintain and preserve these areas [7].

According to the Technical Report on Cerrado Biome Deforestation released in 2009, Minas Gerais state held in 2008, nearly 43.2% of Cerrado remaining areas. The other areas had been removed to make way to any kind of human activity [8]. Even if legally protected by current Brazilian Forest Code, anthropogenic alterations in veredas still occur in despite of their vast distribution. In Triângulo Mineiro region (Minas Gerais state), many of these areas are directly used as natural pastures or drained for subsistence and intensive agriculture [9].

It is generally known that plant community regeneration patterns are determined by a combination of biotic and abiotic factors such as climate conditions, light characteristics, populations' structure and dynamics, interactions with pathogen parasites and the reproductive strategies of the species that compound this community [10]. Nonetheless, the plant species richness might be affected by anthropogenic alterations in the environment. Habitat fragmentation, pasture and edge effects can show a negative relation to diversity in plant communities [11].

Although they are not directly affected, the veredas are highly sensible to environmental disturbances in their surroundings [5]. The edge areas can show significant changes in their floristic composition, according to the characteristics and disturbance levels of the surrounding areas [12]. Moreover, direct alterations in these environments can lead to species diversity reduction and also facilitate biological invasion processes [7].

The biological invasion in veredas by alien species such as the grass *Melinis minutiflora* P. beauv. as well as by native ones like the shrub species *Trembleia parviflora* (D. Don) Cogn., might be related to humidity reduction in veredas soils as a result of anthropogenic actions [13]. The latter mentioned species is able to form high density areas with reduced diversity rates known as "trembleiais areas" [14]. Besides, human activities that cause water table lowering can increase the invasive process by leading this species distribution farther into the vereda area (**Figure 1(c)**) [7].

Disturbed ecosystems can be dominated by native and alien grass species highly productive, which makes the floristic composition in these environments less diverse [15]. Similarly, invasive species belonging to *Pinus* spp. genus change the natural conditions by unbalancing the producer-consumer relationship forming a community dominated by few species. They also form a thick layer of litter which hampers the native species regeneration [16]. These invasive species can disperse through long distances away the plantation area carrying with them compatible invasive ectomycorrhiza fungi [17]. There is contamination by these species in Brazil (**Figure 1(b)**), where *Pinus* spp. plantations are conspicuous [18]. The contamination by these species occurs in open areas such as natural fields, cerrado sensu stricto, sand banks and secondary forests [19].

Once started the biological invasion process its effects on plant community are varied and in some cases difficult to detect because they are idiosyncratic, subtle, and indirect and its likely significant effects might only can be detectable in a long term [20] [21]. Nonetheless, many negative effects of invasive plants had already

Figure 1. General view of woody community in vereda (a), biological invasion by *Pinus caribaea*, underscoring an individual from regeneration (b) and individuals growing in a higher density in the drier and without woody vegetation areas (white arrows in (c)). Photos: Nascimento, A.R.T.

been recorded such as: changes in organic matter deposition in soil [22], changes in fire frequency and/or fire occurrence [23], reduction of diversity rates and basal area of native regenerating species in alien species understory [24], fitness reduction in native species by direct competition [25] or apparent competition [26]. In despite of being scarcely understood, the co-occurrence of distinct invasive species also sets a challenge to the understanding of biological conservation efforts [27].

More than isolating disturbed areas, a successful regeneration process requires in some cases strategies to control invasive species and manage native species in late successional process. The correct intervention though relies on ecological knowledge [28] and monitoring the restoration process in field [28] [29]. In this context studies on regeneration patterns and alien and native species distribution in veredas can provide some important data on water resources and land use in these ecosystems as well as act as guideline for preservation and management [13]. Thus our study aims to investigate the natural regeneration, the coverage of woody vegetation and the potentially invasive species in two veredas sites in secondary successional stage in the core area of Cerrado biome.

2. Materials and Methods

2.1. Study Area

Our study was carried out at two vereda sites located at "Nova Monte Carmelo" farm (**Figure 1**), situated in Minas Gerais state, Brazil. The distance between the studied sites is nearly 12 km. Although the two sites are located at distinct municipalities—site 1 in Estrela do Sul and site 2 in Indianópolis, both fragments are subject to the same anthropogenic pressures as a result of commercial forest plantation at Nova Monte Carmelo farm. Still, the distinct environmental features between both sites can affect the biological invasion process as well as the natural regeneration, since site 1 does not show any vegetation adjacent to vereda, wich makes it more vulnerable to disturbances.

The farm has huge *Pinus* spp. and *Eucalyptus* spp. plantations for commercial purposes. The farm is 58,000-hectare large in which 12,000 ha. are legal reserve areas in distinct regeneration stages, characterized by abandoned pastures and cerrado sensu stricto, semideciduous seasonal forest and veredas fragments. According to the Management Plan of Nova Monte Carmelo farm, the set aside areas for restoration are in natural regeneration process [30]. Nevertheless, the vereda areas in this farm show signs of degradation such as the high density presence of invasive species *Trembleia parviflora* and *Pinus caribaea*. The weather in this region is characterized as Aw, providing hot and rainy summers and dry and cold winters with a 1450 mm average precipitation and annual average temperature between 20˚C and 22˚C [31].

2.2. Data Gathering

We established two distinct size classes to evaluate the natural regeneration process in the two vereda sites: 2 × 2-meter plots (Class I) and 10 × 10-meter plots (Class II). We also established fifteen 4 m^2 plots (Class I) and fifteen 100 m^2 plots (Class II) at each site [32] in the following environments: edge (5 plots), middle (5 plots), and backwards (5 plots). We set these plots randomly throughout transects laid perpendicularly to the vereda's drain line. We identified the environments as edge, middle and backwards according to [1].

We sampled all the regenerating individuals between 15 cm and 1 m high in the smaller plots (4 m^2) and the woody individuals with 1 m high and up to 5 cm D.B.H (Diameter at Breast Height, measured 1.3 m above the ground) in the larger plots (100 m^2). This inclusion criterion for woody vegetation (D.B.H $\geq$ 5 cm) characterizes adult vegetation [33], thus separates the tree layer from the rest of the vegetation.

We identified the sampled woody individuals in situ according to morphological and dendrologic features. When we could not identify these sampled individuals in field we gathered the botanical material and compared it to specialized bibliography and to the botanical material stored at Herbarium Uberlandensis (HUFU). We identified all the collected material according to APG III [34]. We sent the sampled material we could not identify or bearing a doubtful identification to experts.

We used the line intercept method [35] to analyze the vegetation coverage. We sampled fifteen 10-m length lines (five in each environment), the measures were taken directly on the vegetation 1-m above the ground [36]. We set the lines in the 100 m^2 plots parallel to the transects and 2-m away from the edge of the plot.

To analyze the coverage distribution in the distincts plant groups we grouped the data into the following categories: alien grasses, native grasses, herbaceous, woody, lianas, ferns and bare soil.

We also evaluated the distribution of potentially invasive alien species in the two sites. For this purpose we set five 10-m width transects in each of both sites, perpendicular to the vereda's drain line, 100-m away from each other starting at the external edge of vegetation adjacent to vereda up to the watercourse (vereda's backwards environment). We measured the Diameter at Breast Height (D.B.H) for each individual found in the transect area higher than 2.5-m. We measured the Diameter at Soil Height (D.S.H) for those between 1-m and 2.5-m high. We also recorded the height estimate, physiognomy and occurrence zone (edge, middle or backwards) for those found inland the vereda sites.

2.3. Data Analysis

We tested the density, richness, coverage and dendrometric parameters (height and D.B.H) of *Pinus caribaea* (invasive species) for normality using graphic analysis and Kolmogorov-Smirnov test ($p \leq 0.05$). We performed this same procedure to test species richness and density data for native species which we represented by using average and standard deviation. Since the coverage data (total and categorized) did not fir normal distribution, we used Kruskal Wallis non parametrical test. We compared the D.B.H and height averages for *Pinus caribaea* (invasive species) between the two vereda sites using the t-test. We performed ANOVA to compare the averages for trunk diameter and height between the occurrence zones and vegetation types. We performed all the procedures and statistical tests in Systat 10.2 software (Systat Software, Inc, Richmond, Calif).

We investigated the environmental heterogeneity (edge, middle and backwards) inside and amongst the sites by performing a hierarchical cluster analysis according to ward method with the euclidean distance as a clustering coefficient based on the dissimilarity amongst the plots [37]. We created the dendrogram based on Euclidean distance using a presence/absence binary data matrix for the species found in the plots. We used the R software [38] to perform this procedure.

3. Results and Discussion

Based on richness and density estimates the woody layer regeneration is significant (**Figure 2**). The richness did not vary between communities ($U = 67.50$; $p = 0.061$), even though it ranged from 1 to 23 species present in the plots. On the other hand, we found significant difference in the number of individuals per plot estimates ($U = 36.50$; $p = 0.002$) showing a wider variation between areas (1 to 77 individuals). Oppositely to cerrado sensu stricto where the tree-shrub layer is diverse and species rich [39] in vereda vegetation, this layer shows a low diversity, holding a smaller array of woody species which can tolerate the flood conditions and the hydromorphic soils, both characteristical features of these environments.

The cluster analysis identified two distinct groups: one consists basically of the plots located at site 1 and the other one mainly of plots situated at site 2 (**Figure 3**). Therefore we could notice the sites hold distinct floristic compositions even showing similar ecological zonation features. The low similarity between the two sites might be related to distinct disturbance levels and soil humidity conditions, since variations in soil humidity can result in differences in plant community composition either between zones in a vereda area or between veredas areas [5]. Although the formation of distinct floristic zones between edge, middle and backwards plots is not clear we observed a lower similarity between edge and backwards plots when compared to all the other categories—comparison made amongst distinct categories and the same categories.

We did not find any liana species amongst the coverage categories at site 2. In general, the soil coverage did not differ significantly between the two veredas areas except for woody plants category ($U = 45$; $p < 0.05$). The site 2 holds the more significant difference in this category (**Figure 4**). We found significant difference between categories for soil coverage in both sites. Grasses showed the highest coverage when compared to the other categories ($K = 45.06$; $p < 0.001$ for site 1 and $K = 67.47$; $p < 0.001$ for site 2).

When it comes to zonation we did not find any significant difference in coverage amongst zones (edge, middle and backwards) in the categories for site 2. Conversely, Santos and Munhoz [14] identified three distinct floristic groups throughout a humidity gradient in a vereda area at Distrito Federal (DF, Midwest Brazil). The absence of zonation in site 2 might indicates non significant environmental variations in this area, probably related to a flat local topography.

Nonetheless, we observed a zonation tendency in site 1, with a higher native grass coverage at middle and backwards zones ($K = 7.734$; $p < 0.05$). Floristic surveys in vereda vegetation also identify a higher grass diversity at these zones [14], which might indicates adaptive advantages of this plant group to humid areas.

We only observed alien grasses at edge zone in both sites, with an average coverage of 7.60% (±21.35) for site 1% and 0.39% (±1.52) for site 2; this observed coverage is lower when compared to native grass coverage. This species group is usually conspicuous in edge areas, where there is a higher light intensity and wind speed, being this latter responsible for dispersing anemochoric seeds such as *Pinus caribaea* ones. Moreover, the water table lowering (**Figure 1(c)**) can facilitate the colonization of the edge of the veredas by species that could not tolerate the humidity conditions usually found at these environments [7]. Although the average coverage by alien grasses is relatively low, environmental and ecological factors might be maintaining these populations at unstable levels thus any change at the environmental conditions could trigger unpredictable invasive processes

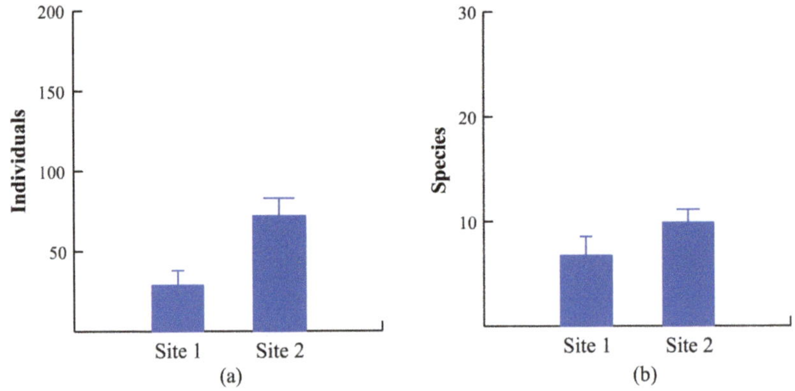

Figure 2. Density estimates (a) and woody species richness (b) per 100 m^2 plot at two veredas in Cerrado biome. Each column represents the average and the bar represents the standard deviation of the estimates (N = 15 samples per area).

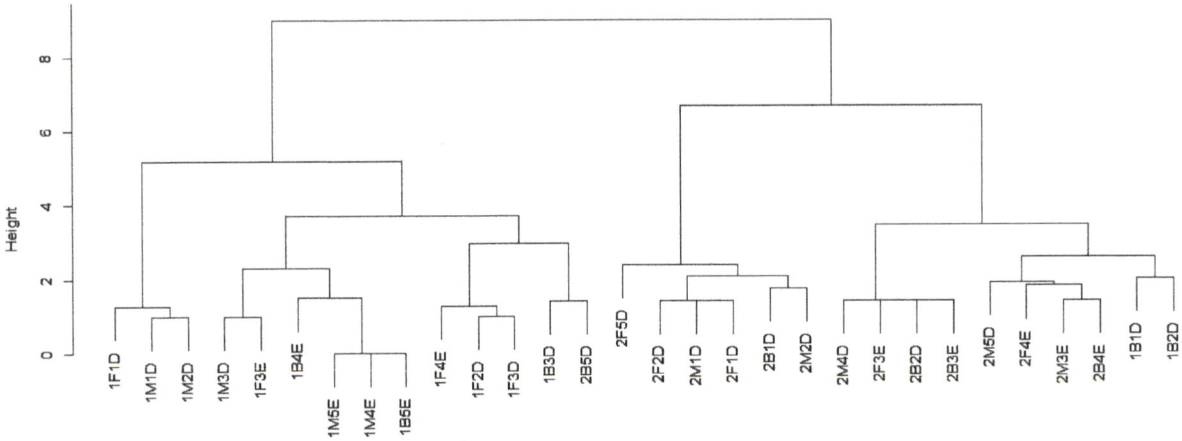

Figure 3. Dendrogram for hierarchical cluster analysis for the two sites using ward's method and the euclidean distance as clustering coefficient. Numbers represent the sites (site 1 and site 2), and letters the zones (E: edge; M: middle; B: backwards).

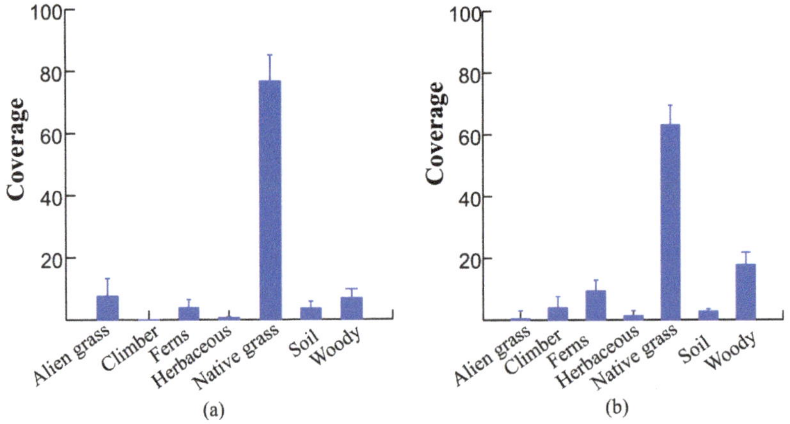

Figure 4. Total soil coverage (%) in distinct categories in two veredas sites: site 1 (a) and site 2 (b) Nova Monte Carmelo farm, Estrela do Sul-MG.

[23] [40] therefore the maintenance of this ecosystem takes monitoring and invasive species management methods.

We only found *Pinus caribaea* individuals in the invasive species survey. We recorded 46 *Pinus caribaea* individuals at site 1 and 116 *Pinus caribaea* individuals at site 2. From the latter group we found 13 individuals inland the vereda and 103 in the surrounding vegetation.

The *Pinus caribaea* individuals showed the highest height (**Figure 5(a)**) at site 2 ($t = -5.873$; $p < 0.01$) and the highest diameter (**Figure 5(b)**) at site 1 ($t = 2.538$; $p = 0.012$). Since the management at site 1 was done using clearcutting these differences might be related to individual's regrowth at this site.

We did not find any significant variation for diameter in *Pinus caribaea* individuals at site 1 amongst edge, middle and backwards zone although we found the highest individuals at the edge of the vereda. We found significant differences for both height and diameter amongst zones in vereda 2 at site 2 (**Table 1**), with the highest individuals found inland (middle and backwards).

We tested the biometrical differences for *Pinus caribaea* individuals at distinct Cerrado physiognomies at site 2 because we only observed adjacent vegetation to vereda at this area. We recorded the invasive species at the 5 analyzed transects allocated at cerrado sensu stricto, semideciduous seasonal forest (SSF) and bordering Cerrado and SSF and Cerrado and vereda. We did not find any significant difference in height ($F = 0.480$; $p = 0.697$) and diameter ($F = 1.491$; $p = 0.221$) amongst the distinct physiognomies. This result might indicate that this invasive species population is able to establish at distinct types of vegetation with distinct environmental features.

The *Pinus caribaea* populations at the two veredas areas studied showed individuals at distinct growing stages, from youngsters to reproductive adult ones, which might play an important role in these ecosystems reinfestation and in the maintenance of these invasive species populations, with the possibility of seed rain and consequently the continuous recruitment of new individuals [41]. The distribution of *Pinus caribaea* individuals at distinct growing stages might be related to the disturbance and the water table level, which can result in distinct plant communities composition, either in veredas zones or between them [5], with places where these individuals establishment might be restrict or absent. Moreover, the environment disturbance levels can be vital to explain the observed variation [42], since the studied areas and the zones within each vereda are subject to distinct

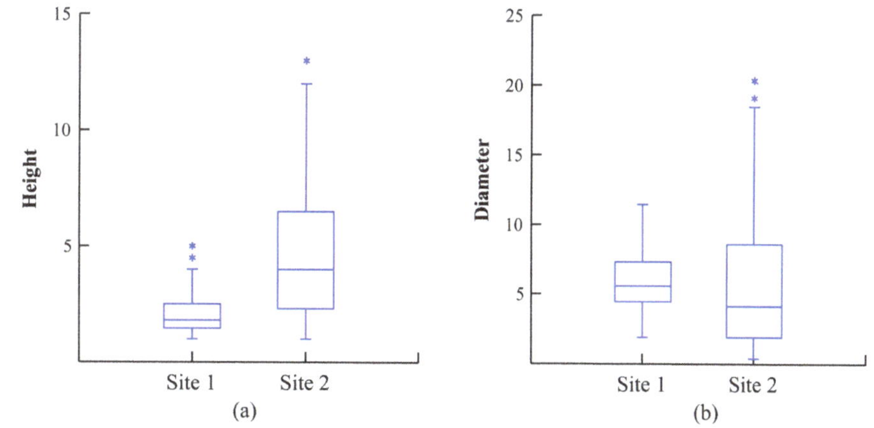

Figure 5. Diameter measured—centimeters, (a) and height—meters of *Pinus caribaea* individuals in both sampling sites at Nova Monte Carmelo farm, Estrela do Sul-MG. The boxes represent 50% of the data, being the central horizontal lines the medians, the superior line represents o 75th percentile and the inferior the 25th percentile. The vertical lines represent the data width, and the asterisks the outliers.

Table 1. Results of ANOVA for height and diameter (on a logarithmic scale) for *Pinus caribaea* individuals found at edge, middle and backwards veredas at two sites.

		Source	Sum-of-squares	df	Mean-square	F-ratio	p
Site 1	Diameter	Zone	0.085	2	0.042	1.508	0.234
		Error	1.066	38	0.028		
	Height	Zone	0.297	2	0.149	4.623	0.016
		Error	1.222	38	0.032		
Site 2	Diameter	Zone	1.323	2	0.662	10.330	0.004
		Error	0.640	10	0.064		
	Height	Zone	0.650	2	0.325	15.456	0.001
		Error	0.210	10	0.21		

levels of anthropogenic disturbance, with the presence or not of native vegetation in the surroundings, proximity to cultivated and fragment areas thus these anthropogenic impacts play an important role in the invasive process [23].

4. Conclusions and Forestry Guidelines

Both sites show a significant natural regeneration potential represented by density and coverage values in spite of the dominance of some species at certain patches, which normally occurs at vereda vegetation. The presence of alien grasses in the veredas' edge even in low densities highlights the vulnerability of these areas to disturbances in the surroundings indicating the necessity of maintaining surrounding vegetation to act as a buffer zone. The invasion of *Pinus caribaea* plus the aforementioned features were recorded as an indicative of disturbance, with individuals at distinct growing levels that reached the area through anemochoric dispersion. A forestry guideline to manage this invasive species at this matrix with a higher alien and native grasses biomass would be the clearcutting followed by poisoning the remaining trunks and roots with herbicid, which besides being an usual forestry procedure can also diminish the reifestation. Girdling the remaining individuals might be necessary. These proceedings have to be done before the seeds production in order to avoid their dispersion to the open areas with a lower woody biomass.

Acknowledgements

We would like to thank Duratex for funding the project of which this study is part. We are also thankful to the graduate Program in Ecology and Natural Resources Conservation and CAPES for the graduate scholarship provided. We are also thankful to the professors Rosana Romero (Melastomataceae), Glein M. Araújo (distinct botany families) and Jimi N. Nakajima (Asteraceae) for helping identify the collected botany material. We would like to thank Cláudio Henrique Eurípedes de Oliveira for his assistance in field.

References

[1] Araújo, G.M., Barbosa, A.A.A., Arantes, A.A. and Amaral, A. (2002) Composição florística de veredas no município de Uberlândia, MG. *Revista Brasileira de Botânica*, **25**, 475-493. http://dx.doi.org/10.1590/S0100-84042002012000012

[2] Brandão, M., Carvalho, P.G. and Baruqui, F.M. (1991) Veredas: Uma abordagem integrada. *Daphne*, **1**, 9-12.

[3] Ribeiro, J.F. and Walter, B.M.T. (2008) As principais fitofisionomias do Bioma Cerrado. In: Sano, S.M., Almeida, S.P. and Ribeiro, J.F., Eds., *Cerrado Ecologia e Flora*, Planaltina, 152-212.

[4] Moreira, S.N., Eisenlohr, P.V., Pott, A., Pott, V.J. and Oliveira-Filho, A.T. (2015) Similar Vegetation Structure in Protected and Non-Protected Wetlands in Central Brazil: Conservation Significance. *Environmental Conservation*, **1**, 1-7. http://dx.doi.org/10.1017/S0376892915000107

[5] Oliveira, G.C., Araujo, G.M. and Barbosa, A.A.A. (2009) Florística e zonação de espécies vegetais em veredas no Triângulo Mineiro, Brasil. *Rodriguésia*, **60**, 1077-1085.

[6] Araújo, G.M., Amaral, F.A., Bruna, E.M. and Vasconcelos, H.L. (2013) Fire Drives the Reproductive Responses of Herbaceous Plants in a Neotropical Swamp. *Plant Ecology*, **214**, 1479-1484. http://dx.doi.org/10.1007/s11258-013-0268-9

[7] Meirelles, M.L., Guimarães, A.J.M., Oliveira, R.C., Araújo, G.M. and Ribeiro, J.F. (2004) Impactos sobre o estrato-herbáceo de areas úmidas do Cerrado. In: Aguiar, L.M.S. and Camargo, A.J.A., Eds., *Cerrado*: *Ecologia e Caracterização*, Planaltina, 41-69.

[8] Ministério do Meio Ambiente (2009) Florestas do Brasil em resumo. Serviço Florestal Brasileiro, Brasília.

[9] Ramos, H.M.V.V., Curi, N., Motta, P.E.F., Vitorino, A.C.T., Ferreira, M.M. and Silva, M.L.N. (2006) Veredas do Triângulo Mineiro: Solos, água e uso. *Ciência e Agrotecnologia*, **30**, 283-293. http://dx.doi.org/10.1590/S1413-70542006000200014

[10] Marimon, B.S., Felfili, J.M., Fagg, C.W., Marimon-Junior, B.H., Umetsu, R.K., Oliveira-Santos, C., Morandi, P.S., Lima, H.S. and Nascimento, A.R.T. (2012) Monodominance in a Forest of *Brosimum rubescens* Taub. (Moraceae): Structure and Dynamics of Natural Regeneration. *Acta Oecologica*, **43**, 134-139. http://dx.doi.org/10.1016/j.actao.2012.07.001

[11] Cagnolo, L., Cabido, M. and Valladares, G. (2006) Plant Species Richness in the Chaco Serrano Woodland from Central Argentina: Ecological Traits and Habitat Fragmentation Effects. *Biological Conservation*, **132**, 510-519.

http://dx.doi.org/10.1016/j.biocon.2006.05.012

[12] Resende, I.L.M., Chaves, L.J. and Rizzo, J.A. (2013) Floristic and Phytosociological Analysis of Palm Swamps in the Central Part of the Brazilian Savanna. *Acta Botanica Brasilica*, **27**, 205-225. http://dx.doi.org/10.1590/S0102-33062013000100020

[13] Felfili, M.J., Silva-Junior, M.C.S., Mendonça, R.C., Fagg, C.W., Filgueiras, T.S. and Mecenas, V.V. (2008) Fitofisionomias e flora. In: Fonseca, O.F., Ed., *Águas Emendadas*, Secretaria de Desenvolvimento Urbano e Meio Ambiente, Brasília, 152-162.

[14] Santos, F.F.M. and Munhoz, C.B.R. (2012) Diversidade de espécies herbáceo-arbustivas e zonação florística em uma vereda no Distrito Federal. *Heringeriana*, **6**, 21-27.

[15] Veldman, J.W. and Putz, F.E. (2011) Grass-Dominated Vegetation, Not Species-Diverse Natural Savanna, Replaces Degraded Tropical Forests on the Southern Edge of the Amazon Basin. *Biological Conservation*, **144**, 1419-1429. http://dx.doi.org/10.1016/j.biocon.2011.01.011

[16] Brown, J.S. and Venable, D.L. (1986) Evolutionary Ecology of Seed-Bank Annuals in Temporally Varying Environments. *The American Naturalist*, **127**, 131-147. http://dx.doi.org/10.1086/284465

[17] Hayward, J., Horton, T.R., Pauchard, A. and Nuñez, M.A. (2015) A Single Ectomycorrhizal Fungal Species Can Enable a *Pinus* Invasion. *Ecology*, **96**, 1438-1444. http://dx.doi.org/10.1890/14-1100.1

[18] Coelho, V.C.M. (2010) Avaliação do manejo da produção econômica de madeira de *Pinus taeda* L. com características qualitativas superiores. Unpublished Master's Thesis, Universida de Federal do Paraná, Curitiba.

[19] Espíndola, M.B., Bechara, F.C., Bazzo, M.S. and Reis, A. (2005) Recuperação ambiental e contaminação biológica: Aspectos ecológicos e legais. *Biotemas*, **18**, 27-38.

[20] Simberloff, D. (2010) Invasions of Plant Communities-More of the Same, Something Very Different, or Both? *The American Midland Naturalist*, **163**, 220-233. http://dx.doi.org/10.1674/0003-0031-163.1.220

[21] Simberloff, D. (2011) How Common Are Invasion-Induced Ecosystem Impacts? *Biological Invasions*, **13**, 1255-1268. http://dx.doi.org/10.1007/s10530-011-9956-3

[22] Witkowski, E.T.F. (1991) Effects of Invasive Alien Acacias on Nutrient Cycling in the Coastal Lowlands of the Cape Fynbos. *Journal of Applied Ecology*, **28**, 1-15. http://dx.doi.org/10.2307/2404109

[23] Rossi, R.D., Figueira, J.E.C. and Martins, C.R.M. (2010) Capim-gordura, invasão biológica, conservação do cerrado e regime de fogo, Minas Gerais, Brasil. *MG Biota*, **3**, 4-27.

[24] Abreu, R.C.R., de Miranda Santos, F.F. and Durigan, G. (2014) Changes in Plant Community of Seasonally Semi-Deciduous Forest after Invasion by *Schizolobium parahyba* at Southeastern Brazil. *Acta Oecologica*, **54**, 57-64. http://dx.doi.org/10.1016/j.actao.2013.03.013

[25] Vila, M. and Weiner, J. (2004) Are Invasive Plant Species Better Competitors than Native Plant Species? Evidence from Pair-Wise Experiments. *Oikos*, **105**, 229-238. http://dx.doi.org/10.1111/j.0030-1299.2004.12682.x

[26] Dangremond, E.M., Pardini, E.A. and Knight, T.M. (2010) Apparent Competition with an Invasive Plant Hastens the Extinction of an Endangered Lupine. *Ecology*, **91**, 2261-2271. http://dx.doi.org/10.1890/09-0418.1PMid:20836448

[27] Kuebbing, S.E., Nuñez, M.A. and Simberloff, D. (2013) Current Mismatch between Research and Conservation Efforts: The Need to Study Co-Occurring Invasive Plant Species. *Biological Conservation*, **160**, 121-129. http://dx.doi.org/10.1016/j.biocon.2013.01.009

[28] Shono, K., Davies, S.J. and Kheng, C.Y. (2006) Regeneration of Native Plant Species in Restored Forests on Degraded Lands in Singapore. *Forest Ecology and Management*, **237**, 574-582. http://dx.doi.org/10.1016/j.foreco.2006.10.003

[29] Ficko, S., Smith, B. and Zeeb, B. (2015) Assisted Revegetation Following Contaminated Site Remediation in the Arctic: Four-Year Case Study of a Former Radar Site. *American Journal of Plant Sciences*, **6**, 1301-1312. http://dx.doi.org/10.4236/ajps.2015.68130

[30] Duratex (2012) Plano de manejo florestal: Versão pública. http://www.duratex.com.br/pt/Download/Plano_de_Manejo_Florestal_Versao_Publica_2012.pdf

[31] Alvares, C.A., Stape, J.L., Sentelhas, P.C., Moraes-Gonçalves, J.L. and Sparevek, G. (2013) Köppen's Climate Classification Map for Brazil. *Meteorologische Zeitschrift*, **22**, 711-728. http://dx.doi.org/10.1127/0941-2948/2013/0507

[32] Pinho Jr., G.V., Nascimento, A.R.T., Valverde, B.T. and Clemente, L.H. (2015) Brazilian Savanna Re-Establishment in a Monoculture Forest: Diversity and Environmental Relations of Native Regenerating Understory in *Pinus caribaea* Morelet. Stands. *Journal of Forestry Research*, **26**, 571-579. http://dx.doi.org/10.1007/s11676-015-0050-z

[33] Felfili, J.M., Carvalho, F.A. and Haidar, R.F. (2005) Manual para monitoramento de parcelas permanentes nos biomas Cerrado e Pantanal. Universidade de Brasília, Brasília.

[34] Chase, M.W. and Reveal, J.L. (2009) A Phylogenetic Classification of the Land Plants to Accompany APG III. *Botan-*

ical Journal of the Linnean Society, **161**, 122-127. http://dx.doi.org/10.1111/j.1095-8339.2009.01002.x

[35] Canfield, R. (1941) Application of Line Interception Method in Sampling Range Vegetation. *Journal of Forestry*, **39**, 388-394.

[36] Bullock, J. (1996) Plants. In: Sutherland, W.J., Ed., *Ecological Census Techniques—A Handbook*, Cambridge University Press, Cambridge, 111-138.

[37] Hair, J.F., Black, W.C., Babin, B.J. and Anderson, R.E. (2005) Análise Multivariada de dados. 5th Edition, Bookman, Porto Alegre.

[38] R Development Core Team (2015) R: A Language and Environment for Statistical Computing. http://www.r-project.org/

[39] Brandão, M. (2000) Cerrado. In: Mendonça, M.P. and Lins, L.V., Eds., *Lista Vermelha das Espécies Ameaçadas de Extinção da Flora de Minas Gerais*, Fundação Biodiversitas, Belo Horizonte, 55-63.

[40] Mack, R.N., Simberloff, D., Lonsdale, M.W., Evans, H., Clout, M. and Bazzaz, F.A. (2000) Biotic Invasions: Causes, Epidemiology, Global Consequences, and Control. *Ecological Applications*, **10**, 689-710. http://dx.doi.org/10.1890/1051-0761(2000)010[0689:BICEGC]2.0.CO;2

[41] Bechara, F.C., Reis, A., Bourscheid, K., Vieira, N.K. and Trentin, B.E. (2013) Reproductive Biology and Early Establishment of *Pinus elliottii* var. *elliottii* in Brazilian Sandy Coastal Plain Vegetation: Implications for Biological Invasion. *Scientia Agricola*, **70**, 88-92. http://dx.doi.org/10.1590/S0103-90162013000200005

[42] Whitmore, T.C. (1990) An Introduction to Tropical Rain Forests. Clarendon Press, Oxford.

24

Finger Millet: An Alternative Crop for the Southern High Plains

Prasanna H. Gowda[1*], P. V. Vara Prasad[2], Sangamesh V. Angadi[3], Umesh M. Rangappa[3], Pradeep Wagle[4]

[1]USDA-ARS Grazinglands Research Laboratory, El Reno, OK, USA
[2]Department of Agronomy, Kansas State University, Manhattan, KS, USA
[3]Agriculture and Science Center, New Mexico State University, Clovis, NM, USA
[4]Department of Microbiology and Plant Biology, and Center for Spatial Analysis, University of Oklahoma, Norman, OK, USA
Email: [*]Prasanna.Gowda@ars.usda.gov

Abstract

In the Southern High Plains, dairies are expanding to take advantage of favorable climatic conditions. Currently, corn (*Zea mays* L.) and forage sorghum [*Sorghum bicolor* (L.) Moench] are the two major crops grown in the region to meet silage demands for the expanding dairy industry, but they have relatively large water requirements of about 840 and 690 mm, respectively, to achieve desirable results. With rising energy costs and declining water levels in the underlying Ogallala Aquifer, crops that use less water, like finger millet (*Eleusine coracana* (L.) Gaertn) could become alternate forage crops for dairies to corn or forage silage. In this study, we evaluated the adaptability of five finger millet accessions to the Southern High Plains and compared nutritional quality of their forage to that of corn and sorghum. Results indicated that finger millet can be grown in the Southern High Plains. Comparison of nutrient composition has shown that the quality of finger millet is relatively higher than that of corn and sorghum in terms of calcium, potassium, and phosphorus levels in their forage. However, potential forage yield of most commonly grown corn and sorghum in the region is higher than that of finger millet. Therefore, finger millet may provide a unique opportunity to improve the dairy-fed silage quality by mixing it with corn or sorghum silage while meeting the growing regional forage demand. Further field research is needed to measure its water requirements in the Southern High Plains.

Keywords

Ogallala Aquifer Region, Semi-Arid, Forage, Texas Panhandle

[*]Corresponding author.

1. Introduction

Declining water resources is a major concern in the Southern High Plains (New Mexico and West Texas, **Figure 1**), where agriculture is the foundation for economic stability. Crop production in the region is dictated by irrigation water, which mostly comes from the Ogallala Aquifer underlying the Southern High Plains [1]. The Ogallala Aquifer, once considered a never-ending source of irrigation water, is declining rapidly in the region which has left many wells dry and crop production more risky. Although New Mexico (5.4%) and Texas (20.4%) account for 26% of total Ogallala Aquifer area, the amount of water that underlies these two states is very small (New Mexico 1.5% and Texas 12%) of the whole Ogallala Aquifer [2]. Irrigation efficiencies have improved considerably in the region with the adoption of center pivot, and more recently, subsurface drip irrigation systems. However, the availability of groundwater is also have decreased. In addition, many other water demands are competing for the groundwater resource. Therefore, new approaches are needed to improve water conservation and water use efficiency of crops in the region.

Southern High Plains has been one of the fastest dairy growing regions in the United States [3]. Animal density for dairies has doubled between 2002 and 2004 [4] in west Texas. Establishment of three large capacity cheese plants and favorable environment for the dairy industry are expected to accelerate dairy growth in eastern New Mexico and west Texas. In addition, west Texas has the highest concentration of beef cattle on feed in the country [4]. This large concentrated animal feeding operation in the region needs large quantities of good quality forages. Water requirement of cattle is minimal compared to the water required to produce large quantities of

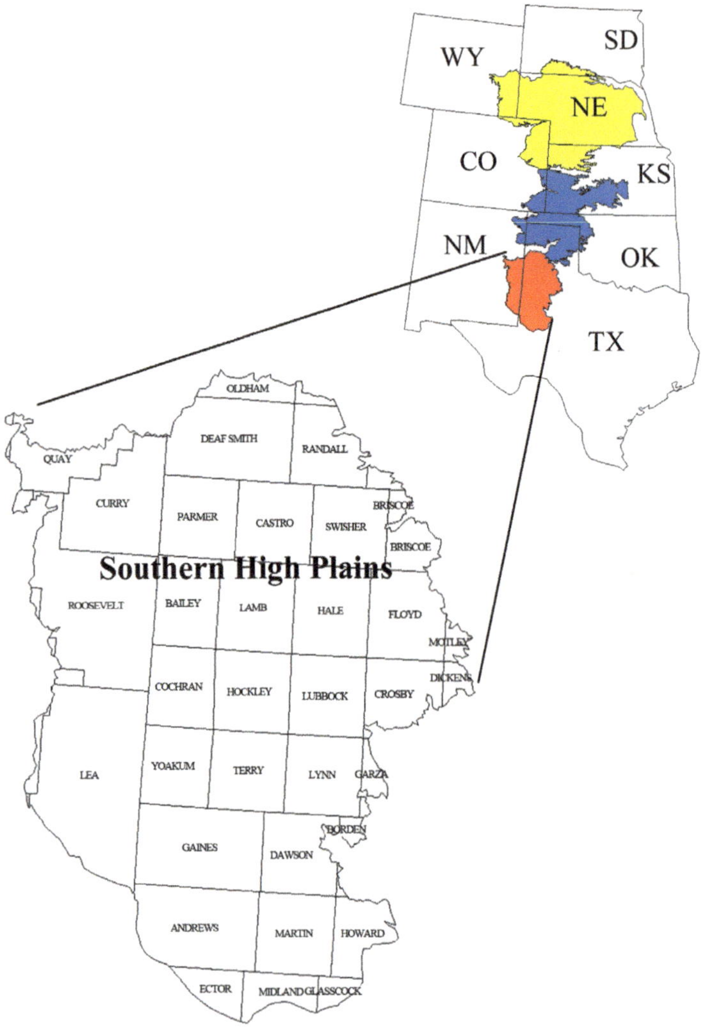

Figure 1. Location of the Southern High Plains in the south-central United States.

good quality silage needed by the cattle. Therefore, to sustain the fast growing animal feeding operations in the region, new crops and cropping systems are needed to maintain quality forage production with less water use.

Currently, corn (*Zea mays* L.) and forage sorghum [*Sorghum bicolor* (L.) Moench] are the two major crops grown in the region to meet the current silage demand by the dairy industry. Corn silage is traditionally used to feed lactating dairy cows which require energy rich feed for maximum milk production. However, corn production requires a large quantity of water (~840 mm), and reducing water resources in the region makes it difficult to sustain corn production. Sorghum is a more drought tolerant crop and has good potential as a silage crop with a reduced water requirement (~690 mm). However, nutritional quality concerns have reduced its acceptability by the industry. Finger millet (*Eleusine coracana* (L.) Gaertn), known to be highly tolerant to pests, diseases, and environmental stresses including drought, is currently grown in arid and semi-arid regions of the world.

Finger millet is a forage grass is also known as Koracan (Nigeria), Ragi (India), Bulo (Uganda), Wimbi (Swahili), and Telebun (Sudan). It is cultivated on more than 36 Mha globally [5] and 2 Mha of that in India in a region receiving rainfall less than or comparable to the Southern High Plains [6]. Although finger millet straw is used for feeding livestock in many Asian and African countries [7], it is mainly grown for food grain for human consumption [8] and also for brewing alcoholic/non-alcoholic beverages [9]. Finger millet exhibits a wide range of adaptability including thriving at higher elevations or higher salinity conditions compared with most other tropical cereals [10]. It is reported to use 10% - 20% less water than sorghum under irrigated conditions in arid conditions of southern India [11]. Forage yield of finger millet under irrigated conditions are nearly double compared to those in rainfed conditions and varies from 4.4 - 8.8 $Mg \cdot ha^{-1}$ under irrigation [10]. Under high input-intensive management, finger millet yields can exceed 15 $Mg \cdot ha^{-1}$, which is reported in the Punjab state of India where finger millet is grown for silage under irrigation and three cuttings are taken annually [12].

Finger millet grain tastes good and is nutritionally rich as it contains high levels of protein and minerals such as calcium, iron, and manganese [8]. It also contains appreciable amount of methionine, an essential amino acid [13] lacking in most food grains. Most of the past research concentrated mainly on nutritional properties of the grain and has been reported mainly from India. However, climatic suitability of finger millet in the United States and nutrient quality of its forage are largely unknown. In this study, we evaluated the climatic adaptability of five finger millet accessions and compared nutritional quality of their forage to that of corn and sorghum in the Southern High Plains.

2. Materials and Methods

This study was conducted during the summer growing season of 2006 in the USDA-ARS Conservation and Production Research Laboratory (CPRL) located at Bushland, Texas [35°11'N, 102°06'W] at 1170 m above mean sea level. Soils in and around Bushland are classified as slowly permeable Pullman clay loam (fine, mixed, superactive, thermic Torrertic Paleustoll) in texture. Average annual precipitation in the region is about 562 mm, while about 835 mm of water is needed to grow grain corn (330 mm growing season rainfall), and 688 mm for grain sorghum (248 mm growing season rainfall) [14].

In this experimental study, five accessions of finger millet were evaluated. Selection of the accessions was based on their ability to produce forage yield and days required to maturity to avoid winter damage. **Table 1** presents plant characteristics of five different finger millet accessions used in this study. Ten thousand seeds for each accession were received from the USDA-ARS Plant Genetic Resources Conservation Unit, Griffin, Georgia. Four accessions (PI271166, PI271167, PI462383, and PI462414) originated from India and the PI307693 originated from South Africa.

In the field experiment, about 5000 seeds for each accession were broadcasted on May 8, 2006 in a 1 × 1 m plot and the 27-days old seedlings were transplanted in clumps on June 5, 2008. Each clump had 4 - 6 seedlings planted at 23 cm (9") apart on 76 cm (30") wide rows. All four were flood irrigated immediately after planting and repeated whenever needed. After 30 days of planting, nitrogen fertilizer was applied at the rate of 125 kg N ha^{-1}. After 120 days of planting, 10 plants were selected randomly from each accession and ground for assessing the forage quality. In addition, silage corn and silage sorghum samples were selected from neighboring plots grown as part of another study. Samples were sent to Dairy One Inc.[1] to analyze for crude protein, neutral detergent fiber (NDF), non-fibrous carbohydrates (NFC), total digestible nutrients (TDN), *in-vitro* true digestibility (IVTD), and *in-vitro* NDF digestibility. Same nutritional qualities were also measured for corn and sorghum fo-

[1]Mention of trade or commercial products in this article is solely for the purpose of providing specific information and does not imply recommendation or endorsement by the US Department of Agriculture.

rages, and were compared and contrasted with that of finger millet accessions.

3. Results and Discussion

For selected five finger millet accessions, number of days from planting to maturity varied from 112 - 122 (**Table 1**). South African accession (PI307693) failed to germinate for unknown reasons. However, excellent germination (more than 90%) was observed with all four Indian accessions. Considering the age of the seeds (see in **Table 1**), it was a great success. This is consistent with observations reported in the literature that finger millet seeds can be stored for a very long time without any treatments. Although the number of days required from planting to maturity was less than or equal to 122 days, none of the four accession senesced within that period. Only one accession (PI462414) senesced before freeze during late October, but exceeded more than 165 days. This unhindered growth with all four accessions may be due to availability of abundant solar radiation on the Southern High Plains combined with managed irrigation and adequate nutrient supply. The remaining three accessions were killed by cold temperatures before senescence. The plant height was consistent with the values reported in the Germplasm Resources Information Network (GRIN) database. Further, the number of tillers was more than 6-8 per plant with PI271166 and PI271167 accessions and 3 - 6 tillers per plant with PI462383 and PI462414 accessions. These results matched with the GRIN classification on tillering capacity for PI271167, PI462383, and PI462414 accessions. However, the observed tillering capacity of PI271166 accession did not match with the GRIN classification (**Table 1**).

Table 2 presents nutritional properties of four finger millet accessions, sorghum, and corn forages grown in the USDA-ARS CPRL, Bushland, Texas. Crude protein levels for finger millet accession PI271167 was significantly higher (14.6%) than that of the remaining three finger millet accessions, corn, and sorghum. Corn silage had the least crude protein levels (7.4%). NDF and TDN were comparable in all three crops. However, the NFC for finger millet accessions (20.1% - 23.5%) was significantly greater than that for sorghum (15.7%) and comparable to corn (21.4%). IVTD and NDFD at 30 hours for all four finger millet forages were equal or better than that for forage sorghum and corn.

Calcium content in finger millet accessions (1.09% - 1.29%) was found to be at least two times greater than that for the corn and sorghum silage (**Table 2**). In case of phosphorus levels, finger millet had 4 to 5 times greater levels than that in corn (0.09%) and 90% - 100% more than that in forage sorghum (0.25%). Calcium and phosphorus are important minerals when formulating diets for dairy animals. Calcium and phosphorus deficiencies are frequently associated with reduction in appetite, retarded growth, decreased efficiency in feed utilization, decreased milk production, and impaired reproduction [15]. Further, greater magnesium content in finger

Table 1. Characterization of five different finger millet accessions.

Trait	Characterization*				
GRIN-ID	PI271166	PI271167	PI307693	PI462383	PI462414
Source	India	India	South Africa	India	India
Maturity (days)**	122	122	120	122	112
Plant height (m)	1.0	1.1	1.1	0.7	0.8
Plant spread (m)	1.0	1.0	1.1	0.6	0.5
Stem size	Large	Large	Large	Medium	Medium
Tillering capacity	Medium	Abundant	Abundant	Medium	Medium
Foliage amount	Abundant	Abundant	Abundant	Medium	Medium
Seed production	Abundant	Medium	Abundant	Abundant	Abundant
Resistance to winter damage	Complete kill	Complete kill	Complete kill	Complete kill	Complete kill
Received year	1961	1961	1965	1981	1981

*GRIN. USDA, ARS, National Genetic Resources Program. *Germplasm Resources Information Network—(GRIN)*. [Online Database] National Germplasm Resources Laboratory, Beltsville, Maryland. Available: http://www.ars-grin.gov/cgi-bin/npgs/html/desc.pl?102007 (29 September 2008).
**Days from planting to first flower.

Table 2. Nutritional properties of four finger millets, forage sorghum, and corn at the USDA-ARS Conservation and Production Research Laboratory, Bushland, Texas.

Parameter*	Finger millet				Forage sorghum	Corn
GRIN ID/plant name	PI 271166	PI 271167	PI 462383	PI462414		
Crude protein (%)	10.8	14.6	9.7	7.9	9.2	7.4
NDF (%)	62.9	58.8	61.6	61.2	66.8	65.8
NFC (%)	20.1	21.1	21.9	23.5	15.7	21.4
TDN (%)	61	62	60	56	58	59
Calcium (%)	1.15	1.29	1.09	1.26	0.59	0.54
Phosphorus (%)	0.40	0.50	0.46	0.40	0.25	0.09
Magnesium (%)	0.31	0.34	0.28	0.30	0.31	0.15
Potassium (%)	4.26	5.08	4.93	3.86	2.34	2.88
Nitrates (%)	0.68	1.01	0.16	0.32	0.14	0.54
Nitrate-nitrogen (PPM)	1256	2270	368	724	319	1224
IVTD 30 hr (% of DM)	69	70	69	66	66	59
NDFD 30 hr (% of NDF)	51	48	50	44	49	37

*NDF: Neutral detergent fiber, NFC: Non fibrous carbohydrates, IVTD: *In vitro* true digestibility, TDN: Total digestible nutrients, and NDFD: *In vitro* NDF digestibility.

millet forages (0.28% - 0.31%) was observed than that in corn (0.15%) and comparable to forage sorghum (0.31%). Deficiencies in magnesium may be related to infertility, anemia or suppressed immune function in dairy cows [15]. Overall, higher levels of calcium, phosphorus, and magnesium in addition to comparable levels of remaining dietary requirements and digestibility indices, indicates that finger millet forage is as an attractive alternative to sorghum or corn forage if not a feed-additive to improve the quality of forage for dairy animals.

4. Conclusion

Comparison of nutritional properties indicated that quality of finger millet forage as feed to dairy livestock was greater than that of both corn and forage sorghum in terms of calcium, potassium, and phosphorus levels, while other nutritional levels were comparable. However, potential forage yield of corn and forage sorghum grown in the region is greater than that of finger millet. Therefore, finger millet may provide a unique opportunity to improve dairy-fed forage quality by mixing it with corn or sorghum while meeting the growing regional forage demand. The complimentary nutritional characteristic makes finger millet a potential candidate for intercropping with corn or sorghum to realize improved water use efficiency while increasing the quality of forage-based diets to dairy cows. However, further research is required to measure water requirements and forage yield of finger millet in the Southern High Plains of the Ogallala Aquifer region.

References

[1] Maupin, M.A. and Barber, N. (2000) Estimated Withdrawals from Principal Aquifers in the United States. US Geological Survey Circular, 1279.

[2] Guru, M. and Horne, J. (2000) The Ogallala Aquifer. The Kerr Center for Sustainable Agriculture, Inc., Poteau, 32.

[3] Hadjigeorgalis, E. (2005) The US Dairy Industry and International Trade in Dairy Products. Department of Agricultural Economics and Agricultural Business, New Mexico State University, Las Cruces.

[4] Texas Agricultural Statistics (2004) 2004 Texas Agricultural Statistics—Texas Agriculture by the Numbers. Texas Field Office, National Agricultural Statistics Service, Austin, 150 p.

[5] Food and Agriculture Organization of the United Nations (FAO) (2003).

[6] Covane, S.T., Jayaraman, N. and Senthil, N. (1995) Heterosis Study in Ragi [*Eleusine corocana* (L) Gaertn]. *Journal of Phytology Research*, **8**, 53-56.

[7] Sumathi, P., Joel, A. and Muralidharan, V. (2005) Combining Ability and Heterosis for Grain Yield and Its Component Traits in Finger Millet under Irrigated Conditions. International Sorghum and Millets Newsletter.

[8] Upadhyaya, H.D., Gowda, C., Pundir, R., Reddy, V.G. and Singh, S. (2006) Development of Core Subset of Finger Millet Germplasm Using Geographical Origin and Data on 14 Quantitative Traits. *Genetic Resources and Crop Evolution*, **53**, 679-685. http://dx.doi.org/10.1007/s10722-004-3228-3

[9] Fernandez, D.R., Vanderjagt, D.J., Millson, M., Huang, Y.-S., Chuang, L.-T., Pastuszyn, A. and Glew, R.H. (2003) Fatty Acid, Amino Acid and Trace Mineral Composition of *Eleusine coracana* (Pwana) Seeds from Northern Nigeria. *Plant Foods for Human Nutrition*, **58**, 1-10. http://dx.doi.org/10.1023/B:QUAL.0000040323.67339.cb

[10] Duke, J.A. (2005) Handbook of Energy Crops: *Eleusine coracana* (L.) Gaertn, Hort. purdue. edu [Consulté le 5 janvier 2005]. www.hort.purdue.edu

[11] Kissan (2006) Irrigation. http://ikisan.com/ap_irrigation.shtml

[12] Bogan, A.V. (1997) Tropical Pasture and Fodder Plants (Grasses and Legumes). Longman, London and New York, 111-126.

[13] The National Academies (1996) Chapter 2. Finger Millet. In: *Lost Crops of Africa: Volume* 1: *Series*. Board on Science and Technology for International Development, Office of International Affairs, National Research Council, 408 p.

[14] Gowda, P.H., Baumhardt, R.L., Esparza, A.M., Marek, T.H. and Howell, T.A (2007) Suitability of Cotton as an Alternative Crop in the Ogallala Aquifer Region. *Agronomy Journal*, **99**, 1397-1403. http://dx.doi.org/10.2134/agronj2006.0275

[15] Harris Jr., B. (2003) Vitamins, Minerals and Dairy Cow Fertility. IFAS Extension, University of Florida, Gainesville, DS 47, 3 p.

25

Genetic Diversity Analysis of *Cotoneaster schantungensis* Klotz. Using SRAP Marker

Yan Ma*, Suqing Qu*, Xiurong Xu*, Tingting Liang, Dekui Zang#

College of Forestry, Shandong Agricultural University, Taian, China
Email: #zangdk@sdau.edu.cn

Abstract

In order to detect the genetic relationship of different *Cotoneaster schantungensis* natural populations, 16 samples were selected for genetic diversity analysis based on SRAP markers. Twelve pairs primers were selected out, 93 bands were amplified, of which 91 bands (97.85%) were polymorphic. At species level, the average value of Nei's genetic diversity (H) was 0.2213, Shannon's information index (I) was 0.3596, whereas at population level, they were 0.1955, 0.3121, the percentage of polymorphic loci was 75.27%. The genetic differentiation coefficient in populations (Gst) was 0.0788, gene flow Nm was 5.8467, indicating that there was 7.88% of the variation among populations and 92.12% of the variance within populations and the variation within populations was relatively high. These results indicate that *C. schantungensis* has high genetic diversity and low differentiation among populations. Although most Cotoneaster species are apomictic, *C. schantungensis*, however, may be an exception.

Keywords

***Cotoneaster schantungensis* Klotz., Apomictic, SRAP, Genetic Diversity, Population**

1. Introduction

Cotoneaster schantungensis Klotz. is a unique species of Shandong, the distribution of it is limited in southern mountain area of Jinan, and it is scattered under *Platycladus orientalis* and *Cotinus coggygria* plantation sporadic sporadically, with very few numbers. It is in state of extreme endangerment [1]-[3]. Genetic diversities of endangered plants can be tested effectively with molecular marker technology, thus the threatening mechanism can be grasped and the corresponding protection measures can be formulated [4]-[14]. Sequence-related ampli-

*These authors contribute equally.
#Corresponding author.

fied polymorphism (SRAP) is a kind of new molecular marker technology based on PCR, it is more convenience, high co-dominance, easy for separation of strips and sequencing, and it doesn't need to know sequence information of species in advance, and is richer in information provided in the aspect of morphologic variation and evolutionary history when compared with AFLP, so it has been widely used at present [15]-[17]. As a kind of extremely small population species, the previous research has been studies only on its living environment and community structure and there is no report on its genetic diversity [18]. Our research makes genetic diversity study on 16 samples of *C. schantungensis* with SRAP marker technology, the result shows that the genetic diversity level of intrapopulational of *C. schantungensis* is high, genetic differentiation is large, which provides scientific basis for the formulation of reasonable and efficient protection measures and enhances the protection of endangered species.

2. Materials and Methods

2.1. Plant Materials

C. schantungensis distributes in low altitude mountain area of limestone in southern Jinan, at present, two distributions, "Fo Yu" and "Red Leaves Valley" are found, and there are only about 300 clumps. In these 16 samples of *C. schantungensis* materials are used for experiment, No. 1 - No. 11 are derived from "Fo Yu" (36°35'N 117°06'E, elevation 300 - 350 m), No. 12 - No. 16 are derived from "Red Leaves Valley" (36°28'N 117°09'E, elevation 500 - 600 m). The sampling interval of individuals is more than 30 meters, and normal growing healthy without diseases and insect pests are selected, the young fresh leaves on the top of it are collected to be put in quick-seal bags with allachroic silica gel, the air in bags is squeezed out, which makes it rapid drying.

2.2. Research Methods

2.2.1. Genome DNA Extraction

Total DNA was obtained from young leaves using improved CTAB method [19]. The quality and concentration of DNA was tested with 0.8% agarose gel electrophoresis and ultraviolet spectrophotometer, then diluted it to 50 ng·μL^{-1} and stored it in −20°C for standby application.

2.2.2. SRAP-PCR Amplification Analysis

SRAP-PCRs were run in 25 μl volumes containing 2.5 μl 10× PCR buffer (5.0 mM $MgCl_2$), 30 ng genomic DNA, 0.3 μM of each primer, 7.5 mM dNTPs and 2 Unit of Taq DNA polymerase. PCR amplifications were performed in Bio-Rad PCR amplifier Cycler using the following cycling parameters: 5 min denaturing at 94°C; 5 cycles of 94°C for 1 min; 35°C for 1 min and 72°C for 1 min; 30 cycles of 94°C for 1 min, 50°C for 1 min and 72°C for 1 min; followed by a final extension of 7 min at 72°C, then store at 4°C.

Twelve pairs of primer combinations, selected from the initial 150 pairs of primer combinations, were used for the study (**Table 1**). Amplified products were electrophoresed on 2% agarose gel at 120 V for 2 hour, stained

Table 1. Primer sequences used for SRAP analysis.

Forward primer	Primer combination sequences	
ME4-EM11	F: 5'-TGAGTCCAAACCGGACC-3'	R: 5'-GACTGCGTACGAATTAGC-3'
ME4-EM12	F: 5-TGAGTCCAAACCGGACC-3'	R: 5'-GACTGCGTACGAATTGCC-3'
ME4-EM7	F: 5'-TGAGTCCAAACCGGACC-3'	R: 5'-GACTGCGTACGAATTCAG-3'
ME4-EM9	F: 5'-TGAGTCCAAACCGGACC-3'	R: 5'-GACTGCGTACGAATTCCA-3'
ME5-EM15	F: 5'-TGAGTCCAAACCGGAAG-3'	R: 5'-GACTGCGTACGAATTATG-3'
ME6-EM12	F: 5-TGAGTCCAAACCGGTAA-3'	R: 5'-GACTGCGTACGAATTGCC-3'
ME6-EM13	F: 5'-TGAGTCCAAACCGGTAA-3'	R: 5'-GACTGCGTACGAATTTAG-3'
ME6-EM6	F: 5'-TGAGTCCAAACCGGTAA-3'	R: 5'-GACTGCGTACGAATTGCA-3'
ME6-EM8	F: 5'-TGAGTCCAAACCGGTAA-3'	R: 5'-GACTGCGTACGAATTCCA-3'
ME6-EM9	F: 5'-TGAGTCCAAACCGGTGC-3'	R: 5'-GACTGCGTACGAATTAAC-3'
ME8-EM5	F: 5'-TGAGTCCAAACCGGTGC-3'	R: 5'-GACTGCGTACGAATTCCA-3'

with ethidium bromide (EB) and then visualized and photographed by using gel documentation system (Beijing LiuyiWD-9403CS type UV Apparatus).

2.2.3. Data Statistics and Analysis

SRAP products obtained were manually scored as present (1) or absent (0). The number and percentage of polymorphic loci, effective number of allele Ne, Nei's genetic similarity coefficient among various materials, Nei's genetic diversity index (H) and Shannon information index (I) was then analyzed by using POPGENE 1.32 [20]. Dice similarity coefficient was calculated and the corresponding cluster analysis was performed using the Unweighted Pair-Group Method with Arithmetic Average (UPGMA) [21].

3. Result and Analysis

3.1. SRAP-PCR Amplification

The selected 12 pairs of primers are used to amplify 16 pieces of materials (**Figure 1**), and then the statistics of amplified primer sites is carried on with binary character coding method (**Table 2**). It amplifies 93 sites, among them there are 91 polymorphic loci, the percentage of polymorphism is as high as 97.85%, and each primer amplifies 7.75 loci and 7.58 polymorphic loci averagely, it indicates that there are rich genetic diversities among *C. schantungensis*, and the genetic background is very complex. The number of polymorphic loci amplified by primer combination ME8-EM9 is the most, there are 12, the number of polymorphic loci amplified by primer combination ME8-EM5 is the least, and there are 3. In the 12 pairs of primers, the percentage of polymorphic loci of primer combination ME6-EM12 is the lowest, it is 75%, the percentage of polymorphic loci of the rest primer combinations are all 100%. The Effective number of alleles is 1.1788 - 1.5849, the average value is 1.3420, and the value of primer combination ME6-EM8 is the largest.

3.2. Genetic Diversity Analysis

Based on the test result of SRAP molecular marker, it shows that there is a big difference in genetic diversity of two populations of *C. schantungensis*. The percentage of polymorphic loci (86.02%), Shannon information index

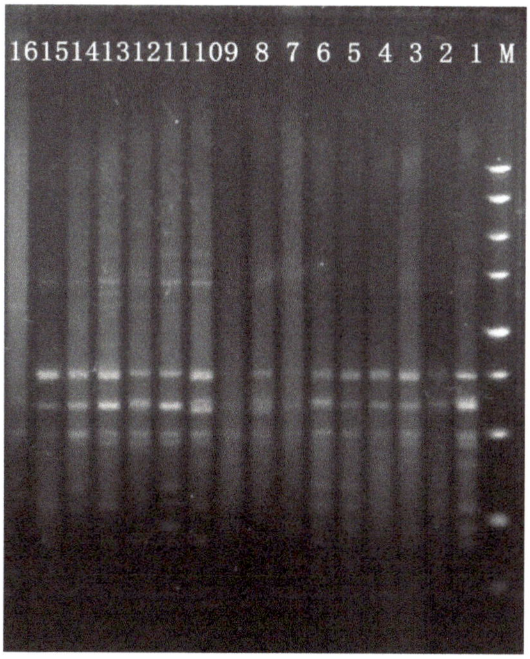

Figure 1. The amplification result of ME8-EM9. No. 1-No. 11: *C. schantungensis* derived from "Fo Yu"; No. 12-No. 16: *C. schantungensis* derived from "Red Leaves Valley" M: DNA Marker 2000.

(0.3550), Nei genetic diversity of the population in "Fo Yu" is much higher than the population in "Red Leaves Valley" (**Table 3**). Because the three parameters could uncover the genetic diversity condition of population effectively, so it indicates that the genetic diversity of population in "Fo Yu" is much higher than population in "Red Leaves Valley". The observation of two parameters of number of alleles and effective number of alleles also proves it.

In the level of species, the number of alleles and effective number of alleles is 1.9785 and 1.3405 respectively; Nei's gene diversity index H is 0.2213, Shannon polymorphic information index I is 0.3596. it indicates that the genetic diversity of *C. schantungensis* presents a high level in the level of population and species.

Table 2. Sequences of 12 pairs of SRAP primers and amplification results.

Primer combination	Number of polymorphic loci (NPL)	Percentage of polymorphic loci (PPL) (%)	Effective number of alleles (Ne)	Nei's gene diversity (H)	Shannon's information index (I)
ME8-EM5	3	100.00	1.1931	0.1585	0.2917
ME8-EM9	12	100.00	1.4581	0.2795	0.4378
ME4-EM11	8	100.00	1.1767	0.1472	0.2746
ME4-EM12	8	100.00	1.3684	0.2624	0.4276
ME4-EM9	8	100.00	1.4909	0.2944	0.4485
ME4-EM7	4	100.00	1.4950	0.3013	0.4603
ME5-EM15	11	100.00	1.2546	0.1643	0.2791
ME6-EM6	9	100.00	1.3521	0.2364	0.3848
ME6-EM8	6	100.00	1.5849	0.3380	0.5037
ME6-EM12	6	75.00	1.3277	0.2061	0.3224
ME6-EM9	10	100.00	1.2240	0.1565	0.2741
ME6-EM13	6	100.00	1.1788	0.1382	0.2528

Table 3. The genetic diversity among different populations of *C. schantungensis* based on SRAP analysis.

Populations	Percentage of polymorphic loci (PPL) (%)	Observed number of allele (Na)	Effective number of alleles (Ne)	Nei's gene diversity (H)	Shannon's information index (I)
Foyu populations	86.02	1.8602	1.3702	0.2260	0.3550
Hongyegu populations	64.52	1.6452	1.2405	0.1650	0.2691
Mean	75.27	1.7527	1.3054	0.1955	0.3121
Species level	97.85	1.9785	1.3405	0.2213	0.3596

3.3. Genetic Differentiation of Population of *C. schantungensis*

Gene differentiation analysis among populations shows that the Genetic differentiation coefficient Gst of two populations of *C. schantungensis* is 0.0788, because the Gst overall level of perennial plant, hybridization plant, insect-pollinated plant is 0.19, 0.22, 0.264 respectively [22]-[24], so the Gst of *C. schantungensis* is obvious low.

The total genetic diversity of population Ht is 0.2123, and most heritable variations exist in the population, and the genetic diversity in population Hs is 0.1955, intra-population occupies 92.1%, among populations only occupies 7.88%. The gene flow among populations estimated by Gst is 5.8467, it shows that there are frequent gene flows between the two populations.

At the same time, the UPGMA with Nei method also shows that the genetic identity of the two populations is 0.9766, the genetic distance is only 0.0236, the genetic similarity among populations is very high, and it is the same as the result that the heritable variation among populations is small.

3.4. The Genetic Distance and Cluster Analysis among Individuals and Populations

Based on genetic similarity coefficient, SRAP marker genetic diversity molecular dendrogram (**Figure 2**) of tested materials is drawn with UPGMA method.

The result shows that the similarity coefficients of all materials are between 0.47 - 0.92, the average is 0.68, and it proves that the genetic relationship of tested materials is very close [25]. It can be seen from dendrogram that 16 pieces of materials can be divided into three classes at the similarity coefficients of 0.38. Individuals among populations have obvious overlaps phenomenon, and they are not classified with geographic features, three materials (15, 16, 12) of "Red Leaves Valley" and all materials of "Fo Yu" are crossed gathered as the first and second class, and No. 13 and No. 14 of "Red Leaves Valley" are solely gathered as the third class. It shows that there is no obvious correlation between genetic distance and geographic space, and it is also not obvious related to distance of two populations and genetic differentiation among populations. At the same time, the genetic distance among individuals of *C. schantungensis* in populations of "Fo Yu" is close, while the distance is far in populations of "Red Leaves Valley", it is speculate that populations in "Fo Yu" are formed by the spread distribution of populations in "Red Leaves Valley".

4. Discussion

The percentage of polymorphic loci is one of the important indexes that reflect variation level in populations [26]. At present, there is no study report of SRAP of Cotoneaster. In species levels, the genetic diversity of *C. schantungensis* is 97.85%, which is higher than *Amygdalus ledebouriana* (P = 92.1%), plants in the same family, but Nei's genetic diversity index (H = 0.2213) and Shannon information index (I = 0.3596) are a little lower than those of *A. ledebouriana* (H = 0.2809, I = 0.4327 [27]. So, although *C. schantungensis* is a unique species, it still has a high basis of genetic diversity. Previous studies also show that narrow distributed species could have higher genetic diversity [28] [29].

Gene flow is very important for evolution of plants, and it has negative correlations with population genetic differentiation [30]. For spermatophyte, the gene flow among populations is mainly produced by foreign genes brought by pollens and seeds [31]. Gene flow among population Nm is less than 1.0, various populations could differentiate intensively, when it is between 1.0 and 4.0, gene flow could give play to its homogenizing function to prevent differentiation among each population caused by genetic drift, when Nm is bigger than 4.0, the population is a random unit [32]. The Nm of *C. schantungensis* is 5.8765, it shows that gene exchange among populations is very frequent, which weakens genetic differentiation among populations, and the genetic distance between two populations is only 0.0236, the genetic identity is high, it also highly inosculates with the value of

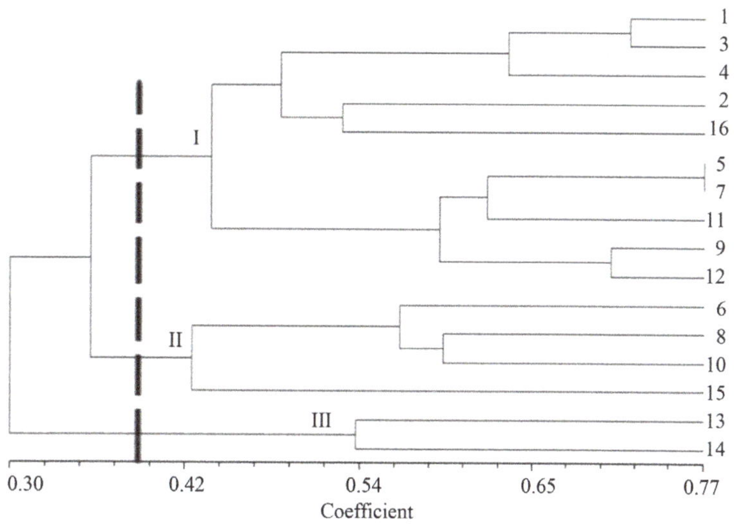

Figure 2. UPGMA cluster analysis of genetic similarity of 16 pieces of tested materials No. 1-No. 11: *C. schantungensis* derived from "Fo Yu"; No. 12-No. 16: *C. schantungensis* derived from "Red Leaves Valley".

gene flow among populations. Although most Cotoneaster are apomixes [33] [34], the research shows that *C. schantungensis* might be an exception.

Acknowledgements

This work was funded by the National Science Foundation of China (NSFC) (30972406) and Shandong agricultural seeds engineering major issue (No. 20106).

References

[1] China Flora Editorial Board of CAS (1974) Flora Reipublicae Popularis Sinicae, Vol. 36. Science Press, Beijing, 149.

[2] Wu, Z.Y., Raven, P.H. and Hong, D.Y., Eds. (2003) Flora of China, Vol. 9. Science Press & St. Louis: Missouri Botanical Garden Press, Beijing, 99.

[3] Li, F.Z. (2004) Shandong Plant Essence. Science Press, Beijing, 296.

[4] Cole, C.T. and Kuchengeuther, M.A. (2001) Molecular Markers Reveal Little Genetic Differentiation among *Aconitum noveboracense* and *A. columbianum* (Ranunculaceae) Populations. *American Journal of Botany*, **88**, 337-347. http://dx.doi.org/10.2307/2657023

[5] Ledig, F.T. and Conkle, M.T. (1983) Gene Diversity and Genetic Structure in a Narrow Endemic, Torrey Pine (*Pinus-torreyana* Parry ex Carrière). *Evolution*, **37**, 79-86. http://dx.doi.org/10.2307/2408176

[6] Lemes, M.R., Gribel, R., Proctor, J. and Grattapaglia, D. (2003) Population Genetic Structure of Mahogany (*Swietenia macrophylla* King, Meliaceae) across the Brazilian Amazon, Based on Variation at Microsatellite Loci: Implications for Conservation. *Molecular Ecology*, **12**, 2875-2883. http://dx.doi.org/10.1046/j.1365-294X.2003.01950.x

[7] Torres, E., Iriondo, J.M. and Pérez, C. (2003) Genetic Structure of a Endangered Plant, *Antirrhinum microphyllum* (Scrophulariaceae): Allozyme and RAPD Analysis. *American Journal of Botany*, **90**, 85-92. http://dx.doi.org/10.3732/ajb.90.1.85

[8] Ge, S. and Hong, D.Y. (1999) Studies of Morphological and Allozyme Variation of the Endangered *Adenophoralobophylla* and Its Widespread Congener *A. potaninii. Journal of Genetics & Genomics*, **26**, 410-417.

[9] Ge, S., Wang, H.Q., Zhang, C.M., *et al.* (1997) Genetic Diversity and Population Differentiation of *Cathaya argyrophylla* in Bamian Mountain. *Act Botanic Sinica*, **39**, 266-271.

[10] Han, C.Y., Sun, W.B. and Gao, L.M. (2004) Analysis of Genetic Diversity of *Trigonobalanus doichangensis* (Fagaceae) by RAPD. *Acta BotanicaYunnanica*, **26**, 513-518.

[11] Ji, Z.X. and Li, J.M. (2007) ISSR Analysis on Genetic Diversity of Endangered Relic Shrub *Sinocanthus chinensis. Chinese Journal of Applied Ecology*, **18**, 247-253.

[12] Su, H.L. and Tang, S.Q. (2004) Genetic Diversity of the Endangered Plant *Abiesziyuanensis* in Two Populations. 24, 414-417.

[13] Xiao, Y.A., He, P., Deng, H.P., *et al.* (2003) Study on Population Genetic Diversity and Genetic Differentiation of *Disanthus Cercidifolius* Maxim var. *Longipes* H.T. Chang in Jinggangshan. *Journal of Southwest China Normal University* (*Natural Science*), **28**, 444-449.

[14] Xie, G.W., Peng, X.Y., Zheng, Y.L., *et al.* (2007) Genetic Diversity of Endangered Plant *Monimopetalumchinense* in China Detected by ISSR Analysis. *Scientia Silvae Sinicae*, **43**, 48-53.

[15] Li, G. and Quiros, C.F. (2001) Sequence-Related Amplified Polymorphism (SRAP), a New Marker System Based on a Simple PCR Reaction: Its Application to Mapping and Gene Tagging in Brassica. *Theoretical and Applied Genetics*, **103**, 455-461. http://dx.doi.org/10.1007/s001220100570

[16] Ferriol, M., Pico, B., de Cordova, P.F. and Nuez, F. (2004) Molecular Diversity of a Germ Plasm Collection of Squash (*Cucurbita moschata*) Determined by SRAP and AFLP Markers. *Crop Science*, **44**, 653-664. http://dx.doi.org/10.2135/cropsci2004.6530

[17] Ferriol, M., Pico, B. and Nuez, F. (2003) Genetic Diversity of a Germplasm Collection of *Cucurbita pepo* Using SRAP and AFLP Markers. *Theoretical and Applied Genetics*, **107**, 271-282. http://dx.doi.org/10.1007/s00122-003-1242-z

[18] Qu, S.Q., Liu, D., Xie, X.M., *et al.* (2012) Study on Composition and Structure Characteristics of *Cotoneaster schantungensis* Community. *Journal of Anhui Agriculture Science*, **40**, 1426- 1430

[19] Kidwell, K.K. and Osborn, T.C. (1992) Simple Plant DNA Isolation Procedures. In: Beckman, J. and Osborn, T.C., Eds., *Plant Genomes*: *Methods for Genetic and Physical Mapping*, Kluwer Academic Publishers, Dordrecht, 1-13. http://dx.doi.org/10.1007/978-94-011-2442-3_1

[20] Qu, S.Q. (2013) Evaluation and Investigation of the Germplasm of the Precious Species *Cotoneaster schantungensis* Klotz. and *Rosa rugosa* Thunb. Shandong Agricultural University, Taian.

[21] Xu, K.X. (1994) Numerical Taxonomy. Science Press, Beijing.

[22] Nybom, H. (2004) Comparison of Different Nuclear DNA Markers for Estimating Intraspecific Genetic Diversity in Plants. *Molecular Ecology*, **13**, 1143-1155. http://dx.doi.org/10.1111/j.1365-294X.2004.02141.x

[23] Zhang, D.Q. and Yang, Y.P. (2008) A Statistical and Comparative Analysis of Genetic Diversity Detected by Different Molecular Markers. *Acta Botanica Yunnanica*, **30**, 159-167.

[24] Nei, M. (1978) Estimation of Average Heterozygosity and Genetic Distance from a Small Number of Individuals. *Genetics*, **89**, 583-590.

[25] Zhang, D.Y. (2003) Plant Life-History Evolution and Reproductive Ecology. Science Press, Beijing, 208-212.

[26] Liu, J.F. and Xiao, W.F. (2008) RAPD Analysis of the Genetic Diversity of a Critically Endangered Plant, *Thuja* Sutchuensis (Cupressaceae). *Acta Agriculture University Jiangxiensis*, **30**, 68-72.

[27] Lü, Z.J., Li, J., Omir, S.T., *et al.* (2010) ISSR Analysis for Genetic Diversity of *Amygdalus ledebouriana* Germplasm from Xinjiang, China. *Journal of Fruit Science*, **27**, 918-923.

[28] Martinez-Palacios, A., Eguiarte, L.E. and Fornier, G.R. (1999) Genetic Diversity of the Endangered Endemic *Agave victoriae-reginae* (Agavaceae) in the Chihuahuan Desert. *American Journal of Botany*, **86**, 1093-1098. http://dx.doi.org/10.2307/2656971

[29] Dolan, R.W., Yahr, R., Menges, E.S. and Halfhill, M.D. (1999) Conservation Implications of Genetic Variation in Three Rare Species Endemic to Florida Rosemary Scrub. *American Journal of Botany*, **86**, 1556-1562. http://dx.doi.org/10.2307/2656793

[30] Grant, V. (1991) The Evolutionary Process: A Critical Study of Evolutionary Theory. Columbia University Press, New York.

[31] Hamrick, J.L. (1987) Gene Flow Distributing of Genetic Variation in Plant Populations. In: Urbanska, K., Ed., *Differentiation Patterns in Higher Plants*, Academy Press, New York, 53-67.

[32] Wright, S. (1931) Evolution of Mendelian Populations. *Genetics*, **16**, 91-176.

[33] Fryer, J. and Hylmo, B. (2001) Captivating Cotoneaster. New Plantsman, **8**, 227-238.

[34] Wu, Z.-Y., Lu, A.-M., *et al.* (2003) The Families and Genera of Angiosperms in China. Science Press, Beijing, 647.

26

Lodging-Resistance Breeding of *Platycodon grandiflorus* Using Distant Hybridization

Yan Ma*, Yan Lin, Huali Zhang, Fang Liang, Liping Sun, Maoliang Wang, Haibo Xin

Beijing Institute of Landscape and Architecture, Beijing Key Laboratory of Green Plants Breeding, Beijing, China
Email: *mayanrong60@163.com

Abstract

The distant hybridization was used in lodging-resistance breeding of *Platycodon grandiflorus.* The parents were *Platycodon grandiflorus* (♀) and *Campanula medium* (♂). 187 seeds of F_1 were harvested by using the inter-generic hybridization in 2009 and 2010, 2 repeat, and 88 normal plants were obtained. The characteristics of leaves, stems and branches were the same as female for F_1 generation, F_2 generation, backcross generation and S_1 generation, and there were some different characteristics with female which were flower colors, branching habit and plant height. A lodging-resistance plant was selected in F_1, which was about 75 cm high, deep pink flower color and developed lateral branches. The DNA groups of each generation were separated by 1% agarose gel electrophoresis and there were not significant differences. Among the each generation many flower colors and forms were obtained.

Keywords

Lodging-Resistance Breeding, *Platycodon grandiflorus*, *Campanula medium*, Distant Hybridization

1. Introduction

Platycodon grandiflorum is a perennial flower, belonging to the family of *Campanulaceae* and genus of *Platycodon*. Chromosome base 2n = 18. *Platycodon grandiflorum* is about 80 - 120 cm high, leaves are alternate or 3 verticillate; flowers are solitary or several flowers in racemes; flower corolla is campanulate form with 5-lobed. Flowers' colors are usually blue, white and light pink. Flowering period is from June to October in each year. *Platycodon grandiflorum* was introduced in Beijing from Tianjin Institute of Landscape and Gardening in 2004. For a perennial flower it has a lot of excellent characters, such as: a long flowering period, more number of

*Corresponding author.

flowers, drought tolerance, cold tolerance, barren tolerance, little pests and diseases, and easy cultivation. So it could be used in the garden, sometimes as cut flowers or potted flowers. But it is easy lodging which has limited the application in landscape. In this research the breeding objective is lodging-resistance.

Campanula medium is biennial flowers, belonging to the family of *Campanulaceae* and genus of *Campanula*. Chromosome base 2n = 34. *Campanula medium* is about 70 - 80 cm high, raceme in apex, campanulate corolla. Flowers' colors are usually blue, white and pink. *Campanula medium* was introduced from Dongsheng Seed Co. Ltd. in 2008.

The bud mutations have been found in *Platycodon grandiflorum* which were the double-color flowers, and the petal numbers more or less than the normal, the double petal flower, and etc. Those characteristic might be induced to generate new variations. The forms of stigma, style, anther and filament of *Platycodon grandiflorum* and *Campanula medium* are similar, and the breeding goal is lodging-resistance of *Platycodon grandiflorum*, so *Platycodon grandiflorum* is chosen as female parent and *Campanula medium* as male parent. In cross breeding, the influence of female parent on hybrid offspring is often stronger than the male parent [1].

In this research, the inter-generic hybridization between *Platycodon grandiflorum* (♀) and *Campanula medium* (♂) has been used in the lodging-resistance breeding and many new characteristics have been obtained in F_1 generation, such as the bright colors, dwarf form, developed lateral branches, lodging-resistance and etc. The superior individual plant was selected from F_1 generation, which was used to self-copulation and the characters of S_1 generation were better than those of F_1. Using this method the new varieties of *Platycodon grandiflorum* that are suitable for horticultural plant could be obtained.

2. Materials and Methods

2.1. Materials

Combined parents were listed in **Table 1**.

2.2. Steps and Methods

2.2.1. Artificial Pollination Steps

1) The stamens of *Platycodon grandiflorum* were castrated before anther dehiscence, and the petals were not removed.

2) The castrated female flowers were set on the anti-powder bag.

3) When the castrated female flowers were opening, the anti-powder bag was removed, and the stigma was pollinated with pollen of *Campanula medium*, and then the bag was recover.

4) The anti-powder bag was taken off at the stigma drying.

5) The fruits were harvested after the seeds were ripe.

6) The seeds were cleaned and numbered.

2.2.2. Inter-Generic Hybridization, Hybridization, Self-Copulation and Backcross Breeding

The inter-generic hybridization tests were carried out between *Platycodon grandiflorum* (♀) and *Campanula medium* (♂) from 2009 to 2010. The stigma was pollinated two times on the day and the next day respectively. The objective plant of F_1 generation was selected to be backcrossed with *Campanula medium* in July 13, 2011, one time pollinated, which was of new color, dwarf form and lodging-resistance. The objective plant of F_1 generation was used to self-copulation as well which was covered with anti-powder net for free pollination in June

Table 1. Parents combination.

Materials	Female (2n = 18, ♀)	Male (2n = 34, ♂)	Generation
Generic cross	*Platycodon grandiflorum*	*Campanula medium*	F_1
Backcross	F_1	*Campanula medium*	Backcross generation
Self-copulation	F_1		S_1
F_1 Group	F_1	F_1	F_2

14, 2012. The F_1 group was pollinated openly to obtain F_2 in 2010. The pollination date, times, and flowers number were listed in **Table 2**, in **Table 2** the codes were set: *Platycodon grandiflorum* [PJ, ♀], *Campanula medium* [CM, ♂], and the F_1 was the selected objective plant from F_1 group.

2.2.3. Cultivation and Characteristics Investigation for F_1, F_2, S_1 and Backcross Generation

The seeds of F_1, F_2, S_1 and backcross generation were sown in plug trays in green houses in winter. Survival seedlings were transferred into pots and cultured as container seedlings, which would be planted in field in next spring. From 2010 to 2015, the characteristics of F_1, F_2, S_1 and backcross generation were investigated after planting which were the lodging-resistance, plant height, flower colors, branches number, lateral branch number and flower number. Comparison analyses between samples were tested by One-Way ANOVA method.

2.2.4. To Separate the DNA Groups of Different Generations

The DNA groups were extracted from F_1, F_2, S_1, backcross generation and parents, using the Plant Genomic DNA Kit which was produced by Tiangen Biotech (Beijing) CO., Ltd.

The PCR technique was used to extend the DNA group, the modified CTAB method was used to extract DNA for RAPD amplification reaction. PCR system: the total volume of basic reaction system were 25 μL, which include: 10 × PCR buffer 2.5 μL, 25 $mmol \cdot L^{-1}$ $MgCl_2$ 2.0 μL, 2.5 $mmol \cdot L^{-1}$ dNTP 2.0 μL, 2.5 $U \cdot L^{-1}$ Taq enzyme 0.5 μL, 5 $\mu mol \cdot L^{-1}$ primer 4.0 μL, template DNA 10ng, sterile ultra pure water up to 25 μL. The primer name is S12 and the sequence is 5'-CCTTGACGCA-3' [2].

PCR reaction program: 94˚C 4 min, 94˚C 30 s, 35˚C 1 min, 72˚C 2 min, 45 cycles, 72˚C, 10 min. The DNA groups were separated by 1% agarose gel electrophoresis.

3. Results and Discussion

3.1. Inter-Generic Hybridization, Self-Copulation and Backcrossed Analysis

The investigation of inter-generic hybridization, self-copulation and backcross were performed from 2009 to 2015. The survey results were listed in **Table 3**, as **Table 3** shown, in inter-generic hybridization 80 flowers were pollinated, 6 seed settings and 187 seeds were harvested, 88 seedlings were obtained, the average seed setting rate was 7.6%, and the average seedling rate was 50.4%. In backcross 15 flowers were pollinated, 15 seed settings and 457 seeds were harvested, 316 seedlings were obtained, the average seed setting rate was 86.7%, and the average seedling rate was 69.1%. In self-copulation 16 seed settings and 530 seeds were harvested, 348 seedlings were obtained, the average seed setting rate was 2.1%, and the average seedling rate was 65.7%.

Table 2. The pollination date, times, and flowers number.

Parents	Emasculation (y-m-d)	Pollination date (y-m-d)	Pollination No.
PJ (♀) × CM (♂)	2009-6-16	2009-6-16, 17	24
PJ (♀) × CM (♂)	2009-7-1	2009-7-2, 3	25
PJ (♀) × CM (♂)	2010-6-24	2010-6-24, 28	31
F_1 (♀) × CM (♂)	2011-7-13	2011-7-14	15
F_1 (♀) × F_1 (♂)	2012-6-14	2012-6-14	779

Table 3. Investigation results of hybridization, backcross and self-copulation.

Parents	Pollination No.	Seed setting No.	Seeds No.	Plants No.	Seed setting rate (%)	Seedling rate (%)
PJ (♀) × CM (♂)	80	6	187	88	7.6	50.4
F_1 (♀) × CM (♂)	15	13	457	316	86.7	69.1
F_1 (♀) × F_1 (♂)	779	16	530	348	2.1	65.7

In this study, compare of the seed seedling rate: self-copulation < inter-generic hybridization < backcross, the reason might be: the self-copulation was the self pollination that the different growth period of pistil and stamen in the same flower caused the low seed setting rate [3]; inter-generic hybridization is of the far phylogenetic relationship, the isolation mechanism were produced during the evolutionary process, so the seed setting rate is very low [4]; backcross was the highest in this test, that backcross method has been proved to be an effective way to solve the problems of distant hybridization which were the hybrid sterility and the isolation from generation to generation so long [5]. In addition, the inter-generic hybridization was repeated 3 times, and each time the living seeds were obtained which could grow into seeding, the further research is needed in the future to prove the reason.

3.2. Characteristics Investigation of F_1, S_1 and Backcross Generation

In this study the objective plant was selected in F_1 group which was the survey and test sample, because of the deep pink flower color, dwarf plant form, more bough number, lateral branching habit, and especially lodging-resistance. The characteristics of the F_1, S_1 and backcross generation were investigated that the results were listed in **Table 4**, as the **Table 4** shown: the plant height was: $S_1 < F_1 <$ backcross generation < female, there was significant difference between each generation and female; the bough number was: $S_1 > F_1 >$ backcross generation > female, there was significant difference between S_1 and F_1 with backcross generation and female; the lateral branches number was: $S_1 > F_1 >$ backcross generation, there was significant difference between S_1 and F_1 with backcross generation; flower number of each branch was: $S_1 > F_1 >$ backcross generation > female, there was significant difference between each generation and female. Especially the selected S_1 was of the developed characteristics. By artificial self-copulation and selection, not only can make the gene homozygous, but also make the good character to be accumulated and strengthened [5].

The reason of lodging resistance might be the developed lateral branch which make the plant gravity center lowing. The increasing flower number of each branch might be the developed lateral branch too. Varieties with moderate plant height, branch point, plant height and number of branches, lower gravity center of plant, compact plant architecture possess high lodging resistance [6].

The objective plant pictures of F_1, S_1, backcross generation and parents were shown in **Figure 1**. The survey results and pictures were shown that the flower color of each generation were different from their parents, but the leaves, stems and flower forms were same with the female's (**Figure 1**) that might be in tow possible, one is the apomixes phenomenon anther is the distant hybridization. The genetic material of apomictic progeny is just as male or female, and the character is just as male or female too [7]. Some people thought that the distant hybrid has not changed necessarily in chromosome number and forms, this problem should be researched in the future [7].

3.3. The Electrophoresis Separation of DNA Groups Comparison

In this test the DNA groups F_1, F_2, S_1, backcross generation and parents were separated by 1% agarose gel electrophoresis. The separated results were shown in **Figure 2**, as the **Figure 2** shown that the separated DNA groups of each plant were represented by each amplified band, 1 was marker, 2 was male, 3 was female, 4 to 15 were F_1, 16 to 22 were F_2, 23 to 27 were backcross generation and 28 to 33 were S_1 (**Figure 2**). Compared with the amplification bands of F_1, F_2, S_1, backcross generation and female there were no difference among them.

Table 4. Comparative analysis of characteristics to different generation (Average for many years).

Generation	Color	High (cm)	Bough No.	Lateral branch No.	Flower No. of each branch
♂	White, Purple	78.3 ± 4.41	1 ± 0.0	0	28.0 ± 1.155
♀	White, Purple	120.0 ± 5.77a	8.0 ± 0.58b	0	22.3 ± 0.33c
F_1	Deep Pink	75.0 ± 2.89b	14.3 ± 0.88a	12.0 ± 0.58a	63.0 ± 0.58a
S_1	Deep Pink	70 ± 2.89b	15.67 ± 0.88a	12.33 ± 1.20a	65.00 ± 1.15a
Backcross	Deep Pink	80 ± 2.00b	10.0 ± 1.00b	8.5 ± 0.50b	47.0 ± 1.00b

Note: Values are mean ± SE, different letters within the same column indicate mean values statistically different at $p < 0.05$ as determined by LSD test.

Figure 1. The pictures of the objective plant of F_1, S_1, backcross generation and parents.

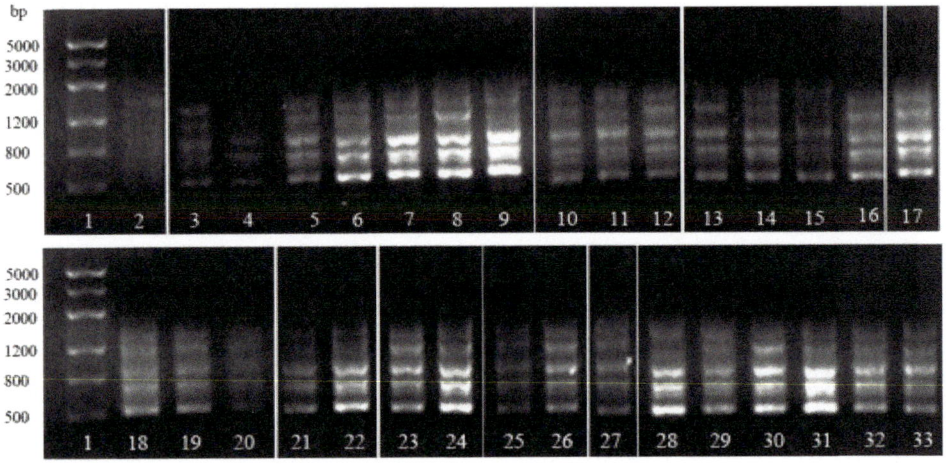

Figure 2. Separation of DNA groups. Marker: 1, Male: 2, Female: 3, F_1: 4 to 15, F_2: 16 to 22, Backcross generation: 23 to 27; S_1: 28 to 33.

In this study, the separated DNA groups have no significant difference and the possible reasons are: the screened primers were not suitable for those DNA groups (10 primers were selected for the experiment, and S12 was the best one); some research shown a multiplex PCR method was developed by which specific fragments with different length in a DNA sample mixture isolated from six transgenic maize could be amplified simultaneously in one PCR reaction, meanwhile, one fragment of maize endogenous reference gene (*zein*) could also be detected [8]; the concentration of agarose gel was not thick enough to make the amplified bands clearly; which was a apomixes phenomenon, apomixes is a phenomenon that the plants produce seeds directly and there is no fusion process of sperm and ovum [7], parthenocarpy is a kind of apomixes that the oocyte is not fertilized, but the fruit can be normally developed under the stimulation of pollen [9]. In the future more studies will be conducted.

4. Inquire into Other Findings

In this study many of new flower colors and flower forms were obtained from F_1 and F_2, which were shown in **Figure 3**. The colors and forms of flower will enrich the variety of *Platycodon grandiflorum*, so the research object should be sustainably developed in the future.

5. Conclusion

The inter-generic hybridization was used between *Platycodon grandiflorum* and *Campanula medium*, and lodging-resistance plant has been obtained from F_1 that the morphological characteristics were similar with female,

Figure 3. The rich flower colors and forms.

as well as other offspring. The DNA groups of each generation and parents were separated by 1% agarose gel electrophoresis, but the results hadn't been got. A lot of flowers' colors and forms have been obtained from the offspring.

References

[1] Cheng, J.S. and Liu, Q.L. (2000) Garden Plant Genetics and Breeding. China Forestry Press, Beijing.

[2] Wang, B.N., Zhang, X., Chen, J.M. and Song, J.Y. (2011) Optimization of RAPD Reaction System on Wild Cymbidium Faber in Qinling Mountain. *Journal of Northwest A&D University(Natural Science Edition)*, **39**, 87-84.

[3] Wu, J.R., Yan, Y.Z., Wu, S.Q., Zhang, Z. and Li, M.S. (2005) The Causes and Settling Method of Low Seed Setting Rate of *Platycodon grandiflorum* by Self-Pollination. *Journal of Agricultural Science Yanbian University*, **27**, 56-60.

[4] Liu, W.D., Zhu, S.N. and Cui Q.X. (2005) Horticultural Crops Genetics and Breeding. China Agricultural Science and Technology Press, Beijing.

[5] Yang, G.S. and Yuan, H.Y. (2009) Principle of Crop Breeding. Science Press, Beijing.

[6] Chen, X.J., Qi, C.K., Pu, H.M., Zhang, J.F., Gao, J.Q. and Fu, S.Z. (2007) Evaluation of Lodging Resistance in Rapeseed (*Brassica napus* L.) and Relationship between Plant Architecture and Lodging Resistance. *Chinese Journal of Oil Crop Sciences*, **29**, 54-57.

[7] Liu, Z.K., Luo, H., *et al.* (2010) Modern Plant Breeding. Science Press, Beijing.

[8] Jin, W.J., Hao, Y., Cheng, H.M., Lu, X.B., Peng, Y.F. and Jia, S.R. (2005) Specific Detection of Alien DNA Sequences in Six Transgenic Maize by Using Multiplex PCR Method. *Journal of Agricultural Biotechnology*, **13**, 562-567.

[9] Hu, H.W. (2010) Crop Genetics and Breeding. Chemical Industry Press, Beijing.

27

The Allelopathic Potential of Hairy Vetch (*Vicia villosa* Roth.) Mulch

Charles M. Geddes, Andrea Cavalieri, Fouad Daayf, Robert H. Gulden

Department of Plant Science, University of Manitoba, Winnipeg, Canada
Email: Rob.Gulden@umanitoba.ca

Abstract

Hairy vetch (*Vicia villosa* Roth.) has been recognized as a good candidate for weed suppressive mulch in organic no-till cropping systems. In our study, the allelopathic potential of hairy vetch, fall rye (*Secale cereale* L.) and winter wheat (*Triticum aestivum* L.) (extracted species) were evaluated at both vegetative and reproductive developmental stages (extract stage) based on the germination and radicle elongation of five response species using aqueous extracts in soil microcosms. Our study found hairy vetch shoot extract to have little allelopathic potential compared to fall rye or winter wheat, both species for which the allelopathic potential is well documented. Interestingly, hairy vetch was the only extracted species to increase in allelopathic suppression of radicle elongation at the reproductive stage when plant biomass is near maximum. This result was conceivably due to the increased concentration of the putative allelochemical cyanamide in reproductive tissue. Chemical inhibition of radicle elongation in this extract, however, was only observed in the domesticated response species wheat (*Triticum aestivum* L.) and canola (*Brassica napus* L.). The allelopathic effect on germination varied and depended on extracted species, extract stage, and response species; whereas the effect on radicle elongation was similar among extracted species, yet unique within response species. This research demonstrated that allelopathy studies should include multiple allelopathic and response species to accurately quantify the magnitude of chemical effects among allelopathic species and to rule out potential phytotoxic chemical defense/detoxification mechanisms that exist in some response species.

Keywords

Cyanamide, Germination, Hormesis, Phytotoxin, Radicle Elongation

1. Introduction

Allelopathy is the chemical effect of one organism, plant or microbe, on another through the release of chemicals into the environment [1]. This effect is most commonly viewed as amensalism, where growth in one plant is inhibited while the other plant remains unaffected. However, allelopathy may also manifest as a neutral/positive (stimulatory) interaction in some situations [1]-[4]. This interaction has often been attributed to hormesis, defined as stimulatory effects on plant growth caused by subtoxic concentrations of allelopathic phytotoxins [4].

Hairy vetch, fall rye, and winter wheat have been utilized as cover crops or mulches for weed suppression in organic no-till agriculture because they allow for an extended window of biomass production in areas restricted to short growing seasons. In Manitoba, Canada, Halde *et al.* (2014) [5] recognized hairy vetch as a good mulch candidate because of high biomass production at the end of the mulch producing growing season (9.1 - 10.7 $Mg \cdot ha^{-1}$) and high biomass retention the following spring (6.0 - 7.6 $Mg \cdot ha^{-1}$) into which subsequent crops could be planted. High mulch biomass combined with high weed control potential make hairy vetch an excellent candidate for weed suppressive mulch.

The differentiation between allelopathy and plant competition is difficult to address through field experimentation because it is challenging to isolate these forms of interference from one another due to lack of suitable control treatments [1] [6] [7]. A controlled laboratory microcosm approach can allow for isolation of the chemical effects of mulch residue, where aqueous extracts can be used to simulate allelochemical leachate. The inclusion of soil in allelopathic microcosms to account for microbial degradation and other interactions with soil is an important factor often overlooked in allelopathy studies [8] [9].

Wheat and rye have been recognized for the suppression of both germination and radicle elongation of several species via the release of benzoxazinoid allelochemicals. The main allelochemicals released by these species and their subsequent derivatives include 2,4-dihydroxy-1,4-benzoxazin-3-one (DIBOA), 2,4-dihydroxy-7-methoxy-1, 4-benzoxazin-3-one (DIMBOA), and 6-methoxybenzoxazolin-2-one (MBOA) as the main allelochemicals along with 2-hydroxy-7-methoxy-1,4-benzoxazin-3-one (HMBOA), 2-hydroxy-1,4-benzoxazin-3-one (HBOA), and benxoxazolin-2-one (BOA) in rye, and MBOA as the main allelochemical in wheat, along with HMBOA, and HBOA [1] [9]-[12].

Results from previous studies on the allelopathic potential of hairy vetch however, have been variable. In a greenhouse pot study, Inderjit and Asakawa, (2001) [13] determined that interference of hairy vetch with radish (*Raphanus sativus* L.) was not likely due to allelopathy, while Ercoli *et al.* (2005; 2007) [14] [15] attributed weed interference in both field and laboratory studies to phytotoxicity of hairy vetch. A series of greenhouse experiments using leached and unleached hairy vetch shoot tissue grown under field conditions suggested that allelopathy may play a role in weed suppression caused by hairy vetch mulch, but only within the first month after hairy vetch termination/mulch formation [16]. These studies, however, lacked osmotic controls and therefore were not clear as to whether emergence of the response species was due to allelopathic or osmotic effects. Inconsistent results surrounding the allelopathic potential of hairy vetch have created a need for further evaluation including effects on other weeds and crops and comparison of the observed effects to other previously documented allelopathic species.

Cyanamide, the putative allelochemical in hairy vetch, was originally thought to be solely a synthetic compound used as an agricultural fertilizer, and was not believed to be biosynthesized by plants [17]. Synthetic cyanamide has also been used as a fungicide, nematocide, herbicide [18], and to release bud and seed dormancy in several species [19]-[21]. The dimer form of cyanamide, dicyandiamide, also has been used as a nitrification inhibitor for agricultural production [22] [23]. In 1991, Maier-Greiner *et al.* [24] first isolated and characterized cyanamide hydratase, an enzyme uniquely responsible for the hydrolysis of cyanamide fertilizer into urea, from the soil borne fungus *Myrothecium verrucaria* (Alb. & Schwein.) Ditmar. This discovery, spurred the question as to how an enzyme evolved to hydrolyze a synthetic compound that had only been synthesized since 1898 [17]. In 2003, Kamo *et al.* [25] discovered cyanamide as an allelochemical in hairy vetch which was the first report of cyanamide as a natural chemical bio-synthesized by plants.

The objectives of this study were to i) determine the allelopathic potential of hairy vetch by comparing it to fall rye and winter wheat throughout several response species; ii) separate the allelopathic from osmotic effects of the extracts evaluated; and iii) quantify the level of cyanamide in hairy vetch extracts. It was hypothesized that leachate from hairy vetch tissue would chemically suppress both germination and radicle elongation of the response species. It was also hypothesized that the allelopathic potential of hairy vetch would be greater than winter wheat, but less than fall rye based on germination and radicle elongation of the response species. Lastly, vegetative stage extracts were hypothesized to have greater allelopathic potential than reproductive stage extracts.

2. Materials and Methods

Canola (cv. Sentry), wheat (cv. Waskada), kochia (*Kochia scoparia* L.), lamb's quarters (*Chenopodium album* L.), and wild oat (*Avena fatua* L.) seeds (collectively the response species), agricultural crop and weed species common to the Canadian Prairies, were imbibed separately in aqueous extracts of vegetative and reproductive stage hairy vetch, fall rye, and winter wheat aboveground biomass. The germination medium was composed of 15 grams of unsterilized silty clay loam soil (Orthic Black Chernozem, Mollisol, 0 - 15 cm depth, 3.4% OM, 7.1 pH, CEC 15.3 meq), collected from the Ian N. Morrison Research Station in Carman, MB, Canada (49°20'43.9"N, 98°58'05.2"W), and stored at 4°C prior to use. This soil medium was added to each petri dish (Fisher 100 × 15 mm) microcosm along with 25 seeds of a specific response species. The microcosms were then irrigated with 6 milliliters of a specific extract, osmotic control, or distilled water (di H_2O). Following the introduction of an extract or control solution, each microcosm was placed in a growth room (16:8 hr, 22:18°C) in the dark for four to seven days. The experiment was a three-way factorial (extracted species × extract stage × response species) with controls, using a randomized complete block design with four blocks. Each of the four experimental blocks consisted of 45 treatments which included: six extract microcosms (three species × two stages), both hypertonic and hypotonic osmotic control microcosms, and a di H_2O control microcosm for each of the five response species (nine treatments per response species). The experiment was also repeated using filter paper (Fisher brand, P8-creped, 9 cm) as a germination medium.

2.1. Biomass and Response Species Collection

Hairy vetch (cv. Unknown), fall rye (cv. Hazlet), and winter wheat (cv. Flourish) biomass, along with response species seed lots, were grown and collected at the Ian N. Morrison Research Station in Carman, Manitoba, Canada. A detailed description of the collection site can be found in Halde *et al.* (2014) [5]. Vegetative shoots of all extracted species were collected at the three-leaf stage (Zadok scale 13), while the reproductive shoots were collected immediately after flowering (Zadok scale 69). Biomass samples were air dried at room temperature, ground using a Wiley Mill (2 mm sieve), then stored at 4°C in air-tight bags for two weeks prior to extraction.

2.2. Extract and Osmotic Control Preparation

In an Erlenmeyer flask, dry ground biomass was submersed in di H_2O (1:10 wt/wt, respectively; determined through preliminary experimentation [3]), then placed on an orbital shaker at 150 rpm for 24 hours at 21°C. The slurry was then filtered through two layers of cheesecloth and stored at 4°C overnight. A Vapor Pressure Osmometer (5500 Vapor Pressure Osmometer, Wescor Inc., Logan, UT) was used to test the osmotic potential of all aqueous plant extracts. Mannitol ($C_6H_{14}O_6$) was added to di H_2O to generate osmotic controls with the same osmotic potential as the highest (−0.712 MPa, hereafter designated "hypotonic", least amount of solutes in solution) and lowest (−1.013 MPa, hereafter designated "hypertonic", greatest amount of solutes in solution) extracts. We only used two osmotic controls because of the narrow range of osmotic potentials among the plant extracts.

2.3. High Performance Liquid Chromatography

Reversed-phase high performance liquid chromatography (HPLC) was conducted using a Waters 2690 Separations Module, fitted with a LiChroCART® 4-4 LiChrospher® 100 RP-18 5 µm guard column and a LiChroCART® 250-4 LiChrospher® 100 RP-18 5 µm separatory column, and equipped with a Waters 996 photodiode array detector to analyze the hairy vetch extracts following the methods of Wang *et al.* (2008) [26]. Briefly, extracts were centrifuged at 4696 g for 40 minutes at 4°C then filtered through a 0.2 µm PTFE syringe filter (Thermo Scientific). A 50 µl sample was injected and the column was eluted at a flow rate of 1 $mL \cdot min^{-1}$ with a mobile phase consisting of A) 0.1% H_3PO_4 in water, and modified with B) HPLC grade acetonitrile using the following gradient: (time [min]/A [%]/B [%]) = 0/100/0, 5/95/5, 14/90/10, 0/80/20, 23/80/20, 30/65/35, 35/65/35, 43/50/50, 48/25/75, 55/0/100, 60/0/100. Cyanamide 99% (NH_2CN, Sigma-Aldrich) was used as an HPLC standard to identify this allelochemical in hairy vetch extracts. The peak specific to cyanamide was identified and quantified through extract spiking, UV absorbance spectra analyses, and the utilization of a concentration curve in ultrapure H_2O with duplicate samples ($y = 630.71x; R^2 = 0.9811$; where: y is peak height, and x is concentration of cyanamide in $\mu g \cdot ml^{-1}$). The cyanamide standard solutions were prepared immediately prior to HPLC analysis to limit the formation of dicyandiamide [22].

2.4. Quantification of Germination and Radicle Elongation

Germination and radicle elongation quantification of canola, wheat, kochia, lamb's quarters, and wild oat microcosms were conducted after 4, 4, 5, 7, and 6 days after imbibition, respectively; due to species-specific time requirements to complete germination. After elutriation, seeds were considered germinated if the radicle had protruded through the seed coat. Average radicle length using three representative seedlings per microcosm were measured using ASSESS 2.0 Image Analysis Software [27].

2.5. Data Preparation

In our study, allelopathy was defined as the chemical effect (inhibition or stimulation) of a plant extract on a response species after accounting for the osmotic effect of the specific plant extract in question. The extract and osmotic effect on germination were corrected for dormancy and viability of response species using the di H_2O control to calculate germination as a percent of that in di H_2O. The effect of the osmotic control (as a percent of the di H_2O control) with the closest osmotic potential to that of the extract was then subtracted from the extract effect (as a percent of that in di H_2O). This resulted in the chemical effect on germination expressed as a percent change relative to the di H_2O control (the "allelopathic effect"). In other words, data are presented as negative percentages indicating suppression and positive percentages indicating stimulation of response variables. Equation (1) was used to calculate the allelopathic effect on germination and radicle elongation within each experimental block:

$$A=\left[\left(\frac{E}{W}\right)\times 100\right]-\left[\left(\frac{O}{W}\right)\times 100\right] \tag{1}$$

where, A = allelopathic effect on germination or radicle elongation, E = number of germinated seedlings or average radicle length per seedling in a specific extract microcosm, W = number of germinated seedlings or average radicle length per seedling in the corresponding di H_2O control microcosm, and O = number of germinated seedlings or average radicle length per seedling in the osmotic control microcosm with an osmotic potential most similar to that of the specific extract.

2.6. Statistical Analyses

Mixed model analysis in SAS 9.3 was used to conduct ANOVA on the allelopathic effect on germination and radicle elongation data as well as the osmotic effect on germination and radicle elongation [28]. The model for testing the allelopathic effect was a complete three-way factorial (extracted species × extract stage × response species) as the controls were used only to determine the allelopathic effect. To test osmotic effects, a two-way factorial (osmotic control × response species) model was used. The residuals were tested for normality, where *Lund's test* was used to remove outliers when necessary to meet the assumptions of ANOVA [29]. Extract stage, extracted species, and response species were considered fixed effects and experimental replication was considered a random effect. Treatment means were separated based on *Fisher's protected LSD* (α = 0.05) using the pdmix 800 macro [30]. *Least squares means* (α = 0.05) were used to determine significant differences between the allelopathic effect and corresponding osmotic effect for both germination and radicle elongation.

3. Results

3.1. Filter Paper Microcosm Experiments

A preliminary experiment conducted on filter paper in petri dish microcosms investigated the difference between shoot and root biomass using the same extracted species and response species as the current study (data not shown) [3]. Extracts were prepared using half the biomass compared to the present study, at a rate of 1:20 wt dry ground biomass/wt di H_2O. These extracts produced a range of allelopathic effects on germination from −72.5% to 3.6% (lamb's quarters in winter wheat shoot extract and wild oat in hairy vetch root extract, respectively) and allelopathic effects on radicle elongation ranging from −48.6% to 96.9% (wild oat in winter wheat shoot extract and lamb's quarters in fall rye shoot extract, respectively) relative to the di H_2O control. This preliminary experiment also found that shoot biomass in all three extracted species consistently resulted in increased allelopathic inhibition of germination and radicle elongation (−43.6% and −14.5%, respectively) when compared to root biomass (−17.6% and −5.3%, respectively). Filter paper microcosms were also used in the present study. These microcosms were prepared using the more concentrated shoot biomass extracts from this study. In the present study, high concentrations of extracts in filter paper microcosms resulted in almost complete inhibition of germination in all treatments (data not shown); for this reason, only the allelopathic effect in soil microcosms data are reported.

3.2. Allelopathic Effect on Germination

Vegetative fall rye was the only extract to chemically suppress germination among all five response species and significant allelopathic effects on germination were only observed sporadically in all other extract and response species combinations. In vegetative fall rye, germination ranged from −27.8% in lamb's quarters to −106.8% in kochia relative to the di H_2O control (**Table 1**). Vegetative hairy vetch and reproductive fall rye extracts, on the

Table 1. The allelopathic effect of vegetative and reproductive stage hairy vetch, fall rye and winter wheat shoot aqueous extracts on the germination of response species; expressed as a change in percent relative to the distilled water control.

Developmental stage	Extracted species	Response species	Allelopathic effect on germination (Δ% relative to di H_2O)[a,b]		
Vegetative	Hairy vetch	Canola	−17.6	a-g	
		Wheat	−13.0	b-e	
		Kochia	0.0	ab	
		Lamb's quarters	4.2	a-f	
		Wild oat	10.3	ac	
	Fall rye	Canola	−55.0	gh	**
		Wheat	−79.6	hi	***
		Kochia	−106.8	j	***
		Lamb's quarters	−27.8	a-g	***
		Wild oat	−35.3	e-g	*
	Winter wheat	Canola	−18.0	b-f	
		Wheat	−32.3	d-g	
		Kochia	−89.2	ij	***
		Lamb's quarters	−47.9	e-h	**
		Wild oat	5.1	ab	
Reproductive	Hairy vetch	Canola	−21.9	b-g	
		Wheat	−4.1	ab	
		Kochia	−10.8	a-e	
		Lamb's quarters	13.9	ab	*
		Wild oat	4.8	ab	
	Fall rye	Canola	−3.3	a-d	
		Wheat	−7.1	ab	
		Kochia	15.5	a	
		Lamb's quarters	−4.2	a-f	
		Wild oat	4.8	ab	
	Winter wheat	Canola	−23.3	b-g	
		Wheat	−42.3	fg	*
		Kochia	−50.9	gh	**
		Lamb's quarters	−12.5	a-g	
		Wild oat	−10.5	a-e	

a) values in a column having the same letter(s) are not statistically different as determined by *Fisher's protected LSD* ($\alpha = 0.05$); b) negative values indicate suppression, whereas positive values indicate stimulation. *, **, *** indicate significant differences between extract and corresponding osmotic control based on *least squares means* ($\alpha = 0.05$) at $P < 0.05$, $P < 0.01$, and $P < 0.0001$, respectively.

other hand, did not have an allelopathic effect on germination of any response species (**Table 1**). Reproductive hairy vetch, which was at the stage of maximum biomass, considered ideal for the formation of mulch, did not chemically suppress germination of any of the response species. This extract, however, caused a 13.9% allelopathic stimulation of germination in lamb's quarters (**Table 1**). Winter wheat extracts inhibited germination in kochia and lamb's quarters to a greater extent at the vegetative than at the reproductive developmental stage. No allelopathic effect on germination of lamb's quarters was observed at the reproductive stage of winter wheat. In contrast, inhibition of wheat germination was greater at the reproductive than at the vegetative developmental stage of the winter wheat extract, where no allelopathic effect on wheat germination was observed at the latter stage. Germination in all other response species and extract combinations was not different from their respective osmotic controls.

3.3. Allelopathic Effect on Radicle Elongation

The allelopathic effect on radicle elongation (negative or positive) tended to be unique to each response species, irrespective of extracted species or developmental stage. In fact, although not always considered an allelopathic effect, radicle elongation was statistically the same in each response species among all extracted species. One exception was lamb's quarters, where a greater degree of stimulation was observed in fall rye extracts than the other extracted species (**Table 2**). The response-species-specific effects on radicle elongation may have been due to common chemical (inhibitory/stimulatory) constituents among all extracts, or potentially species-specific phytotoxin detoxification mechanisms. Interestingly, a stimulation of wild oat radicle elongation was also consistent among all extracted species (**Table 2**). For wild oat, however, this stimulation was not different from the effect of the osmotic control when exposed to hairy vetch and winter wheat extracts. Overall, allelopathic suppression of radicle elongation was observed in canola (−46.9% to −57.7%), wheat (−24.6% to −40.2%), and kochia (−33.8% in winter wheat extracts only) with the greatest and most consistent suppression of radicle elongation observed in the domesticated response species (canola and wheat) (**Table 2**).

3.4. Osmotic Effect on Germination and Radicle Elongation

The range of osmotic potential among the extracts was relatively narrow and the osmotic controls within that range only affected germination in the weedy species, but not the domesticated species (crops). The hypotonic osmotic control, which was most similar to the osmotic potential of the reproductive hairy vetch, fall rye, and winter wheat extracts and the vegetative hairy vetch extract, only reduced germination in wild oat (31% reduction compared to di H_2O) (**Table 3**). The hypertonic osmotic control, used for the vegetative fall rye and winter wheat extracts, suppressed germination of the small seeded species kochia (by about one-third from 52% to 33% total germination) and lamb's quarters (by about half from 26% to 13% total germination) (**Table 3**).

Overall, radicle elongation was more sensitive to osmotic effects than germination. This was particularly obvious at the hypotonic osmotic potential where three species (wheat, lamb's quarters, and wild oat) showed

Table 2. The allelopathic effect of hairy vetch, fall rye, and winter wheat shoot aqueous extracts (throughout both vegetative and reproductive developmental stages) on radicle elongation of the response species; expressed as a change in percent relative to the distilled water control.

Extracted species	Hairy vetch[a,b]			Fall rye[a,b]			Winter wheat[a,b]		
Canola	−57.6	g	***	−46.9	fg	***	−52.4	fg	***
Wheat	−24.6	de	**	−26.6	de	**	−40.2	e-g	**
Kochia	−12.4	d		−7.4	cd		−33.8	d-f	**
Lamb's quarters	21.4	b	**	62.5	a	**	26.3	b	**
Wild oat	9.1	bc		16.1	bc	*	10.5	bc	

[a]values in a column having the same letter(s) are not statistically different as determined by *Fisher's protected LSD* ($\alpha = 0.05$); [b]negative values indicate suppression, whereas positive values indicate stimulation. *, **, *** indicate significant differences between extract and corresponding osmotic control based on *least squares means* ($\alpha = 0.05$) at $P < 0.05$, $P < 0.01$, $P < 0.0001$, respectively.

Table 3. The effect of osmotic controls on germination and radicle elongation of response species; expressed as total germination percentage and average radicle length per seedling.

Osmotic control	Osmotic potential (MPa)	Response species	Germination (%)[a]		Radicle length (mm)[a]	
di H_2O[b]		Canola	93.0	a	55.9	de
		Wheat	99.0	a	200.0	a
		Kochia	52.0	c	22.3	f-h
		Lamb's quarters	26.0	de	41.9	d-f
		Wild oat	88.0	ab	143.8	b
Hypotonic[c]	−0.712	Canola	97.0	a	56.0	d
		Wheat	99.0	a	127.8	b
		Kochia	56.0	c	18.1	gh
		Lamb's quarters	19.0	ef	12.1	h
		Wild oat	57.0	c	51.8	de
Hypertonic[d]	−1.013	Canola	95.0	a	44.3	de
		Wheat	99.0	a	89.3	c
		Kochia	33.0	d	9.6	h
		Lamb's quarters	13.0	f	9.6	h
		Wild oat	76.0	b	35.2	e-g

[a]values in a column having the same letter(s) are not statistically different as determined by *Fisher's protected LSD* ($\alpha = 0.05$); [b]indicates the control used to account for seed-lot dormancy and viability; [c]indicates the osmotic control used for vegetative hairy vetch (−0.792 MPa), and reproductive hairy vetch (−0.792 MPa), fall rye (−0.712 MPa), and winter wheat (−0.841 MPa) extracts; [d]indicates the osmotic control used for vegetative fall rye (−1.013 MPa), and winter wheat (−0.970 MPa) extracts.

altered radicle elongation compared to only one species (wild oat) with altered germination (**Table 3**). Differences between the hypertonic and hypotonic osmotic controls were relatively small among all response species, except in wheat, where radicle elongation was suppressed to a greater extent in the hypertonic (56% reduction in radicle elongation) compared to the hypotonic osmotic control (36% reduction in radicle elongation) four days after imbibition (**Table 3**). However, both of these osmotic controls significantly inhibited radicle elongation of wheat seedlings when compared to di H_2O (200 mm per seedling). The extracts and osmotic controls used in this study spanned a relatively narrow range of osmotic potential ranging from −0.712 to −1.013 MPa (**Table 3**).

3.5. High Performance Liquid Chromatography Analysis

Cyanamide, the putative allelopathic compound biosynthesized by hairy vetch, was identified at a retention time of 2.62 minutes in the hairy vetch extracts (**Figure 1**). Peak height was used to calculate the concentration of cyanamide instead of area under the peak because of the overlap of peaks observed in the aqueous hairy vetch extracts. Cyanamide was present at concentrations of 2159 and 2630 $\mu g \cdot ml^{-1}$ or 215.9 and 263.0 $\mu g \cdot g^{-1}$ DM tissue in the vegetative and reproductive stage hairy vetch extracts, respectively. The 22% increase in cyanamide in the reproductive tissue extract of hairy vetch was consistent with the overall reduction in radicle elongation among all response species caused by the chemical effect of the reproductive hairy vetch extract (**Figure 2**). We did not look for the presence of cyanamide in wheat or rye extracts because cyanamide had not been previously identified as an allelochemical in these species.

4. Discussion

Based on germination and radicle elongation of the response species tested, hairy vetch would not be considered a good candidate species for weed suppressive allelopathic mulch. Hairy vetch did not chemically inhibit germination of any response species, and, in fact, stimulated the germination of lamb's quarters under our conditions. Additionally, the only allelopathic suppression of radicle elongation observed was that of the domesticated response species tested (canola and wheat). These data suggest that the weed suppression observed by Halde *et*

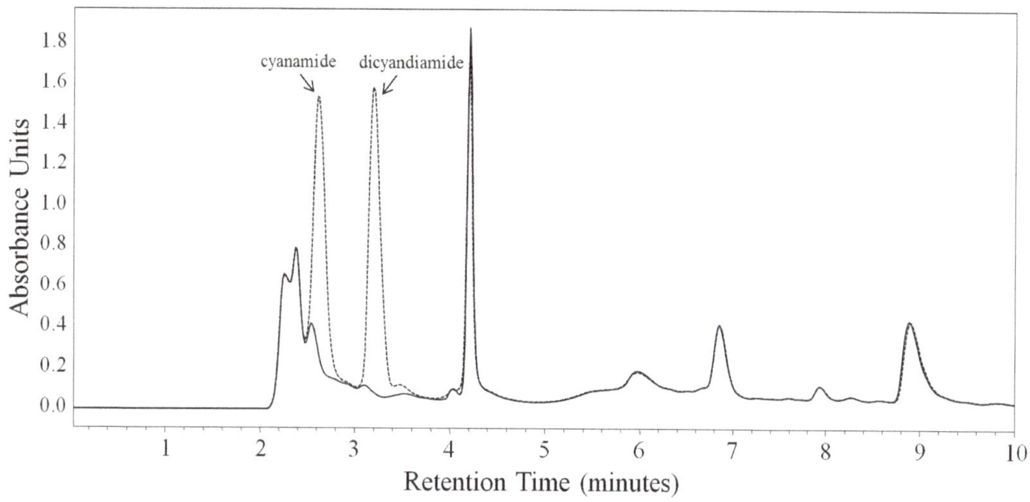

Figure 1. Chromatogram of UV absorbance (MaxPlot 200 to 400 nm) for a 1:3 (v/v) solution of vegetative hairy vetch extract in a) ultrapure H_2O (solid line), or b) 4000 $\mu g \cdot ml^{-1}$ cyanamide solution (dotted line) during HPLC analysis. Cyanamide was identified at a retention time of 2.62 minutes with UV spectra matching that of the cyanamide standard.

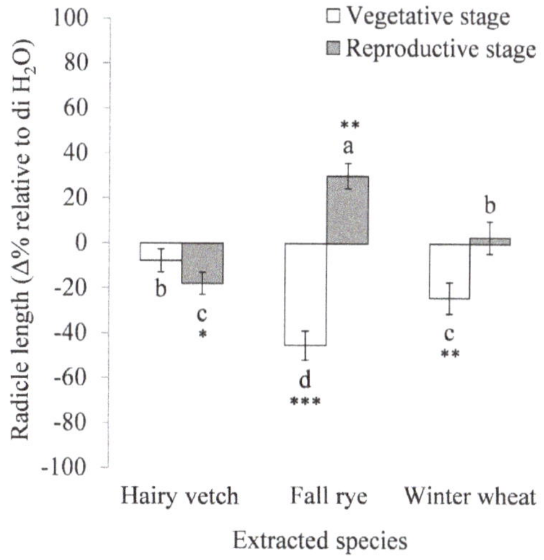

Figure 2. The allelopathic effect of vegetative and reproductive stage shoot aqueous extracts of hairy vetch, fall rye, and winter wheat on radicle elongation throughout all response species tested (canola, wheat, kochia, lamb's quarters, and wild oat). Error bars indicate +/− one standard error. Bars with the same letter are not statistically different as determined by *Fisher's protected LSD* (α = 0.05). Significant differences between extracted species and corresponding osmotic controls based on *least squares means* (α = 0.05) at $P < 0.05$, $P < 0.01$, and $P < 0.0001$ are indicated by *, **, and ***, respectively.

al. (2014) [5] was likely not due to allelopathy, but rather physical weed suppression as a result of high vetch biomass production. Therefore, in opposition to our proposed hypotheses, results from this study delineate hairy vetch as an undesirable candidate species for allelopathic mulch to control the weed species evaluated herein. It is important to note, however, that these data indicate the plausibility of hairy vetch as allelopathic mulch to inhibit volunteer wheat or canola in crop species other than those tested in our study.

Hairy vetch had less allelopathic potential than fall rye or winter wheat. This further suggests that weed interference caused by hairy vetch mulch is unlikely to be caused by allelopathy. However, this result opposes the majority of previous literature [1] [7] [14]-[16] [25]. In a field study, recruitment and biomass of redroot pigweed (*Amaranthus retroflexus* L.) and lamb's quarters were most suppressed by incorporated cover crops of hairy vetch when compared to brown mustard (*Brassica juncea* L.), rye, or bare soil [14]. A subsequent growth room study conducted using aqueous extracts in a filter paper petri dish assay, found that germination of redroot pigweed and common knotgrass (*Polygonum aviculare* L.) was not suppressed by hairy vetch extracts whereas germination of lamb's quarters was suppressed by about 20% [15]. Differences between these laboratory and field studies were attributed to the phytotoxic effect of vetch biomass degradation and allelochemical interactions with soil biota [14] [15]. In our filter paper bioassay, germination of kochia and lamb's quarters was inhibited entirely by the aqueous extracts (data not shown) where more moderate effects were observed in the soil microcosm assay. This further suggests that allelochemical interactions in soil cause decreased allelopathic potential of leachate from plant biomass. Based on the amount of biomass needed to inhibit seedling emergence by 50% (I50), Teasdale *et al.* (2007) [16] demonstrated decreased emergence of velvetleaf (*Abutilon theophrasti* Medic.), green foxtail (*Setaria viridis* L. Beauv.), and smooth pigweed (*Amaranthus hybridus* L.) under hairy vetch mulch one day before and five days after desiccation compared to one month after desiccation. This study determined that correlation of timing between hairy vetch desiccation/mulch formation and seeding of cash crop is imperative when attempting to utilize allelopathy for weed suppression [16]. It must be noted, however, that the absence of osmotic controls in this study did not allow the quantification of the true chemical effect of hairy vetch extracts and mulch leachate.

The putative allelochemical cyanamide was found in both vegetative and reproductive stage hairy vetch extracts at concentrations of 215.9 and 263.0 $\mu g \cdot g^{-1}$ dry weight (DW) tissue. Kamo *et al.* (2003) [25] first found cyanamide in hairy vetch at concentrations of 444, 163, and 89 $\mu g \cdot g^{-1}$ fresh weight (FW) in shoot, endosperm, and root tissue of nine-day-old hairy vetch seedlings, respectively. A later publication reported cyanamide at concentrations of 369 to 498 $\mu g \cdot g^{-1}$ FW in the aboveground biomass of flowering hairy vetch [31]. These researchers also identified cyanamide in four-week-old hairy vetch at concentrations of 629, 256, and less than 20 $\mu g \cdot g^{-1}$ FW in leaf, stem and root tissue, respectively [17]. Differences in chemical concentration between our study and those previously reported could either be due to the use of dried plant tissue, conditions during plant growth, and/or differences in extraction and quantification protocols [32]. For these reasons, a direct comparison of cyanamide concentrations cannot be made among the extracts in our study and those reported previously. It is also important to note that chemical constituents other than cyanamide were not identified in our HPLC analyses and therefore warrant further investigation. With that said, the aqueous extracts of dry tissue evaluated in our study would be reflective of the water soluble allelochemicals leached from dry hairy vetch mulch under conditions in the Northern Great Plains, where, shortly after the spring thaw, a cash crop would be seeded into dry vetch mulch from plants terminated near the end of the previous growing season. Additionally, environmental differences such as length of mulch overwintering, soil type and soil microbiota may also have an effect on allelochemical availability and degradation as well as the concomitant seedling responses to the aforementioned chemicals. Therefore, chemical concentrations alone may not be sufficient to accurately evaluate allelopathic potential in mulch-based systems.

Previous research found monocotyledonous species to be tolerant and dicotyledonous species to be sensitive to cyanamide phytotoxicity [33]. Using Pieper's value, a measure of the average time requirement for germination, Soltys *et al.* (2012) [33] found that 3.0 mM cyanamide solution prolonged germination time of redroot pigweed, tomato (*Solanum lycopersicum* L.), lettuce (*Lactuca sativa* L.), and white mustard (*Sinapsis alba* L.), where germination timing of cleavers (*Gallium aparine* L.), corn (*Zea mays* L.), and wheat was unaffected. This effect was likely attributable to seed weight and thickness of the seed coat, where heavier seeds with thick seed coats generally required more time to germinate [33]. Although we did not measure the average time requirement for germination, this effect may have been observed in the differences among radicle elongation. Our study showed no apparent relationship between growth habit (monocotyledonous or dicotyledonous), seed size or allelopathic effects. In the absence of soil, cyanamide in solution has been shown to affect root growth of onion in a dose-dependent manner from 2 - 10 mM [34]. This coincides with the greater inhibition of radicle elongation and increased cyanamide concentration observed between the vegetative (51.36 mM) and reproductive (62.56 mM) stage hairy vetch extracts in our study, which was well above the minimum biologically active concentrations required. However, the difference in cyanamide concentration between vegetative and reproductive stage extracts was small enough to indicate that there may be other chemical constituents involved in the observed allelopathic effects. Kamo *et al.* (2003) [25] showed that 0.24 mM (100 ppm) cyanamide solution inhibited radicle elongation of lettuce seedlings by 40% in a filter paper petri dish assay. This result, however, did not include the possible effect of cyanamide binding to the soil matrix and/or degradation by soil microbes. Due to the presence of soil in our microcosm assay, the concentration of cyanamide in our extracts would likely not be equal to the concentration being absorbed by the imbibed response species, making any concentration comparison difficult to interpret. However, it is important to consider the effect of allelochemical degradation in soil when comparing among multiple response species with inherently different time requirements to complete germination. Interestingly these data show an observable relationship between length of time from imbibition to microcosm evaluation and allelopathic effect. Canola and wheat, being the first species to germinate, showed the greatest response to allelopathy, whereas wild oat, being the last species to germinate, showed the mildest response to allelopathy overall. This observation indicates a possible relationship between allelopathic effect and allelochemical degradation by soil microorganisms.

The mechanism by which cyanamide alters germination or root elongation has yet to be determined. Cyanamide solution, in a petri-dish assay, decreased the quantity of mitotic cells, suppressed the propagation of meristematic cells, and altered the cytoskeleton arrangement in onion (*Allium cepa* L.) [34]. In a subsequent study on tomato, cyanamide inhibited root growth, likely due to altered cell division, an imbalance of the hormones ethylene and auxin, and differences in expansin gene expression [35]. To date, no research has been conducted on cyanamide mode of action using the species evaluated in this study.

Contrary to hairy vetch, vegetative stage fall rye and winter wheat had the greatest allelopathic effect on germination when compared to the corresponding reproductive stage extracts; likely related to higher hydroxamic acid concentration in vegetative tissue [36]. This shows potential for fall rye or winter wheat to be used as a cover crop incorporated at early growth stages for utilization of allelopathy for weed suppression. We did not evaluate known allelochemicals in fall rye or winter wheat because these putative benzoxazinoid allelochemicals and their derivatives are relatively unstable and not readily available aside from chemical synthesis [9] [12] [37]. Interestingly, reproductive stage fall rye had the greatest stimulatory effect on radicle elongation of response species; likely due to hormesis caused by the decrease of hydroxamic acids in fall rye tissue throughout its development [4] [9] [36]. Reproductive stage fall rye and winter wheat mulches expressed potential for utilization for species-specific growth promotion.

In our study, radicle elongation of both lamb's quarters and wild oat were allelopathically stimulated by all extracts, whereas radicle elongation in the other three response species was suppressed. Coincidentally, lamb's quarters and wild oat also required the longest amount of time before germination could be evaluated. There is a possibility that the stimulation of radicle elongation observed could be due to allelochemical degradation in soil over time; however, differences in species-specific time required to complete germination can create difficulty when trying to correct for degradation of allelochemicals. In a growth room pot experiment, Krogh *et al.* (2006) [12] determined the maximum concentration of the allelochemical MBOA (154 $ng \cdot g^{-1}$ DW soil) in soil one day after the incorporation of wheat (one seedling in 300 g soil). Using first order degradation kinetics, they determined MBOA to have a DT_{50} (time required for the chemical to reach half of the maximum concentration) of 1.3 to 1.6 days in soil. The maximum amount of DIBOA (53 $ng \cdot g^{-1}$ DW soil) in soil one day following incorporation of rye (one seedling in 300 g soil) was also determined [12]. The concentration of HBOA was found to increase as DIBOA and BOA degraded, reaching its maximum concentration (~100 $ng \cdot g^{-1}$ DW soil) in soil four days after the incorporation of rye [12]. Lamb's quarters seedlings have been noted to have a strong detoxification mechanism against benzoxazinoid allelochemicals [9]. In a greenhouse tray experiment, leachate from hairy vetch mulch had a stimulatory effect on the emergence of lamb's quarters seedlings [16]. Additionally, root elongation of lamb's quarters seedlings increased (approximately 120% to 135% of the deionized water control) when imbibed with aqueous extracts of hairy vetch in a filter paper assay [16]. This shows the possibility of allelopathy for species-specific growth promotion.

Radicle elongation seemed to be more sensitive to the osmotic component than the allelopathic component of the plant extracts. Nevertheless, after accounting for the osmotic effect of extracts in this study, radicle elongation still remained a more sensitive measure of allelopathic potential than germination. Although the range of osmotic potential among extracts was relatively narrow, the observed osmotic potentials were hypertonic enough to affect the rate of water uptake by seed [38]. Our study showed that it is imperative to include osmotic controls to account for osmotic potential of extracts in allelopathy studies. Al-Kahari, (1998) [38] found germination and radicle elongation of wheat, after eleven days, to be greatly affected when germinated in 0 to −1.4 MPa solutions of polyethylene glycol 8000, where a decrease in germination (99% to 7%) and root length (15 cm to 0.2 cm) was correlated with an increase in solute concentration. In another study, time to reach 10% germination of wheat increased from 44 h to 124.7 h when germinated in NaCl, Na_2SO_4, KCl, K_2SO_4, $MgCl_2$, $MgSO_4$, or polyethylene glycol 8000 solutions ranging from −0.3 MPa to −1.8 MPa [39]. However, in our study total germination of wheat was not affected by the osmotic controls at −0.712 and −1.013 MPa. Radicle elongation of wheat, on the other hand, was greatly inhibited in both the hypotonic (127.8 mm) and hypertonic (89.3 mm) osmotic controls when compared to di H_2O (200.0 mm). Boyd *et al.* (2004) [40] found that the germination percentage of canola (84%, 42%, and 15%), wheat (83%, 69%, and 40%), and wild oat (42%, 11%, and 1%) decreased with increasing solute concentration in −0.01, −0.5, and −1.00 MPa polyethylene glycol 6000 solutions, respectively. Differences in response species germination between our study and that of Boyd *et al.* (2004) [40] could be due to the presence of soil as a germination medium in our microcosm assay. Many allelopathy assays use germination and radicle elongation as response variables to assess allelopathic potential. However, evaluation of the allelopathic effect on radicle elongation has been shown to be a more sensitive measure than germination [1] [15]. It is possible that the response observed as radicle elongation could be due to an extract induced delay in the time to germination; therefore explaining the sensitivity of the radicle elongation response variable in allelopathy studies.

5. Conclusion

In summary, hairy vetch was determined to be an impractical candidate species for allelopathic mulch based on germination and radicle elongation of the response species in this study. Overall, the allelopathic effect on radicle elongation was more dependent on response species than extracted species. Results from this study suggest that allelopathic effects on radicle elongation are more dependent on the physiological characteristics and detoxification mechanisms of response species rather than the chemical effect of different extracted species. These data indicate that allelopathy studies should not only evaluate one allelopathic species but compare between more than one allelopathic species to determine whether observed effects are caused by the allelochemicals in question or are simply a result of species-specific defense/detoxification mechanisms of several phytotoxic allelochemicals as observed with wild oat and lamb's quarters in this study. Further research should focus on cyanamide degradation in soil, as well as the mode of action of cyanamide on the response species evaluated in this study. Future research should also focus on the stimulation of radicle elongation observed in wild oat and lamb's quarters. This shows the possibility for utilization of allelopathy for species-specific growth promotion. Additionally, agroecologists should focus on the deliberate incorporation of allelopathic interactions in a comprehensive integrated weed management approach. This strategy may help further delay the evolution of herbicide resistance in current conventional cropping systems.

Acknowledgements

We would like to thank the Natural Sciences and Engineering Research Council of Canada for an Industrial Postgraduate Scholarship to CMG and a Discovery Grant to RHG. We would also like to thank Lorne Adam for technical assistance.

References

[1] Rice, E.L. (1984) Allelopathy. Physiological Ecology: A Series of Monographs, Texts, and Treaties. 2nd Edition, Academic Press Inc., Orlando.

[2] Geddes, C.M., Cavalieri, A., Daayf, F. and Gulden, R.H. (2013) The Allelopathic Potential of Hairy Vetch, Fall Rye and Winter Wheat in a Silty Clay Loam Soil. Poster Presentation. CSSS/MAC/ASA, CSSA, and SSSA. Tampa. https://scisoc.confex.com/crops/2013am/webprogram/Paper80586.html

[3] Geddes, C.M., Cavalieri, A. and Gulden, R.H. (2012) The Allelopathic Potential of Hairy Vetch, Fall Rye and Winter Wheat. CWSS/MAC. Poster Presentation. http://umanitoba.ca/faculties/afs/agronomists_conf/media/2013_Geddes-Gulden_Allelopathic_SA2013.pdf

[4] Belz, R.G., Edivaldo, D.V. and Duke, S.O. (2007) Dose/Response Relationships in Allelopathy Research. In: Fujii, Y. and Parvez, M.M., Eds., *Allelopathy: Concepts and Methodology*, Science Publishers, Enfield, 4-29.

[5] Halde, C., Gulden, G.H. and Entz, M.H. (2014) Selecting Cover Crop Mulches for Organic Rotational No-Till Systems in Manitoba, Canada. *Agronomy Journal*, **106**, 1193-1204. http://dx.doi.org/10.2134/agronj13.0402

[6] Blum, U. (2007) Can Data Derived from Field and Laboratory Bioassays Establish the Existence of Allelopathic Interactions in Nature? In: Fujii, Y. and Parvez, M.M., Eds., *Allelopathy: Concepts and Methodology*, Science Publishers, Enfield, 31-38.

[7] Fujii, Y. (2001) Screening and Future Exploitation of Allelopathic Plants as Alternative Herbicides with Special Ref-

erence to Hairy Vetch. *Journal of Crop Production*, **4**, 257-275. http://dx.doi.org/10.1300/J144v04n02_09

[8] Inderjit (2005) Soil Microorganisms: An Important Determinant of Allelopathic Activity. *Plant and Soil*, **274**, 227-236. http://dx.doi.org/10.1007/s11104-004-0159-x

[9] Schulz, M., Marocco, A., Tabaglio, C., Macias, F.A. and Molinillo, J.M.G. (2013) Benzoxazinoids in Rye Allelopathy—From Discovery to Application in Sustainable Weed Control and Organic Farming. *Journal of Chemical Ecology*, **38**, 154-174. http://dx.doi.org/10.1007/s10886-013-0235-x

[10] Burgos, N.R., Talbert, R.E., Kim, K.S. and Kuk, Y.I. (2004) Growth Inhibition and Root Ultrastructure of Cucumber Seedlings Exposed to Allelochemicals from Rye (*Secale cereale*). *Journal of Chemical Ecology*, **30**, 671-689. http://dx.doi.org/10.1023/B:JOEC.0000018637.94002.ba

[11] Copaja, S.V., Villarroel, E., Bravo, H.R., Pizarro, L. and Argandona, V.H. (2006) Hydroxamic Acids in *Secale cereale* L. and The Relationship with Their Antifeedant and Allelopathic Properties. *Verlag der Zeitschriftfür Naturforschung*, **61**, 670-676. http://dx.doi.org/10.1515/znc-2006-9-1010

[12] Krogh, S.S., Mensz, S.J.M., Nielsen, S.T., Mortensen, A.G., Christophersen, C. and Fomsgaard, I.S. (2006) Fate of Benzoxazinone Allelochemicals in Soil after Incorporation of Wheat and Rye Sprouts. *Journal of Agricultural and Food Chemistry*, **54**, 1064-1074. http://dx.doi.org/10.1021/jf051147i

[13] Inderjit and Asakawa, C. (2001) Nature of Interference Potential of Hairy Vetch (*Vicia villosa* Roth) to Radish (*Raphanus sativus* L.) Does Allelopathy Play Any Role? *Crop Protection*, **20**, 261-265. http://dx.doi.org/10.1016/S0261-2194(00)00136-8

[14] Ercoli, L., Masoni, A. and Pampana, S. (2005) Short Communication: Weed Suppression by Winter Cover Crops. *Allelopathy Journal*, **16**, 273-278.

[15] Ercoli, L., Masoni, A., Pampana, S. and Arduini, I. (2007) Allelopathic Effects of Rye, Brown Mustard and Hairy Vetch on Redroot Pigweed, Common Lambsquarter and Knotweed. *Allelopathy Journal*, **19**, 249-256.

[16] Teasdale, J.R., Abdul-Baki, A.A., Park, Y.B. and Rosecrance, R.C. (2007) The Potential for Allelopathy during Decomposition of Hairy Vetch Residue. In: Fujii, Y. and Parvex, M.M., Eds., *Allelopathy: Concepts and Methodology*, Science Publishers, Enfield, 211-225.

[17] Kamo, T., Endo, M., Sato, M., Kasahara, R., Yamaya, H., Hiradate, S., Fujii, Y., Hirai, N. and Hirota, M. (2008) Limited Distribution of Natural Cyanamide in Higher Plants: Occurrence in *Vicia villosa* Subsp. *varia*, *V. cracca*, and *Robinia pseudo-acacia*. *Phytochemistry*, **69**, 1166-1172. http://dx.doi.org/10.1016/j.phytochem.2007.11.004

[18] Kamo, T., Kato, K., Hiradate, S., Nakajima, E., Fujii, Y. and Hirota, M. (2006) Evidence of Cyanamide Production in Hairy Vetch *Vicia villosa*. *Natural Product Research*, **20**, 429-433. http://dx.doi.org/10.1080/14786410500143583

[19] Guevara, E., Jimenez, V.M., Herrera, J. and Bangerth, F. (2008) Effect of Hydrogen Cyanamide on the Endogenous Hormonal Content of Pea Seedlings (*Pisum sativum* L.). *Brazilian Journal of Plant Physiology*, **20**, 159-163. http://dx.doi.org/10.1590/S1677-04202008000200008

[20] Jackson, J.E. and Bepete, M. (1995) The Effect of Hydrogen Cyanamide (Dormex) on Flowering and Cropping of Different Apple Cultivars under Tropical Conditions of Sub-Optimal Winter Chilling. *Scientia Horticulturae*, **60**, 293-304. http://dx.doi.org/10.1016/0304-4238(94)00707-M

[21] Or, E., Vilozny, I., Fennel, A., Eyal, Y. and Ogrodovitch, A. (2002) Dormancy in Grape Buds: Isolation and Characterization of Catalase cDNA and Analysis of Its Expression Following Chemical Induction of Bud Dormancy Release. *Plant Science*, **162**, 121-130. http://dx.doi.org/10.1016/S0168-9452(01)00542-8

[22] Turowski, M. and Deshmukh, B. (2004) Direct Chromatographic Method for Determination of Hydrogen Cyanamide and Dicyandiamide in Aqueous Solutions. *Analytical Letters*, **37**, 1981-1989. http://dx.doi.org/10.1081/AL-120039440

[23] Willison, W. and Anderson, J.M. (1991) Dicyandiamide as an Inhibitor of Denitrification in Coniferous Forest Soils. *Soil Biology & Biochemistry*, **23**, 605-607. http://dx.doi.org/10.1016/0038-0717(91)90071-Q

[24] Maier-Greiner, E.H., Obermaier-Skrobranek, B.M.M., Estermaier, L.M., Kammerloher, W., Freund, C., Wulfing, C., Burkert, U.I., Matern, D.H., Breuer, M., Eulitz, M., Kufrevioglu, O.I. and Hartmann, G.R. (1991) Isolation and Properties of a Nitrile Hydratase from the Soil Fungus *Myrothecium verrucaria* That Is Highly Specific for the Fertilizer Cyanamide and Cloning of Its Gene. *Proceedings of the Natural Academy of Sciences of the United States of America*, **88**, 4260-4264. http://dx.doi.org/10.1073/pnas.88.10.4260

[25] Kamo, T., Hiradate, S. and Fujii, Y. (2003) First Isolation of Natural Cyanamide as a Possible Allelochemical from Hairy Vetch. *Journal of Chemical Ecology*, **29**, 275-283. http://dx.doi.org/10.1023/A:1022621709486

[26] Wang, X., Hadrami, E.I., Adam, L.R. and Daayf, F. (2008) Differential Activation and Suppression of Potato Defence Responses by *Phytophthora infestans* Isolates Representing US-1 and US-8 Genotypes. *Plant Pathology*, **57**, 1026-1037. http://dx.doi.org/10.1111/j.1365-3059.2008.01866.x

[27] Lamari, L. (2008) ASSESS 2.0 Image Analysis Software for Plant Disease Quantification. American Phytopathologi-

cal Society, St. Paul.

[28] Littell, R.C., Miliken, G.A., Stroup, W.W. and Wolfinger, R.D. (1996) SAS System for Mixed Models. SAS Institute, Cary, 633.

[29] Lund, R.E. (1975) Tables for an Approximate Test for Outliers in Linear Models. *Technometrics*, **17**, 473-476. http://dx.doi.org/10.1080/00401706.1975.10489374

[30] Saxton, A. (1998) A Macro for Converting Mean Separation Output to Letter Groupings in Proc Mixed. In: *Proceedings of the 23rd SAS Users Group International*, SAS Institute, Cary, 1243-1246.

[31] Kamo, T., Sato, M., Kato, K., Hiradate, S., Nakajima, E., Fujii, Y. and Hirota, M. (2006) Quantification of Cyanamide Contents in Herbaceous Plants. *Bioscience, Biotechnology, and Biochemistry*, **70**, 2310-2312. http://dx.doi.org/10.1271/bbb.60171

[32] Hiradate, S., Kamo, T., Nakajima, E., Kato, K. and Fujii, Y. (2005) Direct Quantitative Determination of Cyanamide by Stable Isotope Dilution Gas Chromatography-Mass Spectrometry. *Journal of Chromatography A*, **1098**, 138-143. http://dx.doi.org/10.1016/j.chroma.2005.08.035

[33] Soltys, D., Bogatek, R. and Gniazdowska, A. (2012) Phytotoxic Effects of Cyanamide on Seed Germination and Seedling Growth of Weed and Crop Species. *Acta Biologica Cracoviensia Series Botanica*, **54**, 87-92. http://dx.doi.org/10.2478/v10182-012-0025-8

[34] Soltys, D., Rudzinska-Langwald, A., Kurek, W., Gniazdowska, A., Sliwinska, E. and Bogatek, R. (2011) Cyanamide Mode of Action during Inhibition of Onion (*Allium cepa* L.) Root Growth Involves Disturbances in Cell Division and Cytoskeleton Formation. *Planta*, **234**, 609-621. http://dx.doi.org/10.1007/s00425-011-1429-5

[35] Soltys, D., Rudzinska-Langwald, A., Gniazdowska, A., Wisniewska, A. and Bogatek, R. (2012) Inhibition of Tomato (*Solanum lycopersicum* L.) Root Growth by Cyanamide Is Due to Altered Cell Division, Phytohormone Balance and Expansin Gene Expression. *Planta*, **236**, 1629-1638. http://dx.doi.org/10.1007/s00425-012-1722-y

[36] Rice, C.P., Park, Y.B., Adam, F., Abdul-Baki, A.A. and Teasdale, J.R. (2005) Hydroxamic Acid Content and Toxicity of Rye at Selected Growth Stages. *Journal of Chemical Ecology*, **31**, 1887-1905. http://dx.doi.org/10.1007/s10886-005-5933-6

[37] Villagrasa, M., Eljarrat, E. and Barcelo, D. (2009) Analysis of Benzoxazinone Derivatives in Plant Tissues and Their Degradation Products in Agricultural Soils. *Trends Analytical Chemistry*, **28**, 1103-1114. http://dx.doi.org/10.1016/j.trac.2009.07.001

[38] Al-Karaki, G.N. (1998) Response of Wheat and Barley during Germination to Seed Osmopriming at Different Water Potential. *Journal of Agronomy and Crop Science*, **181**, 229-235. http://dx.doi.org/10.1111/j.1439-037X.1998.tb00422.x

[39] Hampson, C.R. and Simpson, G.M. (1989) Effect of Temperature, Salt, and Osmotic Potential on Early Growth of Wheat (*Triticum aestivum*). I. Germination. *Canadian Journal of Botany*, **68**, 524-528. http://dx.doi.org/10.1139/b90-072

[40] Boyd, N. and Van Acker, R. (2004) Seed Germination of Common Weed Species as Affected by Oxygen Concentration, Light, and Osmotic Potential. *Weed Science*, **52**, 589-596. http://dx.doi.org/10.1614/WS-03-15R2

28

Testing Leaf Multispectral Reflectance Data as Input into Random Forest to Differentiate Velvetleaf from Soybean

Reginald S. Fletcher

United States Department of Agriculture, Agricultural Research Service, Crop Production Systems Research Unit, Stoneville, USA
Email: reginald.fletcher@ars.usda.gov

Abstract

Velvetleaf (*Abutilon theophrasti* Medic.) infestations negatively impact row crop production throughout the United States and Canada's eastern provinces. To implement management strategies to control velvetleaf, managers need tools for differentiating it from crop plants. 5 Band, 7 Band, 8 Band, and 16 Band multispectral datasets simulating LANDSAT 3 plus a blue band, LANDSAT 8, WorldView 2, and WorldView 3 spectral bands, respectively were tested as input into the random forest algorithm for velvetleaf soybean [*Glycine max* L. (Merr.)] discrimination. During two separate greenhouse experiments in 2014, leaf reflectance measurements were obtained at the vegetative growth stage of velvetleaf plants and two soybean varieties. The reflectance measurements were collected with a plant contact probe attached to a hyperspectral spectroradiometer. Leaf hyperspectral reflectance measurements were convolved to the four multispectral datasets with computer software. Overall, user's, and producer's accuracies and kappa coefficient were employed to determine classification accuracies. Using the multispectral datasets as input, the random forest algorithm differentiated velvetleaf from the soybean varieties with accuracies ranging from 86.7% to 100%. 7 Band, 16 Band, 8 Band, and 5 Band datasets ranked or tied for the highest accuracies seventeen, sixteen, twelve, and one time, respectively. Kappa coefficients indicated an almost perfect agreement (*i.e.*, kappa value, 0.81 - 1.0) to substantial agreement (*i.e.*, kappa value, 0.61 - 0.80) between reference data and model predicted classes. This study was the first to demonstrate the application of the random forest machine learner and leaf multispectral reflectance data as tools to distinguish velvetleaf from soybean and to identify multispectral band combinations providing the best accuracies. Findings support further application of the random forest machine learner along with remotely-sensed multispectral data as tools for velvetleaf soybean discrimination with future implications for site-specific management of velvetleaf.

Keywords

***Glycine max*, *Abutilon theophrasti*, Machine Learning, Supervised Classification, Ensemble Technique**

1. Introduction

Velvetleaf (*Abutilon theophrasti* Medic.), a broadleaf plant native to China, was introduced into the United States from India as a fiber crop. It escaped cultivation and now has become a problem weed in row crops, especially in corn (*Zea mays* L.) and soybean (*Glycine max* (L.) Merr.) fields throughout the United States and Canada's eastern provinces. The summer annual weed grows to heights ranging from 0.3 to 2.0 m. The plant reproduces from seed and can develop up to 17,000 seeds that may remain viable for up to sixty years. Velvetleaf grows best in warm regions and invades vacant lots, gardens, and cultivated fields. Once established, it is a problem weed for many years to come.

Velvetleaf infestations negatively impact a crop and field in several ways. Velvetleaf plants emerging before or at the same time as crop plants are highly competitive for water and plant nutrients and thus can outgrow the crop. A 25% decrease in crop yield can occur when the velvetleaf plant population is equivalent to 1 plant per 30 cm [1]. Seeds, adult plants, and decaying plant parts contain or produce allelopathic (toxic) chemicals that inhibit water uptake and chlorophyll production of some crop plants, particularly soybean, thus preventing growth. The chemicals enter the soil during rain events.

Producers commonly use preemergence and postemergence measures to manage or control velvetleaf infestations. Detecting and eliminating the plant before seeding is vital because of the long-term dormancy of the seeds and the future problems they may cause. Therefore, field managers need additional techniques besides the common field survey for detecting velvetleaf infestation in crop fields.

Remote sensing technology has gained popularity as a tool for weed detection in agricultural systems [2]-[8]. The technology involves using ground, airborne, or satellite-borne sensors to obtain light reflectance measurements of plant leaves and canopies to differentiate between weed and crop plants. Detecting weeds with remote sensing technologies requires that differences in spectral reflectance exist between weeds and their environment and that the spatial and spectral resolution of remote sensing equipment is sufficient to detect these differences [9].

Soybean weed discrimination has been the focus of several remote sensing studies including velvetleaf as one of the weeds of interest. Reference [10] determined from statistical analysis of multispectral data spanning the visible to near infrared region of the light spectrum that weed-free soybean plots could be distinguished from soybean plus velvetleaf plots, soybean plus mixed weed plots, and soybean plus grass plots. The separation only occurred with red/infrared ratios. None of the soybean plus weed plots, however, could be distinguished from each other with single bands or red/infrared ratios.

Reference [11] obtained mixed results differentiating soybean from velvetleaf and foxtail (*Setaria faberi* Herrm.) in a controlled experiment. At one study site, they reported a classification error less than 17% for the weeds; at the other study site they achieved classification errors of 17% and 39% for foxtail and velvetleaf, respectively. Their study focused on using airborne multispectral imagery collected within the visible green (520 to 600 nm), visible red (630 to 690 nm), and near infrared (760 to 900 nm) wavebands. They concluded that if weed differentiation was not an issue for the weed management program then remote sensing techniques have good potential to differentiate weeds from crops.

Reference [4] demonstrated that a single decision tree approach based on the classification and regression technique could use vegetation indices as input to discriminate between corn, corn and velvetleaf, corn and a mixture of various grass species, corn and mixture of random predominant weed species, soybean, soybean and velvetleaf, soybean and mixture of various grass species, velvetleaf, mixed grass, and mixtures of random predominant weed species. The classification success rate was 85% ± 6%. The study focused on using twenty-four narrowband multispectral bands within the visible and near infrared regions of the light spectrum. From those bands, sixty-five normalized difference vegetation index bands were created and were used as input for classification. Accurate results were achieved; however, the sample size was small at only three plots per treatment.

Based on the above studies, more information is needed on the potential of using remote sensing technology for soybean weed discrimination, especially in the case of velvetleaf. Currently, information is lacking on the comparison of multispectral systems wavebands for soybean velvetleaf discrimination. Additionally, no information exists on including shortwave infrared spectral data to discriminate soybeans from velvetleaf. The shortwave infrared region of the light spectrum (1300 - 2500 nm) is sensitive to the water content in plants [12]. Finally, no information is available on the role that soybean variety may play in differentiating it from velvetleaf.

Another key aspect of using remote sensing technology is the computer algorithm employed to process the data. The success or failure of using the technology is affected by the algorithm selected to analyze the data. In this study, it is proposed to use the random forest machine learner for soybean velvetleaf discrimination. Random forest has gained popularity as a tool to use for classification problems because it is fully automated, and users have the ability to design powerful models with little experience in using the machine learner. Random forest has been ranked as one of the best learners to employ for classification and regression problems [13]. Researchers have successfully used it in genetics, clinical medicine, bioinformatics, agriculture, and remote sensing applications [14]-[17].

Random forest is an ensemble learning method based on the principle that a group of "weak learners" can come together to develop a "strong learner" [18]. Thus, it uses multiple decision trees to make a consensus prediction, hence the name random forest. Each decision tree in the "so-called forest" is derived from a bootstrap sample (*i.e.*, a percentage of the original data is selected for training, and the non-selected data are used for testing) of the original data (sampling with replacement). The splitting of each tree node is determined by the Gini criterion (*i.e.*, a measurement of node purity). For the splitting process, the algorithm selects a subset of the predictor variables at each node and then the best-splitting variable is chosen from that subset. Samples not selected in the bootstrap process for a tree (*i.e.*, approximately 36.8% of the original samples), known as "out-of-bag" (OOB) samples, are used to test the accuracy of the classifier. Random forest assigns an OOB sample to a class by using the decision trees in which the sample was OOB. The votes of each tree are tallied, and the OOB sample is assigned to the class receiving the largest votes. Compared with other machine learners, the random forest algorithm does not need an independent test set because the OOB samples serve as the test set [18]. Random forest also provides a variable importance reading representing the importance of each predictor variable to the model.

Currently, no information is available on using leaf multispectral reflectance data as input into random forest for soybean velvetleaf discrimination. The objective of this investigation was to evaluate leaf multispectral reflectance data as input into the random forest machine learner to differentiate velvetleaf from soybean. Specifically, the study focused on evaluating multispectral data mimicking the spectral bands of satellite sensors to discriminate the velvetleaf from two soybean varieties. Spectral bands of satellite sensors were chosen because the bands are strategically placed in different regions of the light spectrum for land cover mapping, thus providing different spectral band combinations for the model to test for separating velvetleaf from soybean.

2. Materials and Methods

2.1. Plant Descriptions

Two Progeny (P) brand LibertyLink (LL) soybean varieties (P4928LL and P5460LL, Progeny Ag Products, Wynne, Arkansas) and non-glyphosate resistant velvetleaf (United States Department of Agriculture, Agricultural Research Service, Stoneville, MS) were grown for the study. All three plants are characterized as pubescent plants, consisting of gray, light tawny, and white hairs for soybean P4928LL, soybean P5460LL, and velvetleaf, respectively. Soybean P4928LL is characterized as having an indeterminate growth habit (*i.e.*, a continuation of vegetative growth after flowering) and soybean P5460LL as having a determinate growth habit (*i.e.*, vegetative growth completed prior to flowering). The maturity group assigned to soybean P4928LL and soybean P5460LL are 4.9 and 5.4, respectively.

2.2. Greenhouse Experiment

The study was conducted at the United States Department of Agriculture, Agricultural Research Service, Stoneville, MS facility. Data were collected from two separate greenhouse experiments initiated on June 13, 2014, and

August 28, 2014. Soybean and velvetleaf seeds were sown in plugs containing commercial potting mix (Pro-Mix, Ultimate Potting Mix, Quakertown, Pennsylvania). Ten days after germination, thirty plants of each soybean variety and weed species were transplanted to individual 1 L pots filled with the commercial potting mix. Plants were watered at three- to four-day intervals. The potting mix consisted of a slow release nitrogen, phosphorus, and potassium fertilizer. The plants were grown at a temperature and photoperiod of 26.6°C and 14-h, respectively.

2.3. Data Collection

Leaf reflectance measurements were obtained with a full range hyperspectral spectroradiometer (FieldSpec 3, PANalytical Boulder, Boulder, CO). The instrument's fiber optic was attached to a plant probe (PANalytical Boulder, Boulder, CO) equipped with a light source. The plant probe has a 1 cm field of view. A leaf clip (PANalytical Boulder, Boulder, CO) was fastened to the contact probe. This device has a trigger lock/release gripping system designed to hold the leaf in place without removing it from the plant or causing damage to the plant. The leaf clip has a two-sided rotating head. One side of the head contains a black panel face, and the other side has a white panel face. The black and white panels are ideal for reflectance and transmittance measurements, respectively. The former was employed in this study.

The spectroradiometer obtained continuous spectra in the range of 350 - 2500 nm. Its sampling interval and spectral resolution were 1.4 nm and 3 nm, respectively, within the 350 nm to 1000 nm spectral range. The sampling interval and spectral resolution were 2 nm and 10 nm, respectively, within the 1000 nm to 2500 nm spectral range. The proprietary software operating the instrument resampled the reflectance data to 1 nm wavelengths.

Reflectance measurements were collected from the most recently matured leaf of each plant. Soybean has a trifoliate leaf, therefore, the center leaflet of the most recently matured leaf was chosen for data collection. At the selected sample spot of each plant leaf, reflectance measurements were an average of fifteen readings. Leaf reflectance measurements were obtained on June 30, 2014, and September 17, 2014, for the first and second experiments, respectively. For velvetleaf, it is important to identify and treat the plant prior to seeding. Measurements were obtained for all plants during the vegetative growth stage. The instrument was calibrated with a white spectralon panel (PANalytical Boulder, Boulder, CO) at 15-minute intervals.

2.4. Development of Multispectral Bands

The hyperspectral reflectance measurements of the soybean and velvetleaf leaves were resampled to four multispectral datasets (**Table 1**), referred to as 5 Band, 7 Band, 8 Band, and 16 Band. The green, red, and near infrared bands of the 5 Band dataset were replicates of the green, red, and near infrared bands obtained by LANDSAT 1 - 4 multispectral scanners [19]. The blue band was added to the 5 Band dataset to represent a broad region of the blue spectrum. Also, it represented blue light reflectance data obtained by many commercial handheld cameras. The 7 Band, 8 Band, and 16 Band datasets simulated the spectral bands of the LANDSAT 8 Operational Land Imager, WorldView 2 sensors, and WorldView 3 sensors, respectively [19]-[21]. The datasets were unique because they represented different regions of the light spectrum and provided spectral resolutions ranging from 20 to 300 nm. The multispectral bands were created by resampling the original hyperspectral bands using a Gaussian distribution function and the lower and the upper bounds of each satellite sensor spectral bands. The resampled spectral data were created with the hsdar package [22] of the R software [R version 3.2.0 (April 16, 2015) Full of Ingredients].

2.5. Classification Model Development

The conditional inference version of random forest (cforest) was used to create the models evaluated in this study. Reference [15] recommended using cforest instead of the original version of random forest if the prediction variables were highly correlated. Some of the variables were highly correlated in each dataset, thus justifying the use of cforest for model creation. Strong correlation between variables biases the variable importance rankings provided by random forest for classification or regression problems. Cforest implementation of random forest was designed to better handle correlation among variables, thus providing more accurate and unbiased rankings of the variable of importance [15]. The cforest technique utilizes conditional inference trees as base learners, in contrast to random forest, which employs classification and regression trees as base learners [15] [23]

Table 1. Spectral band descriptions and wavelengths of the multispectral datasets used in this study.

Spectral Band	Wavelengths of Each Dataset			
	5 Band[a]	7 Band	8 Band	16 Band
Coastal		430 - 450 nm	400 - 450 nm	400 - 450 nm
Blue	400 - 500 nm	450 - 510 nm	450 - 510 nm	450 - 510 nm
Green	500 - 600 nm	530 - 590 nm	510 - 580 nm	510 - 580 nm
Yellow			585 - 625 nm	585 - 625 nm
Red	600 - 700 nm	640 - 670 nm	630 - 690 nm	630 - 690 nm
Red-edge			705 - 745 nm	705 - 745 nm
Near infrared 1	700 - 800 nm	850 - 880 nm	770 - 895 nm	770 - 895 nm
Near infrared 2	800 - 1100 nm		860 - 1040 nm	860 - 1040 nm
Shortwave infrared 1		1570 - 1650 nm		1195 - 1225 nm
Shortwave infrared 2		2110 - 2290 nm		1550 - 1590 nm
Shortwave infrared 3				1640 - 1680 nm
Shortwave infrared 4				1710 - 1750 nm
Shortwave infrared 5				2145 - 2185 nm
Shortwave infrared 6				2185 - 2225 nm
Shortwave infrared 7				2235 - 2285 nm
Shortwave infrared 8				2295 - 2365 nm

[a]5 Band-simulates LANDSAT 3 spectral bands plus an additional blue band, 7 Band-simulates LANDSAT 8 spectral bands, 8 Band-simulates WorldView 2 spectral bands, and 16 Band-simulates WorldView 3 spectral bands.

[24]. Furthermore, instead of using bootstrap samples to construct its decision trees, cforest utilizes subsampling without replacement for constructing unbiased decision trees for the forest. Finally, the cforest algorithm uses the conditional permutation scheme described by [15] to determine the variable of importance ranking.

The number of samples to evaluate at each split of the tree (*mtry*) and the number of trees to use for creating the model (*ntree*) were the two parameters needed to be set before completing the classification. For this study, the default *mtry* value of 5 was used for each dataset. The default *ntree* value of 500 was employed as the starting point and was adjusted accordingly to obtain consistent variable importance rankings.

The following procedure was used to test the robustness of the models relative to variable importance [15]. A model was created using the default *mtry* and *ntree* values, the variable importance rankings were tabulated, and then the model was rerun using the same *mtry* and *ntree* values and a different starting seed (*i.e.*, the random generator used as a starting point for sampling). The model parameters were accepted if the variable importance ranking was similar between the first and the second runs. If the variable importance rankings were not consistent between runs, then the *ntree* value was increased by 1000, and the model was retested using the same *mtry* and seed values. This process was continued until a stable variable importance ranking was obtained.

2.6. Accuracy Assessment

Classification accuracies of the selected models were determined by evaluating the user's, producer's, and overall accuracies and kappa coefficient [24]. User's accuracy represents the percentage of predicted samples classified correctly. Producer's accuracy characterizes the percentage of reference samples correctly identified. The overall accuracy is a measure of the total number of correctly classified samples divided by the total number of samples. The kappa coefficient quantifies the variation between the observed agreement of the reference data and predicted data and the chance agreement between the two. The accuracy values were tabulated from the "out of bag" samples, those samples not used to train the model. Model development and evaluation were determined with the party package of the R software [25]-[27].

3. Results

3.1. Accuracy Assessment

The accuracy assessment results of the random forest classification for the June 30, 2014, dataset are summarized in **Table 2** for the velvetleaf soybean P4928LL classification. Overall, user's, and producer's accuracies greater than 90% were achieved for all of the multispectral datasets. The highest overall classification accuracy of 96.7% was obtained with the 7 Band, 8 Band, and 16 Band datasets; the lowest overall classification accuracy of 95% occurred for the 5 Band dataset. The same ranking order of the datasets was observed for the kappa coefficients (**Table 2**). The user's and producer's accuracy ranged from 93.3% to 100%. For the velvetleaf class, a tie occurred between the 7 Band and the 16 Band multispectral datasets for the highest user's accuracy; whereas, the 8 Band dataset ranked best in the producer's accuracy (**Table 2**). The 8 Band, and the 7 Band and 16 Band multispectral datasets achieved the greatest user's and producer's accuracies, respectively, for the soybean P4928LL class (**Table 2**).

The random forest classification results of the velvetleaf soybean P4928LL classes are tabulated in **Table 2** for the September 17, 2014, multispectral datasets. The 7 Band dataset obtained the highest measurement accuracies with 93.3%, 0.867, 90.6%, 96.4%, 96.7%, and 90.0% for overall accuracy, kappa coefficient, velvetleaf user's accuracy, soybean P4928LL user's accuracy, velvetleaf producer's accuracy, and soybean P4928LL producer's accuracy, respectively. The other multispectral datasets were tied for second in the measurement accuracies.

Overall, user's, and producer's accuracies and the kappa coefficients are presented in **Table 3** for the June 30, 2014, velvetleaf soybean P5460LL classification. The 7 Band, 8 Band, and 16 Band datasets ranked best in all accuracy categories. Their user's, producer's, and overall accuracies ranged from 96.7% to 100%, and the kappa coefficients were 0.967. The 5 Band dataset obtained the lowest accuracies, with user's, producer's, and overall accuracies ranging from 93.5% to 96.7%. The kappa value was 0.9.

The September 17, 2014, dataset for the velvetleaf soybean P5460LL classification indicated that the 16 Band dataset model was ranked or tied for first in all of the accuracy categories (**Table 3**). The 7 Band and 8 Band dataset models were tied for first for the soybean P5460LL producer's accuracy. They obtained the second highest accuracies for the other categories. The 5 Band dataset ranked last in all of the accuracy assessment categories.

Table 2. Accuracy assessment of the velvetleaf versus soybean P4928LL classification based on leaf multispectral data input into the random forest classifier.

Classification	Date	Accuracy Measurement	Multispectral Dataset[a]			
			5 Band	7 Band	8 Band	16 Band
Velvetleaf-soybean P4928LL	June 30, 2014	User's accuracy velvetleaf	93.5%	96.7%	93.8%	96.7%
		User's accuracy soybean P4928LL	96.6%	96.7%	100%	96.7%
		Producer's accuracy velvetleaf	96.7%	96.7%	100%	96.7%
		Producer's accuracy soybean P4928LL	93.3%	96.7%	93.3%	96.7%
		Overall accuracy	95.0%	96.7%	96.7%	96.7%
		Kappa coefficient	0.900	0.933	0.933	0.933
Velvetleaf-soybean P4928LL	September 17, 2014	User's accuracy velvetleaf	90.3%	90.6%	90.3%	90.3%
		User's accuracy soybean P4928LL	93.1%	96.4%	93.1%	93.1%
		Producer's accuracy velvetleaf	93.3%	96.7%	93.3%	93.3%
		Producer's accuracy soybean P4928LL	90.0%	90.0%	90.0%	90.0%
		Overall accuracy	91.7%	93.3%	91.7%	91.7%
		Kappa coefficient	0.833	0.867	0.833	0.833

[a]Refer to **Table 1** for the spectral band designations of the multispectral datasets.

Table 3. Accuracy assessment of the velvetleaf versus soybean P5460LL classification based on leaf multispectral data input into the random forest classifier.

Classification	Date	Accuracy Measurement	Multispectral Dataset[a]			
			5 Band	7 Band	8 Band	16 Band
Velvetleaf-soybean P5460LL	June 30, 2014	User's accuracy velvetleaf	93.5%	96.7%	96.7%	96.7%
		User's accuracy soybean P5460LL	96.6%	100%	100%	100%
		Producer's accuracy velvetleaf	96.7%	100%	100%	100%
		Producer's accuracy soybean P5460LL	93.3%	96.7%	96.7%	96.7%
		Overall accuracy	95.0%	98.3%	98.3%	98.3%
		Kappa coefficient	0.900	0.967	0.967	0.967
Velvetleaf-soybean P5460LL	September 17, 2014	User's accuracy velvetleaf	87.1%	93.1%	93.1%	93.3%
		User's accuracy soybean P5460LL	89.7%	90.3%	90.3%	93.3%
		Producer's accuracy velvetleaf	90.0%	90.0%	90.0%	93.3%
		Producer's accuracy soybean P5460LL	86.7%	93.3%	93.3%	93.3%
		Overall accuracy	88.3%	91.7%	91.7%	93.3%
		Kappa coefficient	0.767	0.833	0.833	0.867

[a]Refer to **Table 1** for the spectral band designations of the multispectral datasets.

3.2. Model Parameters

For fourteen out of the sixteen classification models, the default *mtry* and *ntree* values were adequate for obtaining stable variable importance readings (**Table 4**). The two exceptions were the random forest models used to complete the velvetleaf soybean P4928LL and the velvetleaf soybean 5460LL classifications based on the 8 Band and 16 Band datasets, respectively, for September 17, 2014. Three thousand five-hundred and 4500 trees were used to complete the classifications of the former and latter, respectively.

3.3. Variable Importance

The variable importance rankings of the random forest models used for the June 30, 2014, velvetleaf versus soybean P4928LL were as follows (**Figure 1**). The green (G) and near infrared two (NIR2) spectral bands were relevant to the model and had similar variable importance scores for the 5 Band dataset. The NIR1, G, and shortwave infrared one (SWIR1) spectral bands were important to the 7 Band dataset model while noticeable differences occurred in their variable importance scores. NIR1 and 2, G, and yellow (Y) spectral bands were needed by the model for the 8 Band dataset; the NIR2 and G spectral bands had similar variable importance scores and appeared in the top tier of variable importance scores. The NIR1 and Y spectral bands had variable importance scores similar to each other and appeared in the second tier of variable importance scores. SWIR1 to 4, NIR1 and 2, G, and Y spectral bands were the most important variables in the 16 Band dataset model. The spectral bands were grouped into six tiers: tier one-SWIR1, tier two-NIR2, tier three-G, tier four-NIR1 and Y, tier five-SWIR3, and tier six-SWIR2 and 4.

Variable importance rankings of the random forest models are shown in **Figure 2** for the September 17, 2014, velvetleaf versus soybean P4928LL classification. The G spectral band was the most important variable in the 5 Band dataset model. The G and blue (B) spectral bands were needed by 7 Band dataset model, with the G band ranked best. The G, Y, and B spectral bands were ranked most useful by the random forest model for the classification with the 8 Band dataset. Noticeable differences occurred in their importance scores. Eight spectral bands including G, Y, B, NIR1 and 2, red (R), SWIR1, and coastal (CA) were ranked important to the 16 Band dataset model. The G spectral band ranked first followed by the Y, B, NIR2, R, NIR1, SWIR1, and CA spectral bands.

Figure 3 illustrates the variable importance rankings of random forest models used in the classification of the June 30, 2014, velvetleaf and soybean P5460LL classes. The G and NIR2 spectral bands were ranked most im-

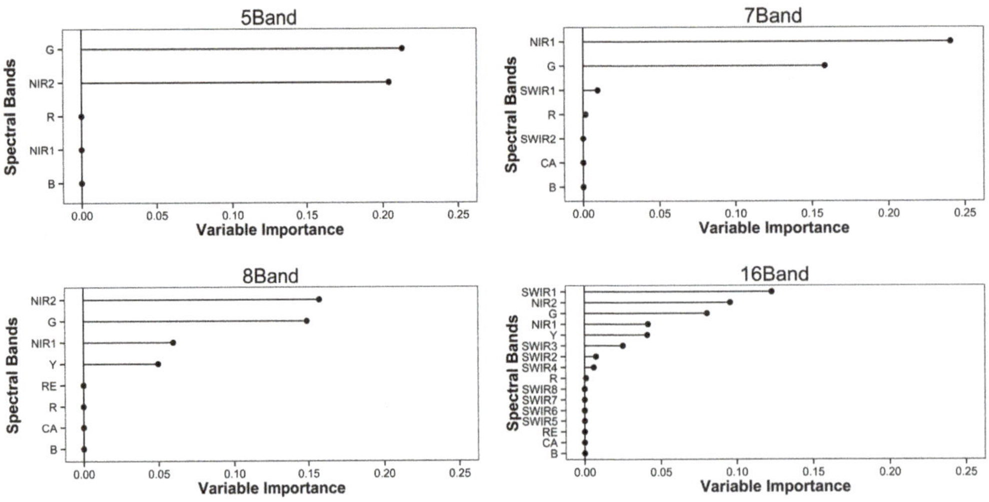

Figure 1. Variable importance rankings per multispectral dataset derived by the random forest model used for the velvetleaf and soybean P4928LL classification, June 30, 2014. CA = coastal, B = blue, G = green, Y = yellow, R = red, RE = red-edge, NIR = near infrared, and SWIR = shortwave infrared.

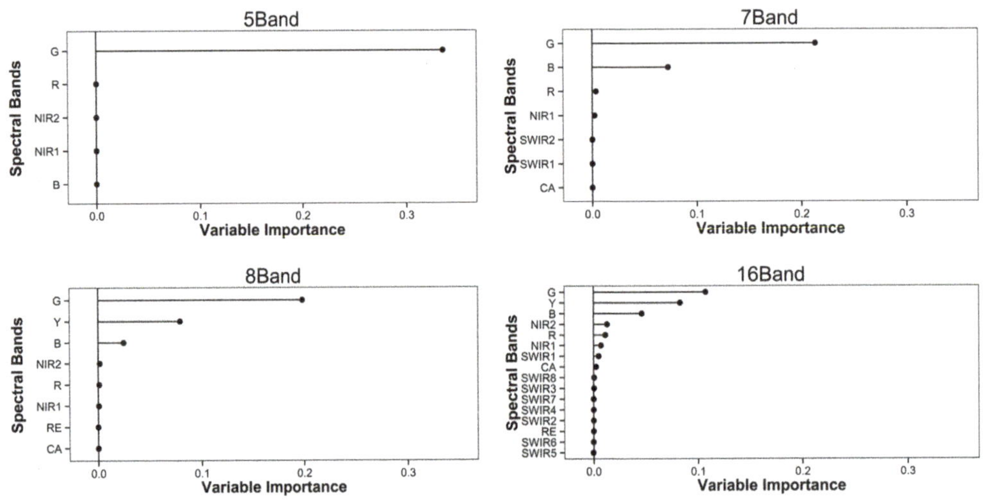

Figure 2. Variable importance rankings per multispectral dataset derived by the random forest model used for the velvetleaf and soybean P4928LL classification, September 17, 2014. CA = coastal, B = blue, G = green, Y = yellow, R = red, RE = red-edge, NIR = near infrared, and SWIR = shortwave infrared.

Table 4. Random forest model parameters used with the multispectral datasets to distinguish velvetleaf from two soybean varieties.

Classification	Dataset[a]	*mtry*[b]	*Ntrees* (June 30, 2014)	*Ntrees* (September 17, 2014)
Velvetleaf-soybean P4928LL	5 Band	5	500	500
Velvetleaf-soybean P4928LL	7 Band	5	500	500
Velvetleaf-soybean P4928LL	8 Band	5	500	3500
Velvetleaf-soybean P4928LL	16 Band	5	500	500
Velvetleaf-soybean P5460LL	5 Band	5	500	500
Velvetleaf-soybean P5460LL	7 Band	5	500	500
Velvetleaf-soybean P5460LL	8 Band	5	500	500
Velvetleaf-soybean P5460LL	16 Band	5	500	4500

[a]Refer to **Table 1** for the spectral band designations of the multispectral datasets. [b]*mtry* = number of randomly preselected variables; *ntrees* = number of trees used in the classification.

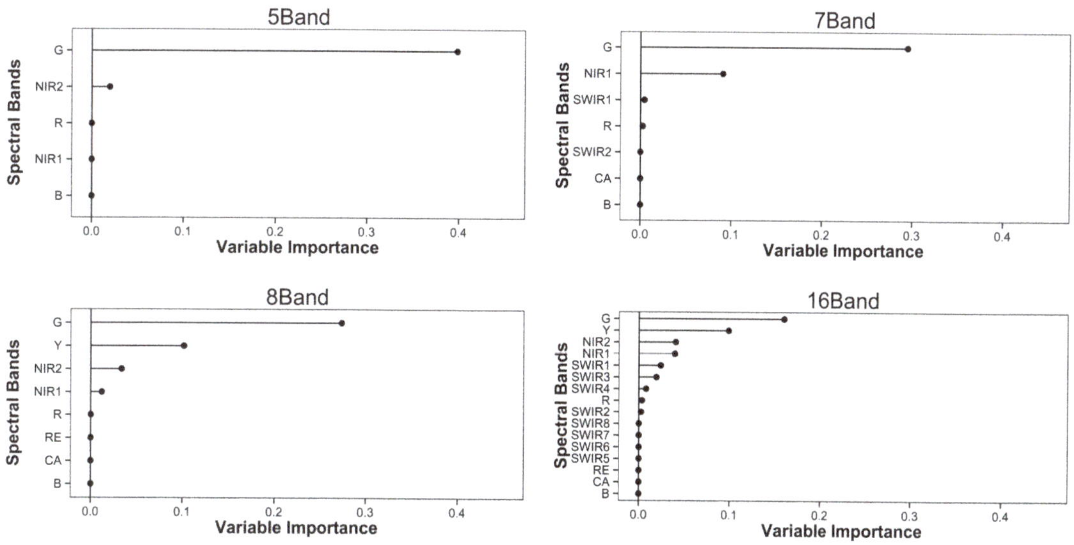

Figure 3. Variable importance rankings per multispectral dataset derived by the random forest model used for the velvetleaf and soybean P5460LL classification, June 30, 2014. CA = coastal, B = blue, G = green, Y = yellow, R = red, RE = red-edge, NIR = near infrared, and SWIR = shortwave infrared.

portant to the model for the 5 Band multispectral dataset. Distinct differences were observed in the scores, with the G band ranked the most important. Essential spectral bands for the 7 Band dataset model in descending order were G, NIR1, SWIR1, and R. The 8 Band dataset random forest model selected the G, Y, and NIR1 and 2 spectral bands as valuable variables for the classification; the rankings appeared in four distinct tiers: tier one-G, tier two-Y, tier three-NIR2, and tier four-NIR1. Five class tiers was observed for the most important rankings for the 16 Band dataset including the G spectral band in tier one, the Y spectral band in tier two, NIR1 and 2 spectral bands in tier three, SWIR spectral bands one and three in tier four, and the SWIR4 band in tier five.

Variable importance scores of the random forest models are shown in **Figure 4** for the September 17, 2014, velvetleaf P5460LL classification. The NIR2 and G spectral bands were the most useful to the model when using the 5 Band dataset, and their scores were nearly identical. The spectral bands critical to the classification model using the 7 Band dataset were as follows in descending order: NIR1, G, and B. There was an obvious difference in the variable importance scores. Four spectral bands were relevant to the model using the 8 Band dataset: NIR1, NIR2, G, and Y. NIR1, NIR2 and G, and Y spectral bands appeared in the first, second, and third tiers of the rankings, respectively. Eight spectral bands were relevant to the model using the 16 Band dataset, and their rankings in descending order were NIR2, NIR1, G, Y, SWIR1, B, SWIR8, and SWIR7.

4. Discussion

The objective of this study was to evaluate leaf multispectral reflectance data as input into the random forest classification algorithm to differentiate soybean from velvetleaf, an invasive weed affecting soybean production throughout the United States and eastern provinces of Canada. The study emphasized using different multispectral band combinations as input into the algorithm to differentiate velvetleaf from two different soybean varieties. The algorithm achieved overall, user's, and producer's accuracies that were greater than 85% for velvetleaf soybean discrimination (**Table 2** and **Table 3**), which was comparable to soybean weed discrimination studies using statistical methods [11] and single decision trees [4] to classify airborne imagery. Kappa values indicated that an almost perfect agreement (*i.e.*, kappa value range 0.81 - 1.0) to substantial agreement (*i.e.*, kappa value range 0.61 - 0.80) occurred between the reference data and predicted data (**Table 2** and **Table 3**). The latter was observed only for the 5 Band dataset for the velvetleaf soybean P5460LL classification occurring on September 17, 2014.

Generally, for all the datasets, the G and NIR spectral bands were ranked as important variables to the models for discriminating velvetleaf from soybean. Plant leaf reflectance and absorption of green light are influenced by leaf chlorophyll content [12], and may have been responsible for soybean velvetleaf differentiation. The inter-

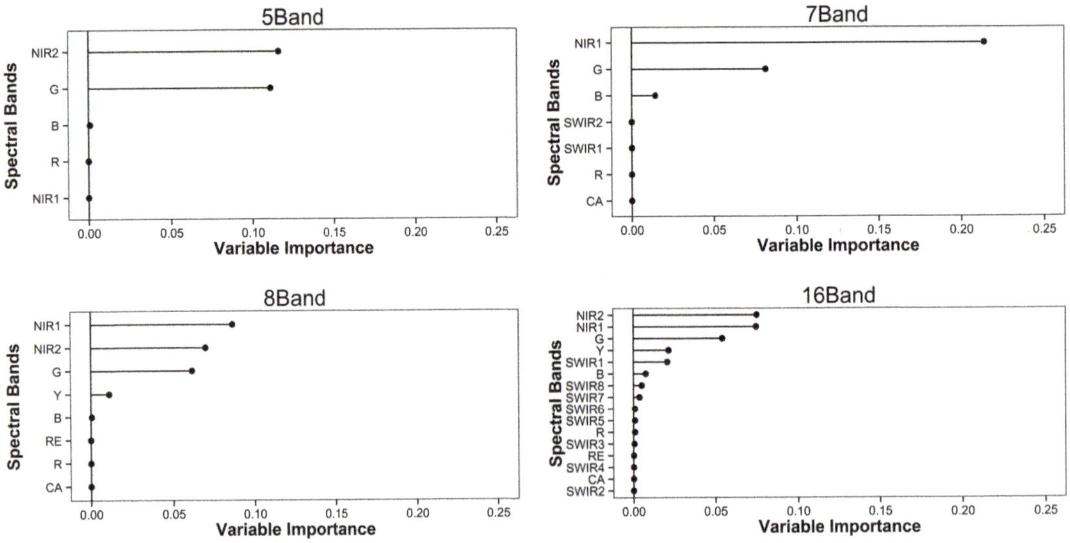

Figure 4. Variable importance rankings per multispectral dataset derived by the random forest model used for the velvetleaf and soybean P5460LL classification, September 17, 2014. CA = coastal, B = blue, G = green, Y = yellow, R = red, RE = red-edge, NIR = near infrared, and SWIR = shortwave infrared.

micellular spaces of plant leaves affect their ability to reflect and absorb near infrared light [12]. Therefore, leaf pigment and internal structure appear to be important components for distinguishing soybean from velvetleaf. Additionally, for the 7 Band and 16 Band datasets, the SWIR bands were important to the models for velvetleaf soybean discrimination; however, the SWIR bands' importance to a model was date specific. The shortwave infrared reflectance of plant leaves is affected by the water content of the leaf tissues [12]. Furthermore, for the 8 Band and 16 Band datasets, the Y spectral band was consistently ranked as an important variable to the models. Plant leaves reflectance of yellow light is also affected by chlorophyll content of the leaves.

With the increase in the number of spectral bands, more variables were ranked important to the random forest models (**Figures 1-4**); however, the increase in the number of bands per se did not always result in an increase in classification accuracy. For example, the number of accuracy test results completed for both dates and soybean varieties equal twenty-four. The 7 Band, 16 Band, 8 Band, and 5 Band datasets ranked or tied for the highest accuracies seventeen, sixteen, twelve, and one time, respectively. The differences in overall, user's, and producer's accuracies ranged from 0% to 6.6%, with the lowest accuracies occurring 95% of the time for the 5 Band dataset. For the kappa coefficients, the 5 Band model ranked last 100% of the time. The lower classification accuracies observed for the 5 Band dataset were most likely a result of the broader bandwidths (*i.e.*, 100 nm or greater). Also, the findings indicated that reliable accuracies generally can be achieved using the default *mtry* and *ntree* values (**Table 4**).

To put this study into perspective, leaf multispectral reflectance data were used as input into the random forest model for differentiating the velvetleaf from the soybean varieties. Leaf reflectance measurements represent pure reflectance measurements. Plant canopy response is affected by leaf angle, leaf positioning in the plant canopy, inter-canopy shadowing, soil background, and intermixing of plant canopies. Those aspects could lead to a different variable importance ranking of the spectral bands for plant canopy studies. Additionally, the study focused on binary classifications of soybean versus velvetleaf. Future studies need to focus on determining the potential of discriminating more than one weed at a time from soybean. Overall, this study provided valuable information on using the machine learning technique and on the influence of using different multispectral band combinations as input into the model for velvetleaf soybean discrimination.

5. Conclusion

This study provided new information on using the random forest algorithm with leaf multispectral reflectance data for differentiating velvetleaf from soybean. It demonstrated that the random forest algorithm could be used with a complement of multispectral datasets to separate velvetleaf from soybean. The best accuracies were achieved with multispectral datasets sensitive to visible (green and yellow spectral bands), near infrared, and shortwave infrared light. Findings support further application of the random forest machine learner along with remotely-sensed multispectral data as tools for velvetleaf soybean discrimination with future implications for site-specific management of velvetleaf.

Acknowledgements and Disclaimer

The author is grateful to Dr. Vijay Nandula for supplying the velvetleaf seed, Mr. Milton Gaston Jr., Mr. Arrington Smith, Ms. Keysha Hamilton, Mr. David Fisher, Ms. Raven Thompson, and Ms. Keyanna Nealon for their assistance in data collection, and Dr. Ken Fisher and Dr. Chenghai Yang for their critical review of the manuscript. Mention of trade names or commercial products in this report is solely for the purpose of providing specific information and does not imply recommendation or endorsement by the US Department of Agriculture.

References

[1] Lanini, W.T. and Wertz, B.A. (2015) Velvetleaf. Penn State Extension. http://extension.psu.edu/pests/weeds/weed-id/velvetleaf

[2] Koger, C.H., Bruce, L.M., Shaw, D.R. and Reddy, K.N. (2003) Wavelet Analysis of Hyperspectral Reflectance Data for Detecting Pitted Morning Glory (*Ipomoea lacunosa*) in Soybean (*Glycine max*). *Remote Sensing Environment*, **86**, 108-119. http://dx.doi.org/10.1016/S0034-4257(03)00071-3

[3] Smith, A.M. and Blackshaw, R.E. (2003) Weed-Crop Discrimination Using Remote Sensing: A Detached Leaf Experiment. *Weed Technology*, **17**, 811-820. http://dx.doi.org/10.1614/WT02-179

[4] Yang, C.C., Prasher, S.O. and Goel, P.K. (2004) Differentiation of Crop and Weeds by Decision-Tree Analysis of Multi-Spectral Data. *Transactions of the ASAE*, **47**, 873-879. http://dx.doi.org/10.13031/2013.16084

[5] Iqbal, J., Owens, P.R. and Ali., I. (2006) Application of Remote Sensing Data to Assess Weed Infestation in Cotton. *Agricultural Journal*, **1**, 186-191.

[6] Gómez-Casero, M.T., Castillejo-González, I.L. and García-Ferrer, A. (2010) Spectral Discrimination of Wild Oat and Canary Grass in Wheat Fields for Less Herbicide Application. *Agronomy for Sustainable Development*, **30**, 689-699. http://dx.doi.org/10.1051/agro/2009052

[7] Nieuwenhuizen, A.T., Hofstee, J.W., van de Zande, J.C., Meuleman, J. and van Henten, E.J. (2010) Classification of Sugar Beet and Volunteer Potato Reflection Spectra with a Neural Network and Statistical Discriminant Analysis to Select Discriminative Wavelengths. *Computers Electronics in Agriculture*, **73**, 146-153. http://dx.doi.org/10.1016/j.compag.2010.05.008

[8] de Castro, A.I., Jurado-Expósito, M., Gómez-Casero, M.T. and López-Granados, F. (2012) Applying Neural Networks to Hyperspectral and Multispectral Field Data for Discrimination of Cruciferous Weeds in Winter Crops. *Science World Journal*, Article ID: 630390. http://dx.doi.org/10.1100/2012/630390

[9] Lamb, D.W. and Brown, R.B. (2001) Remote-Sensing and Mapping of Weeds in Crops. *Journal of Agricultural Engneering Research*, **78**, 117-125. http://dx.doi.org/10.1006/jaer.2000.0630

[10] Goel, P.K., Prasher, S.O., Patel, R.M., Smith, D.L. and Di Tommaso, A. (2002) Use of Airborne Multi-Spectral Imagery for Weed Detection in Field Crops. *Transactions of American Society of Agricultural Engineers*, **45**, 443-449.

[11] Gibson, K.D., Dirks, R., Medlin, C.R. and Johnston, L. (2004) Detection of Weed Species in Soybean Using Multispectral Digital Images. *Weed Technology*, **18**, 742-749. http://dx.doi.org/10.1614/WT-03-170R1

[12] Gausman, H. (1985) Plant Leaf Optical Properties. Texas Tech Press, Lubbock.

[13] Fernández-Delgado, M., Cernadas, E., Barro, S. and Amorim, D. (2014) Do We Need Hundreds of Classifiers to Solve Real World Classification Problems? *Journal of Machine Learning Research*, **15**, 3133-3181.

[14] Gislason, P.O., Benediktsson, J.A. and Sveinsson, J.R. (2006) Random Forests for Land Cover Classification. *Pattern Recognition Letters*, **27**, 294-300. http://dx.doi.org/10.1016/j.patrec.2005.08.011

[15] Strobl, C., Malley, J. and Tutz, G. (2009) An Introduction to Recursive Partitioning: Rationale, Application and Characteristics of Classification and Regression Trees, Bagging and Random Forests. *Psychological Methods*, **14**, 323-348. http://dx.doi.org/10.1037/a0016973

[16] Goldstein, B.A., Polley, E.C. and Briggs, F.B.S. (2011) Random Forest for Genetic Association Studies. *Applications in Genetics and Molecular Biology*, **10**, 1-34. http://dx.doi.org/10.2202/1544-6115.1691

[17] Ok, A.O., Akar, O. and Gungor, O. (2012) Evaluation of Random Forest Method for Agricultural Crop Classification. *European Journal of Remote Sensing*, **45**, 421-432.

[18] Breiman, L. (2001) Random Forests. *Machine Learning*, **45**, 5-32. http://dx.doi.org/10.1023/A:1010933404324

[19] US Geological Survey (2015) Frequently Asked Questions about the Landsat Missions. http://landsat.usgs.gov/best_spectral_bands_to_use.php

[20] Digital Globe (2010) The Benefits of the Eight Spectral Bands of WorldView 2. http://global.digitalglobe.com/sites/default/files/DG-8SPECTRAL-WP_0.pdf

[21] Digital Globe (2014) WorldView 3 Data Sheet. https://dg-cms-uploads-production.s3.amazonaws.com/uploads/document/file/95/DG_WorldView3_DS_forWeb_0.pdf

[22] Lehnert, L.W., Meyer, H. and Bendix, J. (2015) Hsdar: Manage, Analyse and Simulate Hyperspectral Data in R. R Package Version 0.3.0. https://cran.r-project.org/web/packages/hsdar/index.html

[23] Hothorn, T., Buehlmann, P., Dudoit, S., Molinaro, A. and van der Laan, M. (2006) Survival Ensembles. *Biostatistics*, **7**, 355-373. http://dx.doi.org/10.1093/biostatistics/kxj011

[24] Congalton, R. and Green, K. (2009) Assessing the Accuracy of Remotely Sensed Data: Principles and Practices. 2nd Edition, CRC/Taylor & Francis, Boca Raton, 183 p.

[25] Hothorn, T., Hornik, K. and Zeileis, A. (2006) Unbiased Recursive Partitioning: A Conditional Inference Framework. *Journal of Computational and Graphical Statistics*, **15**, 651-674. http://dx.doi.org/10.1198/106186006X133933

[26] Strobl, C., Boulesteix, A.L., Zeileis, A. and Hothorn, T. (2007) Bias in Random Forest Variable Importance Measures: Illustrations, Sources and a Solution. *BMC Bioinformatics*, **8**, 25. http://dx.doi.org/10.1186/1471-2105-8-25

[27] Strobl, C., Boulesteix, A.L., Kneib, T., Augustin, T. and Zeileis, A. (2008) Conditional Variable Importance for Random Forests. *BMC Bioinformatics*, **9**, 307. http://www.biomedcentral.com/1471-2105/9/307 http://dx.doi.org/10.1186/1471-2105-9-307

29

Micronutrient Status in Soil of Central India

Khageshwar Singh Patel*, Santosh Chikhlekar, Shobhana Ramteke, Bharat Lal Sahu, Nohar Singh Dahariya, Reetu Sharma

School of Studies in Chemistry/Environmental Science, Pt. Ravishankar Shukla University, Raipur, India
Email: *patelkhageshwarsingh@gmail.com

Abstract

Two major issues, *i.e.* large crop productions and huge anthropogenic activities (e.g. fuel burning and mineral roasting) disturb the micronutrient balance in the soil of India. In this work, the available and total status of eight micronutrients *i.e.* Fe, Mn, Cu, Zn, Co, Ni, Mo, and S of the soils in the most urbanized area: Raipur area, Chhattisgarh, India (extending over ≈ 2 × 10^4 km^2) is described. The available status of micronutrients *i.e.* Fe, Mn, Cu, Zn, Co, Ni, Mo and SO_4^{2-} in the soils (n = 100) was ranged from 30 - 8253, 205 - 2800, 2.0 - 8.1, 0.7 - 5.0, 2.2 - 31.2, 0.1 - 13.4, 0.1 - 8.9 and 41 - 747 mg/kg with mean value of (at 95% probability) 642 ± 186, 1178 ± 119, 4.3 ± 0.3, 2.3 ± 0.2, 12.8 ± 1.3, 3.9 ± 0.6, 1.5 ± 0.3 and 281 ± 25 mg/kg, respectively. The concentration variations, deficiencies and toxicities of the micronutrients in the soil are discussed.

Keywords

Soil, Micronutrient, Available, Total Content

1. Introduction

The micronutrients *i.e.* Fe, Cu, Zn, Mn, Co, Ni, Mo, and S in soil play a very important role in plant growth, productivity, soil fertility and animal nutrition [1]. The main functions of the micronutrients in living organism are structural components of cell constituents and its metabolically active compounds, in the maintenance of cellular organization, in energy transformation, in enzyme action, etc. [2]. The increment in nutrient supply beyond a certain limit resulting in the decreased yield of plants is often be associated with the production of specific toxic effects [3]. The incidence of micronutrient deficiencies in soil and plants is increasing due to high and multiple plant yields. The quantification of both total and available (active form) of nutrients in soil is important [4]. The main sources of micronutrients in soils are rock weathering and atmospheric deposition in form of dust, precipitates, volatile compounds, etc. The micronutrients in soil occur in different chemical forms *i.e.* water so-

*Corresponding author.

luble, exchangeable, specifically adsorbed, chelated or complexed, secondary clay minerals or oxide, primary minerals, etc. [5].

Their available contents were leached out from soil with various extractants *i.e.* diethylenetriaminepenta acetic acid (DTPA), ammonium bicarbonate-DTPA (AB-DTPA), triethanolamine-DTPA (TEA-DTPA), Mehlich-1 (0.05 N HCl + 0.025 N H_2SO_4), Mehlich-3 (0.2 N CH_3COOH + 0.25 N NH_4NO_3 + O.O15 N NH_4F + 0.013 N HNO_3 + 0.001 M EDTA), acid ammonium acetate-EDTA (AAA-EDTA), $MgCl_2$, HCl (0.05 N), HNO_3 (0.31 N), ammonium acetate, water, etc. [6]. However, most of these extractants were suffered from some shortcoming *i.e.* unable to extract several trace elements present in soil, not always efficient for all nutrients, etc. Thus now a days, multi-nutrient extractants *i.e.* Mehlich-3, AB-DTPA, TEA-DTPA, acid ammonium acetate-EDTA etc. were widely used for the extraction of micronutrients and trace elements from the soils [6]. Among them, AB-DTPA and TEA-DTPA were claimed better extractant than Melich-3 for Zn and Cu, but they extracted only Zn, Cu, Mn, and Fe. However, the AAA-EDTA leached out several nutrients *i.e.* Mn, Fe, Co, Ni, Cu, Zn, Al, Cd, Mo, Cr, Pb, Sr, P, etc. Ammonium oxalate, ammonium acetate, hot water, etc. were reported for the leaching of the available Mo from the soil. Of these oxalate is widely used for the extraction of available Mo from soil but it required prolonged extraction period (≈ 24 hr). The hot water extraction was recommended for leaching of the available Mo from the soil. Calcium chloride, Bray-1 (0.03N NH_4F + 0.025 N HCl), Morgan's reagent (sodium acetate-acetic acid, pH 4.8), deionized water, etc. were reported for the extraction of S from soil. The total content of micronutrients in soil was leached out with acids *i.e.* aqua-regia, $HClO_4$, HF, $HClO_4$ + HNO_3, H_2SO_4 + HCl + HNO_3, etc. [6].

The micronutrient status in surface soils of some parts of India was reported [7]-[23]. However, the information on the levels of micronutrient *i.e.* Co, Ni and Mo in the soil is lacking. In this work, the status of eight micronutrients *i.e.* Fe, Cu, Zn, Mn, Co, Ni, Mo, and S in surface soils of 100 villages of Raipur district is described. The concentration variations, deficiencies and toxicities of the micronutrients in the soil are discussed.

2. Materials and Methods

2.1. Study Area

The most of urbanization and industrialization in central India has been marked nearby capital city, Raipur, Chhattisgarh state, India. Raipur area includes Raipur district (22°33'N - 21°14'N and 82°6' - 81°38'E) and surrounding districts *i.e.* Balodabazar and Gariabandh. They are situated in the fertile plains of Chhattisgarh region of the country. Hundred city, town and villages of Raipur area ($\approx 2.0 \times 10^4$ km^2) were selected for determining the micronutrient status of the soil.

2.2. Sample Collection

Generally, three different types of soil *i.e.* red laterite, gray, yellow soils occurred in this region. Three different types of soil from 100 villages of Raipur block were collected, **Figure 1**. Soils were taken from horizon of 0 - 15 cm depth. A total 300 soil samples were collected in February 2013 as described in the literature [24].

2.3. Analysis of pH and Extraction

The soils were dried, ground and sieved through a 2-mm sieve. All samples were stored in a 500-mL wide mouth polythene bottles for the analysis. A 10.0 g weighed amount of soil was taken in a 100-mL polythene conical flask by mixing with 20 mL deionized water. The mixture was shaken for 6 hrs, and their pH and electrical conductivity (EC) values were measured with the Hanna sensor-HI 991300N.

A mixed solution of reagents (E. Merck) *i.e.* AAAA-EDTA for the extraction of nutrient *i.e.* Mn, Fe, Co, Ni, Cu, Zn and PO_4^{3-} was used by dissolving 38.5 g ammonium acetate, 9.5 g Na_2EDTA and 29 mL acetic acid (17 M) into 1 L deionized water [25]. A 10 g dried and ground soil sample was taken into a 250-mL polyethylene flask with subsequent addition of 100 mL AAAA-EDTA solution. The mixture was equilibrated for 1 hr with a shaker, and solution was filtered through a 0.45 µm glass fiber filter in a 100-mLpolyethylene volumetric flask.

For Mo, a 10 g soil sample was taken into 250-mL conical polyethylene flask by mixing with 100 mL deionized water [26]. It was heated at boiling temperature for 10 min by subsequent filtering the cold solution as above. The activated charcoal and hot water were employed for the extraction of the available content of sulfate. A 10 g soil sample was mixed with 1 g activated charcoal and 100 mL deionized water into a 250-mL conical

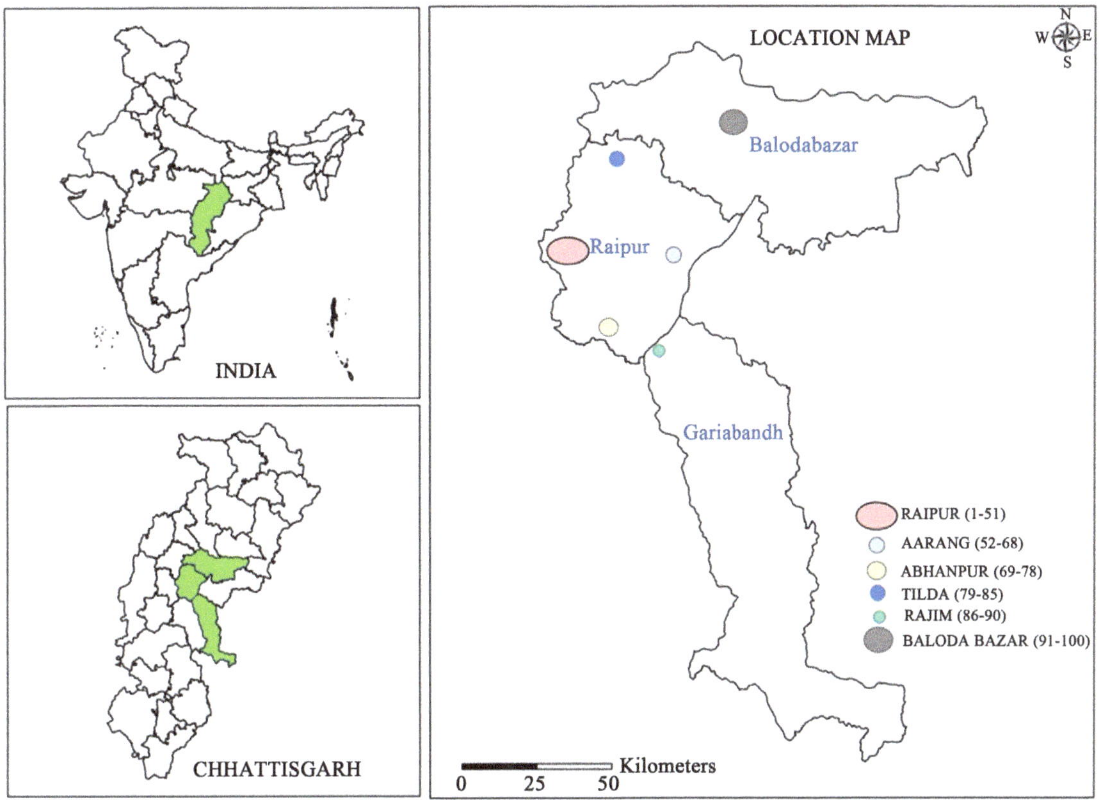

Figure 1. Representation of sampling locations in Chhattisgarh state of the country.

polyethylene flask. The mixture was shaken for a duration of 1 hr by a shaker and solution was filtered through a 0.45 µm glass fiber as above.

The mixed acid (H_2SO_4 + HCl + HNO_3) was used for extraction of the total content of micronutrients *i.e.* Mn, Fe, Co, Ni, Cu and Zn. A 1.0 g dried and powdered soil sample was taken in a 100-mL Teflon beaker. Into it, 30.0 ml of mixed acid solution (H_2SO_4 + HCl + HNO_3) was added. The mixture was heated until white fumes were no longer emitted. The residue was washed with hot dilute hydrochloric acid (0.01 N) and the hot water (50˚C). The mixture was filtered through a 0.45 µm glass fiber as above.

For leaching of total content of sulfate, a 1 g soil sample was mixed with 50 mL solution of acids: (HNO_3 + $HClO_4$) into a 100-mL Teflon beaker by keeping overnight (12 hr). The solution was concentrated to 20 mL by gentle heating, and the cold solution was filtered through a glass fiber as above. The filtrate was evaporated to the dryness by subsequent dissolving with deionized water in a 50-mL polyethylene flask.

2.4. Analysis of Micronutrients

The Flame GBC 932AAwas used for analysis of metals *i.e.* Mn, Fe, Co, Ni, Cu, Zn and Mo in the soil. The Dionex ion chromatography-1100—was employed for analysis of SO_4^{2-} and PO_4^{3-}.

3. Results and Discussion

3.1. Soil Characteristics

The agricultural land includes ≈ 50% area of the total land of the studied area. Three types of soils *i.e.* gray, yellow and red are available in the studied area. The red and yellow soil was originated by weathering of various rocks *i.e.* quartz, feldspars, mica and iron coated quartz formed over different geological periods. The yellow color was ascribed to the higher degree of hydration of the ferric oxide in these soils. The color shaded was also varied from reddish yellow to yellowish brown with often fine textured. The pH and EC of soil was ranged from 4.7 -7.7 and 100 - 900 µS/cm.

3.2. Iron

The levels of available micronutrients in the soils of Raipur area is presented in **Table 1**. Iron comprises about 5% of the earth's crust and is the fourth most abundant element in the lithosphere [27]. The most of the soil iron was found in primary mineral, clays, oxides and hydroxides. The available and total content of Fe in soils of the studied area were varied widely and ranged from 30 - 8253 and 11676 - 40,928 mg/kg with mean value of 642 ± 186 and 19930 ± 5979 mg/kg, respectively. Considering 6 mg/kg as the critical value of Fe, the soils of studied area was found to be contaminated with a very high level of Fe.

3.3. Copper

The concentration of Cu in the earth's crust is averaged 28 mg/kg [27]. The available and total status of Cu in soils of the studied area was ranged from 2.0 - 8.0 and 45 - 69 mg/kg with mean value of 4.3 ± 0.3 and 53 ± 16 mg/kg, respectively. A 0.2 and 50 mg/kg Cu were reported as critical and threshold value for Cu-deficiency and Cu-toxicity to plant growth. Almost all soil of this region was found to be contaminated with sufficient amount of Cu for the healthy growth of plants.

3.4. Zinc

The Zinc content of the lithosphere is 67 mg/kg [27]. Zinc has a strong tendency to combine with sulfide ores, and it occurs most frequently in the lithosphere as sphalerite. The available and total status of Zn in soils of this region was ranged from 0.7 - 5.0 and 27 - 56 mg/kg with mean value of 2.3 ± 0.2 and 38 ± 14 mg/kg, respectively. Critical limit for Zn-deficiency in different type of soils for different crops were ranged from 0.4 to 0.8 mg/kg. A few soils of studied area was found to be deficient in available Zn for the plant growth if the value 0.80 mg/kg was considered as a critical limit. A value of 50 mg/kg Zn was reported as threshold value for the plant toxicity. None of soil of the studied is contaminated Zn at the toxic level.

3.5. Manganese

Manganese concentration in the earth's crust is 1000 mg/kg [27]. The available and total level of Mn in soils of this region lie in the range of 205 - 2800 and 2737 -10,122 mg/kg with mean value 1178 ± 119 and 6889 ± 2274 mg/kg, respectively. A 5.7 and 55 mg/kg were reported as the critical limit for Mn-deficiency and threshold value of Mn-toxicity for plant growth, respectively.

3.6. Cobalt

The average total cobalt concentration in the earth's crust is 40 mg/kg [27]. The available and total concentration of cobalt in soils of this region were varied from 2.2 - 31.2 and 64 - 139 mg/kg with mean value 12.8 ± 1.3 and 119 ± 30 mg/kg, respectively. Considering the 2.5 mg/kg as the critical limit for Co deficiency in soil, almost all soils of this region may be rated as contaminated with sufficient level of Co for plant growth.

3.7. Nickel

The natural abundance of nickel in the earth' crust is 47 mg/kg [27]. The available and total level of Ni in soils of this region were varied from 0.1 - 13.4 and 15 - 70 mg/kg with mean value of 3.9 ± 0.6 and 35 ± 11 mg/kg, respectively. A 0.1 and 50 mg/kg Ni in soil were considered as the critical limit, and threshold value of toxicity for plant growth, respectively. All type of soils were found to be contaminated with a sufficient level of Ni for plant growth.

3.8. Molybdenum

Molybdenum occurs in the soils in extremely small quantities, is usually found in concentrations of less than 1 mg/kg [27]. The available, and total Mo content in soils of the studied area varied from 0.1 - 8.9 and 3.4 - 9.2 mg/kg with mean 1.5 ± 0.3 and 4.6 ± 2.6 mg/kg, respectively. A 0.1 mg/kg was considered as the critical limit for Mo deficiency in soil. A 20.0 mg/kg Mo in soil was considered as threshold value for toxicity. None of the soil in this region was found to contain Mo at the toxic level.

Table 1. Mean (n = 3) micronutrient status of soil, mg/kg.

S. No.	Area	Location	Fe	Cu	Zn	Mn	Co	Ni	Mo	SO_4^{2-}
1	Raipur	Dagniya	333	3.2	0.8	1736	19.7	0.1	1.4	269
2		Jaunda	450	3.4	1.4	1709	6.9	0.1	1.9	410
3		Raipura	151	3.4	2.1	504	4.6	6.0	1.1	210
4		Sarona	157	3.2	1.4	1064	9.9	6.3	1.4	167
5		Atari	326	4.5	1.8	1316	11.9	7.6	0.1	165
6		Ravabhata	323	6.0	4.1	672	12.6	4.1	0.1	199
7		Dhaneli	386	6.3	2.7	840	9.5	4.9	1.9	185
8		Sankara	157	3.4	1.5	924	16.0	3.2	2.9	110
9		Dharsiva	257	4.8	1.4	869	8.1	3.1	8.9	229
10		Deori	280	4.5	2.0	1036	12.3	7.3	7.1	222
11		Sardoo	238	4.3	1.7	2296	21.6	12.3	1.4	249
12		Sarora	208	4.8	1.4	448	18.5	5.2	0.1	324
13		Deopuri	139	3.9	4.5	1764	17.5	2.0	2.5	199
14		Mana	165	4.8	5.0	1400	16.9	11.1	1.5	285
15		Parsatti	321	4.3	2.9	1485	18.3	2.7	1.6	207
16		Labhandi	74	3.9	2.1	1260	18.9	8.0	0.1	585
17		Jora	74	3.6	1.8	504	8.0	0.6	3.2	176
18		Piroda	146	4.1	0.7	1316	12.5	4.3	2.3	276
19		Chherkhedi	55	3.5	2.2	1120	7.7	0.8	2.5	384
20		MandirHasaud	233	4.9	2.1	2800	11.8	1.5	2.0	310
21		Kurud	134	4.1	1.1	350	17.2	0.4	1.2	170
22		Baradera	177	3.5	1.0	476	17.2	0.3	1.5	188
23		Bhanakhedi	263	4.3	2.2	1092	27.0	3.1	1.4	677
24		Chandkhurai	839	4.5	0.7	742	15.3	2.0	0.1	392
25		Badgaon	800	3.4	4.8	420	13.7	7.3	0.1	299
26		Khamtarai	8253	6.0	4.9	1680	13.2	1.7	0.7	747
27		Bothali	935	4.2	1.5	913	14.6	5.3	1.0	436
28		Gullu	427	7.3	3.6	1147	12.3	0.4	0.9	332
29		Bhandarpuri	1004	2.9	1.8	2062	15.0	5.3	0.1	282
30		Boriyakhurd	1832	7.0	3.8	2478	14.1	1.7	1.4	269
31		Dunda	2124	7.0	2.5	2103	15.0	1.3	0.1	254
32		Sejbahar	1428	5.0	2.2	1467	19.2	5.5	0.1	155

Continued

33		Datrenga	1786	4.8	2.0	1006	20.6	4.2	0.1	174
34		Mujgahan	2668	5.7	2.1	888	12.7	3.6	0.1	179
35		Mova	258	3.9	0.7	205	4.3	3.1	2.8	268
36		Daldalseoni	377	5.6	1.7	371	3.2	3.9	2.3	330
37		Saddu	447	7.4	1.0	819	11.8	7.1	2.0	367
38		Sankari	306	3.6	1.5	676	10.1	3.2	2.2	243
39		Dhansuli	235	4.5	1.1	1491	3.2	1.7	1.2	189
40		Lohara	237	4.5	1.4	2746	2.9	0.8	1.7	140
41		Dondekhurd	175	4.5	1.3	680	2.2	1.1	1.3	191
42		Dondekala	251	4.9	1.7	470	5.6	1.5	2.0	358
43		Chataud	258	4.5	1.7	320	3.8	3.1	2.6	343
44		Semaria	132	2.2	2.0	818	4.1	1.3	0.3	163
45		Bhurkoni	44	3.4	1.8	438	2.9	4.9	3.6	162
46		Lalpur	194	4.2	2.4	1217	4.1	0.7	1.6	285
47		Chhapora	122	3.2	2.4	1433	9.5	1.7	1.3	166
48		Donde	226	3.5	1.7	1211	3.4	3.2	1.3	71
49		Saragaon	412	5.6	2.9	321	6.2	3.8	2.2	66
50		Narhada	155	3.4	1.4	1219	7.0	2.5	1.5	41
51		Baronda	429	3.8	1.8	219	13.2	1.1	1.1	101
52	Arang	Godhi	817	4.8	2.5	476	27.9	8.5	0.1	303
53		Paragaon (Arang)	528	3.6	2.2	1792	4.5	1.3	0.6	232
54		Amethi	345	5.6	1.8	717	17.2	9.8	2.0	321
55		Ranisagar	269	4.2	1.1	1408	12.0	1.7	0.4	353
56		Banarasi	482	5.3	2.1	1488	14.8	6.6	0.4	418
57		Bana	1033	4.2	1.7	1379	28.3	3.8	1.9	358
58		Kosrangi	1135	5.5	3.2	984	9.5	3.9	1.2	329
59		Arang	1734	8.1	1.1	907	21.3	4.5	0.7	270
60		Rasani	914	2.0	1.5	644	5.9	1.0	1.5	254
61		Lakholi	1249	2.1	1.5	1515	21.0	1.3	0.7	357
62		Amsena	2240	5.6	2.2	1671	13.9	5.2	0.1	233
63		Chikhali	1189	2.1	2.0	1185	20.9	2.2	0.1	325
64		Kutela	966	2.5	3.2	583	9.0	3.1	0.3	249
65		Kurud	1245	3.1	1.7	290	7.4	1.0	0.9	259
66		Kagdehi	770	2.2	3.1	387	4.5	2.1	1.5	290

Continued

67		Samoda	1685	2.8	2.7	693	6.4	3.4	0.1	118
68		Chaprid	868	2.4	2.8	1909	15.3	3.2	1.2	162
69	Abhanpur	Abhanpur	288	3.2	1.7	2100	13.6	13.4	0.1	426
70		Kendri	218	3.6	2.9	1092	5.5	9.9	0.8	160
71		Manikchauri	130	5.2	1.4	616	18.1	7.8	2.7	325
72		Chhachanpairi	427	3.9	3.5	888	10.9	6.7	1.7	124
73		Kolar	1810	6.4	4.6	1108	26.5	2.0	0.1	343
74		Khopra	367	4.2	1.7	815	9.1	4.5	1.3	275
75		Sakari	609	3.2	0.8	933	3.2	11.2	1.6	252
76		Chandi	378	4.6	0.8	387	8.1	1.3	1.7	262
77		Kachna	250	6.0	2.8	320	4.5	0.7	3.2	385
78		Amasibani	169	6.0	2.0	2301	10.2	1.3	2.3	190
79	Tilda	Tilda	112	6.2	1.8	1652	31.2	7.3	3.4	286
80		Biladi	132	3.4	4.5	2548	6.6	2.8	3.4	395
81		Kota	272	5.5	5.0	1961	11.1	5.5	1.8	332
82		Bahesar	84	4.5	4.6	1485	19.2	6.6	3.1	414
83		Tarashiv	118	4.6	4.6	1540	17.8	5.9	2.9	515
84		Murabangoli	30	5.6	4.9	1988	13.2	2.8	3.9	284
85		Kharora	487	5.7	3.2	1379	15.7	1.1	1.6	253
86	Rajim	Bhelvadih	193	3.6	3.4	1541	13.7	3.2	0.1	285
87		Navapara	169	4.6	1.0	1652	20.2	4.6	0.1	325
88		Kurra	95	4.3	1.3	2045	9.7	2.4	1.6	207
89		Pipraud	125	5.0	3.9	1736	17.4	9.1	2.3	285
90		Rajim	244	2.7	0.8	896	5.3	1.7	1.6	258
91	Balodabazar	Koliyari	500	2.7	0.8	784	13.9	5.2	0.8	213
92		Navagaon	484	4.6	1.0	1933	10.5	6.2	3.3	732
93		Charoda	347	4.6	2.2	934	16.4	7.6	1.8	226
94		Sirri	1089	4.3	2.4	1050	22.3	1.0	2.2	369
95		Kanki	1169	2.1	2.8	539	18.9	1.8	0.5	232
96		Bhaisa	1135	2.0	4.3	1312	4.6	5.9	1.9	274
97		Bhaismudi	1306	2.8	4.3	1302	12.6	4.2	1.3	266
98		Chanpa Jhar	1513	6.0	1.3	1624	21.8	3.6	0.1	272
99		Torla	367	3.4	2.4	1650	19.6	0.6	1.4	324
100		Tama Seoni	694	3.9	2.0	1092	18.5	4.5	0.8	551

3.9. Sulphur

The earth's crust contain about 0.06% Sulphur [27]. It is mostly present as sulfides, sulfates and organic combinations with C and N. The available and total SO_4^{2-} in this region was lie in the range of 41 - 747 and 294 - 1782 mg/kg with mean value of 281 ± 25 and 631 ± 180 mg/kg, respectively. The critical limit of SO_4^{2-} in the soil was reported to be 10 mg/kg. No soils of the studied area were found to be deficient in sulfur level.

3.10. Concentration Variations and Statistics

The range, mean, median, kurtosis and skewness values of micronutrient concentration in soils of 100 villages of the studied area are presented in **Table 2**. The highest value of Fe or SO_4^{2-}, Cu, Zn, Mn, Co, Ni and Mo was observed in site *i.e.* Khamtarai, Arang, Murabangoli, Mandir Hasaud, Tilda, Abanpur and Dharsiva, respectively, **Figure 2**. The content of six micronutrients *i.e.* SO_4^{2-}, Cu, Zn, Co, Ni and Mo are symmetrically distributed in all locations, **Figure 3**. However, large variations in the content of two micronutrients *i.e.* Fe and Mn was seen, may be due to input by the industrial effluents, **Figure 3**. The data for seven micronutrients *i.e.* Cu, Zn, Mn, Co, Ni, Mo and SO_4^{2-} were found to be distributed normally and symmetrically with comparable median and mean values. A large variation in the case of Fe was noticed, may be due to asymmetric distribution of iron minerals in the soil. The available content of the micronutrients in soils of the studied area was found in the following decreasing order: Mn > Fe > SO_4^{2-} > Co > Cu > Ni > Zn > Mo, **Figure 4**. Among them, manganese was found to be at the highest level followed by Fe. However in the case of total levels, a different trend (Fe > Mn > SO_4^{2-} > Co > Cu > Zn > Ni > Mo) was observed, **Figure 4**. A large fraction of S was found to be available for the plant growth, **Figure 5**. However, low to moderate fractions of other micronutrients were available, **Figure 5**. All of them showed a positive correlation between the available and total metal content with the highest and lowest values for Ni and Cu, respectively, **Figure 6**.

The available content of all micronutrients except Zn and SO_4^{2-} was found to decrease as the soil profile was increased from 0 to 120 cm, may be due to strong adsorption of the cations by the geomedia, **Figure 7**. The total content of all micronutrients except SO_4^{2-} was found to increase as the soil depth profile was increased from 0 to 120 cm due to their poor adsorption by the geo-media, **Figure 8**.

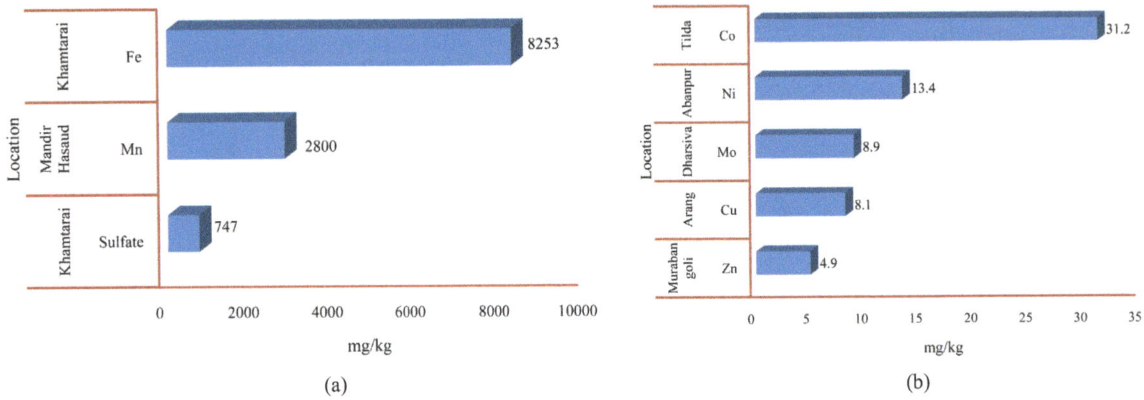

Figure 2. Representation of locations showing maximum micronutrient concentration in studied area.

Table 2. Statistics for distribution of available micronutrients in the soil, (n = 100).

Parameter	Fe	Cu	Zn	Mn	Co	Ni	Mo	SO_4^{2-}
Minimum, mg/kg	30	2.0	0.7	205	2.2	0.1	0.1	41
Maximum, mg/kg	8253	8.1	5	2800	31.2	13.4	8.9	747
Mean, mg/kg	642	4.3	2.3	1178	12.8	3.9	1.5	281
Median, mg/kg	330	4.3	2.0	1100	12.6	3.2	1.4	269
Kurt	41.7	0.1	−0.1	−0.3	−0.3	0.8	9.9	3.3
Skew	5.5	0.4	0.9	0.5	0.4	1.0	2.3	1.4

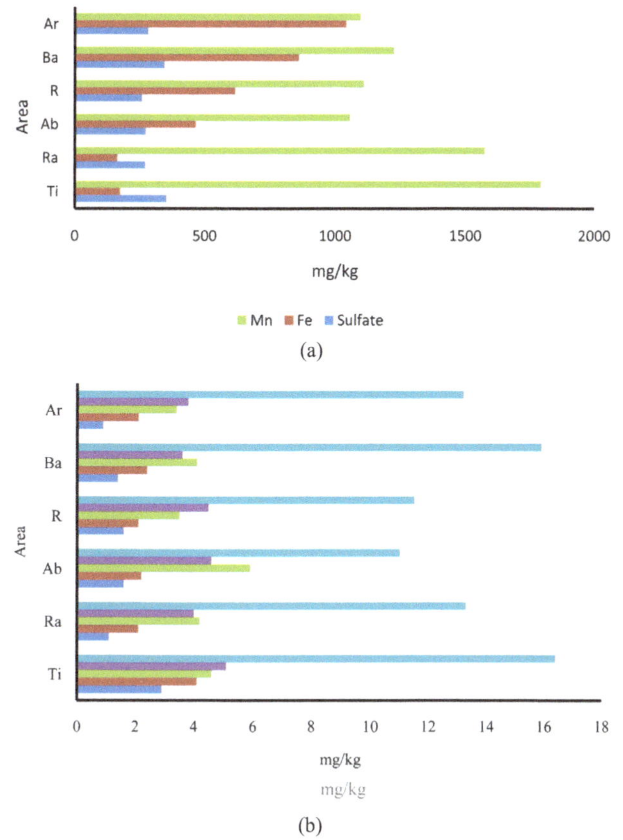

Figure 3. Representation of distribution pattern of micronutrients in surface soil of Raipur area.

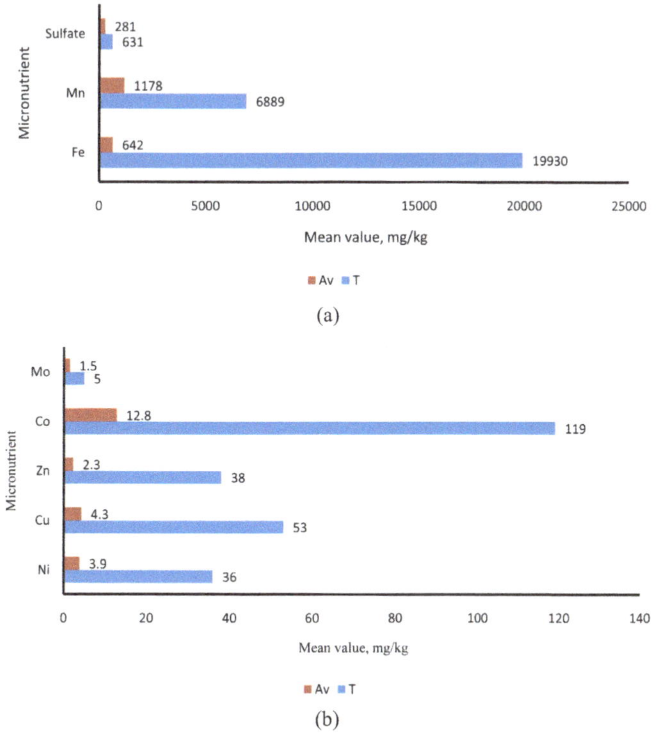

Figure 4. Mean value of total (T) and available (Av) content of micronutrient in the surface soil of Raipur area.

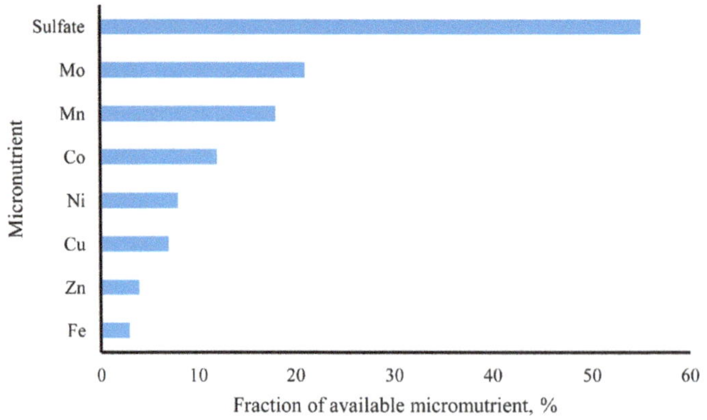

Figure 5. Fraction of available content of micronutrient in surface soil of Raipur area.

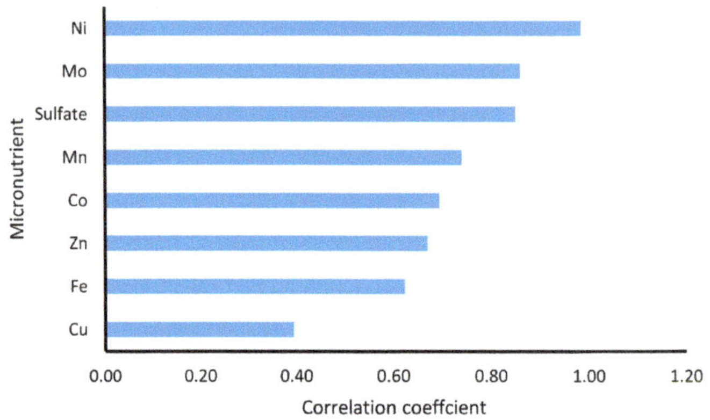

Figure 6. Correlation coefficient of total and available content of micronutrient in surface of Raipur area.

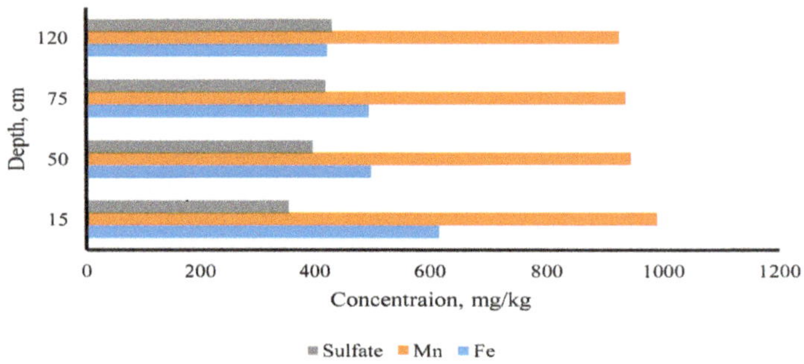

(a)

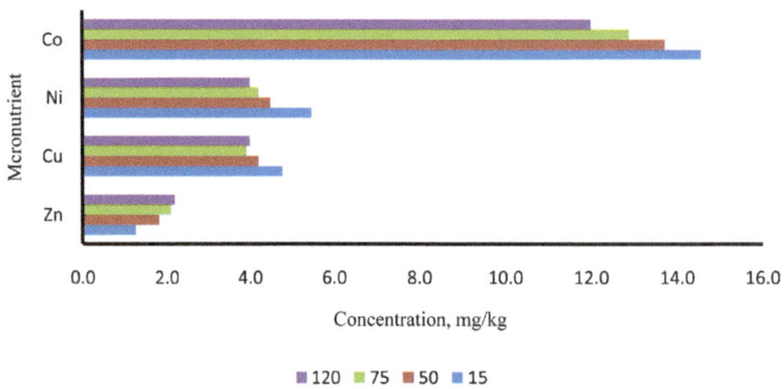

(b)

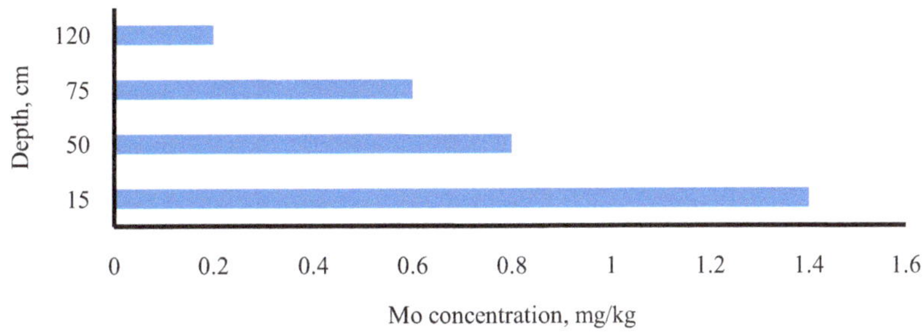

(c)

Figure 7. Depth profile studies of available content of micronutrient in soil of Raipur area.

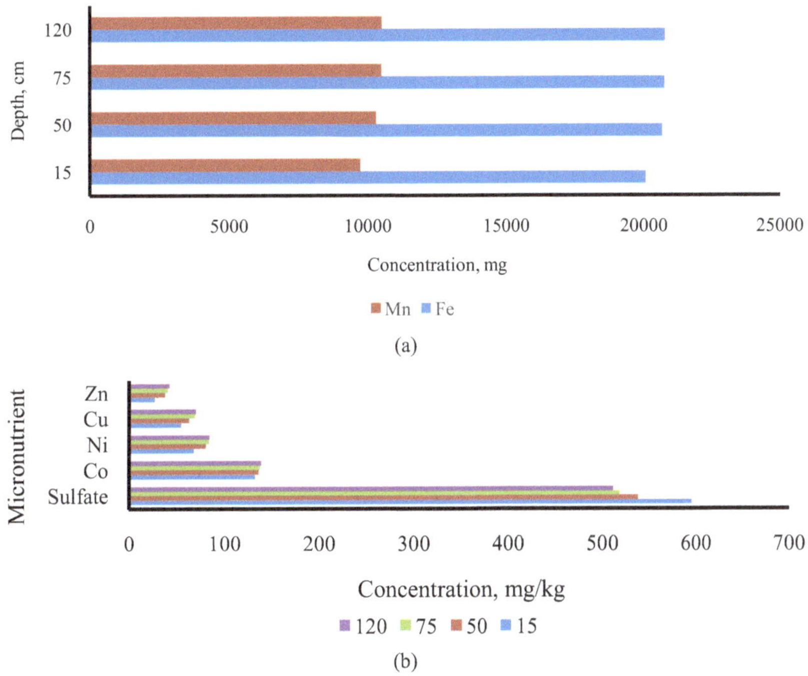

Figure 8. Depth profile studies of total content of micronutrient in soil of Raipur area.

3.11. Micronutrient Deficiency and Toxicity

The micronutrients *i.e.* Fe, Mn, Co, Ni, Cu, Zn, Mo and S are used in very small amounts. The presence of micronutrients below critical limit often causes adverse effects in plant growth and in their yields. For each micronutrient, the critical levels, limit of deficiency, toxicity and optimal growth vary with the genotype and are profoundly affected by plant metabolism and by edaphic and environmental factor that affect the absorption of nutrients. In the studied area, five micronutrients *i.e.* Co, Ni, Cu, Mo and S are found in soils at the sufficient levels. Two micronutrients *i.e.* Fe and Mn are present at toxic levels in all types of soils of this region. In some area, the

Zn deficiency in the soils is observed. The iron chlorosis was commonly seen in the plants of this region which may be due to very high amount of Fe content in soil. The brown or purplish spots on leaves, on lower part of the stem and leaf margins are commonly marked in this region that may be due to manganese toxicity.The most visible zinc deficiency symptoms *i.e.* short internodes and a decrease in leaf size were observed in the plants of this region.

3.12. Comparison of Micronutrient Status

The concentration of available maximum amount of Fe, Mn, Zn, SO_4^{2-} and Cu reported was 12234, 269, 121, 70, and 30 mg/kg, respectively [8] [11] [21]. The levels of micronutrients *i.e.* Fe, Mn, and S were found to be at the highest levels in this region. The content of Cu was present at moderated levels similar to other part of country [7]. The Zn content was present low levels as occurred in other parts of the Country [7]. The level of Mo was present at moderate levels but high content of Fe, Mn, and S may cause Mo deficiency.

4. Conclusion

All types of soils in Raipur area are found to be associated with high levels of Fe, Mn, S; moderate levels of Cu and low level of Mo and Zn. The relative abundance of free form of micronutrients (available content/total content) in soils of this region is found in following decreasing order: S >> Mn > Mo > Cu >> Zn ≈ Fe. The adverse effects *i.e.* chlorosis of young leaves, premature fall of fruits, narcotics, stunned growth of plants/crops in plants of this region are frequently seen may be due to either Fe and Mn toxicities or Zn deficiency or their combination. The Zn deficiency could be corrected by application of Zn compounds e.g. Zinc sulfate, Zinc oxide, Zinc phosphate, etc. in the soil.

Acknowledgements

We are thankful to the Madhya Pradesh Council of Science & Technology, Raipur for granting scholarship to one of the author: SC.

References

[1] Gupta, U.C., Kening, W. and Siyuan, L. (2008) Micronutrients in Soils, Crops and Livestock. *Earth Science Frontiers*, **15**, 110-125. http://dx.doi.org/10.1016/S1872-5791(09)60003-8

[2] Renwick, A.G. and Walker, R. (2008) Risk Assessment of Micronutrients. *Toxicology Letters*, **180**, 123-130. http://dx.doi.org/10.1016/j.toxlet.2008.05.009

[3] Fageria, N.K., Baligar, V.C. and Clark, R.B. (2002) Micronutrients in Crop Production. *Advances in Agronomy*, **77**, 185-268. http://dx.doi.org/10.1016/S0065-2113(02)77015-6

[4] Welch, R.M. and Shuman, L. (1995) Micronutrient Nutrition of Plants. *Critical Reviews in Plant Sciences*, **14**, 49-82. http://dx.doi.org/10.1080/07352689509701922

[5] Mortvedt, J.J., Cox, F.R., Shuman, L.M. and Welch, R.M. (1991) Micronutrients in Agriculture Second Edition Number 4 in the Soil Science Society of America Book Series Chapter 14. Soil Science Society of America, Inc., Madison.

[6] Chatzistathis, T. (2014) Micronutrient Deficiency in Soils & Plants, Micronutrient Solubility and Availability in Soils. Bentham Science Publishers Ltd., Oak Park.

[7] Gupta, A.P. (2005) Micronutrient Status and Fertilizer Use Scenario in India. *Journal of Trace Elements in Medicine and Biology*, **18**, 325-331. http://dx.doi.org/10.1016/j.jtemb.2005.04.003

[8] Gowda, N.K.S., Ramana, J.V., Prasad, C.S. and Singh, K. (2004) Micronutrient Content of Certain Tropical Conventional and Unconventional Feed Resources of Southern India. *Tropical Animal Health and Production*, **36**, 77-94. http://dx.doi.org/10.1023/B:TROP.0000009522.30949.1d

[9] Yadav, B.K. (2011) Micronutrient Status of Soils under Legume Crops in Arid Region of Western Rajasthan, India. *Academic Journal of Plant Sciences*, **4**, 94-97.

[10] Sharma, J.C. and Chaudhary, S.K. (2007) Vertical Distribution of Micronutrient Cations in Relation to Soil Characteristics in Lower Shiwalika of Solan District in North-West Himalayas. *Journal of the Indian Society of Soil Science*, **55**, 40-44.

[11] Venkastesh, M.S., Majumdar, B.K. and Patriam, K. (2003) Status of Micronutrient Cations under Various Land Use Systems of Meghalaya. *Journal of the Indian Society of Soil Science*, **51**, 60-64.

[12] Sharma, R.P., Singh, M. and Sharma, J.P. (2003) Correlation Studies on Micronutrients Vis-a-Vis Soil Properties in Some Soils of Nagaur Districts in Semi-Arid Region of Rajasthan. *Journal of the Indian Society of Soil Science*, **51**, 522-527.

[13] Somasundaram, J., Singh, R.K., Parandiyal, A.K. and Prasad, S.N. (2009) Micronutrient Status of Soils under Different Land Use Systems in Chambal Ravines. *Journal of the Indian Society of Soil Science*, **57**, 307-312.

[14] Yadav, R.L. and Meena, M.C. (2009) Available Micronutrient Status and Their Relationship with Soil Properties of Degana Soil Series of Rajasthan. *Journal of the Indian Society of Soil Science*, **57**, 90-92.

[15] Hundal, H.S., Raj, K., Singh, D. and Machandra, J.S. (2006) Available Nutrient and Heavy Metal Status of Soils of Punjab, North-West India. *Journal of the Indian Society of Soil Science*, **54**, 50-56.

[16] Anil, S., Sharma, P.K., Tur, N.S. and Nayyar, V.K. (2009) Micronutrient Status and Their Spatial Variability in Soils of Muktsar District of Punjab—A GIS Approach. *Journal of the Indian Society of Soil Science*, **57**, 300-306.

[17] Kumar, M. and Babel, A.L. (2011) Available Micronutrient Status and Their Relationship with Soil Properties of Jhunjhunu Tehsil, District Jhunjhunu, Rajasthan, India. *Journal of Agricultural Science*, **3**, 97-106. http://www.ccsenet.org/journal/index.php/jas/article/view/11011 http://dx.doi.org/10.5539/jas.v3n2p97

[18] Prasad, B. and Sinha, N.P. (1982) Changes in the Status of Micronutrients in Soil with Long Term Applications of Chemical Fertilizers, Lime and Manure. *Plant and Soil*, **64**, 437-441. http://dx.doi.org/10.1007/BF02372526

[19] Arunachalam, P., Kannan, P., Prabukumar, G. and Govindaraj, M. (2013) Zinc Deficiency in Indian Soils with Special Focus to Enrich Zinc in Peanut. *African Journal of Agricultural Research*, **8**, 6681-6688. http://eprints.icrisat.ac.in/id/eprint/12416

[20] Patra, P., Mondal, S. and Ghosh, G.K. (2012) Status of Available Sulphur in Surface and Sub-Surface Soils of Red and Lateritic Soils of West Bengal. *International Journal of Plant Animal and Environmental Sciences*, **2**, 276-281.

[21] Singh, A.H., Kumarjit, S.R.K., Singh L.N., Gopimohan, S.N., Nandini, C. and Kumar, S.A.K. (2007) Sulphur Status and forms in Acid Soils of Manipur. *Indian Journal of Agricultural Research*, **41**, 205-209.

[22] Velmurugana, R., Mahendranb, P.P., Wania, S.P., Uttama, K. and Prabhavathic, M. (2013) Molybdenum Status and Critical Limit in the Soil for Green Gram (*Vigna radiata*) Growing in Madurai and Sivagangai Districts of Tamil Nadu, India. *Soil Science and Plant Nutrition*, **59**, 229-236. http://dx.doi.org/10.1080/00380768.2012.763021

[23] Dhamak, A.L., Meshram, N.A. and Waikar, S.L. (2014) Assessment of Micronutrient Indices in Soils of Ambajogai Tahsil of Beed District under Semiarid Agro Ecological Region (Maharashtra). *Journal of Research in Agriculture and Animal Science*, **2**, 40-43. http://www.questjournals.org/

[24] Tan, K.H. (2005) Soil Sampling, Preparation and Analysis. 2nd Edition, CRC Press, Boca Raton. https://www.crcpress.com/Books-in-Soils-Plants-and-the-Environment/book-series/CRCBKSPE

[25] Sippola, J. (1994) Acid Ammonium Acetate-EDTA Universal Extractant in Soil Testing and Environmental Monitoring. *Communications in Soil Science and Plant Analysis*, **25**, 1755-1761. http://dx.doi.org/10.1080/00103629409369149

[26] Zbiral, J. (1992) Determination of Molybdenum in Hot-Water Soil Extracts: Influence of pH and Available Iron on the Molybdenum Content. *Communications in Soil Science and Plant Analysis*, **23**, 817-825. http://dx.doi.org/10.1080/00103629209368631

[27] Rudnick, R.L. and Gao, S. (2003) The Composition of the Continental Crust. In: Holland, H.D. and Turekian, K.K., Eds., *Treatise on Geochemistry*, Vol. 3, The Crust, Elsevier-Pergamon, Oxford, 1-64. http://dx.doi.org/10.1016/b0-08-043751-6/03016-4

Permissions

All chapters in this book were first published in AJPS, by Scientific Research Publishing; hereby published with permission under the Creative Commons Attribution License or equivalent. Every chapter published in this book has been scrutinized by our experts. Their significance has been extensively debated. The topics covered herein carry significant findings which will fuel the growth of the discipline. They may even be implemented as practical applications or may be referred to as a beginning point for another development.

The contributors of this book come from diverse backgrounds, making this book a truly international effort. This book will bring forth new frontiers with its revolutionizing research information and detailed analysis of the nascent developments around the world.

We would like to thank all the contributing authors for lending their expertise to make the book truly unique. They have played a crucial role in the development of this book. Without their invaluable contributions this book wouldn't have been possible. They have made vital efforts to compile up to date information on the varied aspects of this subject to make this book a valuable addition to the collection of many professionals and students.

This book was conceptualized with the vision of imparting up-to-date information and advanced data in this field. To ensure the same, a matchless editorial board was set up. Every individual on the board went through rigorous rounds of assessment to prove their worth. After which they invested a large part of their time researching and compiling the most relevant data for our readers.

The editorial board has been involved in producing this book since its inception. They have spent rigorous hours researching and exploring the diverse topics which have resulted in the successful publishing of this book. They have passed on their knowledge of decades through this book. To expedite this challenging task, the publisher supported the team at every step. A small team of assistant editors was also appointed to further simplify the editing procedure and attain best results for the readers.

Apart from the editorial board, the designing team has also invested a significant amount of their time in understanding the subject and creating the most relevant covers. They scrutinized every image to scout for the most suitable representation of the subject and create an appropriate cover for the book.

The publishing team has been an ardent support to the editorial, designing and production team. Their endless efforts to recruit the best for this project, has resulted in the accomplishment of this book. They are a veteran in the field of academics and their pool of knowledge is as vast as their experience in printing. Their expertise and guidance has proved useful at every step. Their uncompromising quality standards have made this book an exceptional effort. Their encouragement from time to time has been an inspiration for everyone.

The publisher and the editorial board hope that this book will prove to be a valuable piece of knowledge for researchers, students, practitioners and scholars across the globe.

List of Contributors

Qinzhi Yu
Guilin Economic Crops Technology Extending Station, Guilin, China

Congyao Liang
Nanzhang County Agricultural Technology Promotion Center, Xiangyang, China

Xianyu Wang
Agricultural College, Guangxi University, Nanning, China

Yongcheng Du
Institute of Vegetables and Flowers, Chinese Academy of Agricultural Sciences, Beijing, China

Masaharu Masuda and Kenji Murakami
Okayama University, Okayama, Japan

Chitta Ranjan Deb, Sakutemsu L. Jamir and Nangshimeren Sakutemsu Jamir
Department of Botany, Nagaland University, Lumami, India

Vinicius F. Moreira, Ivo J. C. Vieira and Raimundo Braz-Filho
Sector of Natural Products Chemistry, Universidade Estadual do Norte Fluminense Darcy Ribeiro, Campos dos Goytacazes, Brazil

Zamir K. Punja
Department of Biological Sciences, Simon Fraser University, Burnaby, Canada

Mahfuzur Rahman
WVU Extension Service, West Virginia University, Morgantown, USA

Silindile P. Miya and Albert T. Modi
Crop Science, School of Agricultural, Earth and Environmental Sciences, University of KwaZulu-Natal, Scottsville, USA

Xinwei Chen, Chaohe Huangfu and Dianlin Yang
College of Plant Protection, Shenyang Agricultural University, Shenyang, China
Agro-Environmental Protection Institute, Ministry of Agriculture, Tianjin, China

Hongmei Liu and Hui Wang
Agro-Environmental Protection Institute, Ministry of Agriculture, Tianjin, China

Yuxiao Liu
Institute of Information, Tianjin Academy of Agricultural Sciences, Tianjin, China

Tomomi Kakita, Ayami Abe and Takashi Ikeda
School of Agriculture, Meiji University, Kawasaki, Japan

Marcos Schleiden Sousa Carvalho, Luciane Vilela Resende, Gabriele Mikami Costa and Luis Felipe Lima e Silva
Department of Agriculture, Federal University of Lavras, Lavras, Brazil

Maria das Graças Cardoso, Luiz Roberto Marques Albuquerque, Anni Cristini Silvestri Gomes, Thaís Aparecida Sales, Karen Caroline Camargo and Mariana Araújo Espósito
Department of Chemistry, Federal University of Lavras, Lavras, Brazil

Marcos de Souza Gomes
Federal University of Vales de Jequitinhonha e Mucuri, Unaí, Brazil

David Lee Nelson
Federal University of Vales de Jequitinhonha e Mucuri, Diamantina, Brazil

Chinelo A. Ezeabara, C. U. Okeke, J. E. Amadi, A. I. Izundu, Bibian O. Aziagba, P. T. Egboka and C. D. Udechukwu
Department of Botany, Nnamdi Azikiwe University, Awka, Nigeria

Aung Zaw Htwe
Laboratory of Plant Nutrition, Graduate School of Bioresource and Bioenvironmental Sciences, Faculty of Agriculture, Kyushu University, Fukuoka, Japan

Yuchi Saeki
Department of Biochemistry and Applied Biosciences, Faculty of Agriculture, Miyazaki University, Miyazaki, Japan

Kyi Moe
Department of Agronomy, Yezin Agricultural University, Naypyitaw, Myanmar

Takeo Yamakawa
Laboratory of Plant Nutrition, Division of Molecular Biosciences, Department of Bioresource and

Bioenvironmental Sciences, Faculty of Agriculture, Kyushu University, Fukuoka, Japan

Fumiya Kurosaki, Syun Hirohashi, Takahiro Katoh, Futoshi Taura and Jung-Bum Lee
Laboratory of Medicinal Bio-Resources, Graduate School of Medicine and Pharmaceutical Sciences for Research, University of Toyama, Sugitani, Toyama, Japan

Shawn Tay and Jie He
Natural Sciences and Science Education Academic Group, National Institute of Education, Nanyang Technological University, Singapore

Tim Wing Yam
Singapore Botanic Gardens, National Parks Board, Singapore

Zhaolei Zhang
Shandong Changlin Deutz-Fahr Machinery Ltd., Linyi, China

Jianning Yang, Ruxin Li and Jialin Hou
Mechanical and Electrical Engineering College, Shandong Agriculture University, Tai'an, China

Aijun Geng
Mechanical and Electrical Engineering College, Shandong Agriculture University, Tai'an, China
Shandong Provincial Key Laboratory of Horticultural Machineries and Equipments, Tai'an, China

Chhay Ros
Department of Foreign Affairs and Trade, Australian Embassy, Phnom Penh, Cambodia

Peter F. White
Department of Agriculture and Food WA, Baron-Hay Court, S. Perth, Australia

Richard W. Bell
School of Veterinary and Life Sciences, Murdoch University, Murdoch, Australia

María Gabriela Latorre Rapela and María Cristina Lurá
Cátedra de Microbiología General, Facultad de Bioquímica y Ciencias Biológicas, Universidad Nacional del Litoral, Santa Fe, Argentina

Iván Marcipar
Cátedra de Tecnología Inmunológica, Facultad de Bioquímica y Ciencias Biológicas, Universidad Nacional del Litoral, Santa Fe, Argentina

Mustapha Haidar and Elie Shdeed
American University of Beirut, Beirut, Lebanon

Alfred O. Ubalua and Ahamefula I. Ikpeama
Plant Tissue Culture Unit, Biotechnology Research and Development Center, National Root Crops Research Institute (NRCRI) Umudike, Umuahia, Nigeria

Onyinyechi D. Okeagu
Yam Programme, National Root Crops Research Institute (NRCRI) Umudike, Umuahia, Nigeria

Xueyuan Li, Annelie Ahlman and Li-Hua Zhu
Department of Plant Breeding, Swedish Agricultural University, Alnarp, Sweden

Busie E. B. Nsibande
Department of Plant Breeding, Swedish Agricultural University, Alnarp, Sweden
Mushroom Development Unit, National Plant Genetic Resources Centre, Malkerns, Swaziland

Christiane Maria de Oliveira, Maria das Graças Cardoso, Maria Luisa Teixeira, Juliana de Andrade Santiago, Danúbia Aparecida de Carvalho Selvati Rezende, Rafaela Vieira de Souza and Luana Isac Soares
Department of Chemistry, Federal University of Lavras, Lavras, Brazil

Marisa Ionta and Guilherme Álvaro Ferreira da Silva
Institute of Biomedical Sciences, Federal University of Alfenas, Alfenas, Brazil

Marisi Gomes Soares
Department of Chemistry, Federal University of Alfenas, Alfenas, Brazil

David Lee Nelson
Federal University of Vales de Jequitinhonha e Mucuri, Diamantina, Brazil

Marcos Schleiden Sousa Carvalho
Department of Agriculture, Federal University of Lavras, Lavras, Brazil

Dennis C. Gitz III, Zhanguo Xin, Jeffrey T. Baker, Robert J. Lascano and John J. Burke
Cropping Systems Research Laboratory, USDA-ARS, Lubbock, TX, USA

Syed Abdullah Gilani, Fazal Mabood and Javid Hussain
Department of Biological Sciences and Chemistry, University of Nizwa, Birkat Al Mawz, Oman

Zakira Naureen
Department of Biological Sciences and Chemistry, University of Nizwa, Birkat Al Mawz, Oman
COMSATS University, Chak Shehzad, Islamabad, Pakistan

Muhammad Aqeel, Muhammad Nadeem Hassan and Fauzia Y. Hafeez
COMSATS University, Chak Shehzad, Islamabad, Pakistan

Nahla Bouqellah
Department of Biology, Taibah University Madina, Saudi Arabia

N. M. Fakuta and I. F. Ojiekpon
Rubber Research Institute of Nigeria, Gum Arabic Sub-Station, Gashua, Nigeria

I. B. Gashua
Department of Science Laboratory, Technology Federal Polytechnic, Damaturu, Nigeria

O. C. Ogunremi
National Biotechnology Development Agency, Abuja, Nigeria

Danúbia Magalhães Soares
Graduate Degree Program in Ecology and Natural Resources Conservation, Universidade Federal de Uberlândia, Uberlândia, Minas Gerais

André R. Terra Nascimento and Lorena Cunha Silva
Instituto de Biologia, Universidade Federal de Uberlândia, Uberlândia, Minas Gerais

Gastão Viégas de Pinho Júnior
Elementary/High Schools in Uberlândia, Uberlândia, Minas Gerais

Prasanna H. Gowda
USDA-ARS Grazinglands Research Laboratory, El Reno, OK, USA

P. V. Vara Prasad
Department of Agronomy, Kansas State University, Manhattan, KS, USA

Sangamesh V. Angadi and Umesh M. Rangappa
Agriculture and Science Center, New Mexico State University, Clovis, NM, USA

Pradeep Wagle
Department of Microbiology and Plant Biology, and Center for Spatial Analysis, University of Oklahoma, Norman, OK, USA

Yan Ma, Suqing Qu, Xiurong Xu, Tingting Liang and Dekui Zang
College of Forestry, Shandong Agricultural University, Taian, China

Yan Ma, Yan Lin, Huali Zhang, Fang Liang, Liping Sun, Maoliang Wang and Haibo Xin
Beijing Institute of Landscape and Architecture, Beijing Key Laboratory of Green Plants Breeding, Beijing, China

Charles M. Geddes, Andrea Cavalieri, Fouad Daayf and Robert H. Gulden
Department of Plant Science, University of Manitoba, Winnipeg, Canada

Reginald S. Fletcher
United States Department of Agriculture, Agricultural Research Service, Crop Production Systems Research Unit, Stoneville, USA

Khageshwar Singh Patel, Santosh Chikhlekar, Shobhana Ramteke, Bharat Lal Sahu, Nohar Singh Dahariya and Reetu Sharma
School of Studies in Chemistry/Environmental Science, Pt. Ravishankar Shukla University, Raipur, India